Heat Shock Proteins

Volume 14

Series editors
Alexzander A. A. Asea
Professor, Department of Medicine and
Director, Precision Therapeutics Proteogenomics Diagnostic Center
Eleanor N. Dana Cancer Center
University of Toledo College of Medicine and Life Sciences
Toledo, United States of America

Stuart K. Calderwood
Professor and Director, Division of Molecular and Cellular Radiation Oncology
Department of Radiation Oncology
Beth Israel Deaconess Medical Center and Harvard Medical School
Boston, United States of America

Heat Shock Proteins: key mediators of Health and Disease. Heat shock proteins (HSP) are essential molecules conserved through cellular evolution required for cells to survive the stresses encountered in the environment and in the tissues of the developing and aging organism. These proteins play the essential roles in stress of preventing the initiation of programmed cell death and repairing damage to the proteome permitting resumption of normal metabolism. Loss of the HSP is lethal either in the short-term in cases of acute stress or in the long-term when exposure to stress is chronic. Cells appear to walk a fine line in terms of HSP expression. If expression falls below a certain level, cells become sensitive to oxidative damage that influences aging and protein aggregation disease. If HSP levels rise above the normal range, inflammatory and oncogenic changes occur. It is becoming clear that HSP are emerging as remarkably versatile mediators of health and disease. The aim of this series of volumes is to examine how HSP regulation and expression become altered in pathological states and how this may be remedied by pharmacological and other interventions.

More information about this series at http://www.springer.com/series/7515

Alexzander A. A. Asea • Punit Kaur
Editors

HSP70 in Human Diseases and Disorders

Springer

Editors
Alexzander A. A. Asea
Department of Medicine and Precision Therapeutics Proteogenomics Diagnostic Center
Eleanor N. Dana Cancer Center, University of Toledo College of Medicine and Life Sciences
Toledo, OH, USA

Punit Kaur
Department of Experimental Radiation Oncology
The University of Texas MD Anderson Cancer Center
Houston, TX, USA

ISSN 1877-1246 ISSN 1877-1254 (electronic)
Heat Shock Proteins
ISBN 978-3-030-07801-0 ISBN 978-3-319-89551-2 (eBook)
https://doi.org/10.1007/978-3-319-89551-2

Softcover re-print of the Hardcover 1st edition 2018

Printed on acid-free paper

This Springer imprint is published by the registered company Springer International Publishing AG part of Springer Nature.
The registered company address is: Gewerbestrasse 11, 6330 Cham, Switzerland

Preface

The seventy kilo Dalton heat shock protein (HSP70) family is amongst the most studied HSP. The HSP70 family consists of several proteins, including the heat shock cognate 70 (Hsc70 also named Hsp73), which are constitutively expressed whilst others, like the heat shock protein 70 (Hsp70 also named Hsp72), are inducible. It has been proposed that Hsp70 would be requested to amplify the chaperone function carried out under normal conditions by the cognate Hsc70. These proteins are synthesised in response to a variety of stressors including hyperthermia, ischemia, infarct, lesions and seizures or following less drastic metabolic changes as those produced by physical exercise or psychological stress.

The book *HSP70 in Human Diseases and Disorders* provides the most comprehensive review on contemporary knowledge on the role of HSP70 in human diseases and disorders. Using an integrative approach to understanding HSP70 structure, function and immunobiology, the contributors provide a synopsis of novel mechanisms by which HSP70 is involved in the regulation of human diseases and disorders.

To enhance the ease of reading and comprehension, this book has been subdivided into various parts, including Part I, reviews current progress on the role of HSP70 in neuro-oncological disorders; Part II, evaluates the role of HSP70 in circulatory disorders including cardiovascular diseases and kidney disease and Part III, focuses the reader on the role of HSP70 as a novel therapeutic target.

Key basic science and clinical research laboratories from major universities and academic medical hospitals around the world contribute chapters that review present research activity and importantly project the field into the future. The book is a must read for medical students and residents, clinical and basic science researchers, postdoctoral fellows and graduate students in the fields of Medicine, Physiology, Pharmacology, Biotechnology, Molecular Medicine and Pathology.

Toledo, OH, USA — Alexzander A. A. Asea
Houston, TX, USA — Punit Kaur

Contents

Editors Biography

Alexzander A. A. Asea is a highly innovative and accomplished world-renowned clinical and basic research scientist and visionary executive leader who has exceptional experience spearheading clinical and basic science research, training, education, and commercialization initiatives within top-ranked academic biomedical institutes. Prof. Asea's initial findings studying the effects of Hsp72 on human monocytes led to the proposal of a novel paradigm that Hsp72, previously known to be as intracellular molecular chaperones, can be found in the extracellular milieu where it has regulatory effects on immunocompetent cells – a term now called chaperokine. Prof. Asea has authored over 255 scientific publications including peer-reviewed articles, reviews, books, book chapters, editorials, and news headliners in a wide range of biomedical-related disciplines. Prof. Asea is the series editor of the widely successful book series *Heat Shock Proteins* (Springer Nature Publications) and is an editorial board member of 13 other scientific peer-reviewed journals. Currently, Prof. Asea is at the University of Toledo College of Medicine and Life Sciences in Toledo, USA.

Punit Kaur is an expert in onco-proteogenomics, with extensive training and experience in quantitative mass spectrometry imaging, protein chemistry, and biomarker discovery. Dr. Kaur's main research focus is on the use of heat-induced nanotechnology in combination with radiotherapy and chemotherapy in the cancer stem cell therapy. Dr. Kaur has published more than 40 scientific articles, book chapters, and reviews, and currently serves as editorial board member for the *European Journal of Cancer Prevention* and the *Journal of Proteomics and Bioinformatics*. Dr. Kaur is an editor of five books in the highly successful *Heat Shock Proteins* book series by Springer Nature Publishers. Currently, Dr. Kaur is a Visiting Scientist Professor at the University of Texas MD Anderson Cancer Center in Houston, USA.

Part I
Hsp70 in Neuro-oncological Disorders

Heat Shock Protein 70 and Molecular Confession During Neurodegeneration

Komal Panchal, Ajay Kumar, and Anand K. Tiwari

Abstract Molecular chaperones are the group of proteins that participate in the maintenance of cellular homeostasis by regulating several cellular events and protein homeostasis (proteostasis). It has been shown that failure of protein quality control system, formation of protein aggregates and their ectopic accumulation in the neuronal cells is the common pathological hallmark of most of the neurodegenerative diseases such as Alzheimer's disease (AD), Parkinson's disease (PD), Huntington disease (HD), Amyotrophic Lateral Sclerosis (ALS), prion disease and various forms of spinocerebellar ataxia (SCA) etc. Heat shock protein 70 (Hsp70), an evolutionary conserved protein family has been shown to be a key regulator in several neurodegenerative diseases. Hsp70 shows its strong expression in stress condition and is associated with protein folding, refolding of misfolded protein, transport of proteins to different cellular compartments, cell death and cell cycle regulation etc. Several recent studies have suggested that Hsp70 can be a key molecule to address the major pathologies associated with neurodegenerative diseases. This chapter briefly summarizes the Hsp70 and its possible role during neurodegenerative diseases.

Keywords Heat Shock Protein 70 (Hsp70) · Neurodegenerative diseases
Alzheimer's disease (AD) · Parkinson's disease (PD)

Author contributed equally with all other contributors. Komal Panchal and Ajay Kumar

K. Panchal · A. Kumar · A. K. Tiwari (✉)
Genetics & Developmental Biology Laboratory, School of Biological Sciences & Biotechnology, Indian Institute of Advanced Research/IAR, Koba Institutional Area, Gandhinagar, India
e-mail: anandk.tiwari@iar.ac.in

A. A. A. Asea, P. Kaur (eds.), *HSP70 in Human Diseases and Disorders*, Heat Shock Proteins 14, https://doi.org/10.1007/978-3-319-89551-2_1

Abbreviations

17-AAG	17-allylamino-17-demethoxygeldanamycin
17-DMAG	17-(dimethylaminoethylamino)-17- demethoxygeldanamycin
α-Syn	α-synuclein
γPKC	Protein kinase Cγ
$A\beta_{42}$	Amyloid Beta 42
AC	Azure C
AD	Alzheimer's disease
ADP	Adenosine diphosphate
AIF	Apoptosis inducing factor
ALS	Amyotrophic Lateral Sclerosis
Apaf-1	Apoptotic protease activation factor 1
APP	Amyloid precursor protein
AR	Androgen receptor
Ask1	Apoptosis signal-regulating kinase
ATPase	Adenosine tri phosphatease
Bag-1	Bcl-2-associated athanogene-1
Bap	Benzo(a)pyrene
BiP	Binding immunoglobulin protein
CHIP	Carboxy-terminus of HSC70-interacting protein
CMA	Chaperone mediated autophagy
CNS	Central nervous system
DA	Dopaminergic
E. coli	*Escherichia coli*
ER	Endoplasmic reticulum
FMRP	Fragile X mental retardation protein
GBA	Glucocerebrosidase
GFP	Green fluorescent protein
GGA	Geranylgeranyl acetone
Grp75	Glucose-regulated protein
HD	Huntington disease
HOP	HSP70 and HSP90 organizing protein
HSC	Heat shock cognate
HSF1	Heat shock transcription factor-1
HSP	Heat shock proteins
Hsp70	Heat shock protein 70
IDE	Insulin degrading enzyme
JNK	c-Jun N-terminal kinase
LBs	Lewy bodies
LRRK2	Leucine-rich repeat kinase 2
MAPK	Mitogen-activated protein kinases
MB	Methylene blue
MY	Myricetin

NBD	Nucleotide binding domain
ND	Neurodegenerative diseases
NEF	Nucleotide exchange factor
NF-ßB	Nuclear factor-kappaB
NFT	Neurofibrillary tangles
PD	Parkinson's disease
PDB	Protein data bank
PQC	Protein quality control
PTEN	Phosphatase and tensin homolog
rhHSP70	Recombinant human Hsp70
ROS	Reactive oxygen species
SBD	Substrate binding domain
SBMA	Spinal & Bulbar Muscular Atrophy
SCA	Spinocerebellar ataxia
SOD	Superoxide dismutase
Ubl	Ubiquitin-like protein
UCHL-1	Ubiquitin c-terminal hydrolase-1
UPS	Ubiquitin proteasome system
TGF-β1	Transforming growth factor beta 1
TLR4	Toll-like receptor-4
UMN	Upper motor neurons
LMN	Upper motor neurons
NMJ	Neuromuscular junction
FUS	Fused-in-sarcoma
TDP43	TAR DNA binding protein

Introduction

The Heat Shock Proteins (HSP) are a group of proteins categorized under the molecular chaperones family, plays an important role during different cellular events such as protein folding, protein trafficking, autoimmunity and protection from environmental stress etc. (Jaattela and Wissing 1992; Ross and Poirier 2004). The optimum level of HSP in the cell maintains the protein homeostasis, hence represents the "Protein Quality Control" (PQC) system of the cell in different cellular conditions (Sheikh et al. 2013; Meijering et al. 2014; Kakkar et al. 2016; Gidalevitz et al. 2011; Van Drie 2011; Verghese et al. 2012). HSP have been categorized on the basis of their molecular mass such as Hsp100, 90, 70, 60, 40, 27 & 26 etc. Heat shock protein 70 (Hsp70) is one of the most conserved ubiquitously expressed protein, present in the organism from archaebacteria to human (Gupta and Singh 1994; Lindquist and Craig 1988; Hunt and Morimoto 1985). This protein was discovered by Ritossa in 1960 in the fruit fly "*Drosophila*". He examined chromosomal puff in the flies that indicates the transcriptional active region of chromatin, named as "heat shock

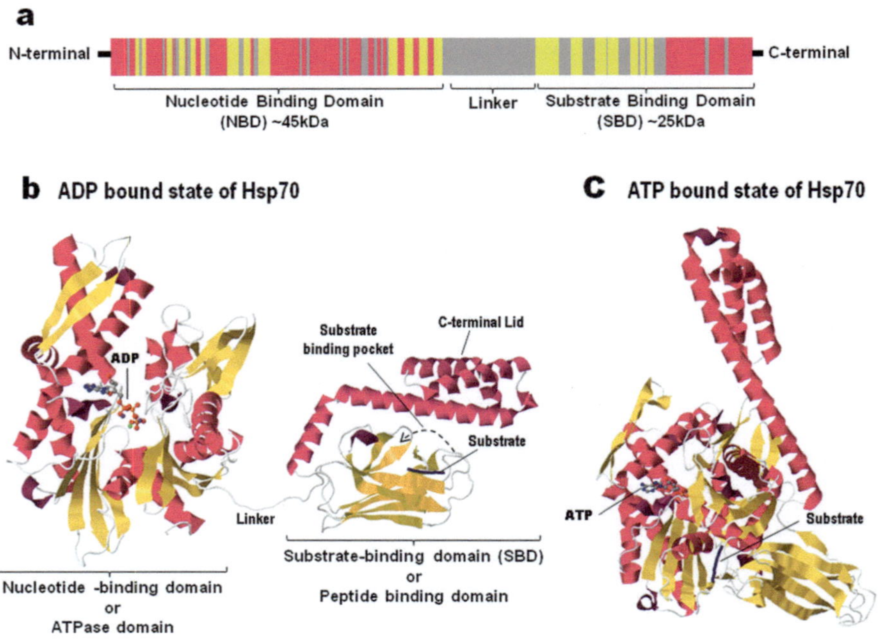

Fig. 1 Hsp70 domain architecture. (**a**) Schematic representation of human Hsp70 domain structure. The N & C terminal domains along with ~45 kDa Nucleotide Binding Domain (NBD), a short Linker and a ~25 kDa Substrate Binding Domain (SBD). (**b**) Three dimensional model of HSP0 showing ADP bound conformational state; that represents the nucleotide-binding domain (Protein Data Bank (PDB) Code: 3HSC) (Flaherty et al. 1990) and substrate binding domain (PDB Code: 1DKZ) (Zhu et al. 1996), are joint together by a flexible linker, the substrate (blue) is locked in substrate binding pocket by lid of substrate binding domain. (**c**) The ATP bound conformation state of HSP70 (PDB Code: 4B9Q) (Kityk et al. 2012) showed the docking of the lid and substrate binding domain showing the folding event of substrate

response" and the protein formed by heat shock was named as "Heat-shock proteins" (HSP) (Ritossa 1964; Lindquist and Craig 1988).

The Hsp70 is ~66–78 kDa proteins possess three functional domains N-terminal ATPase domain (~45 kDa), Substrate binding domain (~25 kDa) and C-terminal domain (Fig. 1) (Hartl 1996; Hendrick and Hartl 1993) ATP hydrolysis in the N-terminal domain is linked to a conformational change in the client binding domain (Vogel et al. 2006). In humans, 13 members of Hsp70 i.e. HSPA1A, HSPA1B, HSPA1L, HSPA2, HSPA3, HSPA4, HSPA5, HSPA5BP1, HSPA6, HSPA7, HSPA8, HSPA9B and HSPA10 are present (Brocchieri et al. 2008). Human Hsp70 is 72% homologues to *Drosophila* (Fruit fly) Hsp70 and 47% to *E.coli* dnak (Jaattela and Wissing 1992; Hunt and Morimoto 1985; Daugaard et al. 2007). *Drosophila* HSP70 family contains two heat inducible (HSP68 and HSP70) and six constitutively expressed members (HSC1–HSC6) (Kumar and Tiwari 2017) (Shopland and Lis 1996); (Rubin et al. 1993). The HSC4 protein (Hsc4p) is the most abundantly produced cytoplasmic Hsc70 members and the HSC3 protein

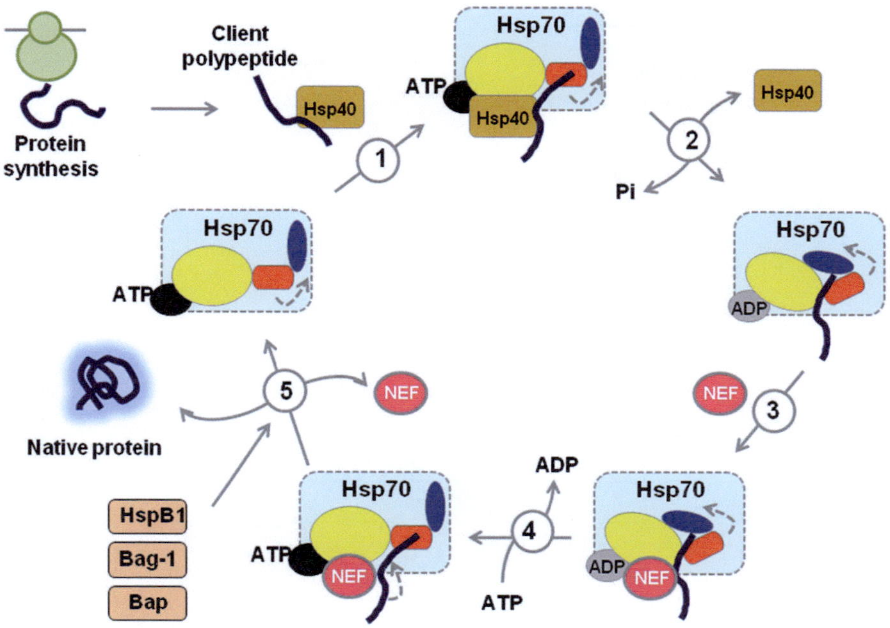

Fig. 2 The chaperone cycle of Hsp70: (*1*) The co chaperone of Hsp70, the Hsp40, first recognized and bind to the client polypeptide or newly synthesized polypeptide chain, later it interact and stimulates the ATPase domain of Hsp70. (*2*) The Hsp70 interact with unfolded client polypeptide chain by its substrate-binding domain (SBD) in an ATP-bound state, in this state Hsp40 stimulate the ATP hydrolysis (ATP = ADP + P*i*) hence dissociate itself from Hsp70 machine, this makes a more tight interaction between Hsp70 and client protein. (*3*) A crucial co-factor, nucleotide exchange factor (NEF) bind with Hsp70 + client protein complex and (*4*) promotes the replacement of ADP to ATP. (*5*) The other co-factors: HspB1, Bag-1 and Bap enter into the cycle and folded client polypeptide and NEF are separate from the Hsp70 molecule, representing the completion of one chaperone cycle

(Hsc3p) is the sole endoplasmic reticulum (ER) resident Hsc70 family member. Hsc3p and Hsc4p are homologous to the mammalian ER resident protein BiP and the cytoplasmic clathrin uncoating ATPase protein respectively (Elefant and Palter 1999). The Hsp70 possess housekeeping function and are associated with variety of cellular processes in normal condition such as it binds with an unfolded protein substrate, maintain it in an extended conformation and stabilized its exposed hydrophobic regions, involved in protein folding through its chaperonin cycle (Fig. 2), refolding, degradation of misfolded proteins, protein trafficking, possess anti-apoptotic activity by inhibiting the translocation of Bax into mitochondria, release of cytochrome c from mitochondria, formation of apoptosome and inhibit activation of initiator caspases, eye development in *Drosophila* (Kumar and Tiwari 2017) etc. (Hartl 1996; Arya et al. 2007; Gething and Sambrook 1992). Under stress condition activation of Hsp70 reduces cellular stress, oxidative stress, promotes cell survival and refolding of the misfolded proteins (Fig. 3 and 4).

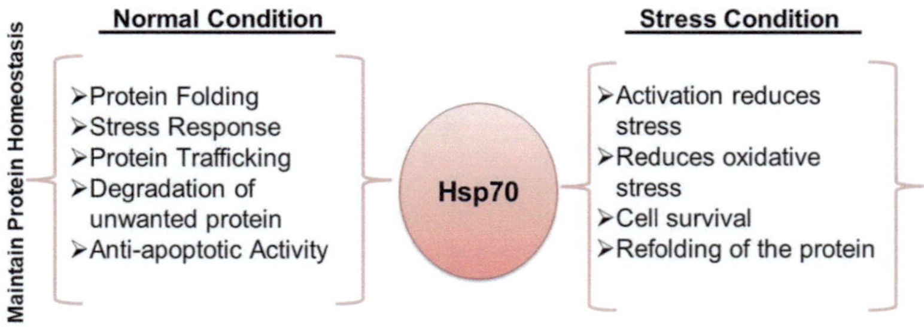

Fig. 3 Molecular Function of Hsp70 in normal and stress condition

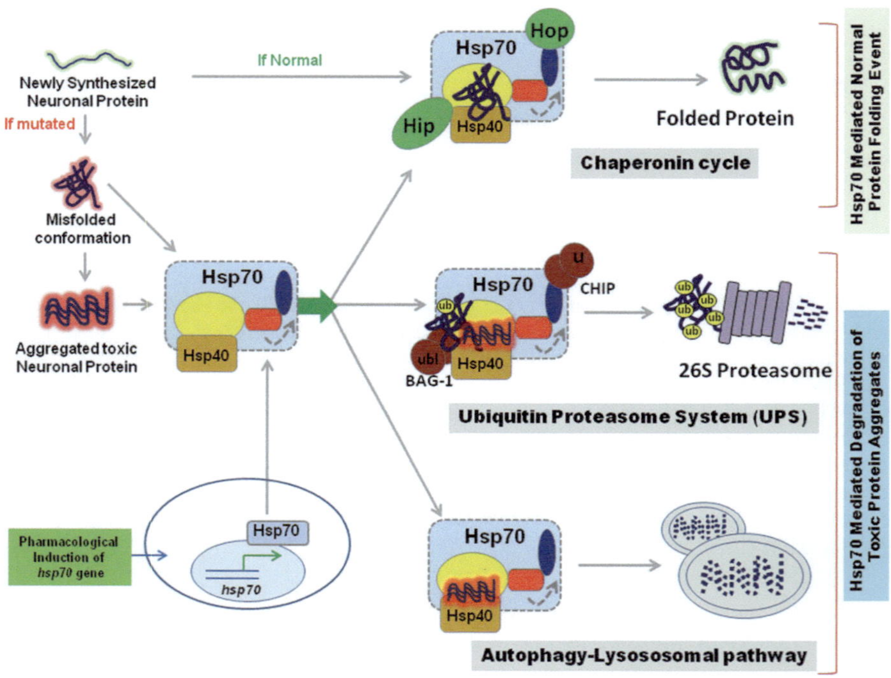

Fig. 4 Molecular functions of Hsp70. Figure showing the normal function of Hsp70 that includes protein folding by chaperonin cycle and degradation of misfolded or protein aggregates by ubiquitin proteasome system (UPS) and autophagy-lysosomal pathways

Hsp70 and Neurodegenerative Diseases

Neurodegenerative diseases are group of disorders that includes brain, peripheral nerves, spinal cord and leads to either functional loss (ataxia like condition) or sensory dysfunction (dementia) (Uttara et al. 2009; Singh et al. 2004). The common neurodegenerative diseases are Alzheimer's disease (AD), Parkinson's disease (PD), Huntington disease (HD), Amyotrophic lateral sclerosis (ALS), Fragile-X-syndrome, Spinocerebellar Ataxia (SCA) etc. (de Diego-Otero et al. 2009; Andersen 2004)

(Table 1). In recent years, these diseases have become a major challenging problem for human beings and need immediate concern. Most of neurodegenerative diseases are associated with degeneration and death of specific neurons due to misfolding or aggregation (cytosolic/nuclear/extracellular) of certain proteins in the brain which are often ubiquitinated and may be associated with chaperones (Ardley and Robinson 2004; Soto and Estrada 2008). The misfolded proteins or aggregates are tagged for refolding by molecular chaperones or degradation by ubiquitin proteasome system (UPS). Generally in a normal cell, this quality control of a protein is well organized and self-regulated for the whole life; however, in an abnormal cell these protein aggregates fail to be targeted or degraded by the proteasome system. Thus, proper cooperation between molecular chaperones and the UPS is an essential for the proper functioning of a protein. Any abnormality in this cooperation results in improper protein folding or aggregate formation that ultimately leading to the neurodegenerative diseases. Several studies have suggested the regulatory role of Hsp70 during neurodegenerative diseases and it was demonstrate that overexpression of Hsp70 in different neurodegenerative disease animal models protects neurons from degeneration, toxicity and aggregate formation (Turturici et al. 2011) (Table 2).

Table 1 List of neurodegenerative diseases

Sr. No.	Neurodegenerative diseases	Culprit protein/s	Molecular functions	Affected areas	References
1.	Alzheimer's disease (AD)	Tau	Hyper-phosphorylation of tau cause microtubule destabilization & hampered the mitochondrial axonal transport	Cortical and hippocampal neurons	Iqbal et al. (2009), Wang et al. (2013b), and Gong and Iqbal (2008)
		Amyloid precursor protein (App)	Amyloidogenic cleavage of App forms $A\beta_{42}$ plaques, hampered vesicular and mitochondrial axonal transport		Chakraborty et al. (2011), Cassar and Kretzschmar (2016), and Moloney et al. (2010)
		Presenilin (Ps1 & Ps 2)	Increase $A\beta_{42}$ plaques formation & accumulation, loss of Photoreceptor cell, degeneration of axonal projections and lethality		Iijima et al. (2008) and Prussing et al. (2013)
		Amyloid β_{42}	Mitochondrial depolarization, apoptosis and neuronal degeneration		Iijima et al. (2008), Iijima-Ando and Iijima (2010), and Starkov and Beal (2008)

(continued)

Table 1 (continued)

Sr. No.	Neurodegenerative diseases	Culprit protein/s	Molecular functions	Affected areas	References
2.	Parkinson's disease (PD)	α-synuclein	Lewy bodies formation, locomotor dysfunction and neuronal loss	Dopaminergic neurons in the brain	Feany and Bender (2000) and Stefanis (2012)
		PTEN-induced putative kinase 1(PINK1)	Apoptosis, abnormal mitochondrial morphology, impaired dopamine release and motor dysfunction, inhibit ETC complex 1 activity		Munoz-Soriano and Paricio (2011) and Morais et al. (2009)
		Parkin	Mitochondrial dysfunction, apoptotic muscle degeneration, abnormal synaptic transmission and neuronal death		Lu and Vogel (2009) and West et al. (2015)
		Protein deglycase DJ-1	Increased ROS, mitochondrial dysfunction, increase lipid peroxidation and neuronal loss		Munoz-Soriano and Paricio (2011), Burchell et al. (2013), and Cookson (2012)
		Leucine-rich repeat kinase 2 (LRRK2)	Locomotor dysfunction, loss of DA neurons, mitochondrial dysfunction, synaptic abnormality and defect in autophagy-lysosomal pathway		Li et al. (2014), Munoz-Soriano and Paricio (2011), and Dachsel et al. (2011)
3.	Huntington's disease (HD)	Huntingtin	Aggregation of the poly-Q expanded Htt protein, aggregates can physically block transport of numerous organelles along the axon, behavioral defect, axonal transport defect and neuronal loss	Striatal neurons	Krench and Littleton (2013), Slow et al. (2005), and Orr and Zoghbi (2007)

(continued)

Table 1 (continued)

Sr. No.	Neurodegenerative diseases	Culprit protein/s	Molecular functions	Affected areas	References
4.	Spinocerebellar ataxia (SCA)	Ataxin-2 (ATXN2)	Pathogenic CAG repeat expansion in the ataxin-2 gene (ATXN2) and cause neuronal toxicity	Midbrain and cerebellum	Laffita-Mesa et al. (2013) and Bonini and Gitler (2011)
5.	Spinal and bulbar muscular atrophy (SBMA)	Androgen receptor (AR)	The *AR* gene mutation that causes abnormal expansion of a DNA segment called a CAG triplet repeat. Weakness in the spinal lower motor neurons. Fasciculations in bulbar, tough and lips muscles.	Nerve cells originate in the spinal cord and the part of the brain that is connected to the spinal cord (the brainstem)	La Spada (2014)
	Amyotrophic lateral sclerosis (ALS)	TAR DNA binding protein (TDP-43)	Neuronal and glial aggregates of phosphorylated and ubiquitinated pTDP-43 protein which leads to degeneration of motor neurons	Bran's motor cortex, upper motor neurons and the lower motor neurons, body muscles and spinal cord	Scotter et al. (2015), Ludolph et al. (2015), and Mackenzie and Rademakers (2008)
		Fused-in-Sarcoma (FUS)	Ubiquitinated FUS forms neuronal and glial cytoplasmic inclusions (NCI) in the region of motor cortex, basal ganglia and spinal cord, as well as dystrophic neuritis, abnormal neurites, astrogliosis, microglial activation, moderate neuronal loss of Betz cells within layer V of motor cortex and motor nuclei of brainstem		Nolan et al. (2016), Sharma et al. (2016), and Rademakers et al. (2010)
		Superoxide dismutase 1 (SOD 1)	Increase oxidative stress and neuronal death		Bunton-Stasyshyn et al. (2015), Banci et al. (2008), and Sau et al. (2007)

Table 2 Hsp70 and its association with neurodegenerative diseases

Sr. No.	Neurodegenerative diseases	Hsp70 functions in disease condition	References
1.	Alzheimer's disease (AD)	It inhibits $A\beta_{42}$ plaques oligomerization	Koren et al. (2009), Ou et al. (2014), and Lu et al. (2014)
		Overexpression of Hsp70 in AD increases Insulin Degrading Enzyme (IDE), an Aβ-degrading enzyme) and TGF-β1 (a cytokine helps in brain injury and inflammation present in the case of AD and also activates phagocytic microglia for Aβ plaques clearance)	
		Hsp70 inhibits tau degradation and enhance its microtubule binding capacity	
		Hsp70 affect both Apaf-1 caspase-dependent and AIF caspase-independent pathways and inhibits apoptosis/neuronal cell death	
2.	Parkinson's disease (PD)	Hsp70 helps in the survival of dopaminergic neurons from toxicity induced by α-syn and Lewy bodies.	Labrador-Garrido et al. (2011), Witt (2009), and Roodveldt et al. (2009)
		Promotes α-syn degradation via Chaperone Mediated Autophagy (CMA) process	
		Hsp70 via CHIP (carboxyl terminus of HSP70-interacting protein) cause ubiquitylation of the unfolded protein followed by proteasomal degradation	
3.	Huntington's disease (HD)	Hsp70 inhibits the conformational change in mutant htt and prevent heterotypic interactions between polyQ expanded htt and transcription factors	Wakabayashi et al. (2007) and Broadley and Hartl (2009)
		Hsp70 cleaves α-Syn in HD patient brain and able to bind with htt and increase its aggregation	
		Overexpression of Hsp70 decreases the polyQ toxicity and the formation of polyQ inclusion body	
4.	Spinocerebellar ataxia (SCA)	Expression of Hsp70 inhibits neurodegeneration induced by ataxin-7	Ogawa et al. (2013), Helmlinger et al. (2004), and Tsai et al. (2005)
		Overexpression of Hsp70 protects cells from cytotoxicity of mutant protein kinase Cγ (γPKC) and neurodegeneration	
		Hsp70 attenuates poly Q induces toxicity in SCA3 fly model and mouse model too	

(continued)

Table 2 (continued)

Sr. No.	Neurodegenerative diseases	Hsp70 functions in disease condition	References
5.	Spinal and bulbar muscular atrophy (SBMA)	Overexpression of Hsp70 decreases the localization of mutant androgen receptor protein in nucleus and rescue the phenotype associated with spinal and bulbar muscular atrophy in transgenic mouse model	Kobayashi et al. (2000), Adachi et al. (2003), and Bailey et al. (2002)
		Overexpression of Hsp70 decreases the mutant AR aggregation and neuronal cell death in neurodegenerative disease model of SBMA	
		Upregulated Hsp70 promotes protrosomal degradation of mutant androgen receptor protein	
6.	Amyotrophic lateral sclerosis (ALS)	Hsp70 decreases the symptoms and increases the lifespan of mouse model of ALS	Gifondorwa et al. (2007), Miyazaki et al. (2016), and Liu et al. (2005)
		Overexpression of Hsp70 increases innervated NMJ numbers.	
		Overexpression of Hsp70 inhibits plaques formation and neuronal toxicity in the $SOD1^{G93A}$ mutant fly	
		Hsp70 helps in the proper folding of mutant SOD1 protein in the case of ALS	
7.	Fragile X-syndrome	Co-expression of Hsp70 rescues the rough eye phenotype in CGG_{100}GFP expressing *Drosophila*	Oh et al. (2015)
		Overexpression of Hsp70 suppresses the CGG_{100}GFP-induced neurodegeneration by enhancing ubiquitin proteasomal system for degradation of protein aggregates and decreases CGG-repeat-elicited toxicity in *Drosophila*	

Hsp70 and Alzheimer's Disease

AD is one of the most common type of irreversible senile dementia occurs mostly after the age of 65 in susceptible individuals (Lu et al. 2014; Iijima-Ando and Iijima 2010). The main pathologies of AD are associated with misfolding of proteins, accumulation of $A\beta_{42}$ plaques, formation of neurofibrillary tangles and neuronal cell death (Prussing et al. 2013; Lu et al. 2014), and is characterized by motor dysfunction, short-term memory loss, sleeping disorders and uptrend agitation (Buchman et al. 2011; Jahn 2013). Previous studies have been shown that aggregation of $A\beta_{42}$

plaques are due to the amyloidogenic cleavage of APP protein by β-secratase and γ-secratase and also by accumulation of neurofibrillary tangles in the brain by tau hyperphosphorylation (Chakraborty et al. 2011; Carmine-Simmen et al. 2009; Moloney et al. 2010; Wittmann et al. 2001). The other pathologies associated with AD are an increased oxidative stress, mitochondrial depolarization and axonal transport defects etc. (Huang et al. 2016; Moreira et al. 2010). Several studies have shown the involvement of Hsp70 and its protective role in AD as discussed below:

Overexpression of Hsp70 Inhibits Aβ Oligomerization and Enhances the Clearance of $A\beta_{42}$ Plaques in AD Brain

As discussed above accumulation of $A\beta_{42}$ plaque is the key pathological condition associated with AD, it is toxic for the survival of the neurons and ultimately causes the onset of disease (Glenner and Wong 1984; Wilhelmus et al. 2007). The accumulation of $A\beta_{42}$ takes place in different steps such as formation of Aβ oligomer, protofibrils and finally the Aβ mature fibrils formation (Fig. 5A) (Upadhaya et al. 2012). It has been shown that Hsp70 can prevent the Aβ toxicity by three ways: *Clearance*, *refolding* and *degradation of Aβ oligomer* (Najarzadegan et al. 2016). Hsp70 induces the clearance of Aβ oligomers by activating TGF-β1, a cytokine, involved in the regulation of brain injury and inflammation. TGF-β1 induces phagocytosis of Aβ oligomers by activating the phagocytic microglia and initiates the degradation of Aβ oligomers (Fig. 5B) (Wyss-Coray et al. 2001; Lu et al. 2014; Tichauer et al. 2014). Furthermore, Hsp70 can also initiate the degradation of Aβ oligomers through activating an $A\beta_{42}$ degrading enzyme also known as insulin degrading enzyme (IDE). Aβ as a substrate tightly bind to IDE with irreversible manner. IDE and Aβ oligomers form stable complex and promote Aβ oligomer's degradation (Fig. 5C) (de Tullio et al. 2013, 2008; Lu et al. 2014). In other mechanism "refolding" the hydrophobic region of $A\beta_{42}$ oligomer is recognized by Hsp70 for refolding and inhibition of Aβ oligomers formation (Evans et al. 2006; Ou et al. 2014; Lu et al. 2014) (Fig. 5D). The ATPase activity of Hsp70 is crucial against the intracellular amyloid formation while the holdase domain of secreted form of Hsp70 (secHsp70) plays an important role in the clearance of extracellular $A\beta_{42}$ in *Drosophila* model of AD (Fernandez-Funez et al. 2016). Moreover, Hsp70 can also facilitate degradation of Aβ oligomes by activating NF-ₖB and p38 MAPK through the Toll-like receptor-4 (TLR4) pathway (Ou et al. 2014; Kakimura et al. 2002). Over expression of Hsp70 is sufficient to reduce the load of Aβ oligomer with a significant reduction in neuronal cell death (Evans et al. 2006).

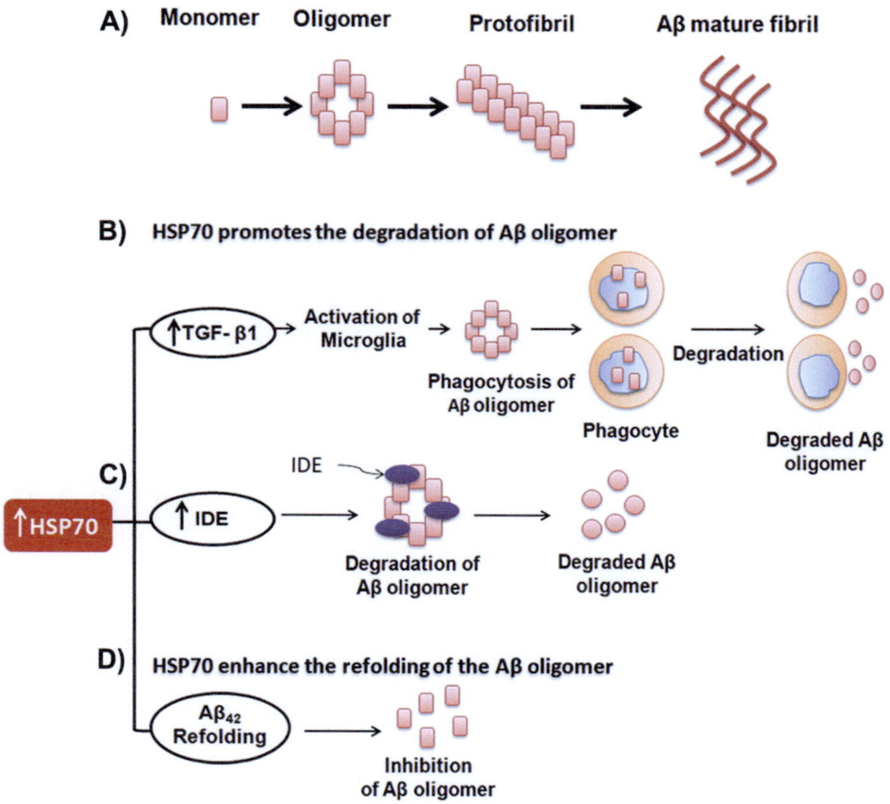

Fig. 5 Hsp70 obviates the $A\beta_{42}$ toxicity. (*A*) $A\beta_{42}$ monomer gets aggregates to form oligomer, protofibrils and mature amyloid beta fibrils. Hsp70 tends to affect the oligomeric form efficiently. Hsp70 neutralizes the effect of $A\beta_{42}$ toxicity by (*B*) activating TGF-β1 and enhancing $A\beta_{42}$ oligomers phagocytosis by the formation of phagocytes and promote degradation of $A\beta_{42}$ oligomers; (*C*) Hsp70 initiates $A\beta_{42}$ oligomers refolding and inhibits the formation of Aβ oligomer; (*D*) Hsp70 activates IDE, an Aβ degrading enzyme which is bind to the Aβ oligomers and degrade it

Hsp70 Inhibits the Accumulation of Neurofibrillary Tangles in AD Brain

Neurofibrillary tangles (NFT) are formed due to the intracellular accumulation of hyperphosphorylated tau protein in the neurons (Iqbal et al. 2009; Prussing et al. 2013). Tau is a microtubule binding protein which provides stability to the microtubules in the neurons and its homeostasis is depended on its expression, phosphorylation and turnover. Hyperphosphorylation of tau causes conformational changes of tau protein which leads to detachment of tau from the microtubules and results in microtubule destabilization and neurodegeneration (Wentzell and Kretzschmar 2010; Iijima-Ando and Iijima 2010; Ballatore et al. 2007; Pooler et al. 2013). Overexpression of Hsp70 stabilized the tau protein and promotes its binding to the microtubules. Hsp70 can also binds to tau aggregates and enhance its degradation by ubiquitin-proteasome/autophagy system and inhibits the tauopathy (Dou

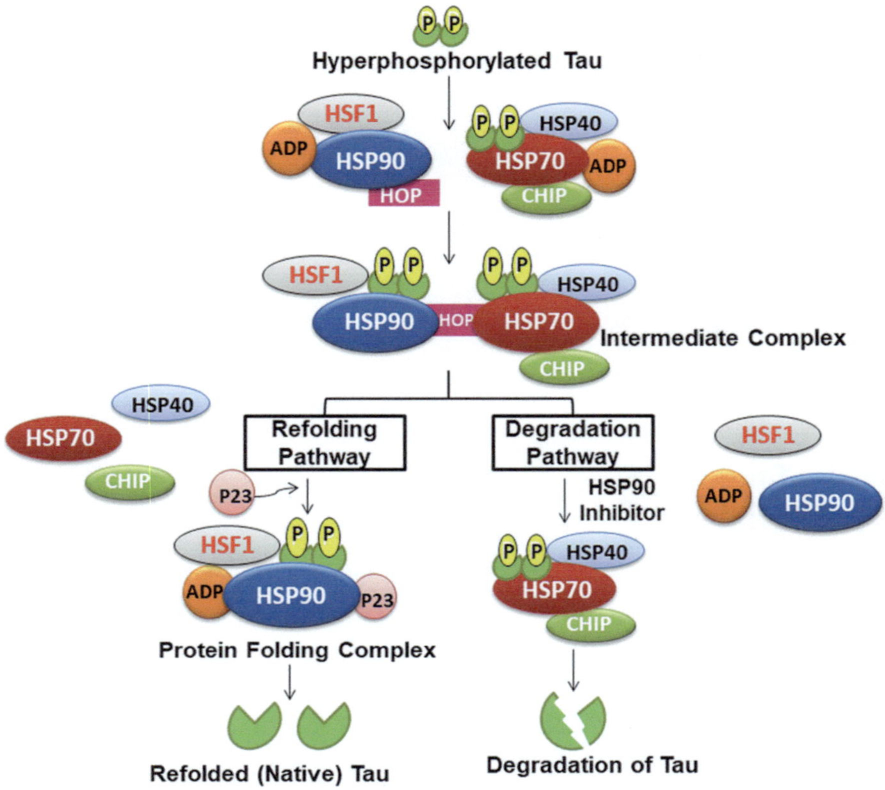

Fig. 6 Hsp70 regulates tau homeostasis. The refolding and degradation of hyperphosphorylated tau is mediated by Hsp70/90 chaperon's complex. In refolding pathway, it prevents degradation of hyperphosphorylated tau with the help of Hsp90, HSF1 and P23 complex and enhances its refolding to form mature tau protein. In degradation pathway Hsp70 degrades the hyperphosphorylated tau with the help of CHIP complex in the presence of Hsp90 inhibitor

et al. 2003; Abisambra et al. 2013; Ou et al. 2014). In AD, expression of Hsp70 inhibits tau toxicity via two pathways such as refolding and degradation of hyperphosphorylated tau protein. Initially, Hsp70 along with CHIP and Hsp40 binds to hyprtphosphorylated tau and make a complex known as "intermediate complex" with Hsp90, HOP (Hsp70 and Hsp90 organizing protein) and HSF1 (Heat shock factor-1). In refolding pathway, Hsp70, Hsp40 and CHIP dissociated from intermediate complex, eventually Hsp90 initiate tau refolding and dephosphorylation with the help of HSF-1 and P23 (a cochaperone that bonds with Hsp90) (Blair et al. 2014; Wang et al. 2014a) (Fig. 6). In degradation pathway, inhibition of Hsp90 by using Hsp90 inhibitors induces the expression of Hsp70 which binds to hyperphosphorylated tau protein and formed a protein complex that is recognized by CHIP (carboxy-terminus of Hsc70-interacting protein), a tau ubiquitin ligase and used for ubiquitination proteasome degradation (Koren et al. 2009; Shimura et al. 2004; Wang et al. 2014a; Blair et al. 2014) (Fig. 6).

Hsp70 Inhibits the Neuronal Cell Death in AD

Neuronal cell death is a prominent pathological feature of AD and it could be due to the accumulation of $A\beta_{42}$ plaques and neurofibrillary tangles which increases the oxidative stress, DNA damage, lipid peroxidation, protein oxidation and release of pro-apoptotic proteins in the cytosol from the mitochondria (Martin 2001; Niizuma et al. 2009; Silva et al. 2014). Previous studies have shown that overexpression of Hsp70 inhibits thermal stress induced neuronal apoptosis by attenuating both caspase dependent and caspase independed pathways via apoptotic protease activation factor 1 (Apaf-1) and apoptosis inducing factor (AIF), respectively (Sabirzhanov et al. 2012b; Ou et al. 2014).

The mechanism via which Hsp70 inhibits caspases depended apoptosis is ATPase domain of Hsp70 binds to Apaf1 and inhibits its oligomerization, therefore, it inhibits the recruitment of pro-caspase-9 to the apoptosome and apoptosis of the neurons (Sabirzhanov et al. 2012b; Beere 2001; Saleh et al. 2000) (Fig. 7a). Peptide-binding domain and EEVD motif of Hsp70 is essential to inhibit the activation of caspase-3 in vitro (Li et al. 2000; Mosser et al. 2000). In caspase independent pathway, Hsp70 attenuates the apoptosis by neutralizing AIF, a mitochondrial intermembrane protein which translocate to the nucleus and induce DNA fragmentation (Sabirzhanov et al. 2012a; Ravagnan et al. 2001) (Fig. 7b). Additionally, Hsp70 also attenuates the expression of death causing signaling molecules such as c-Jun N-terminal kinase (JNK), p38, and apoptosis signal-regulating kinase (Ask1) (Leak 2014; Luo et al. 2008; Li et al. 2010). Mitochondrial depolarization induces release of cytochrome c in the cytosol and Hsp70 plays a vital role in inhibiting the cyto-

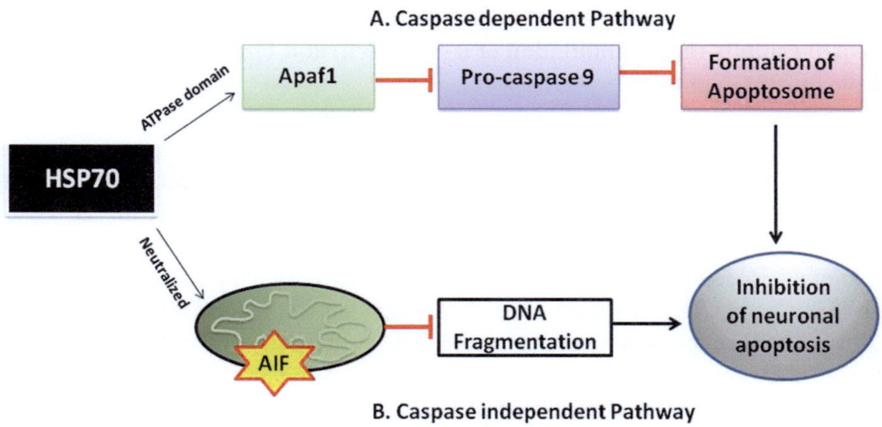

Fig. 7 Hsp70 impedes apoptosis in AD. Hsp70 inhibits apoptosis via A. caspases depended pathway via regulating Apaf-1 and B. caspase independed pathway via affecting AIF. (*A*) Hsp70 regulates Apaf-1 via its ATPase domain and inhibits pro-caspase 9 and subsequently, prevents the formation of apoptosome. (*B*) Furthermore, Hsp70 inhibits AIF and following, inhibition of DNA fragmentation and eventually, via both the pathways, it inhibits the neuronal apoptosis and enhancing neuronal survival

chrome c processing by acting as a downstream molecule and ultimately apoptosis. Peptide-binding domain and EEVD motif of Hsp70 is essential to inhibit the activation of caspase-3 in vitro (Li et al. 2000; Stankiewicz et al. 2005)

Hsp70 and Parkinson's Disease

PD is the second common irremediable neurodegenerative disease which is characterized by progressive loss of dopaminergic neurons, locomotor defects, muscles rigidity, bradykinesia, postural instability and resting tremor (Alexander 2004; DeMaagd and Philip 2015; Magrinelli et al. 2016). In natural population, the sporadic cases of PD occurs predominantly, while the familial form of PD is rare (~5–10%) (Munoz-Soriano and Paricio 2011; Klein and Westenberger 2012). The genetic risk of PD are due to mutations in different genes such as *α-synuclein* (*α-Syn*), *parkin*, *DJ-1*, *phosphatase and tensin homolog (PTEN) induced kinase 1* (*PINK1*), *ubiquitin C-Terminal hydrolase-1* (*UCHL-1*), *leucine-rich repeat kinase 2* (*LRRK2*), *Omi/HtrA2*, *ATP13A2*, and *glucocerebrosidase* (*GBA*) ((Lev et al. 2006; Pickrell and Youle 2015; Li et al. 2014; Stefanis 2012; Munoz-Soriano and Paricio 2011; Maraganore et al. 2004). The pathologies of the PD includes abnormal protein aggregation such as accumulation of Lewy bodies (LBs), which are formed by α-Syn and ubiquitination occur of modified α-Syn in LBs (Tofaris et al. 2003; Wakabayashi et al. 2007). The accumulation of LBs in DA neurons results in mitochondrial dysfunction and increased oxidative stress in the neurons (Beyer et al. 2009; Wakabayashi et al. 2007).

Hsp70 Inhibits the Accumulation of α-Synuclein and Possess Protective Role in PD Condition

The α-Syn present presynaptically in the neurons which is linked to genetic form of PD (Stefanis 2012; Giraldez-Perez et al. 2014). Insoluble α-Syn oligomers form protofibrils which are toxic for the neurons and cause neuronal cell death. This phenomenon is collectively known as "synucleinopathies" characterized by accumulation of misfolded α-Syn protein and formation of intracellular protein bodies known as Lewy bodies (LWs) (Marti et al. 2003; Goedert et al. 2017). Hsp70 plays an important role to prevent the synucleinopathies by promoting proper folding of misfolded α-Syn or through its degradation or via chaperone mediated disaggregation (Ebrahimi-Fakhari et al. 2013; Witt 2010). Previous studies have shown that in PD patient there is an increase of Hsp70 expression in the substantia nigra pars compacta region (SNc) of the brain suggesting the involvement of Hsp70 during disease progression (Ebrahimi-Fakhari et al. 2013; Leak 2014). A study by Klucken et al. (2004) have shown that overexpression of Hsp70 in mice, expressing α-Syn, there is a reduce aggregation of α-Syn and Hsp70 can efficiently bind to α-Syn via its substrate-binding domain to the core hydrophobic region of soluble α-Syn intermediates. Hsp70 initiate open conformational state of α-Syn which inhibits α-Syn to bind with another α-Syn molecule and eventfully formation of oligomers (Witt 2009; Ebrahimi-Fakhari et al. 2013). (Auluck et al. 2002) have shown that Hsp70 increases the clearance of α-Syn accumulation and prevent dopaminergic neuronal loss in PD model of *Drosophila*. Accumulation of α -Syn in LBs are get cleared by the interaction between Hsp70 with C-terminal Hsp70-interacting protein (CHIP) and parkin related ubiquitinylation pathways which promotes proteosomal degradation of misfolded protein in PD (Kalia et al. 2015; Shin et al. 2005). CHIP tends to bind with Hsp70-misfolded protein complex and then promotes the ubiquitylation of the unfolded protein associated with Hsp70 and subsequently promotes its degradation via proteasome and lysosomal pathway (Shin et al. 2005; Donnelly et al. 2013). Previous study has shown that co-expression of CHIP with WT α-Syn in human H4 neuroglial cells results in a significant decrease in the α-Syn oligomerization and eventully, degrades α-Syn (Labrador-Garrido et al. 2011: Witt 2010; Klucken et al. 2004). Witt (2010) have shown that flies expressing human WT α-Syn shows 50% less neurons after 20 days as compare to control flies. Whereas, coexpression of Hsp70 with human WT α-Syn expressing flies shows no loss of fly neurons after 20 days. This experiment reveals that Hsp70 plays a protective role in case of PD.

Furthermore, Chaperone Mediated Autophagy (CMA) is an alternate pathways by which Hsp70 rescue the cell death induced by α-Syn. Hsp70 accelerates the degradation of α-Syn via CMA and thus protect the neuronal death (Witt 2010). Mutation in other PD causing gene "LRRK2" results in early onset of PD and neuronal toxicity. It was shown that CHIP involved in the degradation of LRRK2 with the help of Hsp70 by promoting its ubiquitylation and proteasomal degradation (Ding and Goldberg 2009; Ko et al. 2009).

DJ-1 is also associated with PD, mutation in DJ-1 causes early onset of PD with autosomal recessive inheritance. DJ-1 is present in mitochondria and has protective role in PD through ROS mitigating mechanism. It protects neuronal cells by hydrogen peroxide and 6-hydroxydopamine induced ROS production (Ariga et al. 2013). Lev et al. (2006) have demonstrated that DJ-1 inhibits protein aggregation and neurotoxicity by up-regulating Hsp70 in the cell. In another study, Hsp70 supresses the paraquate induced JNK and caspase activation in *Drosophila* model of PD (Shukla et al. 2014). Simmilarly, in DJ-1$\beta^{\Delta 93}$ flies, a paraquate induced ROS model of PD in *Drosophila*, Hsp70 was upregulated in the adult brain regions (Kumar et al. 2017). Overexpression of WT DJ-1 inhibits A53T human α-Syn toxicity via increasing the expression of Hsp70 (Ariga et al. 2013; Batelli et al. 2008). Therefore, this study suggests that Hsp70 might involve in the DJ-1 mediated inhibition of α-Syn aggregation and toxicity. DJ-1 colocallizes with mitochondrial Hsp70/Grp75 and CHIP in PD. Previous studies have shown that L166P and M26I mutant forms of DJ-1 found in PD were strongly associated with Hsp70 and CHIP as compared to the wild type and other mutants of DJ-1. This association enhances with the treatment of hydrogen peroxides to the cell and concluded that DJ-1 performs anti-oxidant activity to reduce the ROS with the help of Hsp70. In another form of PD, mutation in PTEN-induced putative kinase 1 (*PINK1*) and *Parkin*, an

E3 ubiquitin ligase causes the autosomal recessive form of disease. Previous studies have shown a link of Hsp70 with PINK1 and Parkin in PD (Zheng et al. 2018; Kumar et al. 2012; Yoo et al. 2018). A study by Zheng et al., 2018 demonstrated that Hsp70 stabilizes the PINK1 by decreasing its degradation. In case of Parkin, Hsp70 act as a substarte for ubiquitin ligase activity of Parkin (Moore et al. 2008). Overexpression of Hsp70 ameliorates the effect of parkin null mutation and protect the DA neurons in PD (Zhang et al. 2016). The significance of parkin-mediated ubiquitination of Hsp70 is still a question for debet.

Hsp70 and Huntington's Disease

HD is an inherited neurodegenerative disease caused due to polyglutamine (polyQ) expansion in the huntingtin (*Htt*) gene (Krench and Littleton 2013). It is an autosomal dominant disease caused by single gene mutation and manifests with 100% penetrance (Lewis and Smith 2016; Krench and Littleton 2013). The huntingtin (Htt) protein plays a vital role in the intracellular cascades and in case of HD, it gets mutated and deregulate the basic cellular processes. Mutation in the Htt protein alters the signaling pathways associated with neuronal survival and causes neuronal loss (Krench and Littleton 2013). In HD, the repetition of CAG sequence in the exon 1 of huntingtin gene (*Htt*) generates a protein with an expanded polyglutamine (polyQ) tail at the N terminus. The expansion of polyQ tails results in genetic defect characterized by expansion of glutamine self-aggregation or aggregation with different proteins and formation of inclusion bodies in the affected neurons which leadings to the neuronal death. Apart from HD, another human neurodegenerative diseases caused by polyQ expansion are dentatorubral and pollidoluysian atrophy, spinal and bulbar muscular atrophy and the spinocerebellar ataxias 1, 2, 3, 6, 7, and 17 (Krench and Littleton 2013; Wolfgang et al. 2005).

Hsp70 Inhibits the Toxicity of Huntingtin Protein in HD

Expression of polyQ proteins in cell culture, results in an increase expression of Hsp70 protein (Turturici et al. 2011; Tagawa et al. 2007). To finding out the role of Hsp70 in the HD, researcher used different polyQ models in cell lines, yeast, worm (*C. elegance*), fly (*D. melanogaster*) and Vertebrate (*Zebrafish*) (Kikis 2016; Wolfgang et al. 2005; Muchowski et al. 2000). Hansson et al. (2003) have shown that overexpression of Hsp70 showed moderate effect on neuropathologies associated with HD in R6/2 HD mice. Furthermore, Hsp70 binds with Bag-1 which is associated with inclusion bodies, reduces the pathologies of HD which are associated with polyQ-stretch huntingtin protein (Jana et al. 2005). It was shown that Hsp70 inhibits the formation of mature aggregation of polyQ-stretch of Htt protein with the help of Hsp40 in ATP dependent manner. Hsp70/Hsp40 causes intramolecular conformational changes in polyQ-stretch of huntingtin protein, immediately during, the formation of new aggregates, feasibly while huntingtin present in the monomeric condition. Thus, Hsp70 inhibits the formation of spherical and ring like oligomers's aggregation and promote the stabilization of Huntingtin protein in the monomeric form. Therefore, as a result of Hsp70 action on the polyQ-stretch huntingtin protein, it promotes inhibition of toxic effect of protein aggregation, fibrillary aggregation pathway and prevents accumulation of amorphous aggregates of the huntingtin protein in the HD (Wacker et al. 2004; Schaffar et al. 2004). By the changes of mutant huntingtin conformation by Hsp70, it is no longer involved in the inhibition of essential cellular pathways and machinery such as polyQ-containing transcription factors. Furthermore, Hsp70 inhibits an intra-molecular conformational change of mutant Htt protein. It prevents heterotypic interactions between polyQ expanded htt and transcription factors associated with it. Additionally, Hsp70 helps in the reduction of oxidative stress in HD cellular model. The overexpression of Hsp70 inhibits the intracellular accumulation of iron ions in the cells transiently expressing the polyQ expanded Htt protein and decreases the formation of reactive oxygen species (ROS) (Wyttenbach et al. 2002).

Hsp70 and Spinocerebellar Ataxia

Spinocerebellar Ataxia (SCA) is an autosomal dominant disorder associated with degeneration of cerebellum, often accompanied by central nervous system and brainstem degeneration (and less commonly the peripheral nervous system) (Taroni and DiDonato 2004; Paulson 2009). The pathologies associated with SCA are expanded CAG/polyQ ataxia and non-protein coding repeat expansion ataxia (Paulson 2009). Hsp70 plays a vital role in the neuroprotection in the case of SCA. (Cummings et al. 2001) have shown that overexpression of Hsp70 play a neuroprotective role in SCA1 mice model. (Helmlinger et al. 2004) have demonstrated that Hsp70 attenuates poly Q induced toxicity in SCA3 fly model and in mouse mode. Previous studies have shown that overexpression of Hsp70 in SCA3 poly Q diseased model of *Drosophila* and SBMA mouse model reduced the toxicity of the disease protein without any effect on protein aggregation (Adachi et al. 2003; Cummings et al. 2001; Warrick et al. 1999).

Hsp70 and Spinal and Bulbar Muscular Atrophy

Spinal & Bulbar Muscular Atrophy (SBMA) also known as Kennedy's disease (Kennedy et al. 1968), is an X linked inherited neuromuscular disorder characterized by loss of motor neurons, progressive muscle wasting, muscles weakness, fasciculations and one of the most common genetic causes of infant death. A trinucleotide (CAG) repeat expansion in the androgen receptor (*AR*) gene on the X chromosome occur in the SBMA. This mutation results in an androgen-dependent toxic gain of function in the mutant protein and an expanded polyglutamine tract (Grunseich et al. 2014; Kennedy and Alter 2000). Recent studies have shown that SBMA can be cure by decreasing the amount of polyglutamine androgen receptor (polyQ AR). HSP70 involves in the maintenance of polyQ AR proteostasis. Hsp70 with co-chaperone, Hsp70 interacting protein (Hip) promotes polyQ AR ubiquitination and its clearance. Overexpression of Hsp70 ameliorates the toxicity of the *Drosophila* model of SMA (Wang et al. 2013a). A study by (Adachi et al. 2003) demonstrated that the overexpression of Hsp70 decreases the localization of mutant androgen receptor protein in nucleus and rescue the phenotype associated with spinal and bulbar muscular atrophy in transgenic mouse model. Furthermore, they have shown that overexpression of Hsp70 decreases the mutant AR aggregation and neuronal cell death in neurodegenerative disease model (Kobayashi et al. 2000; Adachi et al. 2003). The treatment with arimoclomol, a well-known HSP co-inducer induced the expression of Hsp70 in spinal and bulbar muscular atrophy AR100 mice so it helps in the delays disease progression. Increased level of Hsp70 promotes proteiosomal degradation of mutant androgen receptor protein (Bailey et al. 2002; Malik et al. 2013). Additionally, Overexpression of human Hsp70 decreases the disease pathology and improves the phenotype of diseased mice (Adachi et al. 2003). Another compound geranylgeranylacetone also known for induction of Hsp70, Hsp90 and Hsp105 via activation of HSF1 (Katsuno et al. 2005; Wang et al. 2014b). Hence, the treatment with geranylgeranylacetone, ameliorates the neuromuscular phenotype in SMA in an AR-97Q mouse model of spinal and bulbar muscular atrophy (Katsuno et al. 2005; Malik et al. 2013).

Hsp70 and Amyotrophic Lateral Sclerosis

Amyotrophic Lateral Sclerosis (ALS) is a diseases associated with degeneration of specific nerve cells regulating muscle movement. ALS is characterized by progressive loss of motor neurone (atrophy) resulting in muscle weakness and inability of motor control (Gifondorwa et al. 2007). It is a disease of upper motor neurons (UMN) and lower (LMN) motor neurons in the brain and spinal cord (Gordon 2013). Copper, zinc *superoxide dismutase 1 (SOD1)* mutation associated with familial ALS. Mutated SOD1 aggragation occur in the mitochondria in the case of familial ALS. HSP70 plays an important role in the decreases the ALS pathology associated with SOD1 mutation (Banci et al. 2008). A study by (Miyazaki et al. 2016) shown that serum level of Hsp70 remain increased throughout the disease period in the patient. So, HSP70 might have a vital role in the ALS. Hsp70 plays a neuroprotective role by inhibited the accumulation of mSOD1 (mutant SOD1) accumulation (Jain et al. 2008). The administration of exogenous recombinant human rhHsp70 in G93A SOD1 mice, increases the life span, induced motor normal function and prolonging motor neurons survival and delayed ALS disease symptoms. By the treatment with exogenous rhHsp70, increasing in the number of innervated neuromuscular junctions as compare to control tissue (Gifondorwa et al.

2007). It also inhibits plaques formation and neuronal toxicity in SOD1^{G93A} mutant flies.

Hsp70 and Fragile X-Syndrome

Fragile X-syndrome is the neurodegenerative disorder caused due to the multiple repeat of CGG trinucleotide (>200 repeats) in 5′ UTR of *FMR1*gene and results in transcriptional silencing and loss of Fragile X Mental Retardation Protein (FMRP) (O'Donnell and Warren 2002; Jin et al. 2003). Studies have suggested that this repeat disorder also has neuronal inclusion bodies that are Hsp70 and ubiquitin positive and overexpression of Hsp70 suppress this phenotype (Jin et al. 2003). Overexpression of Hsp70 suppress the CGG100GFP-induced neurodegeneration by enhancing ubiquitin-Proteasomal system for degradation of protein aggregates and decreases CGG-repeat-elicited toxicity in *Drosophila*. Co-expression of Hsp70 rescues the rough eye phenotype in CGG_{100}GFP expressing *Drosophila* (Oh et al. 2015).

Therapeutic Strategies Using Hsp70

As mentioned above, several studies have suggested the neuroprotective potential of Hsp70 in various diseased conditions. Thus, to find out the mechanistic details how to modulate/control the functions of Hsp70 in diseased conditions is one of the key area for research to explore the possibility of the use of Hsp70 in diseased conditions. The following strategies can be used as potentials ways to modulate/control the Hsp70 in diseased conditions.

Modulation of Endogenous H70 Expression Level

As discussed, Hsp70 plays a vital role in the prevention of the formation of oligomers in the neurodegenerative disease and therefore they inhibit toxicity (Pockley 2001; Paul and Mahanta 2014). The induction of Hsp70 might be a promising approach for the treatment of various neurodegenerative disease such as AD, PD, HD, ALS, SMA and fragile X syndrome (Adachi et al. 2005). The pharmacological induction of Hsp70 is one of the hopeful ways to modulate the expression of Hsp70. As an example some drugs are being used for the treatment of AD which are involved in the induction of endogenous Hsp70 level. A study by (Dou et al. 2003) demonstrated that supplementation of geldanamycin increases the endogenous Hsp70 level and prevent the formation of neurofibrillary tangles in the COS-1 cell (Lu et al. 2014). Several studies have suggested that by controlling the activity of Hsp70 there are probability of regulating tau-based neurodegeneration. It has been shown that Hsp70 activation increases tau stability and inhibition of Hsp70 reduces tau expression (Jinwal et al. 2009). Additionally, celastrol, pentacyclic triterpene act as a Hsp90 inhibitor and Hsp70 inducer and shows neuroprotective effects in polyQ diseased models and in HD (Hay et al. 2004; Cummings et al. 2001; Zhang and Sarge 2007). Geranylgeranylacetone (GGA), an ulcer drug and cyclic isoprenoid has been known for its role in activation of Hsp70. Previous study has shown that oral administration of GGA induced expression of Hsp70 along with Hsp90 and Hsp105 by activating the HSF1. GGA increases the Hsp70 level in CNS and reduces the accumulation of AR protein in the case of SBMA mouse model (Katsuno et al. 2005). There are more chemical compounds such as 17-allylamino-17-demethoxygeldanamycin (17-AAG), 17-(dimethylaminoethylamino)-17- demethoxygeldanamycin (17-DMAG) inhibits the toxicity associated with SBMA and HD, respectively (Waza et al. 2005; Sittler et al. 2001). Arimoclomil (BRX-220), a hydroxylamine derivatives an experimental drug developed by CytRx Corp, is an oral therapeutic medicine for the treatment of amyotrophic lateral sclerosis (ALS). It will help in the inhibition of the toxicity associated with ALS by inducing the expression of Hsp70 (Phukan 2010). BRX-220 also promotes the neuronal survival by increasing the expression of Hsp70 (Vigh et al. 1997; Ousman et al. 2017). A systematic administration of immunosuppressant drugs such as cyclosporine A and FK506 increases the expression of Hsp70 that enhances axonal regeneration and causes the recovery of sciatic nerve crush injury in rat (Gold 1997; Ousman et al. 2017). The other compound herbimycins, a benzoquinone ansamycin antibiotic induces expression of Hsp70 and prevent the pathologies of neurodegeneration (Shaaban et al. 2013). Riluzole is pharmacological drug that is commonly used for the treatment of ALS is also induces the expression of Hsp70 and leads to the survival of primary embryonic spinal cord (Yang et al. 2008; Neef et al. 2011). It has been also noted that pharmacological inhibition of Hsp90 leads to release of Heat Shock Factor 1(HSF1) from it, the free HSF1 can also induced the endogenous Hsp70 (Bagatell et al. 2000; Neckers and Workman 2012; Soga et al. 2013), which can mitigate the neurodegeneration.

Administration of Exogenous Hsp70

Studies have shown that administration of exogenous Hsp70 is helpful in AD treatment. (Bobkova et al. 2015; Kampinga and Bergink 2016) have shown that intranasally-administered Hsp70, efficiently entered into the specific brain regions such as olfactory bulbs, hippocampus, cortex etc. in the mice and helps in reducing Aβ plaque in the mice. Furthermore, exogenous Hsp70 improve spatial memory in mice. (Morozov et al. 2017) have demonstrated that exogenous administration of Hsp70 decreases intracellular proteasomes activity of iso$A\beta_{42}$ treated cells.

Inhibition of Hsp70 ATPase Activity

The ATPase (catalytic) activity of Hsp70 is associated with increasing the affinity of Hsp70 to bind with misfolded, aggregated proteins by changing HSP70 conformation from adenosine triophosphate (ATP) bound form to adenosine diphosphate (ADP) bound form. Previous researches have shown that the inactivation of Hsp70 ATPase activity enhances the capacity of Hsp70 to bind with misfolded, aggregated proteins by promoting its ADP bounded form. Hence, inhibition of ATPase activity increases the proper protein folding, initiate degradation or removal of protein aggregates in AD. Thus, these studies suggested that modulation (activation/inhibition) in ATPase activity of Hsp70 could be beneficial for neurons in the case of neurodegenerative diseases (Repalli and Meruelo 2015). Some molecules that can be used for the inhibition of Hsp70 ATPase activity and keep Hsp70 into its ADP bounded form are MKT-07767, YM-01, methylene blue (MB), and YM-08 (Miyata et al. 2011; Repalli and Meruelo 2015; Abisambra et al. 2013). Other bioactive compounds such as myricetin (MY, flavones) and Azure C (AC, benzothiazines) used to enhance HSP70 activity for elimination of aggregated protein by inhibiting its ATPase activity (Jinwal et al. 2009).

Conclusions

Hsp70 are highly conserved protein and plays a pivotal role during different cellular events. One of the key function is protein folding by which it can regulate the activity of a protein. If a protein is not properly folded it should be refolded or can be going for degradation pathway using ubiquitin proteasome system. Thus, Hsp70 is very important to maintain the cellular homeostasis. As discussed above overexpression of Hsp70 plays a protective role in AD, PD, ALS and Poly Q diseases and there are several molecules that can exogenously induce the expression of Hsp70 and help in safety mechanism. Thus, Hsp70 can be used as a potential therapeutic agent to target various neurodegenerative diseases. Due to advancement of science & technology, onset of neurodegenerative diseases became a major challenging problem for the human beings. As discussed above, protein quality control plays a key role in NDs and this quality control is regulated by molecular chaperones, ubiquitin ligases and autophagy lysosomal pathway etc. In recent year's several efforts have been taken to resolves the issues related with these diseases but still there are no proper cures for any of these diseases. Thus, more and more basic research is required to find out the detailed molecular mechanism of these diseases to use the information for the development of therapeutics targets to control/treat these diseases. As discussed above most of the NDs are due to the misfolding of protein and their accumulation in cell that ultimately generate cellular stress and NDs. Thus, molecules that can refold are initiate the refolding of misfolded proteins or initiate the clearance process can be a key area of research in the search of therapeutic target for the diseases. As described, Hsp70 possess neuroprotective potential thus, identification of mechanistic details of Hsp70 and the molecules that can modulate the activity of Hsp70 will shed a light in the area of use of Hsp70 in different diseased conditions.

Acknowledgements The work in our laboratory was supported by grants from Department of Science & Technology (DST), New Delhi, India (SR/FT/LS-1/2010), Gujarat State Biotechnology Mission (GSBTM) (GSBTM/MD/PROJECTS/SSA/456/2010-2011), and Gujarat Council on Science & Technology (GUJCOST) (GUJCOST/MRP/2015-16/2680) Gujarat. The Laser Scanning Confocal Microscope facility supported by Department of Biotechnology, India is duly acknowledged.

Conflict of Interest The authors declare that they have no conflict of interest.

References

Abisambra, J., Jinwal, U. K., Miyata, Y., et al. (2013). Allosteric heat shock protein 70 inhibitors rapidly rescue synaptic plasticity deficits by reducing aberrant tau. *Biological Psychiatry, 74*, 367–374.

Adachi, H., Katsuno, M., Minamiyama, M., et al. (2003). Heat shock protein 70 chaperone overexpression ameliorates phenotypes of the spinal and bulbar muscular atrophy transgenic mouse model by reducing nuclear-localized mutant androgen receptor protein. *The Journal of Neuroscience, 23*, 2203–2211.

Adachi, H., Katsuno, M., Minamiyama, M., et al. (2005). Widespread nuclear and cytoplasmic accumulation of mutant androgen receptor in SBMA patients. *Brain, 128*, 659–670.

Alexander, G. E. (2004). Biology of Parkinson's disease: Pathogenesis and pathophysiology of a multisystem neurodegenerative disorder. *Dialogues in Clinical Neuroscience, 6*, 259–280.

Andersen, J. K. (2004). Oxidative stress in neurodegeneration: Cause or consequence? *Nature Medicine, 10*(Suppl), S18–S25.

Ardley, H. C., & Robinson, P. A. (2004). The role of ubiquitin-protein ligases in neurodegenerative disease. *Neurodegenerative Diseases, 1*, 71–87.

Ariga, H., Takahashi-Niki, K., Kato, I., Maita, H., Niki, T., & Iguchi-Ariga, S. M. (2013). Neuroprotective function of DJ-1 in Parkinson's disease. *Oxidative Medicine and Cellular Longevity, 2013*, 683920.

Arya, R., Mallik, M., & Lakhotia, S. C. (2007). Heat shock genes – Integrating cell survival and death. *Journal of Biosciences, 32*, 595–610.

Auluck, P. K., Chan, H. Y., Trojanowski, J. Q., Lee, V. M., & Bonini, N. M. (2002). Chaperone suppression of alpha-synuclein toxicity in a Drosophila model for Parkinson's disease. *Science, 295*, 865–868.

Bagatell, R., Paine-Murrieta, G. D., Taylor, C. W., Pulcini, E. J., Akinaga, S., Benjamin, I. J., & Whitesell, L. (2000). Induction of a heat shock factor 1-dependent stress response alters the cytotoxic activity of Hsp90-binding agents. *Clinical Cancer Research, 8*, 3312–3318.

Bailey, C. K., Andriola, I. F., Kampinga, H. H., & Merry, D. E. (2002). Molecular chaperones enhance the degradation of expanded polyglutamine repeat androgen receptor in a cellular model of spinal and bulbar muscular atrophy. *Human Molecular Genetics, 11*, 515–523.

Ballatore, C., Lee, V. M., & Trojanowski, J. Q. (2007). Tau-mediated neurodegeneration in Alzheimer's disease and related disorders. *Nature Reviews. Neuroscience, 8*, 663–672.

Banci, L., Bertini, I., Boca, M., Girotto, S., Martinelli, M., Valentine, J. S., & Vieru, M. (2008). SOD1 and amyotrophic lateral sclerosis: Mutations and oligomerization. *PLoS One, 3*, e1677.

Batelli, S., Albani, D., Rametta, R., Polito, L., Prato, F., Pesaresi, M., Negro, A., & Forloni, G. (2008). DJ-1 modulates alpha-synuclein aggregation state in a cellular model of oxidative stress: Relevance for Parkinson's disease and involvement of HSP70. *PLoS One, 3*, e1884.

Beere, H. M. (2001). Stressed to death: Regulation of apoptotic signaling pathways by the heat shock proteins. *Science's STKE, 2001*, re1.

Beyer, K., Domingo-Sabat, M., & Ariza, A. (2009). Molecular pathology of Lewy body diseases. *International Journal of Molecular Sciences, 10*, 724–745.

Blair, L. J., Sabbagh, J. J., & Dickey, C. A. (2014). Targeting Hsp90 and its co-chaperones to treat Alzheimer's disease. *Expert Opinion on Therapeutic Targets, 18*, 1219–1232.

Bobkova, N. V., Evgen'ev, M., Garbuz, D. G., et al. (2015). Exogenous Hsp70 delays senescence and improves cognitive function in aging mice. *Proceedings of the National Academy of Sciences of the United States of America, 112*, 16006–16011.

Bonini, N. M., & Gitler, A. D. (2011). Model organisms reveal insight into human neurodegenerative disease: Ataxin-2 intermediate-length polyglutamine expansions are a risk factor for ALS. *Journal of Molecular Neuroscience, 45*, 676–683.

Broadley, S. A., & Hartl, F. U. (2009). The role of molecular chaperones in human misfolding diseases. *FEBS Letters, 583*, 2647–2653.

Brocchieri, L., Conway de Macario, E., & Macario, A. J. (2008). Hsp70 genes in the human genome: Conservation and differentiation patterns predict a wide array of overlapping and specialized functions. *BMC Evolutionary Biology, 8*, 19.

Buchman, A. S., Leurgans, S. E., Nag, S., Bennett, D. A., & Schneider, J. A. (2011). Cerebrovascular disease pathology and parkinsonian signs in old age. *Stroke, 42*, 3183–3189.

Bunton-Stasyshyn, R. K., Saccon, R. A., Fratta, P., & Fisher, E. M. (2015). SOD1 function and its implications for amyotrophic lateral sclerosis pathology: New and renascent themes. *The Neuroscientist, 21*, 519–529.

Burchell, V. S., Nelson, D. E., Sanchez-Martinez, A., et al. (2013). The Parkinson's disease-linked proteins Fbxo7 and Parkin interact to mediate mitophagy. *Nature Neuroscience, 16*, 1257–1265.

Carmine-Simmen, K., Proctor, T., Tschape, J., Poeck, B., Triphan, T., Strauss, R., & Kretzschmar, D. (2009). Neurotoxic effects induced by the Drosophila amyloid-beta peptide suggest a conserved toxic function. *Neurobiology of Disease, 33*, 274–281.

Cassar, M., & Kretzschmar, D. (2016). Analysis of amyloid precursor protein function in Drosophila melanogaster. *Frontiers in Molecular Neuroscience, 9*, 61.

Chakraborty, R., Vepuri, V., Mhatre, S. D., et al. (2011). Characterization of a Drosophila Alzheimer's disease model: Pharmacological rescue of cognitive defects. *PLoS One, 6*, e20799.

Cookson, M. R. (2012). Evolution of neurodegeneration. *Current Biology, 22*, R753–R761.

Cummings, C. J., Sun, Y., Opal, P., Antalffy, B., Mestril, R., Orr, H. T., Dillmann, W. H., & Zoghbi, H. Y. (2001). Over-expression of inducible HSP70 chaperone suppresses neuropathology and improves motor function in SCA1 mice. *Human Molecular Genetics, 10*, 1511–1518.

Dachsel, J. C., Wider, C., Vilarino-Guell, C., et al. (2011). Death-associated protein kinase 1 variation and Parkinson's disease. *European Journal of Neurology, 18*, 1090–1093.

Daugaard, M., Rohde, M., & Jaattela, M. (2007). The heat shock protein 70 family: Highly homologous proteins with overlapping and distinct functions. *FEBS Letters, 581*, 3702–3710.

de Diego-Otero, Y., Romero-Zerbo, Y., el Bekay, R., Decara, J., Sanchez, L., Rodriguez-de Fonseca, F., & del Arco-Herrera, I. (2009). Alpha-tocopherol protects against oxidative stress in the fragile X knockout mouse: An experimental therapeutic approach for the Fmr1 deficiency. *Neuropsychopharmacology, 34*, 1011–1026.

de Tullio, M. B., Morelli, L., & Castano, E. M. (2008). The irreversible binding of amyloid peptide substrates to insulin-degrading enzyme: A biological perspective. *Prion, 2*, 51–56.

de Tullio, M. B., Castelletto, V., Hamley, I. W., Martino Adami, P. V., Morelli, L., & Castano, E. M. (2013). Proteolytically inactive insulin-degrading enzyme inhibits amyloid formation yielding non-neurotoxic abeta peptide aggregates. *PLoS One, 8*, e59113.

DeMaagd, G., & Philip, A. (2015). Parkinson's disease and its management: Part 1: Disease entity, risk factors, pathophysiology, clinical presentation, and diagnosis. *P T, 40*, 504–532.

Ding, X., & Goldberg, M. S. (2009). Regulation of LRRK2 stability by the E3 ubiquitin ligase CHIP. *PLoS One, 4*, e5949.

Donnelly, B. F., Needham, P. G., Snyder, A. C., Roy, A., Khadem, S., Brodsky, J. L., & Subramanya, A. R. (2013). Hsp70 and Hsp90 multichaperone complexes sequentially regulate thiazide-sensitive cotransporter endoplasmic reticulum-associated degradation and biogenesis. *The Journal of Biological Chemistry, 288*, 13124–13135.

Dou, F., Netzer, W. J., Tanemura, K., Li, F., Hartl, F. U., Takashima, A., Gouras, G. K., Greengard, P., & Xu, H. (2003). Chaperones increase association of tau protein with microtubules. *Proceedings of the National Academy of Sciences of the United States of America, 100*, 721–726.

Ebrahimi-Fakhari, D., Saidi, L. J., & Wahlster, L. (2013). Molecular chaperones and protein folding as therapeutic targets in Parkinson's disease and other synucleinopathies. *Acta Neuropathologica Communications, 1*, 79.

Elefant, F., & Palter, K. B. (1999). Tissue-specific expression of dominant negative mutant Drosophila HSC70 causes developmental defects and lethality. *Molecular Biology of the Cell, 10*, 2101–2117.

Evans, C. G., Wisen, S., & Gestwicki, J. E. (2006). Heat shock proteins 70 and 90 inhibit early stages of amyloid beta-(1-42) aggregation in vitro. *The Journal of Biological Chemistry, 281*, 33182–33191.

Feany, M. B., & Bender, W. W. (2000). A Drosophila model of Parkinson's disease. *Nature, 404*, 394–398.

Fernandez-Funez, P., Sanchez-Garcia, J., de Mena, L., Zhang, Y., Levites, Y., Khare, S., Golde, T. E., & Rincon-Limas, D. E. (2016). Holdase activity of secreted Hsp70 masks amyloid-beta42 neurotoxicity in Drosophila. *Proceedings of the National Academy of Sciences of the United States of America, 113*, E5212–E5221.

Flaherty, K. M., DeLuca-Flaherty, C., & McKay, D. B. (1990). Three-dimensional structure of the ATPase fragment of a 70K heat-shock cognate protein. *Nature, 346*, 623–628.

Gething, M.-J., & Sambrook, J. (1992). Protein folding in the cell. *Nature, 355*, 33–45.

Gidalevitz, T., Prahlad, V., & Morimoto, R. I. (2011). The stress of protein misfolding: From single cells to multicellular organisms. *Cold Spring Harbor Perspectives in Biology, 3*, a009704.

Gifondorwa, D. J., Robinson, M. B., Hayes, C. D., Taylor, A. R., Prevette, D. M., Oppenheim, R. W., Caress, J., & Milligan, C. E. (2007). Exogenous delivery of heat shock protein 70 increases lifespan in a mouse model of amyotrophic lateral sclerosis. *The Journal of Neuroscience, 27*, 13173–13180.

Giraldez-Perez, R., Antolin-Vallespin, M., Munoz, M., & Sanchez-Capelo, A. (2014). Models of alpha-synuclein aggregation in Parkinson's disease. *Acta Neuropathologica Communications, 2*, 176.

Glenner, G. G., & Wong, C. W. (1984). Alzheimer's disease: Initial report of the purification and characterization of a novel cerebrovascular amyloid protein. *Biochemical and Biophysical Research Communications, 120*, 885–890.

Goedert, M., Jakes, R., & Spillantini, M. G. (2017). The synucleinopathies: Twenty years on. *Journal of Parkinsons disease, 7*, S53–S71.

Gold, B. G. (1997). FK506 and the role of immunophilins in nerve regeneration. *Molecular Neurobiology, 15*, 285–306.

Gong, C. X., & Iqbal, K. (2008). Hyperphosphorylation of microtubule-associated protein tau: A promising therapeutic target for Alzheimer disease. *Current Medicinal Chemistry, 15*, 2321–2328.

Gordon, P. H. (2013). Amyotrophic Lateral Sclerosis: An update for 2013 clinical features, pathophysiology, management and therapeutic trials. *Aging and Disease, 4*, 295–310.

Grunseich, C., Rinaldi, C., & Fischbeck, K. H. (2014). Spinal and bulbar muscular atrophy: Pathogenesis and clinical management. *Oral Diseases, 20*, 6–9.

Gupta, R. S., & Singh, B. (1994). Phylogenetic analysis of 70 kD heat shock protein sequences suggests a chimeric origin for the eukaryotic cell nucleus. *Current Biology, 4*, 1104–1114.

Hansson, O., Nylandsted, J., Castilho, R. F., Leist, M., Jaattela, M., & Brundin, P. (2003). Overexpression of heat shock protein 70 in R6/2 Huntington's disease mice has only modest effects on disease progression. *Brain Research, 970*, 47–57.

Hartl, F. U. (1996). Molecular chaperones in cellular protein folding. *Nature, 381*, 571–579.

Hay, D. G., Sathasivam, K., Tobaben, S., et al. (2004). Progressive decrease in chaperone protein levels in a mouse model of Huntington's disease and induction of stress proteins as a therapeutic approach. *Human Molecular Genetics, 13*, 1389–1405.

Helmlinger, D., Hardy, S., Sasorith, S., et al. (2004). Ataxin-7 is a subunit of GCN5 histone acetyltransferase-containing complexes. *Human Molecular Genetics, 13*, 1257–1265.

Hendrick, J. P., & Hartl, F.-U. (1993). Molecular chaperone functions of heat-shock proteins. *Annual Review of Biochemistry, 62*, 349–384.

Huang, W. J., Zhang, X., & Chen, W. W. (2016). Role of oxidative stress in Alzheimer's disease. *Biomed Rep, 4*, 519–522.

Hunt, C., & Morimoto, R. I. (1985). Conserved features of eukaryotic hsp70 genes revealed by comparison with the nucleotide sequence of human hsp70. *Proceedings of the National Academy of Sciences of the United States of America, 82*, 6455–6459.

Iijima, K., Chiang, H. C., Hearn, S. A., et al. (2008). Abeta42 mutants with different aggregation profiles induce distinct pathologies in Drosophila. *PLoS One, 3*, e1703.

Iijima-Ando, K., & Iijima, K. (2010). Transgenic Drosophila models of Alzheimer's disease and tauopathies. *Brain Structure & Function, 214*, 245–262.

Iqbal, K., Liu, F., Gong, C. X., Alonso Adel, C., & Grundke-Iqbal, I. (2009). Mechanisms of tau-induced neurodegeneration. *Acta Neuropathologica, 118*, 53–69.

Jaattela, M., & Wissing, D. (1992). Emerging role of heat shock proteins in biology and medicine. *Annals of Medicine, 24*, 249–258.

Jahn, H. (2013). Memory loss in Alzheimer's disease. *Dialogues in Clinical Neuroscience, 15*, 445.

Jain, M. R., Ge, W. W., Elkabes, S., & Li, H. (2008). Amyotrophic lateral sclerosis: Protein chaperone dysfunction revealed by proteomic studies of animal models. *Proteomics. Clinical Applications, 2*, 670–684.

Jana, N. R., Dikshit, P., Goswami, A., Kotliarova, S., Murata, S., Tanaka, K., & Nukina, N. (2005). Co-chaperone CHIP associates with expanded polyglutamine protein and promotes their degradation by proteasomes. *The Journal of Biological Chemistry, 280*, 11635–11640.

Jin, P., Zarnescu, D. C., Zhang, F., Pearson, C. E., Lucchesi, J. C., Moses, K., & Warren, S. T. (2003). RNA-mediated neurodegeneration caused by the fragile X premutation rCGG repeats in Drosophila. *Neuron, 39*, 739–747.

Jinwal, U. K., Miyata, Y., Koren, J., 3rd, et al. (2009). Chemical manipulation of hsp70 ATPase activity regulates tau stability. *The Journal of Neuroscience, 29*, 12079–12088.

Kakimura, J., Kitamura, Y., Takata, K., et al. (2002). Microglial activation and amyloid-beta clearance induced by exogenous heat-shock proteins. *The FASEB Journal, 16*, 601–603.

Kakkar, V., Kuiper, E. F., Pandey, A., Braakman, I., & Kampinga, H. H. (2016). Versatile members of the DNAJ family show Hsp70 dependent anti-aggregation activity on RING1 mutant parkin C289G. *Scientific Reports, 6*, 34830.

Kalia, L. V., Kalia, S. K., & Lang, A. E. (2015). Disease-modifying strategies for Parkinson's disease. *Movement Disorders, 30*, 1442–1450.

Kampinga, H. H., & Bergink, S. (2016). Heat shock proteins as potential targets for protective strategies in neurodegeneration. *Lancet Neurology, 15*, 748–759.

Katsuno, M., Sang, C., Adachi, H., Minamiyama, M., Waza, M., Tanaka, F., Doyu, M., & Sobue, G. (2005). Pharmacological induction of heat-shock proteins alleviates polyglutamine-mediated motor neuron disease. *Proceedings of the National Academy of Sciences of the United States of America, 102*, 16801–16806.

Kennedy, W. R., & Alter, M. (2000). Progressive proximal spinal and bulbar muscular atrophy of late onset: A sex-linked recessive trait. *Journal of Clinical Neuromuscular Disease, 2*, 3–5.

Kennedy, W. R., Alter, M., & Sung, J. H. (1968). Progressive proximal spinal and bulbar muscular atrophy of late onset: A sex-linked recessive trait. *Neurology, 18*, 671–671.

Kikis, E. A. (2016). The struggle by Caenorhabditis elegans to maintain proteostasis during aging and disease. *Biology Direct, 11*, 58.

Klein, C., & Westenberger, A. (2012). Genetics of Parkinson's disease. *Cold Spring Harbor Perspectives in Medicine, 2*, a008888.

Klucken, J., Shin, Y., Masliah, E., Hyman, B. T., & McLean, P. J. (2004). Hsp70 reduces alpha-synuclein aggregation and toxicity. *The Journal of Biological Chemistry, 279*, 25497–25502.

Ko, H. S., Bailey, R., Smith, W. W., et al. (2009). CHIP regulates leucine-rich repeat kinase-2 ubiquitination, degradation, and toxicity. *Proceedings of the National Academy of Sciences of the United States of America, 106*, 2897–2902.

Kobayashi, Y., Kume, A., Li, M., Doyu, M., Hata, M., Ohtsuka, K., & Sobue, G. (2000). Chaperones Hsp70 and Hsp40 suppress aggregate formation and apoptosis in cultured neuronal cells expressing truncated androgen receptor protein with expanded polyglutamine tract. *The Journal of Biological Chemistry, 275*, 8772–8778.

Koren, J., 3rd, Jinwal, U. K., Lee, D. C., Jones, J. R., Shults, C. L., Johnson, A. G., Anderson, L. J., & Dickey, C. A. (2009). Chaperone signalling complexes in Alzheimer's disease. *Journal of Cellular and Molecular Medicine, 13*, 619–630.

Kumar, A., & Tiwari, A. K. (2017). Molecular chaperone Hsp70 and its constitutively active form Hsc70 play an indispensable role during eye development of Drosophila melanogaster. *Molecular Neurobiology, 55*, 4345–4361.

Kumar, P., Pradhan, K., Karunya, R., Ambasta, R. K., & Querfurth, H. W. (2012). Cross-functional E3 ligases Parkin and C-terminus Hsp70-interacting protein in neurodegenerative disorders. *Journal of Neurochemistry, 120*, 350–370.

Kumar, A., Christian, P. K., Panchal, K., Guruprasad, B. R., & Tiwari, A. K. (2017). Supplementation of spirulina (Arthrospira platensis) improves lifespan and locomotor activity in paraquat-sensitive DJ-1β Δ93 Flies, a Parkinson's disease model in Drosophila melanogaster. *Journal of Dietary Supplements, 14*, 573–588.

Kityk, R., Kopp, J., Sinning, I., & Mayer, M. P. (2012). Structure and dynamics of the ATP-bound open conformation of Hsp70 chaperones. *Molecular Cell, 48*, 863–874.

Krench, M., & Littleton, J. T. (2013). Modeling Huntington disease in Drosophila: Insights into axonal transport defects and modifiers of toxicity. *Fly (Austin), 7*, 229–236.

La Spada, A. (2014). *Spinal and bulbar muscular atrophy*.

Labrador-Garrido, A., Bertoncini, C. W., & Roodveldt, C. (2011). The HSP70 chaperone system in Parkinson's disease. In: *Etiology and pathophysiology of Parkinson's disease*. InTech.

Laffita-Mesa, J. M., Rodriguez Pupo, J. M., Moreno Sera, R., et al. (2013). De novo mutations in ataxin-2 gene and ALS risk. *PLoS One, 8*, s.

Leak, R. K. (2014). Heat shock proteins in neurodegenerative disorders and aging. *J Cell Commun Signal, 8*, 293–310.

Lev, N., Roncevic, D., Ickowicz, D., Melamed, E., & Offen, D. (2006). Role of DJ-1 in Parkinson's disease. *Journal of Molecular Neuroscience, 29*, 215–225.

Lewis, E. A., & Smith, G. A. (2016). Using Drosophila models of Huntington's disease as a translatable tool. *Journal of Neuroscience Methods, 265*, 89–98.

Li, C. Y., Lee, J. S., Ko, Y. G., Kim, J. I., & Seo, J. S. (2000). Heat shock protein 70 inhibits apoptosis downstream of cytochrome c release and upstream of caspase-3 activation. *The Journal of Biological Chemistry, 275*, 25665–25671.

Li, H., Liu, L., Xing, D., & Chen, W. R. (2010). Inhibition of the JNK/Bim pathway by Hsp70 prevents Bax activation in UV-induced apoptosis. *FEBS Letters, 584*, 4672–4678.

Li, J.-Q., Tan, L., & Yu, J.-T. (2014). The role of the LRRK2 gene in Parkinsonism. *Molecular Neurodegeneration, 9*, 47.

Lindquist, S., & Craig, E. A. (1988). The heat-shock proteins. *Annual Review of Genetics, 22*, 631–677.

Liu, J., Shinobu, L. A., Ward, C. M., Young, D., & Cleveland, D. W. (2005). Elevation of the Hsp70 chaperone does not effect toxicity in mouse models of familial amyotrophic lateral sclerosis. *Journal of Neurochemistry, 93*, 875–882.

Lu, B., & Vogel, H. (2009). Drosophila models of neurodegenerative diseases. *Annual Review of Pathology, 4*, 315–342.

Lu, R. C., Tan, M. S., Wang, H., Xie, A. M., Yu, J. T., & Tan, L. (2014). Heat shock protein 70 in Alzheimer's disease. *BioMed Research International, 2014*, 435203.

Ludolph, A., Drory, V., Hardiman, O., Nakano, I., Ravits, J., Robberecht, W., & Shefner, J. (2015). A revision of the El Escorial criteria - 2015. *Amyotroph Lateral Scler Frontotemporal Degener, 16*, 291–292.

Luo, X., Zuo, X., Zhou, Y., Zhang, B., Shi, Y., Liu, M., Wang, K., McMillian, D. R., & Xiao, X. (2008). Extracellular heat shock protein 70 inhibits tumour necrosis factor-alpha induced proinflammatory mediator production in fibroblast-like synoviocytes. *Arthritis Research & Therapy, 10*, R41.

Mackenzie, I. R. A., & Rademakers, R. (2008). The role of TDP-43 in amyotrophic lateral sclerosis and frontotemporal dementia. *Current Opinion in Neurology, 21*, 693.

Magrinelli, F., Picelli, A., Tocco, P., Federico, A., Roncari, L., Smania, N., Zanette, G., & Tamburin, S. (2016). Pathophysiology of motor dysfunction in Parkinson's disease as the rationale for drug treatment and rehabilitation. *Parkinsons Disease, 2016*, 9832839.

Malik, B., Nirmalananthan, N., Gray, A. L., La Spada, A. R., Hanna, M. G., & Greensmith, L. (2013). Co-induction of the heat shock response ameliorates disease progression in a mouse model of human spinal and bulbar muscular atrophy: Implications for therapy. *Brain, 136*, 926–943.

Maraganore, D. M., Lesnick, T. G., Elbaz, A., et al. (2004). UCHL1 is a Parkinson's disease susceptibility gene. *Annals of Neurology, 55*, 512–521.

Marti, M. J., Tolosa, E., & Campdelacreu, J. (2003). Clinical overview of the synucleinopathies. *Movement Disorders, 18*, 21–27.

Martin, L. J. (2001). Neuronal cell death in nervous system development, disease, and injury (Review). *International Journal of Molecular Medicine, 7*, 455–478.

Meijering, R. A., Henning, R. H., & Brundel, B. J. (2014). Reviving the protein quality control system: Therapeutic target for cardiac disease in the elderly. *Trends in Cardiovascular Medicine, 25*, 243–247.

Miyata, Y., Koren, J., Kiray, J., Dickey, C. A., & Gestwicki, J. E. (2011). Molecular chaperones and regulation of tau quality control: Strategies for drug discovery in tauopathies. *Future Medicinal Chemistry, 3*, 1523–1537.

Miyazaki, D., Nakamura, A., Hineno, A., Kobayashi, C., Kinoshita, T., Yoshida, K., & Ikeda, S. (2016). Elevation of serum heat-shock protein levels in amyotrophic lateral sclerosis. *Neurological Sciences, 37*, 1277–1281.

Moloney, A., Sattelle, D. B., Lomas, D. A., & Crowther, D. C. (2010). Alzheimer's disease: Insights from Drosophila melanogaster models. *Trends in Biochemical Sciences, 35*, 228–235.

Moore, D. J., West, A. B., Dikeman, D. A., Dawson, V. L., & Dawson, T. M. (2008). Parkin mediates the degradation-independent ubiquitination of Hsp70. *Journal of Neurochemistry, 105*, 1806–1819.

Morais, V. A., Verstreken, P., Roethig, A., et al. (2009). Parkinson's disease mutations in PINK1 result in decreased Complex I activity and deficient synaptic function. *EMBO Molecular Medicine, 1*, 99–111.

Moreira, P. I., Carvalho, C., Zhu, X., Smith, M. A., & Perry, G. (2010). Mitochondrial dysfunction is a trigger of Alzheimer's disease pathophysiology. *Biochimica et Biophysica Acta, 1802*, 2–10.

Morozov, A. V., Astakhova, T. M., Garbuz, D. G., Krasnov, G. S., Bobkova, N. V., Zatsepina, O. G., Karpov, V. L., & Evgen'ev, M. B. (2017). Interplay between recombinant Hsp70 and proteasomes: Proteasome activity modulation and ubiquitin-independent cleavage of Hsp70. *Cell Stress & Chaperones, 22*(5), 687.

Mosser, D. D., Caron, A. W., Bourget, L., Meriin, A. B., Sherman, M. Y., Morimoto, R. I., & Massie, B. (2000). The chaperone function of hsp70 is required for protection against stress-induced apoptosis. *Molecular and Cellular Biology, 20*, 7146–7159.

Muchowski, P. J., Schaffar, G., Sittler, A., Wanker, E. E., Hayer-Hartl, M. K., & Hartl, F. U. (2000). Hsp70 and hsp40 chaperones can inhibit self-assembly of polyglutamine proteins into amyloid-like fibrils. *Proceedings of the National Academy of Sciences of the United States of America, 97*, 7841–7846.

Munoz-Soriano, V., & Paricio, N. (2011). Drosophila models of Parkinson's disease: Discovering relevant pathways and novel therapeutic strategies. *Parkinsons Dis, 2011*, 520640.

Najarzadegan, M. R., Ataei, E., & Akbarzadeh, F. (2016). The role of heat shock proteins in Alzheimer disease: A systematic review. *J Syndromes, 3*, 6.

Neckers, L., & Workman, P. (2012). Hsp90 molecular chaperone inhibitors: Are we there yet? *Clinical Cancer Research, 1*, 64–76.

Neef, D. W., Jaeger, A. M., & Thiele, D. J. (2011). Heat shock transcription factor 1 as a therapeutic target in neurodegenerative diseases. *Nature Reviews. Drug Discovery, 10*, 930–944.

Niizuma, K., Endo, H., & Chan, P. H. (2009). Oxidative stress and mitochondrial dysfunction as determinants of ischemic neuronal death and survival. *Journal of Neurochemistry, 109*(Suppl 1), 133–138.

Nolan, M., Talbot, K., & Ansorge, O. (2016). Pathogenesis of FUS-associated ALS and FTD: Insights from rodent models. *Acta Neuropathologica Communications, 4*, 99.

O'Donnell, W. T., & Warren, S. T. (2002). A decade of molecular studies of fragile X syndrome. *Annual Review of Neuroscience, 25*, 315–338.

Ogawa, K., Seki, T., Onji, T., Adachi, N., Tanaka, S., Hide, I., Saito, N., & Sakai, N. (2013). Mutant gammaPKC that causes spinocerebellar ataxia type 14 upregulates Hsp70, which protects cells from the mutant's cytotoxicity. *Biochemical and Biophysical Research Communications, 440*, 25–30.

Oh, S. Y., He, F., Krans, A., Frazer, M., Taylor, J. P., Paulson, H. L., & Todd, P. K. (2015). RAN translation at CGG repeats induces ubiquitin proteasome system impairment in models of fragile X-associated tremor ataxia syndrome. *Human Molecular Genetics, 24*, 4317–4326.

Orr, H. T., & Zoghbi, H. Y. (2007). Trinucleotide repeat disorders. *Annual Review of Neuroscience, 30*, 575–621.

Ou, J. R., Tan, M. S., Xie, A. M., Yu, J. T., & Tan, L. (2014). Heat shock protein 90 in Alzheimer's disease. *BioMed Research International, 2014*, 796869.

Ousman, S. S., Frederick, A., & Lim, E. F. (2017). Chaperone proteins in the central nervous system and peripheral nervous system after nerve injury. *Frontiers in Neuroscience, 11*, 79.

Paul, S., & Mahanta, S. (2014). Association of heat-shock proteins in various neurodegenerative disorders: Is it a master key to open the therapeutic door? *Molecular and Cellular Biochemistry, 386*, 45–61.

Paulson, H. L. (2009). The spinocerebellar ataxias. *Journal of neuro-ophthalmology: The official journal of the North American Neuro-Ophthalmology Society, 29*, 227.

Phukan, J. (2010). Arimoclomol, a coinducer of heat shock proteins for the potential treatment of amyotrophic lateral sclerosis. *IDrugs, 13*, 482–496.

Pickrell, A. M., & Youle, R. J. (2015). The roles of PINK1, parkin, and mitochondrial fidelity in Parkinson's disease. *Neuron, 85*, 257–273.

Pockley, A. G. (2001). Heat shock proteins in health and disease: Therapeutic targets or therapeutic agents? *Expert Reviews in Molecular Medicine, 3*, 1–21.

Pooler, A. M., Phillips, E. C., Lau, D. H., Noble, W., & Hanger, D. P. (2013). Physiological release of endogenous tau is stimulated by neuronal activity. *EMBO Reports, 14*, 389–394.

Prussing, K., Voigt, A., & Schulz, J. B. (2013). Drosophila melanogaster as a model organism for Alzheimer's disease. *Molecular Neurodegeneration, 8*, 35.

Rademakers, R., Stewart, H., Dejesus-Hernandez, M., et al. (2010). Fus gene mutations in familial and sporadic amyotrophic lateral sclerosis. *Muscle & Nerve, 42*, 170–176.

Ravagnan, L., Gurbuxani, S., Susin, S. A., et al. (2001). Heat-shock protein 70 antagonizes apoptosis-inducing factor. *Nature Cell Biology, 3*, 839–843.

Repalli, J., & Meruelo, D. (2015). Screening strategies to identify HSP70 modulators to treat Alzheimer's disease. *Drug Design, Development and Therapy, 9*, 321–331.

Ritossa, F. M. (1964). Experimental activation of specific loci in polytene chromosomes of Drosophila. *Experimental Cell Research, 35*, 601–607.

Roodveldt, C., Bertoncini, C. W., Andersson, A., et al. (2009). Chaperone proteostasis in Parkinson's disease: Stabilization of the Hsp70/α-synuclein complex by Hip. *The EMBO Journal, 28*, 3758–3770.

Ross, C. A., & Poirier, M. A. (2004). Protein aggregation and neurodegenerative disease. *Nature Medicine, 10*(Suppl), S10–S17.

Rubin, D. M., Mehta, A. D., Zhu, J., Shoham, S., Chen, X., Wells, Q. R., & Palter, K. B. (1993). Genomic structure and sequence analysis of Drosophila melanogaster HSC70 genes. *Gene, 128*, 155–163.

Sabirzhanov, B., Stoica, B. A., Hanscom, M., Piao, C. S., & Faden, A. I. (2012a). Over-expression of HSP70 attenuates caspase-dependent and caspase-independent pathways and inhibits neuronal apoptosis. *Journal of Neurochemistry, 123*, 542–554.

Sabirzhanov, B., Stoica, B. A., Hanscom, M., Piao, C. S., & Faden, A. I. (2012b). Over-expression of HSP70 attenuates caspase-dependent and caspase-independent pathways and inhibits neuronal apoptosis. *Journal of Neurochemistry, 123*, 542–554.

Saleh, A., Srinivasula, S. M., Balkir, L., Robbins, P. D., & Alnemri, E. S. (2000). Negative regulation of the Apaf-1 apoptosome by Hsp70. *Nature Cell Biology, 2*, 476–483.

Sau, D., De Biasi, S., Vitellaro-Zuccarello, L., et al. (2007). Mutation of SOD1 in ALS: A gain of a loss of function. *Human Molecular Genetics, 16*, 1604–1618.

Schaffar, G., Breuer, P., Boteva, R., et al. (2004). Cellular toxicity of polyglutamine expansion proteins: Mechanism of transcription factor deactivation. *Molecular Cell, 15*, 95–105.

Scotter, E. L., Chen, H. J., & Shaw, C. E. (2015). TDP-43 proteinopathy and ALS: Insights into disease mechanisms and therapeutic targets. *Neurotherapeutics, 12*, 352–363.

Shaaban, K. A., Wang, X., Elshahawi, S. I., et al. (2013). Herbimycins D-F, ansamycin analogues from Streptomyces sp. RM-7-15. *Journal of Natural Products, 76*, 1619–1626.

Sharma, A., Lyashchenko, A. K., Lu, L., et al. (2016). ALS-associated mutant FUS induces selective motor neuron degeneration through toxic gain of function. *Nature Communications, 7*, 10465.

Sheikh, S., Safia, H.,. E., & Mir, S. S. (2013). Neurodegenerative diseases: Multifactorial conformational diseases and their rherapeutic interventions. *Journal Neurodegenerative Disease, 2013*, 563481.

Shimura, H., Schwartz, D., Gygi, S. P., & Kosik, K. S. (2004). CHIP-Hsc70 complex ubiquitinates phosphorylated tau and enhances cell survival. *The Journal of Biological Chemistry, 279*, 4869–4876.

Shin, Y., Klucken, J., Patterson, C., Hyman, B. T., & McLean, P. J. (2005). The co-chaperone carboxyl terminus of Hsp70-interacting protein (CHIP) mediates alpha-synuclein degradation decisions between proteasomal and lysosomal pathways. *The Journal of Biological Chemistry, 280*, 23727–23734.

Shopland, L. S., & Lis, J. T. (1996). HSF recruitment and loss at most Drosophila heat shock loci is coordinated and depends on proximal promoter sequences. *Chromosoma, 105*, 158–171.

Shukla, A. K., Pragya, P., Chaouhan, H. S., Tiwari, A. K., Patel, D. K., Abdin, M. Z., & Chowdhuri, D. K. (2014). Heat shock protein-70 (Hsp-70) suppresses paraquat-induced neurodegeneration by inhibiting JNK and caspase-3 activation in drosophila model of Parkinson's disease. *PLoS One, 9*, 98886.

Silva, A. R., Santos, A. C., Farfel, J. M., et al. (2014). Repair of oxidative DNA damage, cell-cycle regulation and neuronal death may influence the clinical manifestation of Alzheimer's disease. *PLoS One, 9*, e99897.

Singh, R. P., Sharad, S., & Kapur, S. (2004). Free radicals and oxidative stress in neurodegenerative diseases: Relevance of dietary antioxidants. *Journal Indian Academy Clinical Medicine, 5*, 218–225.

Sittler, A., Lurz, R., Lueder, G., Priller, J., Lehrach, H., Hayer-Hartl, M. K., Hartl, F. U., & Wanker, E. E. (2001). Geldanamycin activates a heat shock response and inhibits huntingtin aggregation in a cell culture model of Huntington's disease. *Human Molecular Genetics, 10*, 1307–1315.

Slow, E. J., Graham, R. K., Osmand, A. P., et al. (2005). Absence of behavioral abnormalities and neurodegeneration in vivo despite widespread neuronal huntingtin inclusions. *Proceedings of the National Academy of Sciences of the United States of America, 102*, 11402–11407.

Soga, S., Akinaga, S., & Shiotsu, Y. (2013). Hsp90 inhibitors as anti-cancer agents, from basic discoveries to clinical development. *Current Pharmaceutical Design, 3*, 366–376.

Soto, C., & Estrada, L. D. (2008). Protein misfolding and neurodegeneration. *Archives of Neurology, 65*, 184–189.

Stankiewicz, A. R., Lachapelle, G., Foo, C. P., Radicioni, S. M., & Mosser, D. D. (2005). Hsp70 inhibits heat-induced apoptosis upstream of mitochondria by preventing Bax translocation. *The Journal of Biological Chemistry, 280*, 38729–38739.

Starkov, A. A., & Beal, F. M. (2008). Portal to Alzheimer's disease. *Nature Medicine, 14*, 1020–1021.

Stefanis, L. (2012). Alpha-synuclein in Parkinson's disease. *Cold Spring Harbor Perspectives in Medicine, 2*, a009399.

Tagawa, K., Marubuchi, S., Qi, M. L., et al. (2007). The induction levels of heat shock protein 70 differentiate the vulnerabilities to mutant huntingtin among neuronal subtypes. *The Journal of Neuroscience, 27*, 868–880.

Taroni, F., & DiDonato, S. (2004). Pathways to motor incoordination: The inherited ataxias. *Nature Reviews. Neuroscience, 5*, 641–655.

Tichauer, J. E., Flores, B., Soler, B., Eugenin-von Bernhardi, L., Ramirez, G., & von Bernhardi, R. (2014). Age-dependent changes on TGFbeta1 Smad3 pathway modify the pattern of microglial cell activation. *Brain, Behavior, and Immunity, 37*, 187–196.

Tofaris, G. K., Razzaq, A., Ghetti, B., Lilley, K. S., & Spillantini, M. G. (2003). Ubiquitination of alpha-synuclein in Lewy bodies is a pathological event not associated with impairment of proteasome function. *The Journal of Biological Chemistry, 278*, 44405–44411.

Tsai, H. F., Lin, S. J., Li, C., & Hsieh, M. (2005). Decreased expression of Hsp27 and Hsp70 in transformed lymphoblastoid cells from patients with spinocerebellar ataxia type 7. *Biochemical and Biophysical Research Communications, 334*, 1279–1286.

Turturici, G., Sconzo, G., & Geraci, F. (2011). Hsp70 and its molecular role in nervous system diseases. *Biochemistry Research International, 2011*, 618127.

Upadhaya, A. R., Lungrin, I., Yamaguchi, H., Fandrich, M., & Thal, D. R. (2012). High-molecular weight Abeta oligomers and protofibrils are the predominant Abeta species in the native soluble protein fraction of the AD brain. *Journal of Cellular and Molecular Medicine, 16*, 287–295.

Uttara, B., Singh, A. V., Zamboni, P., & Mahajan, R. T. (2009). Oxidative stress and neurodegenerative diseases: A review of upstream and downstream antioxidant therapeutic options. *Current Neuropharmacology, 7*, 65–74.

Van Drie, J. H. (2011). Protein folding, protein homeostasis, and cancer. *Chinese Journal of Cancer, 30*, 124–137.

Verghese, J., Abrams, J., Wang, Y., & Morano, K. A. (2012). Biology of the heat shock response and protein chaperones: Budding yeast (Saccharomyces cerevisiae) as a model system. *Microbiology and Molecular Biology Reviews, 76*, 115–158.

Vigh, L., Literati, P. N., Horvath, I., et al. (1997). Bimoclomol: A nontoxic, hydroxylamine derivative with stress protein-inducing activity and cytoprotective effects. *Nature Medicine, 3*, 1150–1154.

Vogel, M., Mayer, M. P., & Bukau, B. (2006). Allosteric regulation of Hsp70 chaperones involves a conserved interdomain linker. *The Journal of Biological Chemistry, 281*, 38705–38711.

Wacker, J. L., Zareie, M. H., Fong, H., Sarikaya, M., & Muchowski, P. J. (2004). Hsp70 and Hsp40 attenuate formation of spherical and annular polyglutamine oligomers by partitioning monomer. *Nature Structural & Molecular Biology, 11*, 1215–1222.

Wakabayashi, K., Tanji, K., Mori, F., & Takahashi, H. (2007). The Lewy body in Parkinson's disease: Molecules implicated in the formation and degradation of alpha-synuclein aggregates. *Neuropathology, 27*, 494–506.

Wang, A. M., Miyata, Y., Klinedinst, S., et al. (2013a). Activation of Hsp70 reduces neurotoxicity by promoting polyglutamine protein degradation. *Nature Chemical Biology, 9*, 112–118.

Wang, J., Wright, H. M., Vempati, P., et al. (2013b). Investigation of nebivolol as a novel therapeutic agent for the treatment of Alzheimer's disease. *Journal of Alzheimer's Disease, 33*, 1147–1156.

Wang, H., Tan, M. S., Lu, R. C., Yu, J. T., & Tan, L. (2014a). Heat shock proteins at the crossroads between cancer and Alzheimer's disease. *Biomedical Research International, 2014*, 239164.

Wang, X., Yuan, B., Dong, W., Yang, B., Yang, Y., Lin, X., & Gong, G. (2014b). Induction of heat-shock protein 70 expression by geranylgeranylacetone shows cytoprotective effects in cardiomyocytes of mice under humid heat stress. *PLoS One, 9*, e93536.

Warrick, J. M., Chan, H. Y., Gray-Board, G. L., Chai, Y., Paulson, H. L., & Bonini, N. M. (1999). Suppression of polyglutamine-mediated neurodegeneration in Drosophila by the molecular chaperone HSP70. *Nature Genetics, 23*, 425–428.

Waza, M., Adachi, H., Katsuno, M., Minamiyama, M., Sang, C., Tanaka, F., Inukai, A., Doyu, M., & Sobue, G. (2005). 17-AAG, an Hsp90 inhibitor, ameliorates polyglutamine-mediated motor neuron degeneration. *Nature Medicine, 11*, 1088–1095.

Wentzell, J., & Kretzschmar, D. (2010). Alzheimer's disease and tauopathy studies in flies and worms. *Neurobiology of Disease, 40*, 21–28.

West, R. J., Elliott, C. J., & Wade, A. R. (2015). Classification of Parkinson's disease genotypes in Drosophila using spatiotemporal profiling of vision. *Scientific Reports, 5*, 16933.

Wilhelmus, M. M., de Waal, R. M., & Verbeek, M. M. (2007). Heat shock proteins and amateur chaperones in amyloid-Beta accumulation and clearance in Alzheimer's disease. *Molecular Neurobiology, 35*, 203–216.

Witt, S. N. (2009). Hsp70 molecular chaperones and Parkinson's disease. *Biopolymers, 93*, 218–228.

Witt, S. N. (2010). Hsp70 molecular chaperones and Parkinson's disease. *Biopolymers, 93*, 218–228.

Wittmann, C. W., Wszolek, M. F., Shulman, J. M., Salvaterra, P. M., Lewis, J., Hutton, M., & Feany, M. B. (2001). Tauopathy in Drosophila: Neurodegeneration without neurofibrillary tangles. *Science, 293*, 711–714.

Wolfgang, W. J., Miller, T. W., Webster, J. M., Huston, J. S., Thompson, L. M., Marsh, J. L., & Messer, A. (2005). Suppression of Huntington's disease pathology in Drosophila by human single-chain Fv antibodies. *Proceedings of the National Academy of Sciences of the United States of America, 102*, 11563–11568.

Wyss-Coray, T., Lin, C., Yan, F., Yu, G. Q., Rohde, M., McConlogue, L., Masliah, E., & Mucke, L. (2001). TGF-beta1 promotes microglial amyloid-beta clearance and reduces plaque burden in transgenic mice. *Nature Medicine, 7*, 612–618.

Wyttenbach, A., Sauvageot, O., Carmichael, J., Diaz-Latoud, C., Arrigo, A. P., & Rubinsztein, D. C. (2002). Heat shock protein 27 prevents cellular polyglutamine toxicity and suppresses the increase of reactive oxygen species caused by huntingtin. *Human Molecular Genetics, 11*, 1137–1151.

Yang, J., Bridges, K., Chen, K. Y., & Liu, A. Y. (2008). Riluzole increases the amount of latent HSF1 for an amplified heat shock response and cytoprotection. *PLoS One, 3*, e2864.

Yoo, L., & Chung, K. C. (2018). The ubiquitin E3 ligase CHIP promotes proteasomal degradation of the serine/threonine protein kinase PINK1 during staurosporine-induced cell death. *Journal of Biological Chemistry, 293*, 1286–1297.

Zhang, C. W., Adeline, H. B., Chai, B. H., Hong, E. T., Ng, C. H., & Lim, K. L. (2016). Pharmacological or genetic activation of Hsp70 protects against loss of Parkin function. *Neurodegenerative Diseases, 16*, 304–316.

Zhang, Y. Q., & Sarge, K. D. (2007). Celastrol inhibits polyglutamine aggregation and toxicity though induction of the heat shock response. *Jouranl Molecular Medicine (Berl), 85*, 1421–1428.

Zheng, Q., Huang, C., Guo, J., Tan, J., Wang, C., Tang, B., & Zhang, H. (2018). Hsp70 participates in PINK1-mediated mitophagy by regulating the stability of PINK1. *Neuroscience Letters, 662*, 264–270.

Zhu, X., Zhao, X., Burkholder, W. F., Gragerov, A., Ogata, C. M., Gottesman, M. E., & Hendrickson, W. A. (1996). Structural analysis of substrate binding by the molecular chaperone DnaK. *Science, 272*, 1606–1614.

Hsp70-Family Proteins and Neurodegenerative Diseases

Zheying Sun, Roy J. Blackburn, Laura J. Blair, and John Koren III

Abstract Neuronal proteostasis is a highly regulated and crucial component of neural function. Unlike other tissues and organ, cell loss due to damage and dysfunctional signaling mechanisms is not an option for the brain. Neurons are thusly dependent on the collective cellular machinery of the molecular chaperones. Hsp70, a molecular chaperone which hydrolyzes ATP to fold proteins into a functional state, has been implicated as both a driver of disease pathogenesis and a therapeutic target for the activities and associations identified in several neurodegenerative diseases. Through interaction studies, genetic models, and small molecule therapeutics which serve as chemical tools, we have gained a greater understanding and appreciation for the role of Hsp70 in several neurodegenerative diseases. This chapter will discuss the studies and tools which elucidated the role of Hsp70 family members in neurodegenerative disorders and will offer perspective into therapeutic interventions which may prove beneficial for treating these diseases.

Keywords Alzheimer's disease · Hsp70 · Molecular chaperones · Neurodegenerative diseases · Parkinson's disease

Abbreviations

α-syn	α-synuclein
Aß	Amyloid beta
AD	Alzheimer's disease
ALS	Amyotrophic lateral sclerosis
AMPA	α-Amino-3-hydroxy-5-methyl-4-isoxazolepropionic acid
BAG-1	Bcl2-associated athanogene-1
CHIP	c-terminus of Hsc70 interacting protein

Z. Sun · R. J. Blackburn · L. J. Blair · J. KorenIII (✉)
Department of Molecular Medicine, College of Medicine, USF Byrd Alzheimer's Institute, University of South Florida, Tampa, FL, USA
e-mail: jkoren@health.usf.edu

A. A. A. Asea, P. Kaur (eds.), *HSP70 in Human Diseases and Disorders*, Heat Shock Proteins 14, https://doi.org/10.1007/978-3-319-89551-2_2

CME	Clatherin-mediated endocytosis
fALS	Familial Amyotrophic lateral sclerosis
GA	Geldanamycin
HD	Huntington's disease
Hsc	Heat shock cognate
Hsp	Heat shock protein
LRRK2	Leucine-rich repeat kinase 2
MAPT, tau	Microtubule associating protein tau
PCR	Polymerase chain reaction
PD	Parkinson's disease
polyQ	Polyglutamine-expansions
SCA	Spinocerebellar ataxia
SNpc	Substantia nigra pars compacta
SOD1	Superoxide dismutase 1
TDP-43	TAR DNA binding protein 43
UPS	Ubiquitin-proteasome system

Introduction

A hallmark pathology of many neurodegenerative disorders is the aberrant protein accumulation. These aggregating proteins have escaped or became immune to proteostasis mechanisms which normally govern their localization and turnover. The causes underlying this breakdown in proteostasis vary as wildly as the accumulating and aggregating proteins themselves. Some, such as the microtubule associating protein tau (*MAPT*, tau) in Alzheimer's disease, aggregate following abnormal patterns of phosphorylation, while others, such as tau in Pick's disease or huntingtin in Huntington's disease, aggregate due to mutations or peptide expansions. Whatever processes and mechanisms serve as the impetus for aggregation, these proteins become abnormally regulated and often present a detrimental impact to neural biology. Hsp70, a molecular chaperone which hydrolyzes ATP to fold proteins into a functional state, with the rest of the molecular chaperone network regulates protein folding, the triage of damaged or misfolded proteins for degradation, and the prevention of protein aggregation. Because of these cellular functions, Hsp70 and other molecular chaperones have been well studied for roles in neurodegenerative disorders; both as pathogenic factors and as potential therapeutic agents. The Hsp70 family of chaperones, ahead of other chaperone families, interacts with several families of chaperones and co-chaperones (namely – Hsp90 and the family of DNAJ proteins) which allow Hsp70 family members to interact with and regulate a wide variety of disordered and damaged proteins. Upon interacting with a mutated or misfolded protein, a molecular "decision" is made following cues from interactions with other molecular chaperones and co-chaperones. This decision can lead to the preservation or degradation of potentially aberrant protein. All too often in neurodegenerative disorders, a preservation of potentially damaging proteins occurs. Though this is not a fixed outcome, many studies have shown that interactions with molecular chaperones are necessary for the formation of pathology; particularly in neurodegenerative diseases. This chapter will detail our understanding of how Hsp70 family members impact various neurodegenerative disease proteins and highlight any attempts to utilize Hsp70 family members as therapeutics or therapeutic targets in the treatment of these disorders.

Alzheimer's Disease and Tauopathies

Alzheimer's disease (AD) is a neurodegenerative disorder and the primary cause of dementia in the United States. Pathologically, AD is characterized by the presence of extra-neuronal plaques comprised of amyloid beta (Aß), and intra-neuronal tangles of the microtubule associating protein tau (*MAPT*, tau). The accumulation of tau is driven by aggressive phosphorylation, thought to be a cellular response to extracellular Aß accumulation. Tau mutations, though typically absent in AD, are common in a group of non-AD neurodegenerative disorders collectively known as tauopathies. These disorders, similar to AD, feature the intracellular accumulation of tau, impaired cognition, and a progressive loss of neurons. To date, there are no FDA approved therapeutics for treating the molecular pathogenesis of AD, and all current approved therapies are palliative. Recently, there have been advances in several small molecule's capable of regulating molecular chaperones which have demonstrated beneficial outcomes for models of tauopathy. Here, we will discuss the roles of heat shock proteins in the regulation of tau and Aß, based on the available literature. Early evidence of a role for Hsp70 in AD was identified when the levels of Hsp70 were found to be elevated in the brains of AD patients (Yoo et al. 1999). Later, it was found that higher levels of Hsp70 and Hsp90 correlated with lower levels of insoluble tau and increased tau-microtubule association (Dou et al. 2003). From these studies, and the availability of commercial Hsp90 inhibitors, much effort was placed on characterizing the role of Hsp90 in tau biology. One interesting phenomenon was the identification of an Hsp90 population in the temporal cortex but not the cerebellum of AD patients which had an elevated affinity for the co-enzyme ATP (Dickey et al. 2007). The elevated affinity of Hsp90 for ATP also meant that Hsp90 in AD brain had an increased affinity for small-molecule Hsp90 inhibitors. Inhibition of Hsp90 reduced soluble tau levels in cells treated with these compounds (Dickey et al. 2007). These findings mirrored a similar phenomenon found in cancer cells and patient tumors which presented Hsp90 with elevated affinity for ATP and a sensitivity of Hsp90 associated oncogenes to Hsp90 inhibition (Kamal et al. 2003). However, prior to encountering and interacting with Hsp90, tau was found to interact with Hsc70 or Hsp70, two cytosolic Hsp70 family members.

Hsp70 and Tau

Hsc70 and Hsp70 regulate the tau 'triage decisions' for degradation by the proteasome or through autophagy mechanisms (Fig. 1). Microtubule destabilization or removal of tau from the microtubules drives an interaction between tau and Hsc70 (Jinwal et al. 2010). Once formed, this complex stabilizes tau, protecting against proteasomal degradation. Additionally, the presence of the Bcl2-associated athanogene-1 (BAG-1), an Hsc70 co-chaperone, prevents proteasomal degradation of tau; this is, however, reversible by the induction or overexpression of Hsp70 or by silencing BAG-1(Elliott et al. 2007). Hsp70, through interactions with CHIP (Hsc70-interacting protein), an E3 ubiquitin ligase, promotes the ubiquitination and of tau. Following ubiquitination, tau is targeted for degradation by the proteasome (Dickey et al. 2007, 2008; Shimura et al. 2004). Together, these data suggest Hsc70 can preserve tau whereas tau interacting with Hsp70 may promote degradation (Fig. 2). Similar findings have been observed with other client proteins of both Hsp70 and Hsc70 (Koren 3rd et al. 2010). This suggests that neuronal levels of Hsc70 and Hsp70, and perhaps most importantly the ratio between these two proteins, may be a critical factor governing the accumulation or clearance of tau. Low cellular levels of Hsp70 and CHIP may control the basal levels of "house-keeping" turnover and degradation of tau, a pathway that is better accessed by tau when more Hsp70 is available to present tau to CHIP. These ideas are consistent with findings that Hsp70 differentially regulates tau isoforms and may function primarily in

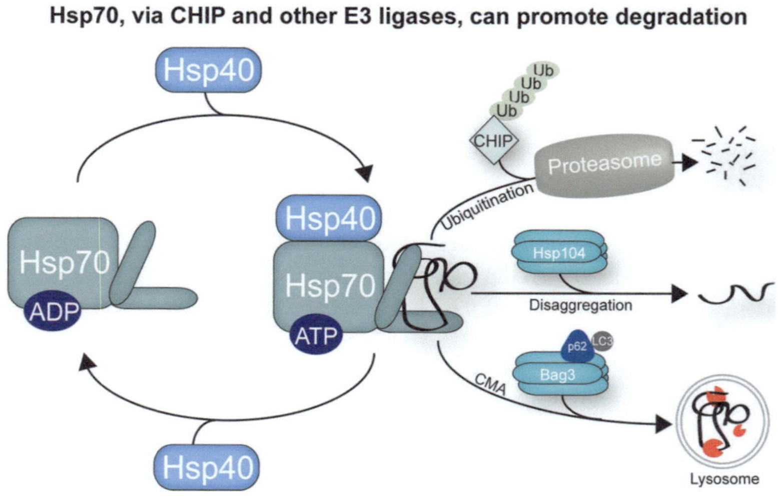

Fig. 1 The Hsp70 family regulates many distinct "fates" for client proteins. The fate of a client following interaction with an Hsp70 family member is often dictated by the (1) member of the Hsp70 family, (2) the repertoire of Hsp70 co-chaperones associating with the Hsp70, and (3) the physical properties of the client. Hsp70 family members have been shown to promote protein degradation though both proteasomal and autophagic mechanisms, reduce or promote the aggregation of distinct clients, and fold proteins following heat shock or other stress events

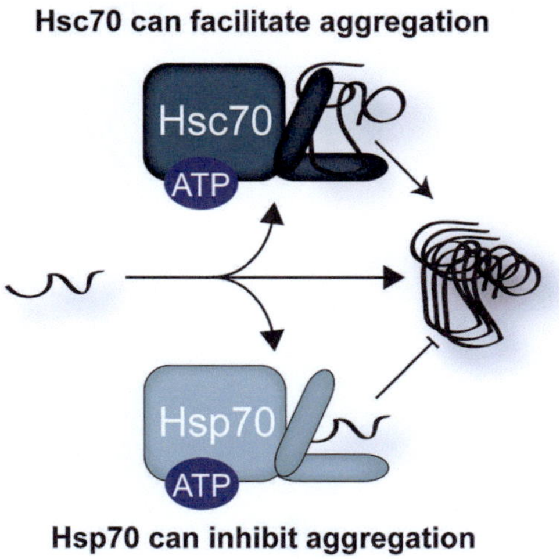

Fig. 2 Hsc70 and Hsp70 are frequently found to promote different "fates" for the same client protein, despite their extreme homology. The constitutive and ubiquitously expressed Hsc70 is often associated with client protein stabilization and aggregation. Whereas Hsp70, which is induced under various stress conditions, is often associated with a clearance of aggregating and mutant proteins; typically through the proteasome degradation machinery

regulating dysfunctional tau isoforms (Voss et al. 2012). Hsp70 family members, through interactions with distinct DNAJ proteins, were also associated with the pathological spread of tau; a phenomenon believed to be a major driver of Alzheimer's disease progression. The complex of DNAJC5 and Hsc70 was found to promote the extracellular release of tau via SNARE and SNAP-23 (Fontaine et al. 2016). This release was not limited to tau, but was also capable of exporting α-synuclein and TDP-43 (TAR DNA Binding protein 43). This mechanism was originally hypothesized to be protective. However, the recent investigations into the mechanisms and outcomes of tau spreading in AD and other tauopathies (Boluda et al. 2015; Guo et al. 2016; Iba et al. 2015) suggests that, though the DNAJC5 and Hsc70 complex may be functioning to protect neurons by forcing the release of pathogenic tau species, DNAJC5 and tau may be working save the cell at the expense of the surrounding tissue and organism.

Currently, small molecules capable of regulating Hsp70 and Hsc70 activity are under investigation as anti-tau therapeutic strategies. Some of these compounds have demonstrated promise results in studies using *in vitro* and mouse models of tauopathy. Work from the Dickey and Gestwicki groups identified and characterized small molecule inhibitors, as well as activators, of Hsc70/Hsp70 (Jinwal et al. 2009). Activation of Hsp70 stabilized tau, possibly bypassing tau through an unproductive folding cycle, or by selectively activating Hsc70 (Fig. 3). Conversely, inhibition of Hsp70 reduced tau levels both *in vitro* and *in vivo*, resulting in the ubiquitination and degradation of tau (Abisambra et al. 2013; Evans et al. 2011; Jinwal et al. 2009; O'Leary 3rd et al. 2010; Thompson et al. 2012). Additionally, the

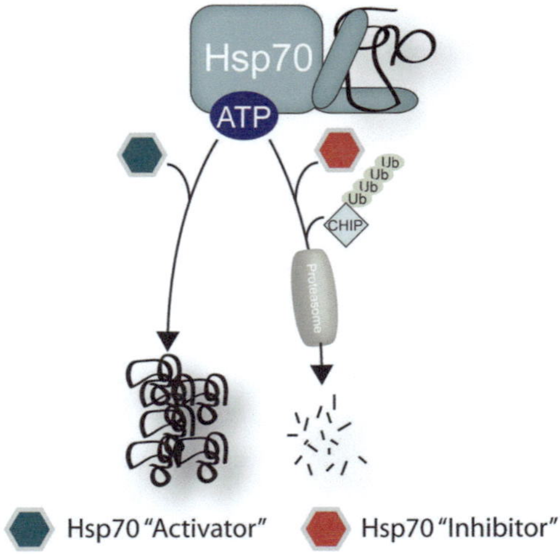

Fig. 3 The fate of an Hsp70 client protein is often dictated by Hsp70 activity. Small molecules capable of enhancing the ATPase activity of Hsp70 have been shown to stabilize and enrich pools of client proteins. While in contrast, small molecule inhibitors of Hsp70 often promote the ubiquitination and degradation of certain client proteins

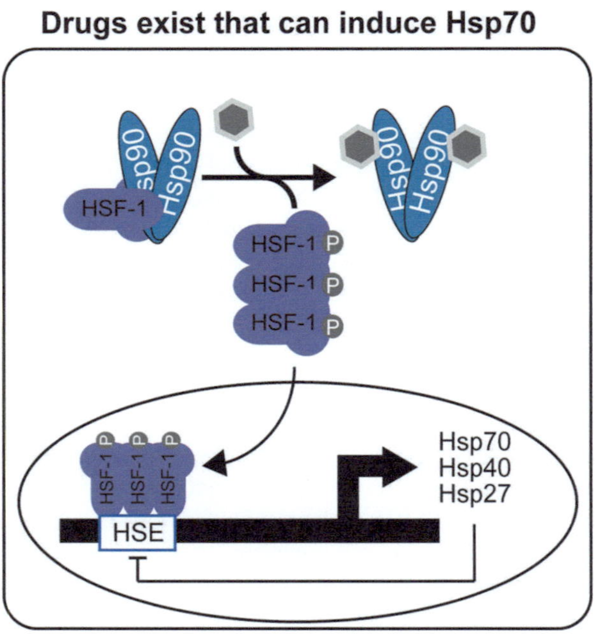

Fig. 4 Small molecules can activate the heat shock response. Some act directly while others, such as ATPase inhibitors of Hsp90, are indirect. These compounds promote the release of HSF1; thus allowing HSF1 to be phosphorylated, trimerize, and translocate to the nucleus

'gold standard' for small molecule inhibitors of Hsp90 is an induction of Hsp70 following treatment. Thus, it is hypothesized that inhibition of Hsp90, which may aid in the clearance of tau, will subsequently drive the induction of Hsp70; thereby increasing neuronal levels of Hsp70 which was shown to promote tau clearance (Fig. 4) as well as provide neuroprotection against apoptosis.

Hsp70 and Amyloid Beta

Though not typically associated with cytosolic chaperones, there is some evidence that Hsp70 can regulate the extracellular Aß. Using a recombinant model, a combination of Hsp70 with the co-chaperone Hsp40 could block early phases of Aß aggregation. This activity was dependent on the ATPase activity of Hsp70 as it could be modified by small molecules capable of enhancing or inhibiting these functions. Though these interactions remain unexplored, it suggests that exogenous or extracellular chaperones from the Hsp70 family, if functional, could regulate the accumulation and pathogenicity of Aß in AD patients.

Parkinson's Disease

Parkinson's disease (PD) is a neurodegenerative disorder caused by decreased dopamine levels in the brain from damaged neurons. PD affects 1% of all people over the age of 60 years old, and includes symptoms such as slow movement, stiffness, and loss of balance. In some cases, PD and progress to generate dementia in afflicted persons. While some medications such as dopamine promoters, and anti-tremor medications can assist in helping PD's symptoms, there is currently no treatment to fully cure the disease (Jankovic 2008). The progression of PD is the result of aggregation of prion-like proteins: alpha-synuclein (α-syn) and parkin, located in the presynaptic terminals of neurons. Accumulating α-syn in neurons causes the degeneration through loss of dendritic spines. These abnormal aggregates of α-syn and Parkin proteins in PD are referred to as Lewy bodies, or lewy neurites depending on their location. PD is also progressed by mutations in the leucine-rich repeat kinase 2 (LRRK2). Mutations in LRRK2 are the most common cause of autosomal dominant PD, and it is found that patients with LRRK2 mutations have accumulation α-syn aggregates, tau aggregates (commonly found in Alzheimer's disease), and/or ubiquitin-positive inclusions (Witt 2010).

Hsp70 and Alpha-Synuclein

The progression of PD is the result of aggregation of the prion-like proteins α-synuclein (α-syn) in the pre-synaptic terminal and dendritic spines. α-syn is a disordered protein which, similar to tau, can form toxic oligomeric species (Lashuel et al. 2002). Accumulating α-syn in promotes neural degeneration through loss of dendritic spines. These abnormal aggregates of α-syn in PD forms aggregate structures collectively known as Lewy bodies, or Lewy neurites depending on their location. A major role in molecular chaperones is to mediate protein misfolding and mitigate accumulation of protein aggregates. It is hypothesized that chaperones are part of a cellular arsenal that protect against α-syn toxicity, and increasing levels of Hsp70 via transgene or chemical activator will protect cells from neurodegeneration. Similar to tau, Hsp70 and CHIP, as well as Hsp90, interact with α-synuclein. Hsp70 has been shown to be critical for regulating the general aggregation of α-syn (Aprile et al. 2015; Auluck et al. 2002; Gao et al. 2015; McLean et al. 2004). Additionally, a publication in 2017 identified a complex between Hsp70 and DNAJB6 to be a stronger inhibitor of α-syn aggregation than Hsp70 alone (Aprile et al. 2017). CHIP, a ubiquitin ligase which frequently, with Hsp70, facilitates the degradation of accumulating proteins via the ubiquitin-proteasome system (UPS), was shown to interact with α-syn and Hsp70 (Tetzlaff et al. 2008). This interaction is significant since CHIP overexpression was observed to reduce high molecular weight oligomers of α-syn. These oligomers were then cleared by both the proteasome and the lysosome (Tetzlaff et al. 2008). Hsp90 also influences α-syn aggregation as determined by *in vitro* studies. In these studies, Hsp90 interacting with monomeric α-syn prevented α-syn from interacting with single-membrane vesicles. Additionally, this interaction promoted α-syn fibril formation in an ATP-dependent manner (Falsone et al. 2009). Unfortunately, only indirect data support the notion that Hsp70 likely binds monomeric forms of soluble α-synuclein: following Hsp70 depletion, α-synuclein reactions resumed at a rate similar to the initial monomer-containing reactions (Luk et al. 2008). While the field continues to explore possible mechanisms of α-synuclein-induced toxicity, *in vitro* studies suggest that, Hsp70 and CHIP likely interact in a dynamic process to regulate α-synuclein fibril assembly and degradation.

In vivo studies, from the study of PD patient tissue or a drosophila model of α-syn accumulation, suggested that overexpression of Hsp70 may be a response to the accumulation of α and an attempt by the cell to inhibit α-syn toxicity. It is assumed that the cellular attempt to mitigate proteotoxic stress in PD models involves recruiting molecular chaperones, including Hsp70. This hypothesis is supported by one of many studies which shows how the overexpression of α-synuclein in the substantia nigra pars compacta (SNpc) of mice resulted in an increased level of Hsp70 mRNA, as assayed by quantitative PCR in neurons recovered by laser capture microscopy (St Martin et al. 2007). Another supporting study showed that time-lapse imaging of fluorescently labeled α-syn demonstrated that aggregate formation and toxicity induced by a C-terminally truncated form of α-syn could be reduced by the co-expression of Hsp70 in living cells (Opazo et al. 2008). In transgenic flies with neuronal-directed expression of α-syn presented features of α-syn accumulation similar to those observed in PD; including adult-onset loss of dopaminergic neurons, filamentous intraneuronal inclusions containing α-syn, and locomotor dysfunction. The study showed that flies expressing wildtype (non-mutated) α-syn in dopaminergic neurons presented a 50% reduction in neurons after 20 days. When wildtype α-syn was co-expressed with human Hsp70, no loss of fly neurons was seen at 20 days. Interestingly, Lewy bodies were still found present within the neurons after the Hsp70 co-expression, which lead to the belief that Hsp70 works by protecting dopaminergic neurons from degenerative cell death caused by α-syn toxicity. The Lewy bodies found present within the neurons after the co-expression of Hsp70 were not toxic, but instead inert. The toxic species of α-syn appeared to be exclusive to soluble forms (Fukuzono et al. 2016).

Hsp70 has since been explored in other model types, including a yeast model of PD. Normally, wildtype α-synuclein, along with α-syn mutations (such as A53T and A30P), will trigger apoptosis in yeast saccharomyces cerevisiae. In this yeast model experiment, cell death was observed with apoptotic markers such as externalization of phosphatidylserine causing membrane asymmetry, release of cytochrome c from the mitochondria, and an accumulation of reactive oxygen species. When co-expressed with Hsp70, α-syn expressing yeast cells showed none of the previously displayed apoptotic markers. Hsp70, however, did not decrease the level of expression of α-syn, indicating that Hsp70 induced protective proteins responsible for inhibiting the α-syn induced apoptosis (Flower et al. 2005). Another PD model that was used to observe the effects of hps70 included a cell culture model of α-syn aggregation and inclusion formation. Co-expression of Hsp70 in this model led to reduced aggregate formation, and lowered total and detergent-insoluble fractions of misfolded α-syn (Klucken et al. 2004). A rat model of PD showed how the induction of Hsp70 mitigated neurotoxicity in acute brain slices, ultimately induced by the mitochondrial toxin rotenone (Tantucci et al. 2009). Studies indicate that activation of the heat shock response, such as through geldanamycin (GA) which induces the expression of Hsp70 via the inhibition of Hsp90, may be therapeutically relevant for the treatment of PD pathology (Fukuzono et al. 2016) (Fig. 4). Introducing Hsp70 has shown to accelerate the degradation of α-syn via CMA and thus protecting cells from neurodegenerative death (Fukuzono et al. 2016). Though Hsp70 can induce the degradation of wildtype α-syn through CMA, mutant forms of α-syn, such as A53T and A30P, are immune to degradation by chaperone-mediated autophagy and may even inhibit the CMA pathway.

Hsp70 and LRRK2

Despite the prevalence and focus on α-syn as a pathological driver of PD, α-syn is not alone in the proteinopathy which is PD. Mutations in LRRK2 are the most common cause of autosomal dominant PD, and it is found that patients with LRRK2 mutations have accumulation a-syn aggregates, tau aggregates (commonly found in Alzheimer's disease), and/or ubiquitin-positive inclusions (Witt 2010). LRRK2 expression in Zebrafish embryos increased GFP ubiquitination. The authors suggested that LRRK2 is more accessible in the presence of Hsp70 and that certain LRRK2 species, such as LRRK2 mono- or oligomers may mediate the accumulation of other proteins. CHIP can bind to multiple LRRK2 domains and promote LRRK2 degradation in HEK293 cells. Further, Hsp90 and CHIP activity levels are determinants of LRRK2-mediated toxicity. CHIP overexpression and increased LRRK2 degradation rates in HEK293 cells and also reduced LRRK2-mediated toxicity in SH-SY5Y cells and HeLa cells.

Mechanisms similar to, yet distinct from, α-synuclein aggregation may govern the function and stability of Leucine-repeat- rich kinase 2 (LRRK2) in Parkinson's disease. The LRRK2 G2019S substitution is the most common sporadic and inherited PD-causing mutation and is associated with dominantly inherited PD. LRRK2 inclusions, similar to α-syn containing Lewy bodies found in Parkinson's patients, also contain molecular chaperones. Specifically, Hsp70 and CHIP, and BAG2 were found to interact with LRRK2. While exploring the interaction between LRRK2 and BAG2, it was discovered that LRRK2 did not interact directly with BAG2 (Lichtenberg et al. 2011). However, overexpression of Hsc70 promoted the interaction of LRRK2 with BAG2, suggesting BAG2 was interacting with LRRK2 specifically through Hsc70. This experiment, however, lead to an experiment which demonstrated that overexpression of Hsc70 enhanced the effects of LRRK2 on protein accumulation but did not influence levels of soluble LRRK2; an effect similar to those observed in tau biology where Hsc70 stabilized tau, but Hsp70 reduced tau accumulation (Lichtenberg et al. 2011) (Fig. 2). When overexpressed with Hsp70, both WT LRRK2 aggregation was decreased. As Hsp70 decreases LRRK2 aggregation and increases its positive effect on protein levels, the accumulation of proteins caused by LRRK2 is due to the higher accessibility in the presence of Hsp70 (Beere et al. 2000).

The characteristics of most PD forms are mostly dependent on α-syn, which brought up the idea that observing the effects of LRRK2 on α-syn levels would lead to a more accurate representative model of LRRK2. This idea lead to one study in which wildtype LRRK2 (WT LRRK2) and a mutant LRRK2 (GSLRRK2) were transfected in HeLa cells and mouse primary neurons to observe its effects on α-syn levels. The mutant gene, GSLRRK2, has been found to be the most common cause of hereditary PD, and is toxic to cells due to its excessive amount of kinase activity. Both WT LRRK2 and GSLRRK2 overexpression led to increased levels of mutated α-syn as well as increased amounts of ubiquitinated proteins (Lichtenberg et al. 2011), indicating that mutant LRRK2 was impairing protein degradation by the proteasome. This data lead to an experiment that showed the co-expression of Hsp70 enhanced the effects of LRRK2 proteasome inhibition and subsequent accumulation of ubiquitinated proteins. However, when overexpressed with Hsp70, both WT LRRK2 and GSLRRK2 aggregation were decreased (Lichtenberg et al. 2011). It was also tested to see if co-expression of Hsp70 affected the LRRK2-mediated accumulation of GFP. The results showed that Hsp70 overexpression did enhance LRRK2 mediated GFP accumulation compared to control. These investigations suggest that Hsp70 allowed LRRK2 to be much more accessible as a promoter of protein aggregation or inhibitor of the proteasome when in monomeric or oligomeric form (Lichtenberg et al. 2011). The mechanism for how Hsp70 impairs the UPS degradation of LRRK2 is unclear. However, this example of "client fate" following an interaction with Hsp70 does suggest that, unlike tau, Hsp70 may not always serve to benefit cells and neurons expressing mutant or aggregating proteins.

HD and PolyQ Expansions

Unlike PD and AD, Huntington's disease (HD) and spinocerebellar ataxia (SCA) are a part of a neurodegenerative disorder family linked by the accumulation of proteins featuring polyglutamine-expansions (polyQ). These diseases (polyQ disorders) are genetically driven autosomal dominant diseases. The common factor is the presence of repeating CAG repeats which can generationally increase in number. Polyglutamine expansion of ataxin proteins causes, a spectrum of progressive neurodegenerative diseases currently attributed to mutations in over 29 genes (Orr 2012). Evidence suggests the Hsp70 family of chaperones and related co-chaperones regulates mutant ataxin biology. Similar to tau and polyQ-expanded huntingtin, Hsp70 and Hsp40 prevented aggregation of mutant ataxin-1 (Cummings et al. 1998). Hsp70 was found to be neuroprotective in several *in vivo* models of polyQ ataxia (Cummings et al. 2001; Warrick et al. 1999). Again similar to tau and α-syn, proteasomal degradation of polyQ-expanded ataxin is regulated by CHIP ubiquitination (Al-Ramahi et al. 2006; Williams et al. 2009); this interaction, under stress conditions, is regulated by Hsp70 (Al-Ramahi et al. 2006). In HD, almost all the cases are caused by inherits of the CAG-expanded huntingtin gene, only 10 percent of the cases are due to other gene mutation (Durr et al. 2012). Increased CAG repeats form an uninterrupted series of glutamines which disrupt the structure of the protein. This disruption, in turn, leads to toxic aggregation.

Huntington's disease patient contains 36-180 glutamine repeats whereas unafflicted persons carry only 6-39 repeats (Muchowski et al. 2000). As Huntington's disease is an inherited disease, genetic testing can be utilized to detect a mutant huntingtin gene (Dayalu and Albin 2015). Akin to other polyQ expansion disorders, both Hsp70 and Hsp40 were shown to limit aberrant huntingtin accumulation and the ubiquitin ligase CHIP was shown to promote the degradation of polyQ huntingtin (Jana et al. 2005; Muchowski et al. 2000). Increasing intracellular CHIP levels ablated the aggregation of huntingtin and reduced huntingtin-driven toxicity. Yeast models showed similar roles for both Hsp70 and Hsp40 when co-expressed in a yeast model of polyQ huntingtin. Both Hsp70 and Hsp40 inhibited the formation of insoluble huntingtin fibrils and, instead, promoted the formation of soluble inclusions. This pattern is similar to what has been observed with tau and α-synuclien: the ability of Hsp70 and related co-chaperones to promote degradation or aggregation formation (Fig. 1). The same result was observed *in vitro* (Muchowski et al. 2000). In 2010, Lotz et al. also noted that Hsp70 inhibited the formation of aggregated huntingtin protein through an interaction with Hsp40 (Lotz et al. 2010). Later in 2015, Monsellier et al performed a characterization of the Hsp70/Hsp40/huntingtin interaction. This group found that Hsp70 interacts with the 17 amino acids in the N-terminal of the huntingtin protein, and, with Hsp40, can inhibit huntingtin protein aggregation (Monsellier et al. 2015). Hsc70 has also been associated with insoluble

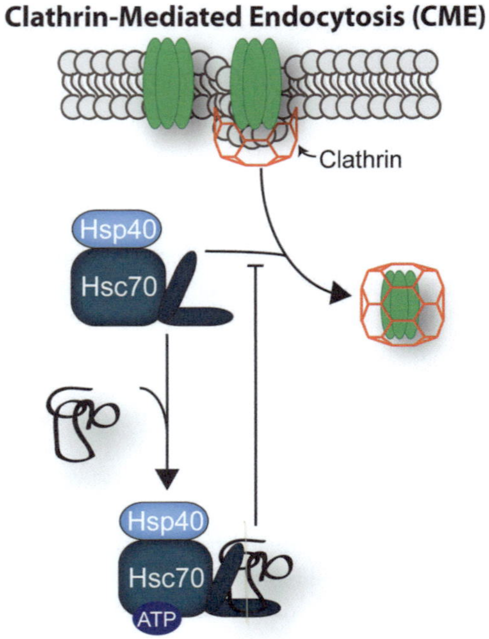

Fig. 5 Sequestration of Hsc70 by aberrant client can impact AMPA internalization, endocytosis, and other clathrin dependent pathways. This function can be restored by increasing the intracellular pool of Hsc70 or through reduction of aberrant client proteins

aggregates of huntingtin in both cell culture models and in HD patient tissue (Jana et al. 2000); this suggesting a role for Hsc70 in the accumulation of polyQ expanded huntingtin distinct from the fate if huntingtin following interaction with Hsp70 (Fig. 2). However, the Morimoto lab discovered an interesting role for Hsc70 in the pathophysiology of polyQ disorders. A previous study found that polyQ expanded proteins, including huntingtin, SOD1, and ataxin, could sequester away a yeast Hsp40 homologue, Sis1p. This sequestration blocked the transportation of proteins into the nucleus for proteasomal degradation; possibly driving toxicities associated with the accumulation of polyQ expanded proteins like huntingtin. Increased expression of Sisp1 restored the inhibited nuclear transport (Park et al. 2013). Hsc70 regulates clatherin-mediated endocytosis (CME); a neuronal process necessary for the internalization of membrane receptors such as AMPA (α-amino-3-hydroxy-5-methyl-4-isoxazolepropionic acid) receptor. The aggregation of polyQ expanded proteins sequestered Hsc70 away from the membrane causing deficits in CME (Fig. 5). Increasing Hsc70 levels restored these deficits (Yu et al. 2014). This phenomenon could lead to many of the cognitive and neuronal deficits observed in polyQ disorders, as well as any other neurodegenerative disorder which features aggregating client proteins of Hsc70.

Amyotrophic Lateral Sclerosis

Amyotrophic lateral sclerosis (ALS), also called Charcot disease or Lou Gehrig's disease, is a neurodegenerative disease caused by the progressive loss of cortical and spinal motor neurons. ALS is a late-onset disease presenting around the age of 60. Initial symptoms can be as innocuous as pain in leg muscles, stumbling when running and difficulty when swallowing. As the disease progresses, weakness in muscles spreads and symptoms become more obviously (Kelly 2013). Because of the innocence and the similar symptoms with other neurodegenerative diseases, it is difficult to diagnose ALS in the early stage. Though most cases are considered sporadic, 5–10% of the cases are familial ALS (fALS) which are inherited from patient's parents (Kiernan et al. 2011; Zarei et al. 2015). Currently, there is no treatment for ALS, Riluzole is the only FDA-approved drug for ALS, however, it only has a modest effect on patient's survival.

Superoxide dismutase 1 (SOD1) is the most common cause of familial ALS. SOD1 protects cells from superoxide free radicals which converts superoxides to oxygen and hydrogen peroxide. In 20% of familial ALS patients, SOD1 is mutated and aggregates into insoluble intracellular inclusions (Rosen et al. 1993) inducing a toxic gain of function, as confirmed by dismutase-inactivated SOD1 animal models (Bruijn et al. 1997). Mutant SOD1 accumulates in spinal cord mitochondria promoting mitochondrial dysfunction, abnormal ATP production, and apoptosis. Mutant SOD1 is regulated by the Hsp70-family. Through interactions with CHIP, Hsp70 can induce the proteasomal degradation of SOD1 via the UPS (Choi et al. 2004; Ishigaki et al. 2007). Hsc70, however, is commonly found to be associated with SOD1 in fALS intracellular aggregates (Watanabe et al. 2001; Zetterstrom et al. 2011). This indicates SOD1 may be differentially regulated by Hsp70 and Hsc70, a situation similar to the tau protein in AD (Fig. 2) (Elliott et al. 2007; Jinwal et al. 2010): Hsc70 interaction with SOD1 protects against proteasomal degradation, facilitating accumulation, whereas SOD1 interacting with Hsp70 promotes the ubiquitination and degradation of SOD1. This hypothesis is supported by data from motor neuron models of mutant SOD1 expression indicated that elevating Hsp70 expression promoted a decrease in aggregated SOD1; this, in turn, reduced toxicity in this model (Durham et al. 1997; Koyama et al. 2006; Roy et al. 1998). Additionally, activation of HSF-1, following the inhibition of Hsp90 with the geldanamycin derivative 17-AAG, also demonstrated cytoprotective effects associated with elevated Hsp70 levels, along with other heat shock proteins (Fig. 4) (Batulan et al. 2006). Direct overexpression, however, of Hsp70 *in vivo* did not recapitulate the *in vitro* results (Liu et al. 2005). However, *in vivo* ALS models did demonstrate positive results when Hsf-1 was activated (Kalmar et al. 2008, 2012; Kieran et al. 2004). These studies treated mouse models of fALS featuring SOD1 mutations with arimoclomol, an inducer of the heat shock response. Arimoclomol treatment resulted in a reduction of SOD1 aggregates, retarded the progression of the mutant SOD1 phenotype, and ultimately increased the lifespan of the treated mutant SOD1 mice. Currently (2017), arimoclomol is under Phase II/III investigation for patients afflicted with fALS caused by mutant SOD1. Thus, increasing Hsp70 levels or activity, in conjunction with the other heat shock proteins, may be an effective strategy for the treatment of mutant SOD1-driven fALS.

Conclusions

The Hsp70 family of molecular chaperones represent a highly adaptive and redundant network. During normal cellular processes, the Hsp70 family, namely Hsc70, work to maintain cellular proteostasis to ensure proper cellular function. However, despite the adaptive nature of this machinery, when a cell or neuron transitions from "normal" to a diseased state, Hsp70 and the other molecular chaperones seem to poorly adjust to the new proteomic conditions of this disease environment. Hsc70 often against the neuron by preserving and possibly facilitating the accumulation of aggregation-prone toxic proteins. These actions are not only not in the best interest of the cell, but also work against the organism. Hsp70, however, appears quite capable of alleviating disease phenotypes initiated by the accumulation of toxic aggregating proteins. These are important considerations when attempting to target the actions of an Hsp70 family member as a treatment for disease. Inhibition of Hsp70 may be beneficial; acting to slow or prevent the accumulation of pathogenic proteins. However, small molecules capable of inhibiting Hsc70 may also be capable of inhibiting Hsp70. Hsp70, as we have discussed, is quite potent and capable of clearing accumulating proteins and is a known regulator of pro-survival (anti-apoptotic) mechanisms. Thus, any ideal therapeutic would be capable of inhibiting Hsc70 while activating or inducing Hsp70. One potential mechanism for this may already exist on the genetic level. Silencing of Hsc70 has been shown to induce the expression of Hsp70 (Koren 3rd et al. 2010). Thus, tools like CRISPR technologies or translation inhibiting morpholinos capable of reducing the intracellular pool of Hsc70 may sufficiently increase Hsp70 levels. It is of critical importance, however, for future studies to elucidate therapeutic risks associated with Hsc70 inhibition or suppression in patients with afflicted with neurotoxic proteinopathies.

Acknowledgements This book chapter is dedicated to the life and memory of Dr. Chad A. Dickey. We apologize to the many authors who have contributed to our understanding of the molecular chaperone and neurodegenerative disease fields, and whose work we have failed to discuss or cite.

References

Abisambra, J., Jinwal, U. K., Miyata, Y., Rogers, J., Blair, L., Li, X., Seguin, S. P., Wang, L., Jin, Y., Bacon, J., et al. (2013). Allosteric heat shock protein 70 inhibitors rapidly rescue synaptic plasticity deficits by reducing aberrant tau. *Biological Psychiatry, 74*, 367–374.

Al-Ramahi, I., Lam, Y. C., Chen, H. K., de Gouyon, B., Zhang, M., Perez, A. M., Branco, J., de Haro, M., Patterson, C., Zoghbi, H. Y., et al. (2006). CHIP protects from the neurotoxicity of expanded and wild-type ataxin-1 and promotes their ubiquitination and degradation. *The Journal of Biological Chemistry, 281*(36), 26714–26724.

Aprile, F. A., Sormanni, P., & Vendruscolo, M. (2015). A rational design strategy for the selective activity enhancement of a molecular chaperone toward a target substrate. *Biochemistry, 54*, 5103–5112.

Aprile, F. A., Kallstig, E., Limorenko, G., Vendruscolo, M., Ron, D., & Hansen, C. (2017). The molecular chaperones DNAJB6 and Hsp70 cooperate to suppress alpha-synuclein aggregation. *Scientific Reports, 7*, 9039.

Auluck, P. K., Chan, H. Y., Trojanowski, J. Q., Lee, V. M., & Bonini, N. M. (2002). Chaperone suppression of alpha-synuclein toxicity in a Drosophila model for Parkinson's disease. *Science, 295*, 865–868.

Batulan, Z., Taylor, D. M., Aarons, R. J., Minotti, S., Doroudchi, M. M., Nalbantoglu, J., & Durham, H. D. (2006). Induction of multiple heat shock proteins and neuroprotection in a primary culture model of familial amyotrophic lateral sclerosis. *Neurobiology of Disease, 24*, 213–225.

Beere, H. M., Wolf, B. B., Cain, K., Mosser, D. D., Mahboubi, A., Kuwana, T., Tailor, P., Morimoto, R. I., Cohen, G. M., & Green, D. R. (2000). Heat-shock protein 70 inhibits apoptosis by preventing recruitment of procaspase-9 to the Apaf-1 apoptosome. *Nature Cell Biology, 2*, 469–475.

Boluda, S., Iba, M., Zhang, B., Raible, K. M., Lee, V. M., & Trojanowski, J. Q. (2015). Differential induction and spread of tau pathology in young PS19 tau transgenic mice following intracerebral injections of pathological tau from Alzheimer's disease or corticobasal degeneration brains. *Acta Neuropathologica, 129*, 221–237.

Bruijn, L. I., Becher, M. W., Lee, M. K., Anderson, K. L., Jenkins, N. A., Copeland, N. G., Sisodia, S. S., Rothstein, J. D., Borchelt, D. R., Price, D. L., et al. (1997). ALS-linked SOD1 mutant G85R mediates damage to astrocytes and promotes rapidly progressive disease with SOD1-containing inclusions. *Neuron, 18*, 327–338.

Choi, J. S., Cho, S., Park, S. G., Park, B. C., & Lee, D. H. (2004). Co-chaperone CHIP associates with mutant Cu/Zn-superoxide dismutase proteins linked to familial amyotrophic lateral sclerosis and promotes their degradation by proteasomes. *Biochemical and Biophysical Research Communications, 321*, 574–583.

Cummings, C. J., Mancini, M. A., Antalffy, B., DeFranco, D. B., Orr, H. T., & Zoghbi, H. Y. (1998). Chaperone suppression of aggregation and altered subcellular proteasome localization imply protein misfolding in SCA1. *Nature Genetics, 19*, 148–154.

Cummings, C. J., Sun, Y., Opal, P., Antalffy, B., Mestril, R., Orr, H. T., Dillmann, W. H., & Zoghbi, H. Y. (2001). Over-expression of inducible HSP70 chaperone suppresses neuropathology and improves motor function in SCA1 mice. *Human Molecular Genetics, 10*, 1511–1518.

Dayalu, P., & Albin, R. L. (2015). Huntington disease: Pathogenesis and treatment. *Neurologic Clinics, 33*, 101–114.

Dickey, C. A., Kamal, A., Lundgren, K., Klosak, N., Bailey, R. M., Dunmore, J., Ash, P., Shoraka, S., Zlatkovic, J., Eckman, C. B., et al. (2007). The high-affinity HSP90-CHIP complex recognizes and selectively degrades phosphorylated tau client proteins. *The Journal of Clinical Investigation, 117*, 648–658.

Dickey, C. A., Koren, J., Zhang, Y. J., Xu, Y. F., Jinwal, U. K., Birnbaum, M. J., Monks, B., Sun, M., Cheng, J. Q., Patterson, C., et al. (2008). Akt and CHIP coregulate tau degradation through coordinated interactions. *Proceedings of the National Academy of Sciences of the United States of America, 105*, 3622–3627.

Dou, F., Netzer, W. J., Tanemura, K., Li, F., Hartl, F. U., Takashima, A., Gouras, G. K., Greengard, P., & Xu, H. (2003). Chaperones increase association of tau protein with microtubules. *Proceedings of the National Academy of Sciences of the United States of America, 100*, 721–726.

Durham, H. D., Roy, J., Dong, L., & Figlewicz, D. A. (1997). Aggregation of mutant Cu/Zn superoxide dismutase proteins in a culture model of ALS. *Journal of Neuropathology and Experimental Neurology, 56*, 523–530.

Durr, A., Gargiulo, M., & Feingold, J. (2012). The presymptomatic phase of Huntington disease. *Revue Neurologique (Paris), 168*, 806–808.

Elliott, E., Tsvetkov, P., & Ginzburg, I. (2007). BAG-1 associates with Hsc70.Tau complex and regulates the proteasomal degradation of Tau protein. *The Journal of Biological Chemistry, 282*, 37276–37284.

Evans, C. G., Jinwal, U. K., Makley, L. N., Dickey, C. A., & Gestwicki, J. E. (2011). Identification of dihydropyridines that reduce cellular tau levels. *Chemical Communication (Cambridge), 47*, 529–531.

Falsone, S. F., Kungl, A. J., Rek, A., Cappai, R., & Zangger, K. (2009). The molecular chaperone Hsp90 modulates intermediate steps of amyloid assembly of the Parkinson-related protein alpha-synuclein. *The Journal of Biological Chemistry, 284*, 31190–31199.

Flower, T. R., Chesnokova, L. S., Froelich, C. A., Dixon, C., & Witt, S. N. (2005). Heat shock prevents alpha-synuclein-induced apoptosis in a yeast model of Parkinson's disease. *Journal of Molecular Biology, 351*, 1081–1100.

Fontaine, S. N., Zheng, D., Sabbagh, J. J., Martin, M. D., Chaput, D., Darling, A., Trotter, J. H., Stothert, A. R., Nordhues, B. A., Lussier, A., et al. (2016). DnaJ/Hsc70 chaperone complexes control the extracellular release of neurodegenerative-associated proteins. *The EMBO Journal, 35*, 1537–1549.

Fukuzono, T., Pastuhov, S. I., Fukushima, O., Li, C., Hattori, A., Iemura, S., Natsume, T., Shibuya, H., Hanafusa, H., Matsumoto, K., et al. (2016). Chaperone complex BAG2-HSC70 regulates localization of Caenorhabditis elegans leucine-rich repeat kinase LRK-1 to the Golgi. *Genes to Cells, 21*, 311–324.

Gao, X., Carroni, M., Nussbaum-Krammer, C., Mogk, A., Nillegoda, N. B., Szlachcic, A., Guilbride, D. L., Saibil, H. R., Mayer, M. P., & Bukau, B. (2015). Human Hsp70 disaggregase reverses Parkinson's-linked alpha-synuclein amyloid fibrils. *Molecular Cell, 59*, 781–793.

Guo, J. L., Narasimhan, S., Changolkar, L., He, Z., Stieber, A., Zhang, B., Gathagan, R. J., Iba, M., McBride, J. D., Trojanowski, J. Q., et al. (2016). Unique pathological tau conformers from Alzheimer's brains transmit tau pathology in nontransgenic mice. *The Journal of Experimental Medicine, 213*, 2635–2654.

Iba, M., McBride, J. D., Guo, J. L., Zhang, B., Trojanowski, J. Q., & Lee, V. M. (2015). Tau pathology spread in PS19 tau transgenic mice following locus coeruleus (LC) injections of synthetic tau fibrils is determined by the LC's afferent and efferent connections. *Acta Neuropathologica, 130*, 349–362.

Ishigaki, S., Niwa, J., Yamada, S., Takahashi, M., Ito, T., Sone, J., Doyu, M., Urano, F., & Sobue, G. (2007). Dorfin-CHIP chimeric proteins potently ubiquitylate and degrade familial ALS-related mutant SOD1 proteins and reduce their cellular toxicity. *Neurobiology of Disease, 25*, 331–341.

Jana, N. R., Tanaka, M., Wang, G., & Nukina, N. (2000). Polyglutamine length-dependent interaction of Hsp40 and Hsp70 family chaperones with truncated N-terminal huntingtin: Their role in suppression of aggregation and cellular toxicity. *Human Molecular Genetics, 9*, 2009–2018.

Jana, N. R., Dikshit, P., Goswami, A., Kotliarova, S., Murata, S., Tanaka, K., & Nukina, N. (2005). Co-chaperone CHIP associates with expanded polyglutamine protein and promotes their degradation by proteasomes. *The Journal of Biological Chemistry, 280*, 11635–11640.

Jankovic, J. (2008). Parkinson's disease: Clinical features and diagnosis. *Journal of Neurology, Neurosurgery, and Psychiatry, 79*, 368–376.

Jinwal, U. K., Miyata, Y., Koren, J., 3rd, Jones, J. R., Trotter, J. H., Chang, L., O'Leary, J., Morgan, D., Lee, D. C., Shults, C. L., et al. (2009). Chemical manipulation of hsp70 ATPase activity regulates tau stability. *The Journal of Neuroscience: The Official Journal of the Society for Neuroscience, 29*, 12079–12088.

Jinwal, U. K., O'Leary, J. C., 3rd, Borysov, S. I., Jones, J. R., Li, Q., Koren, J., 3rd, Abisambra, J. F., Vestal, G. D., Lawson, L. Y., Johnson, A. G., et al. (2010). Hsc70 rapidly engages tau after microtubule destabilization. *The Journal of Biological Chemistry, 285*, 16798–16805.

Kalmar, B., Novoselov, S., Gray, A., Cheetham, M. E., Margulis, B., & Greensmith, L. (2008). Late stage treatment with arimoclomol delays disease progression and prevents protein aggregation in the SOD1 mouse model of ALS. *Journal of Neurochemistry, 107*, 339–350.

Kalmar, B., Edet-Amana, E., & Greensmith, L. (2012). Treatment with a coinducer of the heat shock response delays muscle denervation in the SOD1-G93A mouse model of amyotrophic lateral sclerosis. *Amyotrophic Lateral Sclerosis: Official Publication of the World Federation of Neurology Research Group on Motor Neuron Diseases, 13*, 378–392.

Kamal, A., Thao, L., Sensintaffar, J., Zhang, L., Boehm, M. F., Fritz, L. C., & Burrows, F. J. (2003). A high-affinity conformation of Hsp90 confers tumour selectivity on Hsp90 inhibitors. *Nature, 425*, 407–410.

Kelly, E. B. (2013). *Encyclopedia of human genetics and disease* (Vol. 1). Santa Barbara: ABC-CLIO.

Kieran, D., Kalmar, B., Dick, J. R., Riddoch-Contreras, J., Burnstock, G., & Greensmith, L. (2004). Treatment with arimoclomol, a coinducer of heat shock proteins, delays disease progression in ALS mice. *Nature Medicine, 10*, 402–405.

Kiernan, M. C., Vucic, S., Cheah, B. C., Turner, M. R., Eisen, A., Hardiman, O., Burrell, J. R., & Zoing, M. C. (2011). Amyotrophic lateral sclerosis. *Lancet, 377*, 942–955.

Klucken, J., Shin, Y., Hyman, B. T., & McLean, P. J. (2004). A single amino acid substitution differentiates Hsp70-dependent effects on alpha-synuclein degradation and toxicity. *Biochemical and Biophysical Research Communications, 325*, 367–373.

Koren, J., 3rd, Jinwal, U. K., Jin, Y., O'Leary, J., Jones, J. R., Johnson, A. G., Blair, L. J., Abisambra, J. F., Chang, L., Miyata, Y., et al. (2010). Facilitating Akt clearance via manipulation of Hsp70 activity and levels. *The Journal of Biological Chemistry, 285*, 2498–2505.

Koyama, S., Arawaka, S., Chang-Hong, R., Wada, M., Kawanami, T., Kurita, K., Kato, M., Nagai, M., Aoki, M., Itoyama, Y., et al. (2006). Alteration of familial ALS-linked mutant SOD1 solubility with disease progression: Its modulation by the proteasome and Hsp70. *Biochemical and Biophysical Research Communications, 343*, 719–730.

Lashuel, H. A., Petre, B. M., Wall, J., Simon, M., Nowak, R. J., Walz, T., & Lansbury, P. T., Jr. (2002). Alpha-synuclein, especially the Parkinson's disease-associated mutants, forms pore-like annular and tubular protofibrils. *Journal of Molecular Biology, 322*, 1089–1102.

Lichtenberg, M., Mansilla, A., Zecchini, V. R., Fleming, A., & Rubinsztein, D. C. (2011). The Parkinson's disease protein LRRK2 impairs proteasome substrate clearance without affecting proteasome catalytic activity. *Cell Death & Disease, 2*, e196.

Liu, J., Shinobu, L. A., Ward, C. M., Young, D., & Cleveland, D. W. (2005). Elevation of the Hsp70 chaperone does not effect toxicity in mouse models of familial amyotrophic lateral sclerosis. *Journal of Neurochemistry, 93*, 875–882.

Lotz, G. P., Legleiter, J., Aron, R., Mitchell, E. J., Huang, S. Y., Ng, C., Glabe, C., Thompson, L. M., & Muchowski, P. J. (2010). Hsp70 and Hsp40 functionally interact with soluble mutant huntingtin oligomers in a classic ATP-dependent reaction cycle. *The Journal of Biological Chemistry, 285*, 38183–38193.

Luk, K. C., Mills, I. P., Trojanowski, J. Q., & Lee, V. M. Y. (2008). Interactions between Hsp70 and the hydrophobic core of alpha-synuclein inhibit fibril assembly. *Biochemistry, 47*, 12614–12625.

McLean, P. J., Klucken, J., Shin, Y., & Hyman, B. T. (2004). Geldanamycin induces Hsp70 and prevents alpha-synuclein aggregation and toxicity in vitro. *Biochemical and Biophysical Research Communications, 321*, 665–669.

Monsellier, E., Redeker, V., Ruiz-Arlandis, G., Bousset, L., & Melki, R. (2015). Molecular interaction between the chaperone Hsc70 and the N-terminal flank of huntingtin exon 1 modulates aggregation. *The Journal of Biological Chemistry, 290*, 2560–2576.

Muchowski, P. J., Schaffar, G., Sittler, A., Wanker, E. E., Hayer-Hartl, M. K., & Hartl, F. U. (2000). Hsp70 and hsp40 chaperones can inhibit self-assembly of polyglutamine proteins into amyloid-like fibrils. *Proceedings of the National Academy of Sciences of the United States of America, 97*, 7841–7846.

O'Leary, J. C., 3rd, Li, Q., Marinec, P., Blair, L. J., Congdon, E. E., Johnson, A. G., Jinwal, U. K., Koren, J., 3rd, Jones, J. R., Kraft, C., et al. (2010). Phenothiazine-mediated rescue of cognition in tau transgenic mice requires neuroprotection and reduced soluble tau burden. *Molecular Neurodegeneration, 5*, 45.

Opazo, F., Krenz, A., Heermann, S., Schulz, J. B., & Falkenburger, B. H. (2008). Accumulation and clearance of alpha-synuclein aggregates demonstrated by time-lapse imaging. *Journal of Neurochemistry, 106*, 529–540.

Orr, H. T. (2012). Cell biology of spinocerebellar ataxia. *The Journal of Cell Biology, 197*, 167–177.

Park, S. H., Kukushkin, Y., Gupta, R., Chen, T., Konagai, A., Hipp, M. S., Hayer-Hartl, M., & Hartl, F. U. (2013). PolyQ proteins interfere with nuclear degradation of cytosolic proteins by sequestering the Sis1p chaperone. *Cell, 154*, 134–145.

Rosen, D. R., Siddique, T., Patterson, D., Figlewicz, D. A., Sapp, P., Hentati, A., Donaldson, D., Goto, J., O'Regan, J. P., Deng, H. X., et al. (1993). Mutations in Cu/Zn superoxide dismutase gene are associated with familial amyotrophic lateral sclerosis. *Nature, 362*, 59–62.

Roy, J., Minotti, S., Dong, L., Figlewicz, D. A., & Durham, H. D. (1998). Glutamate potentiates the toxicity of mutant Cu/Zn-superoxide dismutase in motor neurons by postsynaptic calcium-dependent mechanisms. *The Journal of Neuroscience: The Official Journal of the Society for Neuroscience, 18*, 9673–9684.

Shimura, H., Schwartz, D., Gygi, S. P., & Kosik, K. S. (2004). CHIP-Hsc70 complex ubiquitinates phosphorylated tau and enhances cell survival. *The Journal of Biological Chemistry, 279*, 4869–4876.

St Martin, J. L., Klucken, J., Outeiro, T. F., Nguyen, P., Keller-McGandy, C., Cantuti-Castelvetri, I., Grammatopoulos, T. N., Standaert, D. G., Hyman, B. T., & McLean, P. J. (2007). Dopaminergic neuron loss and up-regulation of chaperone protein mRNA induced by targeted over-expression of alpha-synuclein in mouse substantia nigra. *Journal of Neurochemistry, 100*, 1449–1457.

Tantucci, M., Mariucci, G., Taha, E., Spaccatini, C., Tozzi, A., Luchetti, E., Calabresi, P., & Ambrosini, M. V. (2009). Induction of heat shock protein 70 reduces the alteration of striatal electrical activity caused by mitochondrial impairment. *Neuroscience, 163*, 735–740.

Tetzlaff, J. E., Putcha, P., Outeiro, T. F., Ivanov, A., Berezovska, O., Hyman, B. T., & McLean, P. J. (2008). CHIP targets toxic alpha-synuclein oligomers for degradation. *The Journal of Biological Chemistry, 283*, 17962–17968.

Thompson, A. D., Scaglione, K. M., Prensner, J., Gillies, A. T., Chinnaiyan, A., Paulson, H. L., Jinwal, U. K., Dickey, C. A., & Gestwicki, J. E. (2012). Analysis of the tau-associated proteome reveals that exchange of Hsp70 for Hsp90 is involved in tau degradation. *ACS Chemical Biology, 7*(10), 1677–1686.

Voss, K., Combs, B., Patterson, K. R., Binder, L. I., & Gamblin, T. C. (2012). Hsp70 alters tau function and aggregation in an isoform specific manner. *Biochemistry, 51*, 888–898.

Warrick, J. M., Chan, H. Y., Gray-Board, G. L., Chai, Y., Paulson, H. L., & Bonini, N. M. (1999). Suppression of polyglutamine-mediated neurodegeneration in Drosophila by the molecular chaperone HSP70. *Nature Genetics, 23*, 425–428.

Watanabe, M., Dykes-Hoberg, M., Culotta, V. C., Price, D. L., Wong, P. C., & Rothstein, J. D. (2001). Histological evidence of protein aggregation in mutant SOD1 transgenic mice and in amyotrophic lateral sclerosis neural tissues. *Neurobiology of Disease, 8*, 933–941.

Williams, A. J., Knutson, T. M., Colomer Gould, V. F., & Paulson, H. L. (2009). In vivo suppression of polyglutamine neurotoxicity by C-terminus of Hsp70-interacting protein (CHIP) supports an aggregation model of pathogenesis. *Neurobiology of Disease, 33*, 342–353.

Witt, S. N. (2010). Hsp70 molecular chaperones and Parkinson's disease. *Biopolymers, 93*, 218–228.

Yoo, B. C., Seidl, R., Cairns, N., & Lubec, G. (1999). Heat-shock protein 70 levels in brain of patients with Down syndrome and Alzheimer's disease. *Journal of Neural Transmission. Supplementum, 57*, 315–322.

Yu, A., Shibata, Y., Shah, B., Calamini, B., Lo, D. C., & Morimoto, R. I. (2014). Protein aggregation can inhibit clathrin-mediated endocytosis by chaperone competition. *Proceedings of the National Academy of Sciences of the United States of America, 111*, E1481–E1490.

Zarei, S., Carr, K., Reiley, L., Diaz, K., Guerra, O., Altamirano, P. F., Pagani, W., Lodin, D., Orozco, G., & Chinea, A. (2015). A comprehensive review of amyotrophic lateral sclerosis. *Surgical Neurology International, 6*, 171.

Zetterstrom, P., Graffmo, K. S., Andersen, P. M., Brannstrom, T., & Marklund, S. L. (2011). Proteins that bind to misfolded mutant superoxide dismutase-1 in spinal cords from transgenic amyotrophic lateral sclerosis (ALS) model mice. *The Journal of Biological Chemistry, 286*, 20130–20136.

Heat Shock Protein70 in Neurological Disease

Pinar Ortan, Ozden Yildirim Akan, and Ferda Hosgorler

Abstract The HSP70 is a chaperon protein that is expressed during stress conditions that participates in many biological processes, including protein trafficking, nascent polypeptide folding and the refolding of the wrong proteins and cleaning of the misfolded ones. The expression is increased during various pathological conditions such as cerebral ischemia, neurodegenerative diseases, epilepsy, and trauma. They are found in both intracellular and extracellular compartments. HSP70 exhibits different functions in accordance with its location. Intracellular HSP70 exerts cytoprotective functions as a chaperone protein, whereas extracellular HSP70 exerts immunomodulatory functions that trigger immunological responses. They play an auxiliary role in antigen presentation in the appearance of immunological response in multiple sclerosis. Epilepsy is thought to have emerged as a stressor. HSP overexpression is proposed as a potential therapy for neurodegenerative diseases characterized by the accumulation or aggregation of abnormal proteins. In this chapter, we wanted to summarize the recent studies on the role of HSP70 in neurological disorders.

Keywords Alzheimer disease · Heat shock protein 70 · Hsp70 · Neurological disorders · Neuroprotection

Abbreviations

Aβ	Amiloid beta
ALS	Amyotrophic Lateral Sclerosis
CJD	Creutzfeldt-Jakob disease
DA	Dopamine

P. Ortan (✉)
Division of Neurology, SB University, Izmir Bozyaka Training and Research Hospital, Alsancak Izmir, Turkey

O. Y. Akan
Division of İnternal Medicine, SB University, Izmir Bozyaka Training and Research Hospital, Bornova Izmir, Turkey

F. Hosgorler
Division of Physiology, Dokuz Eylul University Medical Faculty, Inciralti/Izmir, Turkey

A. A. A. Asea, P. Kaur (eds.), *HSP70 in Human Diseases and Disorders*, Heat Shock Proteins 14, https://doi.org/10.1007/978-3-319-89551-2_3

EAE	Experimental allegic encephalomyelitis
FFI	Fatal familial insomnia
GSS	Gerstmann-Sträussler-Scheinker syndrome
HD	Huntington disease
HSP	Heat shock protein
LRRK2	Leucine-rich repeat kinase-2
MG	Myasthenia gravis
MS	Multiple sclerosis
MTS	Mesial temporal sclerosis
PD	Parkinson's disease
PINK1	PTEN-induced putative kinase 1
polyQ	Poly-glutamine
PrPC	Cellular prion associated proteins
PrPSc	Disease associated prion proteins
SNCA	Alpha-synuclein
TDP-43	Tar DNA binding protein 43
UPS	Ubiquitin-proteasome system
vCJD	Variant Creutzfeldt-Jakob disease

Introduction

Heat shock protein (HSP) is a survival protein that acts as a molecular chaperone. When the metabolism is in danger, HSP70 expression increases in order to remove the unwanted, unfolded proteins, to repair of the damaged proteins and to help the synthesis of new polypeptides. HSP70 binds to protein substrates to stabilize them to avoid denaturation and apoptosis, assists the maintenance of cellular integrity. In the central nervous system, it has been found that HSP is produced in many cell types, including neurons, glia and endothelial cells. In this chapter we review the role of HSP70 in common neurological disorders and discuss the therapeutic interventions.

HSP70 in Alzheimer Disease

Alzheimers Disease is the most common cause of fatal neurodegenerative disease characterized by progressive memory loss, language disorders, cognitive function impairment. Advancing age, mitochondrial DNA mutations, oxidative stress are the factors that facilitate the development of the disease. The amyloid plaques caused by β Amyloid (Aβ) peptide aggregation inside the cell and the neurofibrillary tangles formed by phosphorylated tau aggregation outside the cell are the pathologic hallmarks of the disease. Hippocampus and cerebral cortex that is responsible for memory and cognition are the most affected parts (Ciechanover and Kwon 2015). Efforts to treat the disease is focused on the accumulation of Aβ and hyperphosphorylated tau proteins. Pathological processes resulting incorrectly folded proteins leads to the clinical manifestation of the disease (Desler et al. 2017). In the normal cell life, the nascent (newly formed) proteins about 30%of them are misfolded and they are destroyed quickly in the cell. Misfolded and aggregated proteins pass from quality control but some mutant proteins escape from proteolysis and form intracellular inclusions and extracellular plaques (Pratt et al. 2015). As the proteasome activities diminish at the older ages, the accumulation of protein aggregate gets easier.

Protein folding is required for a functional and stable protein. Synthesized polypeptide chains pass from thermodynamically unstable, α-helix form to a stable three-dimensional tertiary structure. The most important reason of misfolding of proteins is mutations rather than posttranslational protein modifications, oxidative stress, environmental conditions like pH and heat (Ho et al. 2015). Aβ proteins can accumulate as oligomer, and aggregate as amyloid fibril forms. HSP70 can selectively recognize the Aβ oligomers which are the most toxic form (Whyte et al. 2017). Therefore, HSP70 overexpression inhibits Aβ aggregation that may lead to the clinical improvement. On the other hand, HSP 70 has an important role on tau hemostasis by helping degradation of tau by either proteasomal pathway or ubiquitination (Patterson et al. 2011). HSP70 prevents cell death by weakening the caspase-dependent and independent pathways (Sabirzhanov et al. 2012). For this reason, treatment approaches to elevate the HSP70 levels are under investigation. Geldamycin which is a HSP70 inducer prevents the formation of neurofibrillary tangles by inhibiting tau accumulation (Lu et al. 2014; Hung and Fu 2017). We think neuroprotective therapy approaches on the chaperone system based studies must be executed.

HSP70 in Parkinson's Disease

Parkinson's disease (PD) is second most common movement disorder, and it affects nearly 1% of the population over the age of 60. PD is characterized mainly by progressive and selective loss of dopaminergic neurons in the substantia nigra pars compacta, with the subsequent dopamine (DA) decline in the nigrostriatal pathway, and by the presence of intracytoplasmic fibrillar α-Syn protein aggregates (Lewy Bodies) in the remaining nigral neurons (Halliday et al. 2011). Increased levels of α-synuclein or α-synuclein-containing aggregates are also characteristic of other neurodegenerative diseases, including Lewy body dementia, multiple system atrophy and AD. This group of diseases termed "synucleinopathies". Motor impairments, bradykinesia, rigidity, and resting tremor are clinical characteristic features of PD. These impairments are result of loss of dopaminergic neuron loss in substantia nigra. Although the mechanism is not fully understood it is hypothesized that disease arise from combination of genetic and environmental factors. Oxidative stress associated with mitochondrial dysfunction, proteolytic stress due to dysfunction of the ubiquitin-proteasome system (UPS), and local inflammation are pathogenic pathways that have been concerned. Although the majority of cases of PD appear to be sporadic when the age of symptom onset is younger than 50 years genetic factors play more role. Glucocerebrosidase, Alpha-synuclein (SNCA), leucine-rich repeat kinase-2 (LRRK2), Parkin, PTEN-induced putative kinase 1 (PINK1; PARK6) mitochondrial DJ-1 (PARK7) are the genes that was described which has role in PD (Krüger et al. 1998).

Correct protein folding is essential for proteins' biological functions. Heat shock proteins are critical elements of the cellular response to unfolded proteins. HSP are involved in promoting proper protein folding and preventing aggregation, as well as promoting ubiquitination and degradation of misfolded proteins. Under certain pathological conditions the protein quality control machinery is not sufficient to prevent the accumulation of misfolded proteins and these accumulations may lead to neurodegenerative disorders including PD (Kalmar and Greensmith 2017). The role of molecular chaperones in PD was first suggested by the detection of HSP90, HSP70, HSP60, HSP40, and HSP27 molecules in lewy bodies. In 1991 Namba Y et al. worked on brain tissues, which were obtained from the autopsy of patients who had neurodegenerative diseases including PD. They performed immunohistochemical studies on brain tissues from patients with various neurodegenerative conditions by using specific polyclonal antibody to HSP 70 and found HSP 70 association with abnormal cytoplasmic inclusions, which are characteristic for neurodegenerative diseases (Namba et al. 1991).

Fiszer U et al. worked on CSF of patients PD. They measured IgG levels against anti-HSP 70 molecules and found significantly increased levels in patients with PD compared to patients who had non-inflammatory neurological diseases (Fiszer et al. 1996). It has been shown that neuron cells, are postmitotic cells and they are susceptible to misfolded proteins (Muchowski and Wacker 2005). The misfolded aggregates are immunoreactive for ubiquitin, and most have been reported to contain molecular chaperones and components of the proteasome (Davies et al. 1997). Molecular chaperones and components of the proteasome can also be found in aggregates formed in transgenic animal models and transfected cell cultures (Suhr et al. 2001). Sheng Chen and Ian R. Brown demonstrated intermediate Hsc70 levels in neurons of the substantia nigra affected as PD in a rat model (Chen and Brown 2007). In this way determining these proteins, suggest that protein aggregates are recognized as targets and cellular protein quality control mechanisms are activated in an attempt to prevent their accumulation (Kazemi-Esfarjani and Benzer 2002). α-Syn is a 140-amino acid neuronal protein probably involved in regulating cell differentiation, synaptic plasticity, and dopaminergic neurotransmission.It has been demonstrated that HSP70 overexpression reduced α-Syn accumulation and toxicity in both mouse and Drosophila models of PD. (Klucken et al. 2004; Auluck et al. 2002). Other experiments using heat shock induced expression of chaperones demonstrate that HSP70 supplies protection against cytotoxicity of PD-inducing pesticide rotenone and reduced alpha synuclein aggregation in a cellular model (Zhou et al. 2004). In the light of this information's Huang et al. showed that HSP70 inhibits as fibril formation via preventing the formation of prefebrillar α-syn, binding with these species to inhibit nuclei formation, as well as binding with nuclei to retard fibril elongation (Huang et al. 2006).

Roodveldt et al. showed that HSP70 depletion can be a direct result for the presence of aggregation-prone polypeptides. They found that a nucleotide-dependent interaction between HSP70 and αSyn, which leads to the aggregation of HSP70, with the presence of ADP along with αSyn. Such a co-aggregation phenomenon could be prevented in vitro by the co-chaperone Hip (ST13). Their findings indicated that Hip utilizes stabilization of HSP70 and helps chaperone mediated amyloid formation inhibition. Another finding of this study was that ADP-bound HSP70 has a very high tendency to co-aggregate with α-syn, suggesting that chaperone depletion favored under certain conditions could be an important feature in the onset and progression of amyloid disorder (Roodveldt et al. 2009). DnaK/DnaJ/GrpE of HSP70 system model used by Ahamad et al. By this system they showed that HSP70 inhibits αSyn fibrillar assembling but cannot activate refolding process (Ahmad 2010). For all these reasons misfolded proteins are considered a common therapeutic target in PD and many studies have focused on the neuroprotective role of HSP.

HSP70 in Amyotrophic Lateral Sclerosis

Tar DNA binding protein 43 (TDP-43) neuronal cytoplasmic inclusion aggregations are thought to be the key in some neurodegenerative disease (amyotrophic lateral sclerosis, frontotemporal lobar degeneration) Inhibition or the clearance of this toxic aggregation is, therefore, a strategy for therapeutic intervention. The ubiquitin-proteasome system or the autophagy pathway participates in resolving potentially detrimental protein aggregates. HSP might refold TDP-43 and return it to its natural physiological state. Either transgenic TDP-43 mouse model or sporadic ALS patients reduced HSP70 levels are assessed. Strategies against HSP activation be may be an important therapeutic challenge for the TDP-43 proteinopathies. Arimoclomol that is a co-inducers of heat shock protein 70 and 90 expressions under cellular stress, is neuroprotective in a number of neurodegenerative disease models, including familial Amyotrophic Lateral Sclerosis (ALS). Superoxide Dismutase 1 transgenic mice (an animal model of ALS), Arimoclomol has effects on the prevention of neuronal loss and the promotion of motor neuron survival. The therapeutic potential of Arimoclomol is currently under investigation in Phase II/III clinical trials for familial ALS patients with SOD1 mutations. The HSP70 co-inducer drugs like Arimoclomol may be a hope for ALS and other neurodegenerative disorders.

HSP in Huntington Disease

Huntington disease (HD) is a non curable, adult-onset, autosomal dominant inherited disorder associated with cell loss within a specific subset of neurons in the basal ganglia and cortex. The disease named by the physician who described it as hereditary chorea in 1872 (Huntington 1872). Involuntary movements, dementia, and behavioral changes are HD characteristic hallmarks. Formation of intracellular inclusions composed primarily of the ubiquitous protein huntingtin, the subsequent death of striatal medium spiny neurons and cortical pyramidal neurons are responsible for disease symptoms and signs (Vonsattel and DiFiglia 1998). Electron microscopy reveals both cytoplasmic and nuclear abnormalities, including the presence of large neuronal intra nuclear inclusions or aggregates similar to those in other polyglutamine disorders. The aggregates are also found in dystrophic neurites. The genetic basis of HD is the expansion of a cysteine-adenosine-guanine repeat encoding a polyglutamine tract in the N-terminus of the protein product called huntingtin. This leads to a mutant protein that contains an expanded poly-glutamine (polyQ) sequence (Ho and Hocaoglu 2011). The length of the polyQ sequence determines the age of disease onset and severity (Zoghbi and Orr 2000). In HD pathologic examination we can see cytoplasmic and nuclear aggregates, including large neuronal inclusions like other polyglutamine disorders (Davies et al. 1998).

PolyQ-expanded Htt cooperate differently with the proteostasis network compared to other disease-associated proteins. Bersuker K et al. show that the heat shock response is not activated by mutant huntingtin gene even in cells selected for the highest expression levels and for the presence of inclusion bodies containing aggregated protein (Bersuker et al. 2013). Tagawa K et al. worked on tree types of neuron and found different response to mutant polyQ proteins. They established that HSP70 was up regulated only by mutant htt and selectively in the granule cells within the cerebellum. They also found that granule cells, that are unresponsive to degeneration in HD pathology, lost their resistance by suppressing HSP70 with siRNA, on the other hand cortical neurons, affected in human HD, gained resistance by over expressing HSP70. This indicates that induction levels of HSP70 are a critical factor for determining vulnerabilities to mutant htt among neuronal subtypes (Tagawa et al. 2007). The recent studies show that HSP70 and its co-chaperone HSP40 inhibits huntingtin exon 1 fragment aggregation and modifies the structural, seeding, and infectious properties of the resulting fibrils in a polyQ-independent manner (Monsellier et al. 2015).

HSP70 in Stroke

The effect of HSP70 is studied in many brain ischemia models. In a global cerebral ischemia model, like the brain damage after cardiac arrest, HSP70 mRNA expression increased in hours. HSP70 expression was observed within astrocytes, neurons by glial transfer to protect the neurons from damage. Ischemia in rodents showed neuroprotection after different kinds of nerve injuries in astrocyte cultures of HSP70 transgenic mice. HSP70 was used as a therapeutic agent in a focal ischemia model which was formed with 4-min middle cerebral artery occlusion. Twenty minutes before occlusion some mice were given intravenous HSP70. They reported that the ischemic zone was little about the group treated with HSP70 (Shevtsov et al. 2014). Gomez-Choco M et al. studied the existence of HSP70 in lymphoid tissue of acute stroke patients and exhibited that highly immunoreactive HSP70 was associated with smaller infarct size and better outcome (Gomez-Choco et al. 2014). Geldanamycin is a HSP inhibitor that is emerged as a therapeutic agent with the ability of upregulation of HSP70. Pretreated astrocytes and glial cells with geldamycin were more preserved from cell death (Kacimi and Yenari 2015). Another experiment on a model of brain ischemia was done with a protein called Dynamin that triggers apoptosis leads to caspase-dependent cell death. Kim et al. showed that HSP70 transgenic mice had a better outcome and dynamin inhibitor dynasore also protected the mice from the stroke. They drew attention to HSP70 and dynamin interaction may be worthy and target strategies on dynamin inhibition may protect the brain from the ischemic stroke (Kim et al. 2016).

HSP70 in Epilepsy

Epilepsy is a paroxysmal disorder that is characterized by repeated seizures, which may be idiopathic, symptomatic or cryptogenic in origin. The repeated seizures induce apoptosis in the neurons within the brain especially hippocampus. Stress proteins might be assumed to counteract the pathology of increased neuronal excitation. HSP70 is the most studied neuroprotective chaperon on epilepsy. Seizures are triggered by hyperexcitability that is caused by the excessive calcium influx into the cells. HSP70 has an influence of the decrease of the calcium influx to the neurons thus prevents brain cells from seizure-induced apoptosis. Consequently, HSP70 might be involved in endogenous cellular preservation during seizures. Many animal experiments demonstrate overexpression of HSP70 exerts protective effects during kindling. According to recent studies HSP70, transgenic mice were more resistant to kindling with chemical agents. Seizure threshold and survival during kindling were higher in HSP70 transgenic mice as compared to wild-type mice (Ammon-Treiber et al. 2007). Thus, overexpression of HSP70 exerts protective effects on seizure severity and overall survival during PTZ kindling and decreases the development of kindling. Mesial temporal sclerosis (MTS) is the most known cause of intractable epilepsy that can be treated by hippocampal resection. In the surgery material of human MTS cases, Kandratavicius et al. showed increased expression of HSP70 in the hippocampal formation. After successful surgery, they saw decreased levels of HSP70 and HSP90. Surgical excision of the hippocampus with more HSP expression showed poorer outcome compared with the hippocampus with less HSP expression (Kandratavicius et al. 2014). There was a positive correlation between seizure frequency, duration of epilepsy and HSP70 expression as in many studies.

HSP 70 in Migraine

Migraine is the most common primary headache that affects many people in the world. The World Health Organization estimates the prevalence of migraine %14. It is characterized by recurrent episodes of headache, mostly unilateral throbbing or pulsatile intensified with movement and accompanied by nausea or vomiting. The pathogenesis is explained by the neurovascular theory which is a primarily neurogenic process, and secondary change takes place in cerebral perfusion. In aura phase, Neuronal hyperexcitability usually begins in the occipital cortex and spreading cortical depression affects the cortex and activates the trigeminovascular system by stimulation of nociceptive neurons. The sterile inflammation is accompanied with vasodilatation that causes the headache. In our paper, we examined serum HSP27 levels during the migraine attack and during remission phase and compared with control subjects. We couldn't show any difference between attack and remission. We showed a positive correlation of HSP27 with headache severity scores during the migraine attack (Coban et al. 2011). Similarly, Yön et al. found no statistical difference in chronic migraine patients (Yon et al. 2016).

HSP70 in Myasthenia Gravis

Myasthenia gravis (MG) is an autoimmune neuromuscular disease. There is a limited number of studies into the role of HSP70 in MG. Munakata, et al. studied serum levels of MG patients just before and after treatment. They found a significant increase in the patient group. After treatment, in the group which responded therapy, the HSC71 levels decreased. In the therapy-resistant group, they found no change. They revealed that heat shock cognate protein 71 elevations may be a useful marker for the disease prognosis (Munakata et al. 2008). Helgeland et al. showed anti-HSP70 level elevation in Myasthenia Gravis and Guillain Barre syndrome (Helgeland et al. 2010).

HSP70 in Multiple Sclerosis

Multiple sclerosis (MS) is an autoimmune disease characterized by inflammation, demyelination, and axonal injury in central nervous system that leads disability in one-third of the patients. Between the ages of 20–40, it is the most common cause of disability after trauma. Myelin sheaths, oligodendrocytes and less frequently axon and the nerve cell itself are damaged by inflammation (Lucchinetti et al. 2000). The hallmarks of MS pathology are multifocal demyelination that is characterized by inflammation and gliosis. These areas are seen as MS plaques, which are diagnostic for MS. According to the disease status, MS plaques may be active, including the more inflammatory infiltrates, T lymphocytes, macrophages fewer B-lymphocytes, plasma cells, immunoglobulins, and complements (Lucchinetti et al. 2000). Critical approaches to MS therapy focus on the protection of oligodendrocytes and neurons by immunomodulation of T-cells against myelin. As a consequence, the activation of myelin-specific CD4 T-cells by secretion of Inflammatory cytokines(such as IFN-γ, TNF-α, IL-1β, and IL-6) leads to release of adhesion molecules from the vascular endothelial cells, help the migration of lymphocytes across the blood-brain barrier. CD8+ T cells, B cells, antibodies, natural killer cells, join the complex immune response and responsible for the axonal injury (Mansilla et al. 2014).

The exact role of HSP70 in MS pathogenesis is still debated as the effect changes with the localization of the protein. While the intracellular HSP70 is neuroprotective, the HSP70 released into the extracellular space seems to be an antigen-adjuvant and may have a role in antigen presenting. Selmaj demonstrated the colocalization of T-cell receptor gamma delta cells with HSP65 and HSP70 in MS lesions postmortem (Selmaj et al. 1991). Boiocchi C et al. studied genetic polymorphism of HSP70 159 relapsing-remitting and 36 secondary progressive MS patients and compared with 586 healthy controls. They hypothesized that over expression of HSP70 as in AA genotype may have less inflammation, and better prognosis (Boiocchi et al. 2016).

Experimental allergic encephalomyelitis (EAE) is an animal model of MS. After the injection of a myelin protein (myelin associated protein, myelin basic protein, proteolipid protein, etc.), a cell-mediated immune reaction against myelin develops and causes disease like MS (Hernandez-Pedro et al. 2013). Studies with EAE showed HSP70 over expression seems to be beneficial for the recovery from the disease. Conversely, the HSP70 knockout mice were significantly resistant to EAE development (Mansilla et al. 2014). Talla et al. demonstrated that mitochondrial HSP70 elevation in retinal ganglion cells, which preserves vision by preventing neuronal apoptosis, and axonal damage in the EAE model with mice. HSP70 can also act via an anti-inflammatory mechanism by inducing the expression of anti-inflammatory cytokines and inhibiting inflammatory ones (Talla et al. 2014). In Boiocchi's study, Genotyping of HSP70–2 + 1267 A/G polymorphism was found in 195 MS patients. In addition, HSP70-2 protein content in vitro from PBMC was meaningfully lower in MS patients with GG genotype compared to AA genotype, indicating an implication of the G allele of HSP70-2 gene polymorphism in the development of MS (Boiocchi et al. 2016). Caussi et al. found that humoral response to HSP70 was meaningfully high in a large group of MS patients (Cassu et al. 2013). Significantly elevated antibody titers against HSP70 proteins were not only found in peripheral blood but also in the cerebrospinal fluid sample of MS patients (Chiba et al. 2006). The mitochondrial dysfunction has been proposed to be the key of neurodegeneration in MS. There are studies targeting on mitochondrial HSP70 enhancement as a treatment of choice. Gene therapy to enhance the HSP70 in mitochondria is under investigation.

HSP70 in Prion Disease

Prion diseases are neurodegenerative diseases that have long incubation periods and once clinical symptoms destructively progress. Five human prion diseases are currently recognized: kuru, Creutzfeldt-Jakob disease (CJD), variant Creutzfeldt-Jakob disease (vCJD also known as new variant CJD), Gerstmann-Sträussler-Scheinker syndrome (GSS), and fatal familial insomnia (FFI) (Prusiner 2001). In prion diseases cellular prion associated proteins(PrPC) are changed into disease associated prion proteins(PrPSc) and those misfolded proteins accumulate within the neural tissue (Budka 2003). This accumulation gives rise to common features to prion diseases including neuronal loss, proliferation of glial cells, absence of an inflammatory response, and the presence of small vacuoles within the neurons, which produces a spongiform appearance. For understanding role of chaperons in prion multiplication yeast eukaryotic models have been used. Transmission, self propagating and aggregation are properties of yeast prions (PSI+) (King et al. 1997). Jones et al. worked with yeast cells and showed that a mutation in the SSA1 HSP70 allele (SSA1–21p) significantly impaired PSI+ self-replication and propagation (Jones et al. 2004). Propagation of Saccharomyces cerevisiae [PSI+] prion is impaired by factors that regulate HSP70 substrate binding. Diedrich et al. show elevation in protein levels of inducible HSP70 in active astrocytes with a C57BL6 mice model injected with 22L strain of scrapie (Diedrich et al. 1993). In addition a significant increase in HSP70 RNA expression demonstrated in mice brain that infected with scrapie forms (Kenward et al. 1994). Tamguney et al. conducted a study on 20 potential genes candidates that could regulate the replication of prions in mice infected by scrapie or cow 301V prions. They worked in a mice infected with scrapie or cow 301V prions model potential genes that may regulate replication of prions examined and overexpression of human HSP70 did not show effect on prion disease onset (Tamguney et al. 2008). To our knowledge, the mechanism of the aggregation of those proteins and interventions with chaperons is not exactly understood.

Conclusions

HSP70 has an important role on the immunopathogenesis of multiple neurological conditions. It has neuroprotective effects in various diseases. We think that, treatment of many neurological diseases with HSP70 overexpression, will be possible in the future.

Acknowledgements We would like to thank the editorial staff for the opportunity of being able to be among the authors of the book.

References

Ahmad, A. (2010). DnaK/DnaJ/GrpE of HSP70 system have differing effects on alpha-synuclein fibrillation involved in Parkinson's disease. *International Journal of Biological Macromolecules, 46*(2), 275–279.

Ammon-Treiber, S., Grecksch, G., Angelidis, C., et al. (2007). Pentylenetetrazol kindling in mice overexpressing heat shock protein 70. *Naunyn-Schmiedeberg's Arch Pharmacol, 375*, 115–112.

Auluck, P. K., Chan, H. Y., Trojanowski, J. Q., et al. (2002). Chaperone suppression of alpha-synuclein toxicity in a Drosophila model for Parkinson's disease. *Science, 295*(5556), 865–868.

Bersuker, K., Hipp, M. S., Calamini, B., et al. (2013). Heat shock response activation exacerbates inclusion body formation in a cellular model of Huntington disease. *The Journal of Biological Chemistry, 288*, 23633–23638.

Boiocchi, C., Monti, M. C., Osera, C., et al. (2016). Heat shock protein 70-hom gene polymorphism and protein expression in multiple sclerosis. *Journal of Neuroimmunology, 298*, 189–193.

Budka, H. (2003). Neuropathology of prion diseases. *British Medical Bulletin, 66*, 121–130.

Cassu, D., Masala, S., Frau, J., et al. (2013). Anti Mycobacterium avium subsp. Paratuberculosis heat shock protein 70 antibodies in sera of Sardinian patients with multiple sclerosis. *Journal of the Neurological Sciences, 355*(1–2), 131–133.

Chen, S., & Brown, I. R. (2007). Neuronal expression of constitutive heat shock proteins: Implications for neurodegenerative diseases. *Cell Stress & Chaperones, 12*(1), 51–58.

Chiba, S., Yokota, S., Yonekura, K., et al. (2006). Autoantibodies against HSP70 family proteins were detected in the cerebrospinal fluid from patients with multiple sclerosis. *Journal of the Neurological Sciences, 241*(1–2), 39–43.

Ciechanover, A., & Kwon, Y. T. (2015). Degradation of misfolded proteins in neurodegenerative diseases: Therapeutic targets and strategies. *Experimental & Molecular Medicine, 47*, 147.

Coban, P., Çe, P., Erkizan, O., & Gedizlioglu, M. (2011). Heat shock protein 27 in migraine patients. *Journal of Neurological Sciences [Turkish], 28*(1), 28–34.

Davies, S. W., Turmaine, M., Cozens, B. A., et al. (1997). Formation of neuronal intranuclear inclusions underlies the neurological dysfunction in mice transgenic for the HD mutation. *Cell, 90*(3), 537–548.

Davies, S. W., Beardsall, K., Turmaine, M., et al. (1998). Are neuronal intranuclear inclusions the common neuropathology of triplet-repeat disorders with polyglutamine-repeat expansions? *Lancet, 351*, 131.

Desler, C., Lillenes, M. S., Tønjum, T., et al. (2017). The role of mitochondrial dysfunction in the progression of Alzheimer's disease. *Current Medicinal Chemistry*. https://doi.org/10.2174/0929867324666170616110111. [Epub ahead of print].

Diedrich, J. F., Carp, R. I., & Haase, A. T. (1993). Increased expression of heat shock protein, transferrin, and beta 2-microglobulin in astrocytes during scrapie. *Microbial Pathogenesis, 15*, 1–6.

Fiszer, U., Fredrikson, S., & Członkowska, A. (1996). Humoral response to HSP 65 and HSP 70 in cerebrospinal fluid in Parkinson's disease. *Journal of the Neurological Sciences, 139*(1), 66–70.

Gómez-Chocoa, M., Doucerain, C., Urra, X., et al. (2014). Presence of heat shock protein 70 in secondary lymphoid tissue correlates with stroke prognosis. *Journal of Neuroimmunology, 270*(1–2), 67–74.

Halliday, G. M., Holton, J. L., Revesz, T., et al. (2011). Neuropathology underlying clinical variability in patients with synucleinopathies. *Acta Neuropathologica, 122*, 187–204.

Helgeland, G., Petzold, A., Hoff, J. M., et al. (2010). Anti-heat shock protein 70 antibody levels are increased in myasthenia gravis and Guillain-Barré syndrome. *Journal of Neuroimmunology, 225*(1–2), 180–183.

Hernández-Pedro, N. Y., Espinosa-Ramirez, G., de la Cruz, V. P., et al. (2013). Initial immunopathogenesis of multiple sclerosis: Innate immune response. *Clinical and Developmental Immunology*. Article ID 413465, 15 pages.

Ho, A. K., & Hocaoglu, M. B. (2011). Impact of Huntington's across the entire disease spectrum: The phases and stages of disease from the patient perspective. *Clinical Genetics, 80*(3), 235–239.

Ho, S. L., Poon, C. Y., Lin, C., et al. (2015). Inhibition of β-amyloid aggregation By albiflorin, aloeemodin and neohesperidin and their neuroprotective effect on primary hippocampal cells against β-amyloid induced toxicity. *Current Alzheimer Research, 12*(5), 424–433.

Huang, C., Cheng, H., Hao, S., et al. (2006). Heat shock protein 70 inhibits alpha-synuclein fibril formation via interactions with diverseintermediates. *Journal of Molecular Biology, 364*(3), 323–336.

Hung, S. Y., & Fu, W. M. (2017). Drug candidates in clinical trials for Alzheimer's disease. *Biomedical Science, 24*(1), 47.

Huntington, G. (1872). *Med Surg Report 26*, 320.

Jones, G., Song, Y., Chung, S., et al. (2004). Propagation of *Saccharomyces cerevisiae* [PSI+] prion is impaired by factors that regulate HSP70 substrate binding. *Molecular and Cellular Biology, 24*(9), 3928–3937.

Kacimi, R., & Yenari, M. A. (2015). Pharmacologic heat shock protein 70 induction confers cytoprotection against inflammation in gliovascular cells. *Glia, 63*(7), 1200–1212.

Kalmar, B., & Greensmith, L. (2017). Cellular chaperones as therapeutic targets in ALS to restore protein homeostasis and improve cellular function. *Frontiers in Molecular Neuroscience, 10*, 251. https://doi.org/10.3389/fnmol.2017.00251.

Kandratavicius, L., Hallak, J. E., Carlotti, C. G., et al. (2014). Hippocampal expression of heat shock proteins in mesial temporal lobe epilepsy with psychiatric comorbidities and their relation to seizure outcome. *Epilepsia, 55*, 1834–1843.

Kazemi-Esfarjani, P., & Benzer, S. (2002). Suppression of polyglutamine toxicity by a Drosophila homolog of myeloid leukemia factor 1. *Human Molecular Genetics, 11*(21), 2657–2672.

Kenward, N., Hope, J., Landon, M., et al. (1994). Expression of polyubiquitin and heat-shock protein 70 genes increases in the later stages of disease progression in scrapie-infected mouse brain. *Journal of Neurochemistry, 62*, 1870–1877.

Kim, J. Y., Kim, N., Zheng, Z., et al. (2016). 70kDa heat shock protein downregulates dynamin in experimental stroke: A new therapeutic target? *Stroke, 47*(8), 2003–2011.

King, C. Y., Tittmann, P., Gross, H., et al. (1997). Prion-inducing domain 2-114 of yeast Sup35 protein transforms in vitro into amyloid-like filaments. *Proceedings of the National Academy of Sciences of the United States of America, 94*(13), 6618–6622.

Klucken, J., Shin, Y., Masliah, E., et al. (2004). HSP70 reduces alpha-synuclein aggregation and toxicity. *The Journal of Biological Chemistry, 279*(24), 25497–25502.

Krüger, R., Kuhn, W., Müller, T., et al. (1998). Ala30Pro mutation in the gene encoding alpha synuclein in Parkinson's disease. *Nature Genetics, 18*(2), 106–108.

Lu, R. C., Tan, M. S., Wang, H., et al. (2014). Heat shock protein 70 in Alzheimer's disease. *BioMed Research International, 2014*, 435203. https://doi.org/10.1155/2014.

Lucchinetti, C., Brück, W., Parisi, J., et al. (2000). Heterogenity of multiple sclerosis lesions: Implications for the pathogenesis of demyelination. *Annals of Neurology, 47*(6), 707–717.

Mansilla, M. J., Costa, C., Eixarch, H., et al. (2014). HSP70 regulates immune response in experimental autoimmune encephalomyelitis. *PLoS One, 9*(8). https://doi.org/10.1371/journal.pone.0105737.

Monsellier, E., Redeker, V., Ruiz-Arlandis, G., et al. (2015). Molecular interaction between the chaperone Hsc70 and the N-terminal flank of huntingtin exon 1 modulates aggregation. *The Journal of Biological Chemistry, 290*(5), 2560–2576.

Muchowski, P. J., & Wacker, J. L. (2005). Modulation of neurodegeneration by molecular chaperones. *Nature Reviews. Neuroscience, 6*(1), 11–22.

Munakata, S., Chen, M., Aosai, F., et al. (2008). The clinical significance of anti-heat shock cognate protein 71 antibody in myasthenia gravis. *Journal of Clinical Neuroscience, 15*(2), 158–165.

Namba, Y., Tomonaga, M., Ohtsuka, K., et al. (1991). HSP 70 is associated with abnormal cytoplasmic inclusions characteristic of neurodegenerative diseases. *Nō to Shinkei, 43*(1), 57–60.

Patterson, K. R., Ward, S. M., Combs, B., et al. (2011). Heat shock protein 70 prevents both tau aggregation and the inhibitory effects of preexisting tau aggregates on fast axonal transport. *Biochemistry, 50*(47), 10300–10310.

Pratt, W. B., Gestwicki, J. E., Osawa, Y., et al. (2015). Targeting proteostasis through the protein quality control function of the HSP90/HSP70-based chaperone machinery for treatment of adult onset neurodegenerative diseases. *Annual Review of Pharmacology and Toxicology, 55*, 353–371.

Prusiner, S. B. (2001). Shattucklecture – neurodegenerative diseases and prions. *The New England Journal of Medicine, 344*(20), 1516–1526.

Roodveldt, C., Bertoncini, C. W., Andersson, A., et al. (2009). Chaperone proteostasis in Parkinson's disease: Stabilization of the HSP70/alpha-synuclein complex by Hip. *The EMBO Journal, 28*(23), 3758–3770.

Sabirzhanov, B., Stoica, B. A., Hanscom, M., et al. (2012). Over-expression of HSP70 attenuates caspase-dependent and caspase-independent pathways and inhibits neuronal apoptosis. *Journal of Neurochemistry, 123*(4), 542–554.

Selmaj, K., Brosnan, C. F., & Raine, C. S. (1991). Immunology. *Proceedings of the National Academy of Sciences of the United States of America, 88*, 6452–6456.

Shevtsov, M. A., Nikolaev, B. P., Yakovleva, L. Y., et al. (2014). Neurotherapeutic activity of the recombinant heat shock protein HSP70 in a model of focal cerebral ischemia in rats. *Drug Design Development and Therapy, 8*, 639–650.

Suhr, S. T., Senut, M. C., Whitelegge, J. P., et al. (2001). Identities of sequestered proteins in aggregates from cells with induced polyglutamine expression. *The Journal of Cell Biology, 153*(2), 283–294.

Tagawa, K., Marubuchi, S., Qi, M. L., et al. (2007). The induction levels of heat shock protein 70 differentiate the vulnerabilities to mutant huntingtin among neuronal subtypes. *The Journal of Neuroscience, 27*(4), 868–880.

Talla, V., Porciatti, V., Chiodo, V., et al. (2014). Gene therapy with mitochondrial heat shock protein 70 suppresses visual loss and optic atrophy in experimental autoimmune encephalomyelitis. *Investigative Ophthalmology & Visual Science, 55*(8), 5214–5226.

Tamguney, G., Giles, K., Glidden, D. V., et al. (2008). Genes contributing to prion pathogenesis. *The Journal of General Virology, 89*, 1777–1788.

Vonsattel, J. P., & DiFiglia, M. (1998). Huntington disease. *Journal of Neuropathology and Experimental Neurology, 57*(5), 369–384.

Whyte, L. S., Lau, A. A., Hemsley, K. M., et al. (2017). Endo-lysosomal and autophagic dysfunction: A driving factor in Alzheimer's disease? *Neurochemistry, 140*(5), 703–717.

Yon, M. I., Titiz, A. P., Bilen, S., et al. (2016). Elevated interictal serum HSP-70 levels as an indicator of neurodegeneration for chronic migraine. *The Journal of the Pakistan Medical Association, 66*(6), 677–681.

Zhou, Y., Gu, G., Goodlett, D. R., et al. (2004). Analysis of alpha-synuclein-associated proteins by quantitative proteomics. *The Journal of Biological Chemistry, 279*(37), 39155–39164.

Zoghbi, H. Y., & Orr, H. T. (2000). Glutamine repeats and neurodegeneration. *Annual Review of Neuroscience, 23*, 217–247.

Involvement of Heat Shock Protein 70 (Hsp70) in Gastrointestinal Cancers

Dipamoy Datta, Suparna Banerjee, Anupama Ghosh, Soumyajit Banerjee Mustafi, Prosenjit Sen, and Sanghamitra Raha

Abstract Intracellular protein homeostasis is largely controlled by Heat shock proteins (Hsp). Heat shock proteins (Hsp) impart an age-old defense mechanism for all forms of life on earth. Misfolded proteins are refolded with the aid of Hsp and proteins which are damaged beyond repair are eliminated with assistance from Hsp. Hsp are known as molecular chaperones for their cytoprotective roles. In cancer cells the Hsp are frequently overexpressed and are assumed to be associated with tumor formation. Hsp demonstrate specific affinity to particular classes of oncogenic peptides and client proteins in cancer cells, and are able to stabilize mutated oncogene proteins. They play a key regulatory role in prevention of apoptotic cell death during tumorigenesis and thereby enhance cell growth and proliferation. They may also promote chemoresistance in cancer cells. Here we present the current knowledge on the role of molecular chaperones in particular heat shock protein 70 (Hsp70) in human gastrointestinal cancers along with their therapeutic targeting. This review will focus on the role of Hsp 70 and related chaperones in several gastrointestinal cancers such as pancreatic, gastric, and liver cancers.

Keywords Chaperone · Gastric cancer · Gastrointestinal cancer · Heat shock protein 70 (Hsp 70) · Liver cancer · Pancreatic cancer

D. Datta
Department of Biotechnology, Visva-Bharati, Santiniketan, India

S. Banerjee · P. Sen
Department of Biological Chemistry, Indian Association for the Cultivation of Science, Kolkata, India

A. Ghosh
Division of Plant Biology, Bose Institute, Kolkata, India

S. Banerjee Mustafi
University of Oklahoma Health Sciences Center, Oklahoma City, OK, USA

S. Raha (✉)
Department of Biotechnology, Visva-Bharati, Santiniketan, India

Kolkata, India

A. A. A. Asea, P. Kaur (eds.), *HSP70 in Human Diseases and Disorders*, Heat Shock Proteins 14, https://doi.org/10.1007/978-3-319-89551-2_4

Abbreviations

α-SMA	α-smooth muscle actin protein
CHIP	*c*arboxyl-terminus of *H*sp70 interacting Protein
CSC	cancer stem cells
GC	gastric cancer
GI	gastrointestinal
HCC	hepatocellular carcinoma
HNSCC	head and neck squamous cell cancer
HOP	Hsp70/Hsp90 organizing protein
MAPK	mitogen activated protein kinase

Introduction

Ferruccio Ritossa in 1962 first discovered Heat shock proteins (HsP) or stress proteins (Ritossa 1962). He observed temperature induced puffing patterns in salivary gland chromosomes of *Drosophila melanogaster* larvae. The increase in temperature stimulated the expression of proteins with molecular masses of 26 and 70 kDa (Tissieres et al. 1974). As such, the original definition of HsP was based on their enhanced expression in response to cellular insults, such as raised temperature, oxidative stress, chemical exposure and irradiation (Young et al. 2004). Under normal physiological conditions, a complete set of functionally competent proteins are maintained in the cell. When exposed to cellular stressors, disturbance of the intracellular milieu induces a stress response in the cell, which inhibits the activity of many housekeeping genes while activating stress genes (Gao et al. 2004). This leads to increased levels of stress protein and their chaperones in the cell in a concerted effort to maintain protein homeostasis.

Chaperoning System

In 2008 the 'chaperoning system' has been projected as a concept to include all molecular chaperones, co-chaperones, and cofactors of an organism (Macario and Conway de Macario 2009). The concept visualizes a physiological system encompassing all chaperones and their functionally-related molecules and structures, in all tissues, organs, and biological fluids. This physiological system is essential for the control of protein homeostasis and maintenance of a complete set of proteins in all fluids, cells, and tissues, with the correct and functional conformation (Macario and Conway de Macario 2009). Many vital biological processes such as antigen presentation, hormone receptor assembly, formation of complexes with a variety of ligands and unrelated to protein homeostasis also show direct involvement of the chaperoning system (Macario et al. 2010). The scientific discipline that deals with the chaperoning system is called chaperonology which also includes the study of the genomic sequences of chaperone genes (*e.g.* by applying chaperonomics) (Brocchieri et al. 2007), the study of the diseases that involve chaperones as causative factors (*i.e.*chaperonopathies) (Macario and Conway de Macario 2005), and the use of chaperones (molecules and genes) for the treatment of chaperonopathies (*i.e.*chaperonotherapy) (Macario et al. 2010).

Classification of Heat Shock Proteins

Hsps are classified into six major families according to their molecular mass in kilodaltons (Schlesinger 1990) (a) the large molecular weight Hsp of100–110 kDa, (b) the Hsp90 family of 83–90 kDa, (c) the Hsp70 family of 66–78 kDa, (d) the Hsp60 family of 55–64 kDa, (e) the Hsp40 family of 35–54 kDa, and (f) the small Hsp of 8–34 kDa. The family members have functional homologs in different compartments of the cell. There is a high degree of homology between the Hsp counterparts of different organisms (e.g., the Escherihia coli DnaK and the human Hsp70 share approximately 50% sequence identity). However, no noticeable sequence homology between different families of Hsp could be perceived (e.g., Hsp60 and Hsp70).

The Hsp70 system collaborates with extended peptide segments of proteins as well as partially folded proteins to avert aggregation, modification of folding pathways, and modulation of activity. When not in a state of interaction with a substrate peptide, Hsp70 is typically in an ATP bound state. A very weak intrinsic ATPase activity is possessed by Hsp70. As newly synthesized proteins emanate from the ribosomes, the substrate binding domain of Hsp70 identifies sequences of hydrophobic amino acid residues, and interacts with them. After a peptide binds on the binding domain of Hsp 70, ATPase activity of Hsp70 is stimulated and the slow rate of ATP hydrolysis is enhanced. On ATP hydrolysis to ADP, the binding pocket of Hsp70 closes, thus, the now-trapped peptide chain is tightly bound (Mashaghi et al. 2016).

Chaperones Act in Multi Chaperone Complexes

Although chaperones are relatively abundant, they very rarely function alone (Smith et al. 1995). They typically create large multiprotein complexes that contain other chaperones, co chaperones and various accessory proteins. Chaperone assisted folding is a complex multistep process based on non-covalent interactions between chaperones and their substrates, called “clients”. The folding cycle of Hsp90 is propelled by ATP hydrolysis which aids conformational changes and the recruitment of different co-chaperones. The mechanism of the Hsp90 folding cycle was described for the maturation of steroid-hormone receptors (SHR) by Smith et al. (Smith et al. 1995). The chaperone cycle starts when the newly synthesized or denatured client protein associates with Hsp70 (heat-shock protein of 70 kDa), Hsp40 (heat-shock protein of 40 kDa) and the adapter HIP (Hsp70-interacting protein) to form an early complex. Then adapter protein HOP (Hsp70/Hsp90-organising protein), that binds both Hsp70 and Hsp90 chaperones simultaneously, shifts the client protein to Hsp90 dimer and displaces Hsp40 to form an intermediate complex. In an ATP-dependent manner, the Hsp90 dimer binds the client protein and Hsp70, HOP and HIP are replaced by cochaperones p23 and CYP40 (cyclophilin 40) to complete the mature complex. Hormone binding to SHR in the mature complex leads to a conformational change of SHR driven by ATP hydrolysis. Finally, SHR is dissociated and transferred to the nucleus to regulate gene transcription. The spectrum of folded clients is also influenced by association of Hsp90 with different co-chaperones. For example, Cdc37 (cell division cycle 37) is a co-chaperone which binds to the N-terminal domain of Hsp90 and facilitates the recruitment of various kinases to the Hsp90 machinery (Smith et al. 1995).

Hsp70 binds tightly to partially synthesized peptide sequences (incomplete proteins), therefore forbidding them from aggregation and being relinquished as non-functional. After the synthesis of the entire protein, a nucleotide exchange factor (such as BAG-1 and HspBP1) prompts ADP release and binding of fresh ATP, opening the binding pocket. The protein can fold free on its own, or can be transported to other chaperones for further processing. Peptides can be transferred from Hsp70 to Hsp90 with the help of HOP (Hsp70/Hsp90 Organizing Protein) as HOP can bind to both Hsp70 and Hsp90 concomitantly. Hsp70 also facilitates in transmembrane relocation of proteins, by ossifying them in a partially folded state. Hsp70 proteins can protect cells from thermal or oxidative stress. These stresses normally act to impair proteins, fostering partial unfolding culminating in aggregation. By transiently binding to hydrophobic residues exposed by stress, Hsp70 intercepts these partially denatured proteins from aggregating, and permits them to refold. Low ATP is an attribute of heat shock and strengthened binding is seen as aggregation repression, while recovery from heat shock inculcates substrate binding and nucleotide cycling (Wegele et al. 2004).

Hsp70 is able to take part in disposition of damaged or defective proteins. These altered proteins are directed to the cell's ubiquitination and proteolysis pathways through interaction with CHIP (*C*arboxyl-terminus of *H*sp70 *I*nteracting *P*rotein) – an E3 ubiquitin ligase (Lüders et al. 2000). Finally, in addition to improving overall protein integrity, Heat-shock protein 70 (Hsp70) can block apoptosis at several levels, and is thought to play a carcinogenic role related to its antiapoptotic activity (Lüders et al. 2000; Murphy 2013). One distinctive feature of apoptosis is the release of cytochrome c, which then employs Apaf-1 and dATP/ATP into an apoptosome complex. This complex then cleaves procaspase-9, leading to caspase-9 activation and eventually induction of apoptosis via caspase-3 activation. The arrival of procaspase 9 to this complex is blocked by Hsp70. It renders procaspase-9 binding less favorable by inducing a conformational change. Hsp70 is shown to interplay with endoplasmic reticulum stress sensor protein IRE1alpha thereby safeguarding the cells from ER stress induced apoptosis. This interaction prolonged the splicing of XBP-1 (a transcription factor) mRNA thereby actuating transcriptional upregulation of targets of spliced rescuing the cells from apoptosis (Gupta et al. 2010). HBP 21, a chaperone of heat shock protein 70, acts as tumor suppressor in hepatocellular carcinoma. But, due to allele loss and aberrant methylation when HBP21 is down regulated, HSP70 confers resistance to apoptosis by blocking Bax translocation to mitochondria (Yang et al. 2012) which is required for cytochrome *c* release. Thus, HBP21 can promote cell apoptosis caused by unfavorable conditions and exert negative influence on Hepatocellular carcinoma (HCC) pathogenesis.

Chaperones maintain protein homeostasis not only by maturation of newly synthesised proteins and stabilization of unstable proteins, but also by recognitionand transport of defective proteinsto the degradation pathway. This needs the recruitment of another co-chaperone, CHIP (an E3 ubiquitin ligase), into the Hsp90 chaperone machinery (Murata et al. 2001). It was shown that CHIP suppresses tumor progression in human breast cancer by enhancing the degradation of several oncogenic proteins (Kajiro et al. 2009). Moreover, knockdown of CHIP in breast cancer cells results in rapid tumor growth and metastatic phenotypes in mice. The mechanisms regulating the protein folding/degradation balances involve chaperone binding to CHIP and HOP that depends on a phosphorylation state of Hsp90 and Hsp70 C-termini (Muller et al. 2013). The phosphorylation of these chaperones prevents binding to CHIP and enhances binding to HOP. Proliferating cells express lower levels of CHIP and higher levels of HOP, Hsp70 and Hsp90 compared to non-proliferating cells (Ruckova et al. 2012). Decreased CHIP expression in proliferative cells supports its proposed tumor suppressor properties, while overexpression of HOP may contribute to excessive Hsp90 activity and stabilization of client proteins in cancer cells. These reports reflect elevated protein folding environment in cancer cells regulated by the action of co-chaperone expression and chaperone modifications.

Cancer Stem Cells (CSC) in GI Cancer: Role of Hsp70

Our understanding of phenotypic plasticity of gastrointestinal cancer cells has considerably evolved in the past decade (Bekaii-Saab and El-Rayes 2017; Brungs et al. 2016). Evidences from several malignancies indicate that the cancer cells acquire stem-like phenotype often after chemotherapy (Monteiro and Fodde 2010; Hu et al. 2012). This has been substantiated both in functional assays (ALDH1 activity) and by the expression of several stem cell markers (CD44, CD133, NANOG, SOX2, BMI1, ALDH1, cMYC, KLF4 etc. (Hadjimichael et al. 2015). The hypothesis that a small sub-population of the tumor bulk maintains a stem-like phenotype and accounts for the therapeutic refractoriness, dormant behaviour with subsequent metastasis and recurrence has now been experimentally established using genetic lineage tracing of cancer cells in mouse model. CSC is at the hierarchical apex of tumorcells with the potency of self-renewal, clonal growth and differentiation into different cell types which characterizes the tumor bulk (Alcolea and Jones 2013; Kretzschmar and Watt 2012; Schepers et al. 2012).

Gastric CSCs were first described in 2007 by Yang et al. (Yang et al. 2007). The gastric cells that were isolated from patient tumors showed several phenotypic characteristics of stem and progenitor cells as determined by their enhanced clonogenic potential and differentiation capacity, high susceptibility to spontaneous immortalization and the expression of stem cell inversely correlated with patient survival abrogating the expression of GRP78 marker OCT4, and Nanog. Work by Takashi et al. further showed that CD44 can be a reliable cell surface marker for isolating gastric cancer stem cells (Takaishi et al. 2009). Gastric cancer cell lines MKN-45, MKN-74, and NCI-N87 had a significant subpopulation of high CD44 expressing cells which can be FACS sorted into CD44$^+$ and CD44$^-$ population. CD44$^+$ cells formed tumor spheres and showed higher tumorigenic ability when injected into stomach and skin of severe combined immunodeficient (SCID) mice. Further, The CD44$^+$ gastric cancer cells differentiated to give rise to CD44(−) cells. CSCs are also critical for the formation and maintenance of liver metastasis derived from colorectal cancers (de Sousa e Melo et al. 2017). The role Hsp70 in the malignant progression of highly metastatic human colorectal cancer (CRC) cell lines suggested a strong correlation with the advanced clinical stages and positive lymph node involvement (Hwang et al. 2003). Particularly in colorectal cancer, CD44–hyaluronan (HA) binding, leads to the assembly of a signalling complex that includes Hsp70 and the co-chaperone CDC37, which promotes phosphorylation and activation of the ERBB2–ERBB3 receptor tyrosine kinases (RTKs). The CD44–ERBB2–ERBB3 complex provides a strong stimulus for cyclooxygenase 2 (COX2) transcription and thus trigger the anti-apoptotic machinery in the CSC population (Misra et al. 2015).

GRP78 (BiP/HSPA5), is an important member of the Hsp70 superfamily that is evolutionarily conserved across a wide range of species from yeast to humans, and plays a crucial role in the maintenance of the embryonic stem cells (Wang et al. 2010). Wu et al. have recently demonstrated a significantly high expression of GRP78 in CSC population of the Head and neck squamous cell cancer (HNSCC) (Wu et al. 2010). Depleting GRP78 reduced the self-renewal ability, expression of stemness genes, tumorigenic potential both in cell based and mouse xenograft models and induced apoptotic cell death, thus providing a potential therapeutic avenue. Studies conducted on 86 cases of resected gastric cancer demonstrated a positive correlation of GRP78 with lymph node metastasis. Compared to the adjacent non tumorigenic region of the gastric mucosa GRP78 was specifically overexpressed in the tumor specimens including both primary tumors and metastatic lymph nodes (Zhang et al. 2006). Although this study did not focus on CSC population of the gastric tumors, but given the elevated expression of GRP78 both in primary and secondary metastatic sites of gastric tumors, and the similarity in phenotypic characteristics observed in HNSCC, a putative role of GRP78 in the gastric cancer can be envisioned and warrants further study.

Most preclinical and clinical studies on gastro-intestinal carcinoma have indicated that maximum therapeutic advantage in terms of tumor regression, inhibition of metastasis and recurrence, along with resenstization of persistent tumors to second line chemotherapy could be achieved by the combinatorial therapy targeting

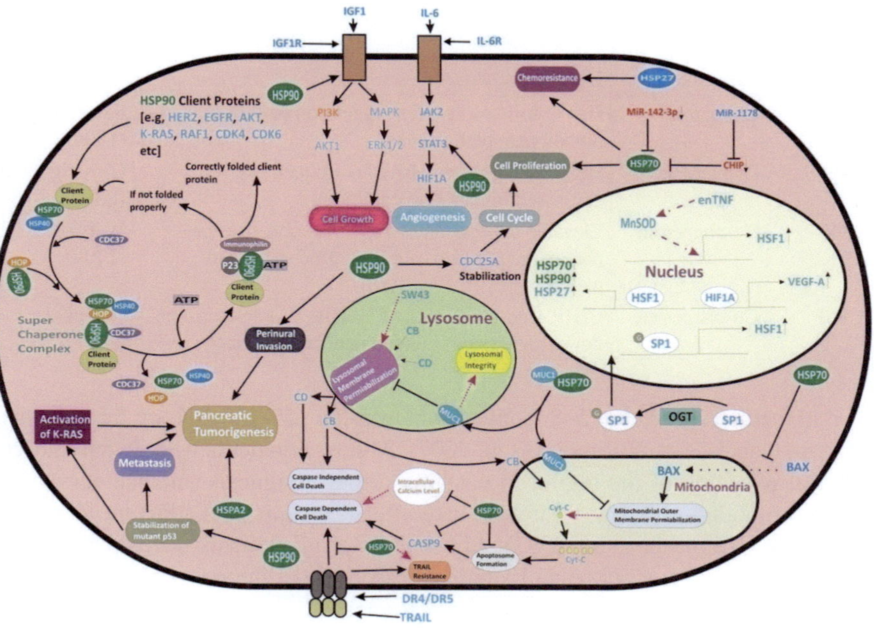

Fig. 1 A literature extracted representation of chaperoning in pancreatic cancer pathogenesis

both the non-stem-like differentiated tumor bulk and the CSC population (Takaishi et al. 2009; Vries et al. 2010). In lymphoma, siRNA-mediated attenuation of HSF1 targeted both the CSC and the non CSC population arguing strongly in favor of Hsp90/Hsp70 inhibitors for treating gastrointestinal cancers that are frequently plagued by chemoresistance (Newman et al. 2012). Promising data was obtained using small molecule inhibitors of Hsp70 and Hsp90 in the in vitro studies conducted on colorectal cancer (Massey et al. 2010). Treatment of colon cancer cells with the Hsp70 inhibitor VER-155008 or Hsp90 inhibitor 17AAG alone, did not exhibit any potent therapeutic effect, but a combinatorial therapy induced more than 90% cell death. Similar results were also obtained by combining another Hsp90 inhibitorVER-82,160 with VER-155008 in HCT116 colon carcinoma cell lines (Massey et al. 2010). It is likely that while Hsp90 inhibitor induces a more cytotoxic effect in the non CSC population, ablation of the CSC population is achieved by the inhibition of Hsp70 function (Fig. 1).

Pancreatic Cancer: An Overview

Pancreatic ductal adenocarcinoma is one of the most aggressive cancers and is the fourth most-frequent tumour-related cause of death in the Western world (Siegel et al. 2012). The prognosis for patients with this disease is extremely poor, with an

Table 1 Important chaperones and co-chaperones in human pancreatic cancer

Protein name	Type of molecule	Evidence	Reference
Hsp70	Chaperone	Hsp70 follows an over-expression pattern in human pancreatic cancer	Lee et al. (1994)
Hsp90	Chaperone	Heat shock protein Hsp90 is frequently over-expressed in pancreas carcinomas	Giessrigl et al. (2012)
HOP	Co-chaperone	Elevated expression of HOP was observed in pancreatic tumor tissue compared to normal pancreas	Walsh et al. (2009)
CHIP	Co-chaperone	CHIP, a co-chaperone of HSP70, was found to be down-regulated in human pancreatic cancer	Cao et al. (2015)

overall median survival of 5–8 months and fewer than 5% have a long-term survival of more than 5 years (Bilimoria et al. 2007). Although patients who do not undergo surgical resection of the tumor do not have a long prognosis (median survival of about 6 months), the 5-year survival rate after resection plus adjuvant therapy is over 20% (Bilimoria et al. 2007). About 70% of all patients develop metastatic recurrence even after successful surgical resection (Bilimoria et al. 2007; Oettle et al. 2007). Therefore, pancreatic adenocarcinoma is considered to be a systemic disease in the majority of patients at the time of diagnosis. Consequently, adjuvant chemotherapy is the standard of care after surgical resection, and has been shown to further increase long-term survival rates by about 10% (Oettle et al. 2007).

The increased expression of one or more Hsp above the level observed in normal tissues is a common feature of human cancers, both solid tumours and haematological malignancies (Kimura et al. 1993). A number of molecular chaperones have been found to be over expressed in human pancreatic cancer (detailed in the Table 1).

Role of Chaperones in Human Pancreatic Cancer Pathogenesis

This may be an adaptive response by pancreatic cancer cells to maintain protein homeostasis and promote cell survival in an unfavorable environment, as well as to stimulate cell proliferation and inhibit cell death (given in Table 2). Increased amounts of Hsps allow pancreatic cancer cells to tolerate changes from within, such as potentially lethal mutations that have a role in oncogenesis. Chaperones, such as Hsp90, are known to be highly expressed in most tumor cells, including pancreatic cancer cells (Giessrigl et al. 2012). Hsp90 also acts as protectors for genetic aberrations found in cancer, facilitating the performance of mutated proteins in promotion of malignancy while conferring cellular tolerance to the imbalanced signaling produced by these oncoproteins (Zhang et al. 2011). Indeed, Hsp are seen to participate in the six essential alterations in cell physiology proposed by Hanahan and Weinberg to define cancerous growth (Hanahan and Weinberg 2000) and described below:

Table 2 The role of Hsp70 chaperones in pancreatic cancer pathogenesis

Protein name	Type of molecule	Affected oncogenic process	Evidence	Reference
Hsp70	Chaperone	Cell proliferation	Hsp70 plays an important role in pancreatic ductal adenocarcinoma (PDAC) associated cell proliferation	MacKenzie et al. (2013)
		Apoptosis	Hsp70 mediates TRAIL resistance in human pancreatic cancer	Monma et al. (2013)
		Apoptosis	Hsp70 prevents activation and translocation of BAX into mitochondrial membrane and thereby protects pancreatic cancer cells from apoptosis	MacKenzie et al. (2013)
		Apoptosis	Hsp70 inhibits caspase-9 recruitment and apoptosome formation in human pancreatic cancer	Aghdassi et al. (2007)
		Apoptosis	Hsp70 promotes lysosomal integrity and subsequently suppress caspase independent apoptosis in a MUC1 responsive ways	Banerjee et al. (2012)
		Chemo-resistance	Hsp70, constitutively over-expressed in pancreatic cancer cells and provides chemoresistance	Hyun et al. (2013)

Self-Sufficiency in Generating Growth Signals in Cancer Cell

Hsp90 is needed to stabilize the fragile structures of many transcription factors and protein kinases that are involved in normal cellular growth pathways. This molecular chaperone is also required to maintain signaling molecules in an active conformation so as to allow rapid triggering by growth signals. In cancer, Hsp90 maintains the activities of the EGFR, HER2 (proto-oncogenic trans-membrane receptor), protein kinases Akt, c-Src and Raf-1(important signaling molecules), CDK4, CDK6 (cell cycle regulator) to promote pancreatic tumour growth, proliferation and survival (Zhang et al. 2008). It also stabilizes the conformations of mutant proteins such as p53, thus allowing these mutated molecules to accumulate within the pancreatic cancer cell (Zhang et al. 2011).

Insensitivity to Anti-growth and Proliferative Signals

Hsp70 has been shown to bind to p53 and other tumor suppressor proteins. Mutation in the p53 protein is one of the most common events in cancer development (Lee et al. 1994). However, although Hsp70 has been shown to accumulate in large amounts in association with mutant p53 protein in cancer cells, there is no conclusive evidence to indicate that increased Hsp levels are necessary to inactivate tumour suppressor molecules for malignant transformation to take place (Lee et al. 1994). On the contrary, Hsp90 plays a very critical role in stabilization of mutant p53 in pancreatic tumor and thereby largely accounts for the enhanced level of oncogenicity (Zhang et al. 2011).

Resistance to Cell Death or Avoidance of Apoptosis

A number of studies showed that elevated levels of Hsp can protect malignant cells against apoptosis generated by therapy (Joly et al. 2010). The apoptotic mechanism follows two pathways, the intrinsic and the extrinsic. For both intrinsic and extrinsic pathways, the ultimate executers of apoptosis are proteases, called caspases, which are activated enzymatically in response to an apoptotic stimulus (Cappello et al. 2002). Due to their cytoprotective role, Hsp have been found to play extremely complex roles in the regulation of apoptosis. They are implicated in both caspase-dependent and independent apoptotic pathways, as well as in the maintenance and activation of anti-apoptotic mediators. Hsp70, which is over expressed in human pancreatic cancer, plays a master regulatory role in inhibition or neutralization of pancreatic cancer associated apoptosis, through imparting TRAI (TNF-related apoptosis-inducing ligand) resistance (Monma et al. 2013), preventing BAX translocation into mitochondria (MacKenzie et al. 2013), inhibiting caspase-9 recruitment and apoptosome formation (Aghdassi et al. 2007) and by providing lysosomal and mitochondrial integrity (Banerjee et al. 2012).

Molecular Chaperones as Anti-cancer Targets in Pancreatic Cancer

There is currently much effort being made to develop more effective Hsp inhibitors for use in pancreatic cancer treatment (Table 3). A large number of chemotherapy combinations has been tested in patients with advanced pancreatic cancer. Only one combination showed significant improvement of survival, however also increased toxicity. Survival of patients with pancreatic cancer could not be improved by the use of these inhibitors. That is why the most important challenge is to define the appropriate way in which to deploy chaperone inhibitors in the treatment of pancreatic cancer.

Table 3 List of potential therapeutic drugs against molecular chaperones in human pancreatic cancer

Name of drug	Molecular targets	Evidence	Reference
Triptolide	Hsp70	Triptolide causes pancreatic cancer cell death in vitro and in vivo by induction of apoptosis and its mechanism of action is mediated via the inhibition of Hsp70	MacKenzie et al. (2013)
PFT-miu	Hsp70	HSP70 and autophagy inhibitor pifithrin-miu enhances the antitumor effects of TRAIL on human pancreatic cancer	Monma et al. (2013)
GO-201	Hsp70	GO-201 was effective in reducing tumor burden in pancreatic cancer mouse model	Banerjee et al. (2012)
Triazole Nucleoside Analog	HSF1, Hsp70, Hsp90, Hsp27	This compound targets heat shock response pathways by down-regulation of heat shock transcription factor 1 and consequential down-regulation of multiple heat shock proteins HSP27, HSP70 and HSP90	Xia et al. (2012)

Gastric Cancer: An Overview

Gastric Cancer and Hsp70

Among most commonly diagnosed type of cancers in the world Gastric cancer (GC) is the fifth (Bray et al. 2012; IARC (International Agency for Research on Cancer) 2012). Although incidence rates have been declining since the 1990s, significant geographical variations have been observed in the occurrence pattern of GC with some Asian countries and Central and Eastern Europe still showing higher incidence (Bertuccio et al. 2009; Ferro et al. 2014). Anti-sense nucleotides against Hsp70 have been shown to cause cell death in gastric adenocarcinoma cells (Zhao and Shen 2005). In patient samples of gastric tumors, high Hsp 70 levels were associated with poor prognosis (Lee et al. 2013). Reduced levels of Hsp 70 and increased apoptosis were observed in gastric adenocarcinoma cells after triptolide treatment (Arora et al. 2017). On the other hand, *Helicobacter pylori,* which is among many of those agents that lead to incidences of gastric cancer in human, may bring about a reduced level of Hsp70. Although the detailed pathology of *H. pylori* induced gastric cancer is not well demonstrated till date, few studies indicate the ability of the expression is one such effect that could be evidenced in patients exhibiting *H. pylori* infection. A study conducted by Axsen WS and colleagues (Axsen et al. 2009) showed that C57B/L mice infected with human *H. pylori* undergo reduced expression of not only *hsp70* but also *hsf1*, a transcription factor regulating the expression of *hsp70* gene (Axsen et al. 2009). A similar study with human epithelial cell line infected with *H. pylori* also demonstrated altered proliferation of the cultured cells accompanied by reduced expression of *hsp70* (Tao et al. 2014). This altered proliferation is predominantly represented by reduced growth rate. Accordingly, in a separate study, the fate of cultured gastric epithelial cells exposed *H. pylori* infection was investigated in detail. This study demonstrated increased expression of apoptosis inducing factor and cytosolic cytochrome c leading to increased instances of apoptotic cell death events. Besides, an upregulation of p21 and associated cell cycle modulation could also explain S-phase arrest of the infected cells and subsequently reduced growth rate (Liu et al. 2011). All these studies therefore quite successfully established a negative correlation between *H. pylori* infection and expression of host *hsp70* gene. However, one of the studies led by Ding S. Z. and colleagues provided important molecular insights into the underlying mechanism of *H. pylori* infection mediated *hsp70* down-regulation (Ding et al. 2010). This study showed an epigenetic component in the regulation of *hsp70* expression upon *H. pylori* infection. Phosphorylation on serine 10 of histone H3 in the promoter region of *hsp70* gene was shown to condense chromatin thereby inhibiting *hsp70* transcription resulting in reduced levels of Hsp70 protein.

Besides altered transcript levels of Hsp70 observed in gastric cancer patients, in many cases differential spatial distribution of the protein within the cell could also be noticed. Like increased translocation of Hsp70 to the nucleus that helps protecting the cell from stress induced nuclear damage. One of the nuclear import carriers of Hsp70 is an evolutionarily conserved protein called Hikeshi (Kose et al. 2012). In a study led by Yanoma T *et al.*, the expression pattern of Hikeshi was investigated in 207 gastric cancer tissue samples. This study showed an elevated expression of this nuclear import protein in gastric cancer tissue samples compared to normal cells. Although apparently this signifies an increased nuclear translocation of Hsp70 as well, no such observation could be noticed at normal temperature. At elevated temperature however, the increased expression of Hikeshi could be well-correlated to increased migration of Hsp70 from cytoplasm to nucleus. Accordingly Hikeshi inhibition in combination with hyperthermia has been suggested as a potential therapeutic tool for refractory gastric cancer (Yanoma et al. 2017). Other therapeutic tools targeting Hsp70 for treatment of gastric cancer includes recently demonstrated cytomegalovirus protein UL138, overexpression of which has been found to induce apoptosis in gastric cancer cell lines (Chen et al. 2016). This is however, in contrary to the common belief that human cytomegalovirus (HCMV) plays an 'oncomodulatory' role in neoplastic processes stimulating neoplastic growth (Michaelis et al. 2009).

Hsp70 being an important component of signaling cascades regulating inflammatory responses, therefore constitute a crucial limiting factor affecting successful *H. pylori* infection. Polymorphisms within the gene sequence controlling the vigor of immune response therefore can be expected to expose individuals to risk of *H. pylori* infection induced acute gastritis followed by onset of gastric cancer differentially. Oswaldo Partida Rodriguez and colleagues carried out studies in this direction and found out that in a Mexican population a single nucleotide polymorphism in *hsp70-1* gene at position +190 is associated with predisposition to gastric cancer and duodenal ulcer (Partida-Rodrıguez et al. 2010). Similarly the BB genotype of hsp70-2 is found to be associated with gastric pre malignant condition in *H. pylori* infected older patients in a Japanese population (Tahara et al. 2009).

Liver Cancer (Hepatocellular Carcinoma): An Overview

Role of Extracellular Hsp70 in the Progression of Tumor Cell Invasion and Metastasis Leading to Hepatocellular Carcinoma (liver cancer)

Tumor cell invasion and metastasis may be integrated with RhoA. RhoA-mediated regulation of this process takes place through the activation of cytoskeletal proteins that promote myosin interaction with F-actin, leading to contractility. Enhanced contractility of tumor cell greatly aids tumor cell movement and migration ability (Valtcheva et al. 2013). RhoA is an essential member of the rho gene family, which is a homolog of Ras protein (Strutt et al. 1997). Heat shock protein 70 and peptide complexes (eHSP70/HSP70-PCs) regulate Hepatocellualar Carcinoma (HCC) cell migration that occurs via regulation of RhoA expression (Yi et al. 2017).

An array of biological functions of tumor cells is modulated by extracellular heat shock protein 70 and peptide complexes (eHsp70/Hsp70-PCs). The expressions of Hsp70, E-cadherin, α-SMA and phosphorylated-p38 MAPK may characterize malignant potential and could categorize the extent of liver cancer. eHsp70/Hsp70-PCs play a focal role in the EMT of hepatocellular carcinoma through the mediation of the p38/MAPK pathway (Li et al. 2013). EMT is an intricate process that refers to the transmogrification of epithelial cells to stromal cells, where the polarity of epithelial cell fizzles out, consorted by escalated migration and invasion. EMT could be identified by the loss of epithelial cell markers (cadherin and E-cadherin) expression and/or overexpression of mesenchymal cell markers (e.g., α- smooth muscle actin protein, α-SMA) (Gheldof and Berx 2013). The occurrence of liver cancer is closely related to its tumor microenvironment (Gao et al. 2012). A number of components of the tumor microenvironment such as hepatocyte growth factor (HGF), fibroblast growth factor (FGF), and platelet-derived growth factor (PDGF) could be involved with EMT (Jung et al. 2007). Presence of Hsp70 and Hsp70 peptide complexes (Hsp70/Hsp70-PCs) has also been observed in the tumor microenvironment (Schildkopf et al. 2011). Moreover, eHsp70/Hsp70-PCs could also influence a variety of tumor cell functions, such as proliferation and invasion. (Wu et al. 2012; Walsh et al. 2011). A recent study has shown that eHSP70/HSP70-PCs carried out their action through Jun-terminal kinase JNK1/2 signaling pathway in hepatocarcinoma (Zhe et al. 2016). Studies have also demonstrated that the stress-activated mitogen activated protein kinases (MAPK) c-Jun NH2-terminal kinase (JNK) and p38 MAPK greatly influence hepatocarcinogenesis. Both these kinases have important cellular functions and considerable amount of cross-talk between them. Therefore, it is difficult to firmly indicate a specific drug target associated with these pathways (Nakagawa and Maeda 2012).

Role of Hsp70 as DAMP (Damage-Associated Molecular Pattern Molecules): Implications in Liver Cancer (HCC)

In the tumor microenvironment, tumor cells destructed by immune cells release intracellular molecules. Some of these molecules are inflammatory mediators referred to as damage-associated molecular pattern molecules (DAMPs) (Tsai et al. 2014). DAMPs play crucial roles in triggering immune responses and activating repair mechanisms to produce both antitumor and protumor effects. Hsp 70 is a well-characterized DAMP in chronic inflammation (Kang et al. 2013). Stress-inducible Hsp70 functions as a cytoprotective protein when cells are subjected to stressful stimuli. Although, necrotic cells can passively release stress-inducible Hsp70 (Gogate et al. 2012), Hsp70 is also actively released when tumor cells suffer from extrinsic stress. Inside the tumor microenvironment, extracellular Hsp70 can bind to TLR2 and TLR4 expressed by tumor cells and can promote immune tolerance, cancer advancement and promulgation of the tumor microenvironment (Horibe et al. 2014). However, Hsp70 also takes part in averting apoptosis by means of autophagy and can promote a pro-survival mechanism. Eminently, Hsp70 augmented autophagy by means of c-Jun N-terminal kinase (JNK) phosphorylation and Beclin-1 upregulation. Several studies have indicated that inhibition of the autophagy regulator Beclin-1 in tumor cells inhibited stress induced autophagy and high-mobility group protein B1 (HMGB1) release (Schmitt et al. 2007). Numerous tumor cells widely express HMGB1 and upon necrotic cell death it can be secreted or released. Migrating growth cones and malignant cells show high HMGB1 expression. HMGB1 binds tissue-type plasminogen activator and plasminogen, stimulating production of plasmin and tissue invasion (Gong 2013). Secreted HMGB1effectuates responses to infection and injury by achieving high affinity binding with several receptors including receptor for advanced glycation end products (RAGE), TLR2 (Toll-like receptor 2) and TLR4 (Toll-like receptor 4), hence invigorating tumor invasion and metastasis (Chen and Yu 2016). Extracellular Hsp70/Hsp70-PCs can stimulate the HCC cell proliferation through TLR2 and TLR4 activation and subsequent activation of the intracellular JNK1/2/MAPK signaling pathway. Recent studies show, MAPK signaling pathway may mediate signaling through TLRs (Wu et al. 2013). MAPK (p38, ERK1/2, and JNK1/2) signaling pathways can control a number of cellular functions in cancer cells. Previous studies have revealed that cyclin D1 plays a principal role in the normal regulation of the cell cycle (Gong 2013). Cyclin D1 is regulated positively by MAPK, thus exerting influence on tumor cell proliferation (Qiu et al. 2014). Cyclin D1 over-expression could shorten the time required for G1 to S phase transition and elevates cell cycle rate, resulting in rapid and uncontrolled cell proliferation. HCC demonstrates clear evidence of Cell cycle imbalance Overexpression of cyclin D1 has been clearly linked to the occurrence and growth of hepatocarcinoma (Zhang et al. 2015, b). Among different signaling pathways induced by Hsp70, the promotion of tumor growth by activation of the NF-κB cascade has been identified in several tumor cells (Wu et al. 2012). Thus, Hsp70 can enhance both the NF-κB and JNK-Beclin-1/HMGB1 pathways in tumor cells (Zhe et al. 2016). Activation of NF-κB was significant for the effect of Hsp70. H70sp induced a positive feedback loop associating Beclin-1/HMGB1 production, causing re-phosphorylation of NF-κB. This triggered a promotion of cell invasion ability of the cancer cells, a critical aspect of tumor progression (Gong 2013).

Regulatory Effects of Glucose on Heat Shock Response

In the hepatocellular carcinoma cell lines, Glucose can activate HSF1 transcription activity and augment the gene expression of alpha B-crystallin, Hsp70, and oncogenic proteins CSK2, and RBM23 (Mendillo et al. 2012). Glucose can stimulate HSF1 hyper-phoshorylation at serine 326 by activation of the mTOR pathway. The mTOR pathway is upregulated in most tumors and takes part in modulating tumor protein synthesis, cell proliferation, and autophagy by targeting various substances. Inhibition of mTORC1 with rapamycin can significantly suppress the Hsp90 and Hsp70 protein synthesis in breast cancer cell lines (Chou et al. 2012). mTOR C1 kinase can interact with and stimulate the phosphorylation of HSF1/S326 leading to activation of HSF1. Glucose-mTOR is one of the signaling pathways that maintain HSF1 activation in tumor cells. HSF1-mediated heat shock response can in turn regulate the glucose metabolism as well as the proteostasis involved in cell proliferation, metastasis, and therapeutic resistances. The glucose-mTOR-HSF1 pathway is a prospective target for tumor therapy (Ma et al. 2015).

Role of Hsp 70 as Biomarker of Hepatocellular Carcinoma (HCC)

Hepatocellular carcinoma (HCC) is one of the predominant malignancies in the world (Di Tommaso and Roncalli 2017). Many causative factors are connected to HCC occurrence and development, such as chronic hepatitis B virus (HBV) and hepatitis C virus (HCV) infection, consumption of alcohol, and cirrhosis (Gehrmann et al. 2014). It has been observed that tumor suppressor protein (p53) and heat shock protein 70 may exert vital influence in chronic liver diseases. Hsp70 and p53 are frequently observed to be over-expressed in HCC biopsy, especially in advanced HCC (Chuma et al. 2003). They are believed to be putative biomarkers for HCC diagnosis, and proper amalgamation of these 2 markers could improve diagnostic accuracy. It has been reported that both the mRNA and protein levels of Hsp70 increase markedly more in advanced HCC than in early HCC (Di Tommaso et al. 2007). Up-regulation of serum Hsp70 is observed in both liver cirrhotic and HCC patients (Shevtsov and Multhoff 2016).

Conclusions

Over the last few years new vistas leading towards a clearer insight of the molecular basis of GI cancers have opened up. Our knowledge was substantially aided by a better understanding of the genetic basis of the disease and the availability of new information on the role of molecular signaling pathways. Still, many aspects of the functioning of chaperones in disease progression remain unexplored. This review aims to provide a summary of the current knowledge of the role of the Hsp70 and related chaperones in human GI tumorigenesis. Since their initial discovery, significant input has been given towards developing molecular chaperone inhibitors as anticancer agents. The question whether the inhibitors of heat shock proteins would be of any significant help to combat GI cancers still remains to be clearly answered.

Acknowledgements D. Datta acknowledges financial support from Visva-Bharati University.

References

Aghdassi, A., Phillips, P., Dudeja, V., et al. (2007). Heat shock protein 70 increases tumorigenicity and inhibits apoptosis in pancreatic adenocarcinoma. *Cancer Research, 67*(2), 616–625.

Alcolea, M. P., & Jones, P. H. (2013). Tracking cells in their native habitat: Lineage tracing in epithelial neoplasia. *Nature Reviews Cancer, 13*, 161–171.

Arora, N., Alsaied, O., Dauer, P., et al. (2017). Downregulation of Sp1 by Minnelide leads to decrease in HSP70 and decrease in tumor burden of gastric cancer. *PLoS One, 12*(2), e0171827.

Axsen, W. S., Styer, C. M., & Solnick, J. V. (2009). Inhibition of heat shock protein expression by *Helicobacter pylori*. *Microbial Pathogenesis, 47*, 231–236.

Banerjee, S., Majumdar, N., Dudeja, V., et al. (2012). MUC1c regulates cell survival in pancreatic cancer by preventing lysosomal permeabilization. *PLoS One, 7*(8), e43020.

Bekaii-Saab, T., & El-Rayes, B. (2017). Identifying and targeting cancer stem cells in the treatment of gastric cancer. *Cancer, 123*(8), 1303–1312.

Bertuccio, P., Chatenoud, L., Levi, F., et al. (2009). Recent patterns in gastric cancer: A global overview. *International Journal of Cancer, 125*(3), 666–673.

Bilimoria, K. Y., Bentrem, D. J., Ko, C. Y., et al. (2007). National failure to operate on early stage pancreatic cancer. *Annals of Surgery, 246*, 173–180.

Bray, F., Jemal, A., Grey, N., et al. (2012). Global cancer transitions according to the human development index (2008–2030): A population-based study. *The Lancet Oncology, 13*, 790–801.

Brocchieri, L., Conway de Macario, E., & Macario, A. J. L. (2007). Chaperonomics, a new tool to study ageing and associated diseases. *Mechanisms of Ageing and Development, 128*, 125–136.

Brungs, D., Aghmesheh, M., Vine, K.,. L., et al. (2016). Gastric cancer stem cells: Evidence, potential markers, and clinical implications. *Journal of Gastroenterology, 51*(4), 313–326.

Cao, Z., Xu, J., Huang, H., et al. (2015). MiR-1178 promotes the proliferation, G1/S transition, migration and invasion of pancreatic cancer cells by targeting CHIP. *PLoS One, 10*(1), e0116934.

Cappello, F., Bellafiore, M., Palma, A., et al. (2002). Defective apoptosis and tumorigenesis: Role of p53 mutation and Fas/FasL system dysregulation. *European Journal of Histochemistry, 46*, 199–208.

Chen, L., & Yu, J. (2016). Modulation of toll-like receptor signaling in innate immunity by natural products. *International Immunopharmacology, 37*, 65–70.

Chen, W., Lin, K., Zhang, L., et al. (2016). The cytomegalovirus protein UL138 induces apoptosis of gastric cancer cells by binding to heat shock protein 70. *Oncotarget, 7*, 5630–5645.

Chou, S. D., Prince, T., & Gong, J. (2012). mTOR is essential for the proteotoxic stress response, HSF1 activation and heat shock protein synthesis. *PLoS One, 7*, e39679.

Chuma, M., Sakamoto, M., Yamazaki, K., et al. (2003). Expression profiling in multistage hepatocarcinogenesis: Identification of HSP70 as a molecular marker of early hepatocellular carcinoma. *Hepatology, 37*, 198–207.

de Sousa e Melo, F., Kurtova, A. V., Harnoss, J. M., et al. (2017). A distinct role for Lgr5+ stem cells in primary and metastatic colon cancer. *Nature, 543*(7647), 676–680.

Di Tommaso, L., & Roncalli, M. (2017). Tissue biomarkers in hepatocellular tumors: Which, when, and how. *Frontiers in Medicine, 4*, 10.

Di Tommaso, L., Franchi, G., Park, Y. N., et al. (2007). Diagnostic value of HSP70, glypican 3, and glutamine synthetase in hepatocellular nodules in cirrhosis. *Hepatology, 45*, 725–734.

Ding, S. Z., Fischer, W., Kaparakis-Liaskos, M., et al. (2010). Helicobacter pylori-induced histone modification, associated gene expression in gastric epithelial cells, and its implication in pathogenesis. *PLoS One, 5*(4), e9875.

Ferro, A., Peleteiro, B., Malvezzi, M., et al. (2014). Worldwide trends in gastric cancer mortality (1980–2011), with predictions to 2015, and incidence by subtype. *European Journal of Cancer, 50*, 1330–1344.

Gao, H., Wang, Y., Liu, X., et al. (2004). Global transcriptome analysis of the heat shock response of Shewanella oneidensis. *Journal of Bacteriology, 186*,(22), 7796–7803.

Gao, Q., Zhao, Y. J., Wang, X. Y., et al. (2012). CXCR6 upregulation contributes to a proinflammatory tumor microenvironment that drives metastasis and poor patient outcomes in hepatocellular carcinoma. *Cancer Research, 72*, 3546–3556.

Gehrmann, M., Cervello, M., Montalto, G., et al. (2014). Heat shock protein 70 serum levels differ significantly in patients with chronic hepatitis, liver cirrhosis, and hepatocellular carcinoma. *Frontiers in Immunology, 5*, 307.

Gheldof, A., & Berx, G. (2013). Cadherins and epithelial-to-mesenchymal transition. *Progress in Molecular Biology and Translational Science, 116*, 317–336.

Giessrigl, B., Krieger, S., Rosner, M., et al. (2012). Hsp90 stabilizes Cdc25A and counteracts heat shock-mediated Cdc25A degradation and cell-cycle attenuation in pancreatic carcinoma cells. *Human Molecular Genetics, 21*(21), 4615–4627.

Gogate, S. S., Fujita, N., Skubutyte, R., et al. (2012). Tonicity enhancer binding protein (TonEBP) and hypoxia-inducible factor (HIF) coordinate heat shock protein 70 (Hsp70) expression in hypoxic nucleus pulposus cells: Role of Hsp70 in HIF-1α degradation. *Journal of Bone and Mineral Research, 27*, 1106–1117.

Gong, W. (2013). Invasion potential of H22 hepatocarcinoma cells is increased by HMGB1-induced tumor NF-κBsignaling via initiation of HSP70. *Oncology Reports, 30*(3), 1249–1256. https://doi.org/10.3892/or.2013.2595.

Gupta, S., Deepti, A., Deegan, S., et al. (2010). HSP72 protects cells from ER stress-induced apoptosis via enhancement of IRE1alpha-XBP1 signaling through a physical interaction. *PLoS Biology, 8*, e1000410.

Hadjimichael, C., Chanoumidou, K., & Papadopoulou, N. (2015). Common stemness regulators of embryonic and cancer stem cells. *World Journal Stem Cells, 7*(9), 1150–1184.
Hanahan, D., & Weinberg, R. A. (2000). The hallmarks of cancer. *Cell, 100*, 57–70.
Horibe, T., Torisawa, A., Kohno, M., et al. (2014). Synergetic cytotoxic activity toward breast cancer cells enhanced by the combination of Antp-TPR hybrid peptide targeting Hsp90 and Hsp70- targeted peptide. *BMC Cancer, 14*, 615.
Hu, X., Ghisolfi, L., Keates, A. C., et al. (2012). Induction of cancer cell stemness by chemotherapy. *Cell Cycle, 11*(14), 2691–2698.
Hwang, T. S., Han, H. S., Choi, H. K., et al. (2003). Differential, stage-dependent expression of Hsp70, Hsp110 and Bcl-2 in colorectal cancer. *Journal of Gastroenterology and Hepatology, 18*(6), 690–700.
Hyun, J. J., Lee, H. S., Keum, B., et al. (2013). Expression of heat shock protein 70 modulates the chemoresponsiveness of pancreatic cancer. *Gut Liver, 7*(6), 739–746.
IARC (International Agency for Research on Cancer). (2012). *GLOBOCAN 2012*. Lyon: IARC. URL: http://globocan.iarc.fr/Pages/fact_sheets_cancer.aspx.
Joly, A. L., Wettstein, G., Mignot, G., et al. (2010). Dual role of heat-shock proteins as regulator of apoptosis andinnate immunity. *Journal of Innate Immunity, 2*, 238–247.
Jung, J. W., Hwang, S. Y., Hwang, J. S., et al. (2007). Ionising radiation induces changes associated with epithelialmesenchymal transdifferentiation and increased cell motility of A549 lung epithelial cells. *European Journal of Cancer, 7*, 1214–1224.
Kajiro, M., Hirota, R., Nakajima, Y., et al. (2009). The ubiquitin ligase CHIP acts as an upstream regulator of oncogenic pathways. *Nature Cell Biology, 11*(3), 312–319.
Kang, Y., Jung, W. Y., Lee, H., et al. (2013). Prognostic significance of heat shock protein 70 expression in early gastric carcinoma. *Korean Journal Pathology, 47*, 219–226.
Kimura, E., Enns, R. E., Alcaraz, J. E., et al. (1993). Correlation of the survival of ovarian cancer patients with mRNA expression of the 60kDa heatshock protein Hsp60. *Journal of Clinical Oncology, 11*, 891–898.
Kose, S., Furuta, M., & Imamoto, N. (2012). Hikeshi, a nuclear import carrier for Hsp70s, protects cells from heat shock-induced nuclear damage. *Cell, 149*, 578–589.
Kretzschmar, K., & Watt, F. M. (2012). Lineage tracing. *Cell, 148*(1–2), 33–45.
Lee, C. S., Montebello, J., Rush, M., et al. (1994). Overexpression of heat shock protein (hsp) 70 associated with abnormal p53 expression in cancer of the pancreas. *Zentralblatt für Pathologie, 140*(3), 259–264.
Lee, H. W., Lee, E. H., Kim, S. H., et al. (2013). Heat shock protein 70 (HSP70) expression is associated with poor prognosis in intestinal type gastric cancer. *Virchows Archiv, 463*(4), 489–495.
Li, H., Li, Y., Liu, D., et al. (2013). Extracellular HSP70/HSP70-PCs promote epithelial-mesenchymal transition of hepatocarcinoma cells. *PLoS One, 8*(12), e84759.
Liu, W. L., Chen, Y., & Lu, G. F. (2011). Down-regulation of HSP70 sensitizes gastric epithelial cells to apoptosis and growth retardation triggered by H. pylori. *BMC Gastroenterology, 11*, 145.
Lüders, J., Demand, J., & Höhfeld, J. (2000). The ubiquitin-related BAG-1 provides a link between the molecular chaperones Hsc70/Hsp70 and the proteasome. *The Journal of Biological Chemistry, 275*, 4613–4617.
Ma, W., Zhang, Y., Mu, H., et al. (2015). Glucose regulates heat shock factor 1 transcription activity via mTOR pathway in HCC cell lines. *Cell Biology International, 39*, 1217–1224.
Macario, A. J. L., & Conway de Macario, E. (2005). Sick chaperones, cellular stress and disease. *The New England Journal of Medicine, 353*, 1489–1501.
Macario, A. J. L., & Conway de Macario, E. (2009). The chaperoning system: Physiology and pathology. In A. Gerbino, G. Crescimanno, & G. Zummo (Eds.), *Experimental medicine reviews, Plumelia* (Vol. 2/3, pp. 9–21).
Macario, A. J. L., Cappello, F., Zummo, G., et al. (2010). Chaperonopathies of senescence and the scrambling of interactions between the chaperoning and the immune systems. *Annals of the New York Academy of Sciences, 1197*, 85–93.

MacKenzie, T. N., Majumdar, N., Banerjee, S., et al. (2013). Triptolide induces the expression of miR-142- 3p: A negative regulator of heat shock protein 70 and pancreatic cancer cell proliferation. *Molecular Cancer Therapeutics, 12*(7), 1266–1275.

Mashaghi, A., Bezrukavnikov, S., Minde, D. P., et al. (2016). Alternative modes of client binding enable functional plasticity of Hsp70. *Nature, 539*, 448–451.

Massey, A. J., Williamson, D. S., Browne, H., et al. (2010). A novel, small molecule inhibitor of Hsc70/Hsp70potentiates Hsp90 inhibitor induced apoptosis in HCT116 colon carcinoma cells. *Cancer Chemotherapy and Pharmacology, 66*(3), 535–545.

Mendillo, M. L., Santagata, S., Koeva, M., et al. (2012). HSF1 drives a transcriptional program distinct from heat shock to support highly malignant human cancers. *Cell, 150*, 549–562.

Michaelis, M., Doerr, H. W., & Cinat, J. (2009). The story of human cytomegalovirus and cancer: Increasing evidence and open questions. *Neoplasia, 11*, 1–9.

Misra, S., Hascall, V. C., Markwald, R. R., et al. (2015). Interactions between hyaluronan and its receptors (CD44, RHAMM) regulate the activities of inflammation and cancer. *Frontiers in Immunology, 6*, 201.

Monma, H., Harashima, N., Inao, T., et al. (2013). The HSP70 and autophagy inhibitor pifithrin-μ enhances the antitumor effects of TRAIL on human pancreatic cancer. *Molecular Cancer Therapeutics, 12*(4), 341–351.

Monteiro, J., & Fodde, R. (2010). Cancer stemness and metastasis: Therapeutic consequences and perspectives. *European Journal of Cancer, 46*(7), 1198–1203.

Muller, P., Ruckova, E., Halada, P., et al. (2013). C-terminal phosphorylation of Hsp70 and Hsp90 regulates alternate binding to co-chaperones CHIP and HOP to determine cellular protein folding/degradation balances. *Oncogene, 32*(25), 3101–3110.

Murata, S., Minami, Y., Minami, M., et al. (2001). CHIP is a chaperone- dependent E3 ligase that ubiquitylates unfolded protein. *EMBO Reports, 2*(12), 1133–1138.

Murphy, M. E. (2013). The HSP70 family and cancer. *Carcinogenesis, 34*, 1181–1188.

Nakagawa, H., & Maeda, S. (2012). Molecular mechanisms of liver injury and hepatocarcinogenesis: Focusing on the role of stress-activated MAPK. *Pathology Research International, 2012*, 172894.

Newman, B., Liu, Y., Lee, H. F., et al. (2012). HSP90 inhibitor 17-AAG selectively eradicates lymphoma in stem cells. *Cancer Research, 72*(17), 4551–4561.

Oettle, H., Post, S., Neuhaus, P., et al. (2007). Adjuvant chemotherapy with gemcitabine vs observation patients undergoing curative-intent resection of pancreatic cancer: A randomized controlled trial. *JAMA, 297*, 267–277.

Partida-Rodrıguez, O., Torres, J., Flores-Luna, L., et al. (2010). Polymorphisms in TNF and HSP-70 show a significant association with gastric cancer and duodenal ulcer. *International Journal of Cancer, 1126*(8), 1861–1868.

Qiu, C., Xie, Q., Zhang, D., et al. (2014). GM-CSF induces cyclin D1 expression and proliferation of endothelial progenitor cells via PI3K and MAPK signaling. *Cellular Physiology and Biochemistry, 33*, 784–795.

Ritossa, F. (1962). A new puffing pattern induced by temperature shock and DNP in Drosophila. *Cellular and Molecular Life Sciences, 18*, 571–573.

Ruckova, E., Muller, P., Nenutil, R., et al. (2012). Alterations of the Hsp70/Hsp90 chaperone and the HOP/CHIP co-chaperonesystem in cancer. *Cellular & Molecular Biology Letters, 17*(3), 446–458.

Schepers, A. G., Hugo, J., Snippert, H. J., et al. (2012). Lineage tracing reveals Lgr5+ stem cell activity in mouse intestinal adenomas. *Science, 337*(6095), 730–735.

Schildkopf, P., Frey, B., Ott, O. J., et al. (2011). Radiation combined with hyperthermia induces HSP70-dependent maturation of dendritic cells and release of pro-inflammatory cytokines by dendritic cells and macrophages. *Radiotherapy and Oncology, 101*, 109–115.

Schlesinger, M. J. (1990). Heat shock proteins. *The Journal of Biological Chemistry, 265*, 12111–12114.

Schmitt, E., Gehrmann, M., Brunet, M., et al. (2007). Intracellular and extracellular functions of heat shock proteins: Repercussions in cancer therapy. *Journal of Leukocyte Biology, 81*, 15–27.

Shevtsov, M., & Multhoff, G. (2016). Heat shock protein–peptide and HSP-based immunotherapies for the treatment of cancer. *Frontiers in Immunology, 7*, 171.

Siegel, R., Naishadham, D., & Jemal, A. (2012). Cancer statistics, 2012. *CA: A Cancer Journal for Clinicians, 62*, 10–29.

Smith, D. F., Whitesell, L., Nair, S. C., et al. (1995). Progesterone receptor structure and function altered by geldanamycin, an hsp90-binding agent. *Molecular and Cellular Biology, 15*(12), 6804–6812.

Strutt, D. I., Weber, U., & Mlodzik, M. (1997). The role of RhoA in tissue polarity and frizzled signalling. *Nature, 387*, 292–295.

Tahara, T., Shibata, T., Arisawa, T., et al. (2009). The BB genotype of heat-shock protein (HSP) 70-2 gene is associated with gastric pre-malignant condition in H. pylori-infected older patients. *Anticancer Research, 29*, 3453–3458.

Takaishi, S., Okumura, T., Shuiping, T., et al. (2009). Identification of gastric cancer stem cells using the cell surface marker CD44. *Stem Cells, 27*(5), 1006–1020.

Tao, L. P., Zou, H., & Huang, Z. M. (2014). Effects of Helicobacter pylori and heat shock protein 70 on the proliferation of human gastric epithelial cells. *Gastroenterology Research and Practice, 2014*, 79342.

Tissieres, A., Mitchell, H. K., & Tracy, U. M. (1974). Protein synthesis in salivary glands of *Drosophila melanogaster*: Relation to chromosome puffs. *Journal of Molecular Biology, 84*, 389–398.

Tsai, S. Y., Jesus, A., Segovia, J. A., et al. (2014). DAMP molecule S100A9 acts as a molecular pattern to enhance inflammation during influenza A virus infection: Role of DDX21-TRIF-TLR4-MyD88 pathway. *PLoS Pathogens, 10*, e1003848.

Valtcheva, N., Primorac, A., & Jurisic, G. (2013). The orphan adhesion G protein-coupled receptor GPR97 regulates migration of lymphatic endothelial cells via the small GTPases RhoA and Cdc42. *The Journal of Biological Chemistry, 288*, 35736–35748.

Vries, R. G. J., Huch, M., & Clevers, H. (2010). Stem cells and cancer of the stomach and intestine. *Molecular Oncology, 45*, 373–384.

Walsh, N., O'Donovan, N., Kennedy, S., et al. (2009). Identification of pancreatic cancer invasion-related proteins by proteomic analysis. *Proteome Science, 7*, 3.

Walsh, N., Larkin, A., Swan, N., et al. (2011). RNAi knockdown of Hop (Hsp70/Hsp90 organising protein) decreases invasion via MMP-2 down regulation. *Cancer Letters, 306*, 180–189.

Wang, M., Ye, R., Barron, E., et al. (2010). Essential role of the unfolded protein response regulator GRP78/BiP in protection from neuronal apoptosis. *Cell Death and Differentiation, 17*, 488–498.

Wegele, H., Müller, L., & Buchner, J. (2004). Hsp70 and Hsp90—A relay team for protein folding. *Reviews of Physiology, Biochemistry and Pharmacology, 151*, 1–44.

Wu, M. J., Jan, C., Tsay, Y. G., et al. (2010). Elimination of head and neck cancer initiating cells through targeting glucose regulated protein78 signaling. *Molecular Cancer, 9*, 283.

Wu, F. H., Yuan, Y., Li, D., et al. (2012). Extracellular HSPA1A promotes the growth of hepatocarcinoma by augmenting tumor cell proliferation and apoptosis-resistance. *Cancer Letters, 317*, 157–164.

Wu, T. T., Tai, Y. T., Cherng, Y. G., et al. (2013). GATA-2 transduces LPS-induced il-1β gene expression in macrophages via a toll-like receptor 4/MD88/MAPK-dependent mechanism. *PLoS One, 8*, e72404.

Xia, Y., Liu, Y., Rocchi, P., et al. (2012). Targeting heat shock factor 1 with a triazole nucleoside analog to elicit potent anticancer activity on drug-resistant pancreatic cancer. *Cancer Letters, 318*(2), 145–153.

Yang, Y. C., Wang, S. W., Hung, H. Y., et al. (2007). Isolation and characterization of human gastric cell lines with stem cell phenotypes. *Journal of Gastroenterology and Hepatology, 22*(9), 1460–1468.

Yang, X., Wang, J., Zhou, Y., et al. (2012). Hsp70 promotes chemoresistance by blocking Bax mitochondrial translocation in ovarian cancer cells. *Cancer Letters, 321*, 137–143.

Yanoma, T., Ogata, K., Yokobori, T., et al. (2017). Heat shock-induced HIKESHI protects cell viability via nuclear translocation of heat shock protein 70. *Oncology Reports, 38*(3), 1500–1506.

Yi, Z., Li, Y., Liu, D., et al. (2017). Extracellular HSP70/HSP70-PCs regulate hepatocarcinoma cell migration and invasion via RhoA. *Oncology Letters, 13*, 1095–1100.

Young, J. C., Agashe, V. R., Siegers, K., et al. (2004). Pathways of chaperone-mediated protein folding in the cytosol. *Nature Reviews. Molecular Cell Biology, 5*, 781–791.

Zhang, J., Jiang, Y., Jia, Z., et al. (2006). Association of elevated GRP78 expression with increased lymph node metastasis and poor prognosis in patients with gastric cancer. *Clinical & Experimental Metastasis, 23*, 401.

Zhang, T., Hamza, A., Cao, X., et al. (2008). A novel Hsp90 inhibitor to disrupt Hsp90/Cdc37 complex against pancreatic cancer cells. *Molecular Cancer Therapeutics, 7*(1), 162–170.

Zhang, L., Pang, E., Loo, R. R., et al. (2011). Concomitant inhibition of HSP90, its mitochondrial localized homologue TRAP1 and HSP27 by green tea in pancreatic cancer HPAF-II cells. *Proteomics, 11*(24), 4638–4647.

Zhang, H., Gao, H., Liu, C., et al. (2015a). Expression and clinical significance of HSPA2 in pancreatic ductal-adenocarcinoma. *Diagnostic Pathology, 10*, 13.

Zhang, Y., Guo, X., Li, Z., et al. (2015b). A systematic investigation based on microRNA-mediated gene regulatory network reveals that dysregulation of microRNA-19a/Cyclin D1 axis confers an oncogenic potential and a worse prognosis in human hepatocellular carcinoma. *RNA Biology, 12*, 643–657.

Zhao, Z. G., & Shen, W. L. (2005). Heat shock protein 70 antisense oligonucleotide inhibits cell growth and induces apoptosis in human gastric cancer cell line SGC-7901. *World Journal of Gastroenterology, 11*(1), 73–78.

Zhe, Y., Li, Y., Liu, D., et al. (2016). Extracellular HSP70-peptide complexes promote the proliferation of hepatocellular carcinoma cells via TLR2/4/JNK1/2MAPK pathway. *Tumour Biology, 37*, 13951–13959.

Heat Shock Protein 70 and Cancer

Tuoen Liu and Shousong Cao

Abstract Heat shock proteins (HSP) constitute a large family of proteins involved in protein folding and maturation and the expressions of HSP are induced by heat shock or other stressors. The major groups, which are classified based on their molecular weight, include HSP27, HSP40, HSP60, HSP70, HSP90, and large HSP (HSP110 and glucose-regulated protein 170). The human HSP70 family is consists of 13 members and five of them have a strong association with cancer. HSP play a significant role in cellular proliferation, differentiation, survival, apoptosis, and carcinogenesis. In this chapter, we thoroughly discussed the roles of HSP70s in cancer biology and pharmacology. The HSP70 proteins have important functions in the molecular mechanisms leading to cancer development, progression, and metastasis. They may also have potential clinical use as biomarkers for cancer diagnosis or assessing disease progression, and as therapeutic targets for cancer therapy. Understanding of the functions and molecular mechanisms of HSP70 proteins is critical for enhancing the accuracy of cancer diagnosis as well as for developing more effective and less toxic chemotherapeutic agents.

Keywords Biomarker · Cancer · Carcinogenesis · Heat shock protein 70 · Therapeutic target

Abbreviations

5-FU	fluorouracil
ADP	adenosine diphosphate

T. Liu (✉)
Department of Biomedical Sciences, West Virginia School of Osteopathic Medicine, Lewisburg, WV, USA
e-mail: tliu@osteo.wvsom.edu

S. Cao
Department of Pharmacology, School of Pharmacy, Southwest Medical University, Luzhou, Sichuan, China

A. A. A. Asea, P. Kaur (eds.), *HSP70 in Human Diseases and Disorders*, Heat Shock Proteins 14, https://doi.org/10.1007/978-3-319-89551-2_5

AIF	apoptosis-inducing factor
AKT	protein kinase B
ALL	acute lymphoid leukemia
AML	acute myelogenous leukemia
AMPK	AMP-activated protein kinase
APAF-1	apoptotic protease activating factor 1
ATF	activating transcription factor
ATP	adenosine triphosphate
ATPase	adenosine triphosphatase
BAG-1	Bcl-2 associated athanogene 1
BAX	Bcl-2 associated X
Bcl-2	B-cell lymphoma-2
Bcl-xL	B-cell lymphoma-extra-large
BIK	Bcl-2 interacting killer
BIP	binding immunoglobulin protein
B-RAF	v-raf murine sarcoma viral oncogene homolog B
CDK	cyclin-dependent kinase
CHIP	carboxyl-terminus of HSP70-interacting protein
CML	chronic myeloid leukemia
C-RAF	v-raf murine sarcoma viral oncogene homolog C
DMC	demethoxycurcumin
EGFR	epidermal growth factor receptor
EMT	epithelial-mesenchymal transition
ER	endoplasmic reticulum
ERK	Ras/extracellular signal-regulated kinase
FDA	Food and Drug Administration
GRP	glucose-regulated protein
HBV	hepatitis B virus
HDAC	histone deacetylases
HER2	human epidermal growth factor receptor 2
HIV	human immunodeficiency virus
HSC	heat-shock cognate protein
HSE	heat shock element
HSF	heat shock factor
HSP	heat shock protein
HUGO	Human Genome Organization
IL	interleukin
JAK	Janus kinase
MAPK	mitogen-activated protein kinase
MCL-1	myeloid cell leukemia 1
MEK	mitogen-activated protein kinase kinase
mTOR	mechanistic target of rapamycin
MTX	methotrexate
NF-κB	nuclear factor NF-κB
NSCLC	non-small cell lung cancer;

oxLDL	oxidative modified low density lipoprotein
PES	phenylethynesulfonamide
PFT	pifithrin
PI3K	phosphatidylinositol-4, 5-bisphosphate 3-kinase
PSA	prostate-specific antigen
SAHA	suberanilohydroxamic acid
SK1	sphingosine kinase 1
STAT	signal transducer and activator of transcription
TGF	transforming growth factor
TNF	tumor necrosis factor;
TNIP1	TNF-α-induced protein 3-interacting protein 1
UBXN2A	UBX Domain Protein 2A

Introduction

Heat shock proteins (HSP) are a group of proteins that function to reverse or inhibit denaturation or unfolding of cellular proteins in response to stress or high temperature. Traditionally, HSP have also been known as molecular chaperones because of their physiological and protective roles in cells. They facilitate protein folding and maintenance of natural structure and function when cells are exposed to homeostatic challenges such as extreme temperature, anoxia, hypoxia, heavy metals, drugs, or other chemical agents that may induce stress or protein denaturation (Macario and Conway de Macario 2007; Liu et al. 2012). HSP are generally classified according to their molecular weights with the majority of them belonging to the groups of HSP27(HSPB1), HSP40, HSP60, HSP70, HSP90, and large HSP [HSP110 and glucose-regulated protein 170 (GRP170)] (Ciocca and Calderwood 2005). Heat shock factors (HSF) act as inducible transcriptional regulators of HSP and they are required for the expression of the majority of HSP. Heat shock elements (HSE) are *cis*-acting sequences located upstream of HSP genes where HSF bind to and induce HSP gene expression (Akerfelt et al. 2010). Except for the small HSP group, other HSP family members including HSP70 proteins are ATP-dependent proteins with adenosine triphosphatase (ATPase) activity (Bepperling et al. 2012). The HSP70 proteins are an ATP binding chaperones with intrinsic ATPase activity which hydrolyzes ATP into ADP. Hydrolysis of ATP initiates the conformational change of HSP70 proteins and further causes substrate binding to them (Fig. 1) (Jakob et al. 1996; Sullivan and Pipas 2002).

The nomenclature of human HSP is based on the system assigned by the Human Genome Organization (HUGO) Gene Nomenclature Committee and uses the Entrez Gene database from the National Center of Biotechnology Information. Specifically, as shown in Table 1, the human HSP70 family consists of thirteen members which are encoded by the HSPA genes (Kampinga et al. 2009). HSP70 proteins have a highly conserved domain structure including a ~44 kDa amino-terminal ATPase

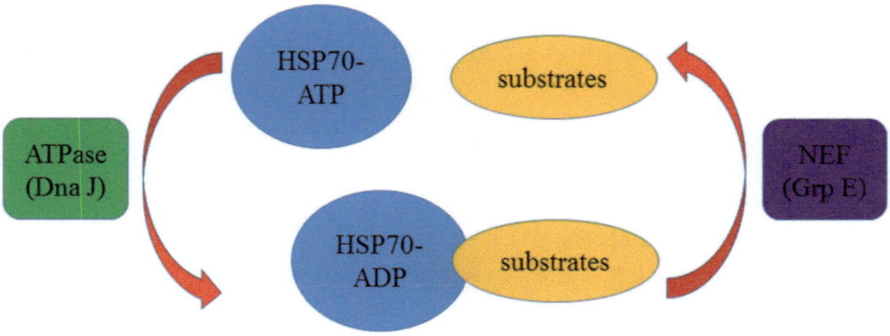

Fig. 1 The model of HSC70 binds to the substrates and releasing cycle. HSP70 has a low affinity with the substrates in the ATP-bound state. After hydrolysis of ATP by ATPase, HSP70 binds to the substrates with a high affinity in the ADP bound state. Some co-chaperones such as Dna J homologues enhance the ATPase activity of HSP70. Nucleotide exchange factors such as GrpE enhance the dissociation of bound ADP from HSP70 to allow the binding of ATP and reset the cycle

Table 1 Human HSP70 (HSPA) family members

HSPA	Gene name	Protein name	Old/common name(s)	Human gene ID
1	*HSPA1A*	HSPA1A	HSP72; HSPA70–1; HSPA1	3303
2	*HSPA1B*	HSPA1B	HSP70–2	3304
3	*HSPA1L*	HSPA1L	Hum70t; hsp-hom	3305
4	*HSPA2*	HSPA2	Heat-shock 70KDa protein-2	3306
5	*HSPA5*	HSPA5	GRP78; BIP; MIF2	3309
6	*HSPA6*	HSPA6	HSP70B'; heat shock 70 protein 6	3310
7	*HSPA7*	HSPA7	Heat shock 70 protein 7	3311
8	*HSPA8*	HSPA8	HSC70; HSC71; HSP71; HSP73	3312
9	*HSPA9*	HSPA9	Mortalin; GRP75; HSPA9B; MOT; MOT2; mtHSP70; PBP74; mot-2	3313
10	*HSPA12A*	HSPA12A	FLJ13874; KIAA0417	259,217
11	*HSPA12B*	HSPA12B	RP23-32 L15.1; 2700081N06Rik	116,835
12	*HSPA13*	HSPA13	Stch	6782
13	*HSPA14*	HSPA14	HSP70–4; HSP70L1; MGC131990	51,182

domain, a ~18 kDa substrate binding domain, and a ~10 kDa carboxyl-terminal domain (Liu et al. 2012). The chaperone activity of HSP is regulated by their co-chaperones including HSP40, Bcl-2 associated athanogene 1 (BAG-1), and carboxyl-terminus of HSP70-interacting protein (CHIP) (Wang et al. 2014a, b). In general, HSP70 proteins play an important role in cancer development, progression, and metastasis and they are often observed at abnormally high expression in cancer cells. HSP70 proteins promote carcinogenesis as a survival factor due to their tumor-associated expression and anti-apoptotic effect (Rérole et al. 2011). HSP70 proteins can protect cancer cells from tumor necrosis factor (TNF)-induced cytotoxicity, indicating that HSP70s may increase the oncogenic potential in certain cancer cells by an escaping mechanism from the immune system (Jäättelä et al. 1992). The

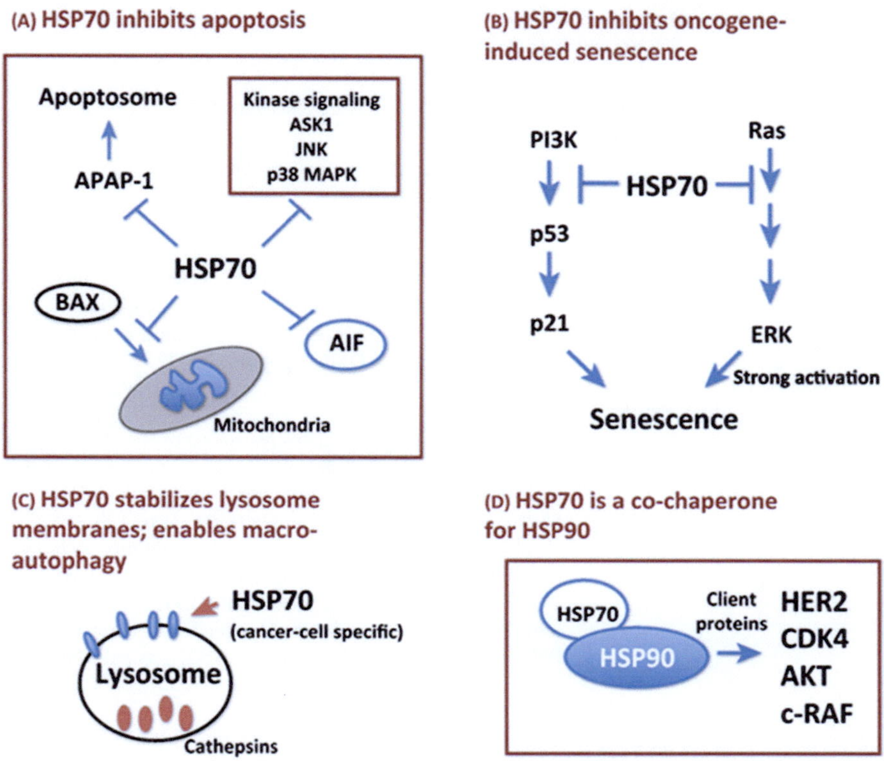

Fig. 2 Proposed roles of HSP70 in regulation of cancer development and as the potential therapeutic targets for anticancer drugs. (**a**) Inhibition of both intrinsic and extrinsic apoptotic pathways by HSP70. HSP70 inhibits apoptosis through (1) binding to and inhibition of translation of BAX from cytosol to mitochondria, (2) prevention of the recruitment and transportation of APAF-1 to apoptosome, (3) binding to and inhibition of the activity of kinases involved in stress signaling, and (4) binding to and disruption of the sensitivity and function of AIF. (**b**) Control of cell senescence by HSP70. HSP70 inhibits cell senescence through (1) interruption of p53-dependent regulation and (2) antagonization of p53-independent regulation. (**c**) Stabilization of lysosome function and regulation of autophagy by HSP70. HSP70 promotes cell survival through (1) stabilization of lysosomes by binding to endo-lysosmal lipid bisphosphate, (2) inhibition of lysosomal membrane permeabilization and (3) stimulation of autophagy. (**d**) Regulation of HSP90 client proteins by HSP70. HSP70 works as a co-chaperone of HSP90 via delivery of client proteins to HSP90. The HSP90 client proteins delivered by HSP70 include HER2, CDK4, AKT and C-RAF

proposed roles of HSP70s in cancer development and as potential therapeutic targets for anti-cancer agents are illustrated in Fig. 2 (Wu et al. 2017).

HSP70 proteins have crucial functions in mediation of protein folding, maintenance of protein homeostasis, and enhancement of cell survival following a multitude of stresses (Murphy 2013). Five important and well studied members in the HSP70 family have a strong association with cancer: the stress inducible HSP70s include HSPA1A (also known as HSP72, mainly located in the cytosol, ~72 kDa) and HSPA6 (also known as HSP70B', mainly located in the cytosol, ~71 kDa); the

Table 2 The roles of major HSP70 family members in cancer development, progression, metastasis, diagnosis, therapy and drug resistance

Gene/protein	Cancer development	Biomarker	Drug resistance and target
HSPA1A/ HSP72	↑Tumor cell	↑ In oral, liver, prostate, colorectal, lung, uterine and cervical cancers, melanoma	Mediates resistance of cisplatin, oxalipatin, borezomib
	oncogenesis	↓ In renal cancer	Target of HSP70 general inhibitors (fisetin, cantharidin, apoptozole, MKT-077, PES, pifithrin-μ, quercetin, DMC, apoptin), HS-72
	proliferation		
	migration		
	metastasis		
	↓ Apoptosis		
HSPA5/ GRP78	↑Tumor-genesis	↑ In ER stress, breast, lung, prostate, overian, gastric, liver, esophageal, renal, and endometrial cancers, melanoma, glioma, fibrosarcoma	Mediates resistance of BRAF inhibitors (vemurafenib and dabrafenib), cisplatin, sorafenib, doxorubicin
	↑Autophagy		Target of HKH40A, OSU-CG-5, carfizombi, SAHA, gefitinib, clarithromycin, celecoxib, antibodies (C38, C107, anti-CDT, SAM-6), natural compounds (epigallocathchingallate, subtilase toxin AB5, versipelostatin, prunustatin A, isoliquirtigenin), peptides
	↓ Apoptosis		
HSPA6/ HSP70B'	↓ Apoptosis	↑ In breast, liver, and colon cancers, lymphoma	Target of MG-132, HSP90 inhibitors tanepimycin and radicicol
	Macrophage survivial		
HSPA8/ HSC70	↑Tumor cell survival	↑ In colon and esophageal cancers	Mediates resistance of MTX
			Target of 15-deoxyspergualin
HSPA9/ Mortalin	↑Migration, invasion, EMT, metastasis	↑ In breast and liver cancers	Mediates resistance of cisplatin
	↓ Apoptosis		Target of MKT-077, embelin, UBXN2A

constitutively expressed HSP70s include HSPA5 (also known as GRP78 or BIP, mainly located in the endoplasmic reticulum (ER), ~78 kDa), HSPA8 (also known as HSC70, mainly located in the cytosol, ~73 kDa), and HSPA9 (also known as mortalin, GRP75 or mtHSP70, mainly located in the mitochondria, ~75 kDa) (Arispe and De Maio 2000; Kampinga et al. 2009). In this chapter, we will focus on discussing the five major HSP70 members of *HSPA1A*/HSP72, *HSPA5*/GRP78, *HSPA6*/HSP70B', *HSPA8*/HSC70, and *HSPA9*/mortalin. Their roles related to cancer characters are summarized in Table 2.

Role of *HSPA1A*/HSP72 in Cancer Development and Diagnosis

The HSP72 protein is coded by the *HSPA1A* gene and has malignant behaviors as evidenced by the decrease of cell proliferation, migration and invasion, increase of apoptosis after knockdown of HSP72 in cervical squamous cell carcinoma cells (Yoshidomi et al. 2014). HSP72 promotes the survival of glioblastoma cells through the inhibition of the degradation of activating transcription factor 5 (ATF5) (Li et al. 2011). HSP72 also inhibits oncogene-induced senescence pathways in either a p53-depdendent (via PI3K) or a p53-independent (via Ras/ERK) manner in multiple types of cancer cells including breast, colon, lung, ovarian and pancreatic cancers (Gabai et al. 2009). In addition, the study of HSP72 in autophagy and cancer showed that HSP72 functions as a survival protein by stabilizing the integrity of lysosomes in cancer cells (Nylandsted et al. 2004).

HSP72 is overexpressed in various cancers and the high expression is correlated with increased tumor grade and poor prognosis. The overexpression of HSP72 is observed in oral cancer, liver cancer, prostate cancer, colorectal cancer, lung cancer, uterine cervical cancer, and melanoma (Murphy 2013; Wu et al. 2017). Cai et al. studied tumor tissues of 507 patients with nasopharyngeal cancer and they found that the expression patterns of HSP72 are correlated with the outcomes of the patients as: high levels of HSP72 in the membranes and cytoplasm are associated with increased survival, while high levels of HSP72 in the nuclei are correlated with poor survival (Cai et al. 2012). The study by Wang et al. showed that the expression of HSP72 in the cytoplasm of esophageal cancer cells was significantly associated with remote metastasis (Wang et al. 2010). However, HSP72 was found to be under-expressed in renal cancer cells (Ramp et al. 2007). HSP72 is a prognostic marker for cholangiocarcinoma, chondroscarcoma, bladder cancer, acute myelogenous leukemia (AML), breast cancer, endometrial cancer, cervix/uterus cancer, pancreatic cancer, heat and neck cancer and colorectal cancer (Kocsis et al. 2011; Boonjaraspinyo et al. 2012; Bayer et al. 2014; Trieb et al. 2016; Wu et al. 2017).

Role of *HSPA5*/GRP78 in Cancer Development and Diagnosis

The HSPA5 protein, commonly known as GRP78, is coded by the *HSPA5* gene. It is the HSP70 family protein primarily presented in the ER, thus it is important in regulation of the cellular response to ER stress. GRP78 is responsible for maintaining the normal function of ER during protein translocation, assisting protein folding and assembly, and targeting misfolded proteins for degradation (Ma and Hendershot 2004]. Cancer cells are subject to ER stress and GRP78 is a survival factor in cancer development. GRP78 plays an important role in initiation of cancinogenesis, prevention of apoptosis and autophagy, and enhancement of drug resistance (Lee 2007).

GRP78 is a co-factor for Cripto signaling via both TGF-β and Src/MAPK/PI3K pathways to promote oncogenesis in somatic stem cells (Gray and Vale 2012). GRP78 binds to prostate-specific antigen (PSA) and upregulates the expression of PSA in prostate cancer cells, promoting cell survival via the activation of ERK, p38 MAPK, and PI3K pathways (Misra et al. 2011). GRP78 prevents ER-stress induced apoptosis in normal and cancer cells. GRP78 interacts to and inhibits pro-apoptotic BIK and BAX proteins in ER to protect human breast cancer cells from estrogen starvation-induced apoptosis (Fu et al. 2007). GRP78 also protects cancer cells from topoisomerase inhibitors-induced apoptosis via binding to and suppression of caspase-7 (Reddy et al. 2003). Inhibition of GRP78 by polyclonal antibodies upregulates p53 activity and promotes apoptosis, via the inhibition of Ras/MAPK and PIK3/AKT pathways and activation of NF-κB (Misra et al. 2009). Another important function of GRP78 in cancer development is the regulation of ER stress-related autophagy in cancer cells (Li et al. 2008). For examples, GRP78 level is elevated during ER-stress mediated autography, which is induced by drug and irritation treatments in nasopharyngeal carcinoma cells (Song et al. 2013). The polymethoxyflavone derivative nobiletin displays anticancer effect by inducing apoptosis, however, the apoptotic effect is compromised by protective autophagy induced by GRP78 in human gastric cancer cells SNU-16 (Moon and Cho 2016). Studies also have showed that treatment of serine/threonine-protein kinase B-RAF inhibitors vemurafenib and debrafenib activated GRP78 regulated ER stress response to induce protective autophagy and further induced drug resistance, while autophagy inhibitor hydroxychloroquine can overcome the resistance to B-RAF inhibitors in melanoma cells and tumor xenografts in animal model (Ma et al. 2014).

GRP78 is overexpressed in multiple cancers including breast cancer, lung cancer, prostate cancer, ovarian cancer, gastric cancer, liver cancer, esophageal cancer, renal cancer, endometrial cancer, melanoma, glioma, and fibrosarcoma. Overexpression of GRP78 is a negative predictor of survival in high-risk patients with endometrial cancer (Ulianich and Insabato 2014). In addition, cell surface GRP78 may be a potential marker for good prognosis and response to chemotherapy in breast cancer (Yerushalmi et al. 2015).

Role of *HSPA6*/HSP70B′ in Cancer Development and Diagnosis

The *HSPA6* gene is located in chromosome 1q in the human genome but not presents in the genomes of rat and mouse (Khalouei et al. 2014). It encodes a 71-kDa HSP70 family protein, also known as HSP70B', with the characteristics of unique induction (Khalouei et al. 2014). The expression of HSPA6 is low to undetectable under normal physiological condition but it is highly stress induced in most cancer cells and tissues (Leung et al. 1990). Ramired et al. reported that the expressions of HSP including HSPA6, HSP72, and HSP40s were reduced following overexpression of TNF-α-induced protein 3-interacting protein 1 (TNIP1) in human epidermal keratinocytes, suggesting the potential role of HSPA6 in TNF-mediated inflammation and apoptosis (Ramired et al. 2015). The study by Smith showed that the expression of HSPA6 can be induced by oxidative modified low density lipoprotein (oxLDL) in the human promonocytic lymphoma cells U937 which originated from resident macrophages. Knockdown of HSPA6 by small interfering RNA resulted in a decrease of the level of mRNA and phosphorylation of sphingosine kinase 1 (SK1), as well as an increase in the production of IL-10 in U937 cells. The results indicate that HSPA6 has functions in activation and stimulation of macrophage survival via the regulation of SK1 activity and IL-10 production (Smith 2010). A study by Regeling et al. also showed that the extract from cigarette induced HSPA6 expression in human intestinal epithelial cells DLD-1. HSPA6 interacts with and stabilizes the anti-apoptotic protein BCL-XL in DLD-1 cells, suggesting the anti-apoptotic role of HSPA6 (Regeling et al. 2016). The expression of HSPA6 is higher in the tumor tissues of liver cancer compared to normal tissues and it is associated with earlier recurrence and poor outcome in the patients with HBV-related early-stage hepatocellular carcinoma (Yang et al. 2015).

Role of *HSPA8*/HSC70 in Cancer Development and Diagnosis

The HSC70 (also called HSP73) protein is a 73 kDa heat shock cognate protein which is coded by the *HSPA8* gene. The basic structure of human HSC70 has three parts: a 44 kDa amino-terminal ATPase domain (1–384 residues), also known as the ATP-binding domain, a 18 kDa peptide (substrate) binding domain (385–543 residues), and a 10 kDa carboxyl-terminal domain (544–646 residues), also known as the variable or "lid" domain (Liu et al. 2012). Studies have found that the concentrations of HSC70 were much higher in the cells during proliferation, particularly in S-phase, compared to differentiated cells in rat glioma C6 cells, suggesting the role of HSC70 in promotion of cell proliferation (Helmbrecht and Rensing 1999). HSC70 regulates the functions of tumor-related genes and proteins. It binds and links to the non-phosphorylated tumor suppressor retinoblastoma (pRb) protein and mutant form of p53 and p73 for degradation (Liu et al. 2012). More recent studies showed that HSC70 promoted cell survival by inhibition of the degradation of Ras-related protein Rab-1A under stress conditions in colon adenocarcinoma cells (Tanaka et al. 2014).

HSC70 is overexpressed in colon cancer and esophageal cancer (Kubota et al. 2010; Moghanibashi et al. 2013). Mutation or deletion of HSC70 is detected in the tissues from patients with breast cancer, suggesting HSC70 is a target of somatic mutation and deletion in breast cancer (Bakkenist et al. 1999). The 1541–1542 delGT heterozygous genotype of HSC70 is associated with decreased risk in lung cancer (Rusin et al. 2004). In addition, HSC70 has been identified in the neuroblastoma cells cultured in conditioned media, suggesting it may be a potential tumor marker (Sandoval et al. 2006).

Role of *HSPA9*/Mortalin in Cancer Development and Diagnosis

The *HSPA9* gene is located in the human chromosome 5q31 interval coding for mortalin. It is involved in myeloid malignancies and is commonly deleted in patients with myelodysplasticsyndrome (MDS) and AML (Chen et al. 2011). Mortalin (HSPA9) binds to and sequesters p53 in the cytoplasm to prevent the translocation of p53 into nucleus in human colorectal adenocarcinoma cells, indicating its role in regulation of cell cycle and apoptosis (Gestl and Anne Böttger 2012). Inhibition of mortalin induces apoptosis in hematopoietic progenitor cells and cancer cells (Chen et al. 2011; Lu et al. 2011; Liu et al. 2017). Mortalin contributes to carcinogenesis by activation of MAPK/MEK/ERK pathway in ovarian cancer and medullary thyroid cancer cells (Starenki et al. 2015; Hu et al. 2016). Overexpression of mortalin increases the migration, invasion, epithelial–mesenchymal transition (EMT), and metastasis in breast cancer cells, in which the process is regulated by PI3/AKT and Janus kinase (JAK)-signal transducer and activator of transcription (STAT) signaling pathways (Na et al. 2016). Mortalin is overexpressed in breast cancer and liver cancer (Chen et al. 2014; Jin et al. 2016). The overexpression of mortalin is associated with early tumor recurrence and metastasis in liver cancer (Yi et al. 2008). Elevated levels of mortalin can be used as a prognostic marker for worse prognosis and poor survival in patients with gastric cancer and colorectal cancer (Rozenberg et al. 2013; Ando et al. 2014).

Role of HSP70 Proteins as the Therapeutic Targets for Anticancer Drug Discovery and Development

HSP70 proteins promote tumorigenesis and their expressions are elevated in various cancers. HSP70 proteins are also involved in mediating drug resistance in cancer therapy. Due to their important roles in cancer biology, the HSP70 molecular chaperones, especially HSP72 and GRP78, are promising drug targets for cancer therapy. Therefore, identification, characterization and development of HSP70 inhibitors are of high interest for novel anticancer drug discovery and development.

Some compounds target and inhibit general HSP70 family proteins. Kim and colleagues reported that fisetin, a dietary flavonoid, can induce apoptosis in colon cancer HCT-116 cells by inhibition of the activity of HSF1 via blocking the binding of HSP72 promoter. Fisetin down-regulates HSP72/BAG3 and induces apoptosis via decreasing the levels of BCL-2, BCL-XL and myeloid cell leukemia 1 (MCL-1) proteins in human colon cancer HCT-116 cells (Kim et al. 2015). The same group also found that the natural compound cantharidin induces cancer cell death through the inhibition of HSP72 and BAG3 by binding of HSF1 to the promoter of HSP72 (Kim et al. 2013). Apoptozole, a small molecule inhibitor of the ATPase region of HSP72, induces apoptosis in cancer cells and displays antitumor activities against tumor xenografts in mouse models, through disruption of the interaction between HSP70 and APAF-1 (Ko et al. 2015). MKT-077, as an analog of the allosteric HSP70 inhibitor, is effective against human breast cancer MDA-MB-231 and MCF-7 cells (Li et al. 2013). Pifithrin-μ (PFT-μ), also known as 2-phenylethynesulfonamide (PES) inhibits the activity of HSP72 via binding to the substrate-binding domain of HSP70 (Balaburski et al. 2013). As a potent and selective HSP72 inhibitor, PFT-μ has been proved to have anti-tumor activities against multiple cancers including AML, acute lymphoid leukemia (ALL), and prostate cancer (Kaiser et al. 2011; Sekihara et al. 2013). Quercetin is a HSP72 inhibitor (also inhibiting HSP27 and HSP90) and inhibits the proliferation of human cervical cancer Hela cells via down-regulation of AMP-activated protein kinase (AMPK)-induced HSP72 and epidermal growth factor receptor (EGFR) (Jung et al. 2010). Demethoxycurcumin (DMC) is a kind of curcumin. Similar to quercetin, DMC also inhibits cell proliferation via AMPK-induced down-regulation of HSP72 and EGFR in prostate cancer cells (Hung et al. 2012). Apoptin is a nonstructural viral protein encoded by the VP3 gene of chicken anemia virus. Apoptin inhibits HSP72 transcription and induces apoptosis in a HSP70-dependent manner in liver cancer HepG2 cells. Inhibition of HSP72 expression is regulated by the ability of apoptin to promote HSF1 trimer depolymerization and inhibits HSF1-mediated HSP70 transcription (Yuan et al. 2013).

Studies have also shown that HSP72 is a therapeutic target for certain anti-cancer drugs and regulates drug resistance. For example, knockdown of HSP72 enhances the sensitivity to cisplatin treatment in cervical cancer cells (Yoshidomi et al. 2014). HSP72 protects human gastric cancer cells from oxalipatin induced apoptosis (Takahashi et al. 2016). Overexpression of HSP72 promotes bortezomib resistance, and inhibition of HSP72 enhances bortezomib-induced cell death in human bladder cancer cells (Qi et al. 2013). Howe et al. identified HS-72 as an allosteric selective inhibitor of HSP72. HS-72 reduces the ATP affinity of HSP72 to inhibit tumor growth and prolong the survival of tumor-bearing mice in an animal model of HER2+ breast cancer (Howe et al. 2014). Study by Zhang et al. showed that HSP72 significantly promoted the proliferation of human gammasigma T cells *in vitro*, indicating that HSP72 could be a potential target for adjuvant therapy of gammasigma T cells in cancer immunotherapy (Zhang et al. 2005).

GRP78 is involved in mediation of drug resistance and may be a potential therapeutic target for anticancer drugs. GRP78 and its downstream target AKT are important in regulation of cisplatin resistance in ER stress-tolerant human lung cancer cells (Lin et al. 2011). Inhibition of GRP78 by small-interference RNA sensitizes cisplatin- and doxorubicin-induced apoptosis in drug-resistant human melanoma cells (Jiang et al. 2009). GRP78 was identified as a positive regulator for sorafenib resistance acquisition in liver cancer cells (Chiou et al. 2010). The anti-HIV drug nelfinavir enhances the cytotoxicity of doxorubicin *in vitro* and potentiates the antitumor efficacy of doxorubicin *in vivo* via up-regulation of GRP78 in doxorubicin-resistant breast cancer MCF-7 cells, suggesting a role of GRP78 in mediation of doxorubicin resistance (Chakravarty et al. 2016). In addition, GRP78 is also a therapeutic target for the exertion of anti-tumor activity of anticancer drugs.

Nanoparticles conjugated with GRP78 antibody inhibit carcinogenesis in liver cancer cells and promote the delivery of 5-fluorouracil (5-FU) to the cell surface with higher expression of GRP78 to enhance the anti-tumor efficiency (Zhao et al. 2014). GRP78 is identified as a specific and direct target of natural compound isoliquirtigenin which inhibits the proliferation of breast cancer cells via the blockage of β-catenin signaling (Wang et al. 2014a, b). HKH40A belongs to bisimidazoacridones and has antitumor activity. The mechanisms of action for its antitumor activity include downregulation of GRP78 and upregulation of ATF6 in different types of cancer cells (Kosakowska-Cholody et al. 2014). OSU-CG-5 is a novel energy restriction mimetic agent. It inhibits the proliferation of human colorectal cancer cells via up-regulation of GRP78 and inhibition of p-mTOR and p-p70S6 kinase (Arafa et al. 2014). Combination of carfilzomib (a proteasome inhibitor) and suberanilohydroxamic acid (SAHA, an HDAC inhibitor) synergistically promotes ER stress with up-regulation of GRP78 in non-small cell lung cancer (NSCLC), suggesting GRP78 may be a potential therapeutic target for the combination therapy in the treatment of NSCLC (Hanke et al. 2016). The combination treatment of gefitinib (an EGFR inhibitor) and clarithromycin (a macrolide antibiotic) effectively enhances cytotoxity via up-regulation of GRP78 in NSCLC cells (Sugita et al. 2015). The COX-2 inhibitor celecoxib can inhibit cell growth and induce apoptosis via down-regulation of GRP78 expression in human leukemia cells (Sobolewski et al. 2015). Some important GRP78 inhibitors including antibodies, natural compounds and peptides are listed in a review article by Roller and Maddalo (Roller and Maddalo 2013). We modified and updated them in Table 2.

The expression of HSPA6 can be induced by anticancer agents. MG-132 is an inhibitor of proteasome used for the treatments of multiple myeloma and lymphoma clinically. Treatment of MG-132 can induce the expression of HSPA6 on the surface of human colon cancer HT-29 and CRL-1809 cells (Noonan et al. 2008). HSP90 inhibitors tanespimycin and radicicol significantly induced the expression of HSPA6 in breast cancer MCF-7 cells, suggesting that HSPA6 may be a specific target for HSP90 inhibition (Kuballa et al. 2015).

Liu et al found that HSC70 located in the cell membrane works as a transporter of methotrexate (MTX) in L1210 leukemia cells, and abnormal tyrosine phosphorylation of HSC70 leads to the occurrence of MTX resistance in the cells (Liu et al. 2015). Significant upregulation of HSC70 is observed in CD34+ chronic myeloid leukemia (CML) cells and treatment of HSC70-specific inhibitor 15-deoxyspergualin decreases cell proliferation, suggesting HSC70 can be a target for cancer therapy (Liu et al. 2012).

Mortalin mediates drug resistance in cancer therapy. For instance, inhibition of the expression of mortalin inhibits cell growth and reverses cisplatin resistance in ovarian cancer cells (Yang et al. 2013). MKT-077 shows anti-tumor efficacy through binding to mortalin and interruption of the interaction between mortalin and p53. MKT-077 also reverses the resistance and sensitizes the cellular response to complement-dependent cytotoxicity in human leukemia K562 cells and human colon cancer HCT-116 cells (Wadhwa et al. 2000; Pilzer et al. 2010). Embelin, a natural quinone, is found in the fruits of Embelia ribes. It inhibits the transcription of mortalin and abrogates the interactions of mortalin-p53 in cancer cells. Embelin targets mortalin leading to the activation of p53 and inhibition of the metastatic signaling in human breast cancer cells (Nigam et al. 2015). Recent studies have showed that UBXN2A, a UBX-domain containing protein, is a potential mortalin inhibitor and can sensitize cancer cell response to chemotherapeutic agents such as 5-FU. UBXN2A also promotes cell death by interfering with the interaction of mortalin-p53 in colon cancer cells (Sane et al. 2016).

Conclusions

Here we have provided an overview of the complex relationship between cancer and the major HSP70 genes/proteins including *HSPA1A*/HSP72, *HSPA5*/GRP78, *HSPA6*/HSP70B', *HSPA8*/HSC70, and *HSPA9*/mortalin. As summarized in Table 2, HSP70 proteins play an important role in tumorigenesis and may be used as potential clinical biomarkers for the diagnosis and predicting prognostic outcome of the patients with cancer. HSP70 proteins are molecular chaperones and their expressions and functions are activated during stress stimulation. Most of the HSP70 proteins have similar functions in carcinogenesis, prevention of apoptosis and conferring of drug resistance. In addition, HSP70 proteins may be used as therapeutic targets for cancer therapy, prompting discovery and development of novel chemotherapeutic agents. Among HSP70 proteins, GRP78 is the target of most of the novelly developed anti-cancer drugs. These drugs have proven to be effective in cancer cells *in vitro* and tumor xenografts in animal models *in vivo*, and some of them are being tested in clinical trials. A variety of anticancer drugs with HSP targets have been approved by the USA FDA for cancer treatment. For example, sorafenib (Nexavar®), a kinase inhibitor that reduces the expression of GRP78 in cancer cells, was approved by the USA FDA for treatment of renal cell carcinoma (2005), hepatocellular carcinoma (2007) and locally recurrent or metastatic, progressive, differentiated thyroid carcinoma (2013) (Roberts et al. 2015). Ruxolitinib (Jakafi®) is a JAK inhibitor that also decreases the expression of HSP70 and HSP90 in cancer cells and animal models. It was approved by the USA FDA for the treatment of intermediate or high-risk myelofibrosis (2011) and polycythemia vera (2014) (Tavallai et al. 2016). However, none of the specific HSP inhibitors has been approved by the USA FDA for the treatment of patients with cancer so far. We propose three major possible reasons for why no HSP inhibitors have been approved by the USA FDA in clinical use: (1) HSP are very critical for the survival of the cells. Once the activities of HSP are inhibited, the basic functions of cancer cells as well as normal cells will be severely affected. We still do not fully understand the different requirement in the amount of HSP between cancer cells and normal cells, which may provide a necessary therapeutic window for discovery and development of more efficacious and less toxic novel anticancer HSP inhibitors. (2) HSP inhibitors can cause severe organ-specific toxicities (liver or ocular toxicity) which are not easy to be managed. The functions of HSP are essential for both normal and tumor cells. It is difficult to identify the functions of HSP that are only presented in cancer cells but not in normal cells. To identify the functions of HSP only being specific for cancer cells may overcome HSP inhibitors induced organ-specific toxicities. (3) Some HSP inhibitors may lack convincing anticancer activity. Different HSP family members interact and collaborate with each other in the signaling network to regulate cellular functions in the cells. Inhibition of one HSP can stimulate the overexpressions of other HSP members, which compensates the inhibitory effects caused by a single HSP inhibitor. Therefore, combination of different HSP inhibitor may be a good strategy to enhance the anticancer efficacy in cancer therapy. In conclusion, HSP70 proteins have a significant role in cancer development and progress. Understanding of the functions and molecular mechanisms of HSP70 proteins is critical for enhancing the accuracy of cancer diagnosis and for the development of more effective chemotherapeutic agents.

Acknowledgements This work was supported by the West Virginia School of Osteopathic Medicine faculty startup funding (T. Liu); and the Distinguished Professor Research Startup Funding (S. Cao) from Southwest Medical University, Lanzhou, Sichuan, China.

References

Akerfelt, M., Morimoto, R. I., & Sistonen, L. (2010). Heat shock factors: Integrators of cell stress, development and lifespan. *Nature Reviews. Molecular Cell Biology, 11*(8), 545–533.

Ando, K., Oki, Z., Zhao, Y., et al. (2014). Mortalin is a prognostic factor of gastric cancer with normal p53 function. *Gastric Cancer, 17*(2), 255–262.

Arafa, e.-S. A., Abdelazeem, A. H., Arab, H. H., et al. (2014). OSU-CG5, a novel energy restriction mimetic agent, targets human colorectal cancer cells in vitro. *Acta Pharmacologica Sinica, 35*(3), 394–400.

Arispe, N., & De Maio, A. (2000). ATP and ADP modulate a cation channel formed by Hsc70 in acidic phospholipid membranes. *The Journal of Biological Chemistry, 275*(40), 30839–30843.

Bakkenist, C. J., Koreth, J., Williams, C. S., et al. (1999). Heat shock cognate 70 mutations in sporadic breast carcinoma. *Cancer Research, 59*(17), 4219–4221.

Balaburski, G. M., Leu, J. J., Beeharry, N., et al. (2013). A modified HSP70 inhibitor shows broad activity as an anticancer agent. *Molecular Cancer Research, 11*(3), 219–229.

Bayer, C., Liebhardt, M. E., Schmid, T. E., et al. (2014). Validation of heat shock protein 70 as a tumor-specific biomarker for monitoring the outcome of radiation therapy in tumor mouse models. *International Journal of Radiation Oncology, Biology, Physics, 88*(3), 694–700.

Bepperling, A., Alte, F., Kriehuber, T., et al. (2012). Alternative bacterial two-component small heat shock protein systems. *Proceedings of the National Academy of Sciences of the United States of America, 109*(50), 20407–20412.

Boonjaraspinyo, S., Boonmars, T., Kaewkes, S., et al. (2012). Down-regulated expression of HSP70 in correlation with clinicopathology of cholangiocarcinoma. *Pathology Oncology Research, 18*(2), 227–237.

Cai, M. B., Wang, X. P., Zhang, J. X., et al. (2012). Expression of heat shock protein 70 in nasopharyngeal carcinomas: Different expression patterns correlate with distinct clinical prognosis. *Journal of Translational Medicine, 10*, 96.

Chakravarty, G., Mathur, A., Mallade, P., et al. (2016). Nelfinavir targets multiple drug resistance mechanisms to increase the efficacy of doxorubicin in MCF-7/Dox breast cancer cells. *Biochimie, 124*, 53–64.

Chen, T. H., Kambal, A., Krysiak, K., et al. (2011). Knockdown of Hspa9, a del(5q31.2) gene, results in a decrease in hematopoietic progenitors in mice. *Blood, 117*(5), 1530–1539.

Chen, J., Liu, W. B., Jia, W. D., et al. (2014). Overexpression of Mortalin in hepatocellular carcinoma and its relationship with angiogenesis and epithelial to mesenchymal transition. *International Journal of Oncology, 44*(1), 247–255.

Chiou, J. F., Tai, C. J., Huang, M. T., et al. (2010). Glucose-regulated protein 78 is a novel contributor to acquisition of resistance to sorafenib in hepatocellular carcinoma. *Annals of Surgical Oncology, 17*(2), 603–612.

Ciocca, D. R., & Calderwood, S. K. (2005). Heat shock proteins in cancer: Diagnostic, prognostic, predictive, and treatment implications. *Cell Stress & Chaperones, 10*(2), 86–103.

Fu, Y., Li, J., & Lee, A. S. (2007). GRP78/BiP inhibits endoplasmic reticulum BIK and protects human breast cancer cells against estrogen starvation-induced apoptosis. *Cancer Research, 67*(8), 3734–3740.

Gabai, V. L., Yaglom, J. A., Waldman, T., & Sherman, M. Y. (2009). Heat shock protein Hsp72 controls oncogene-induced senescence pathways in cancer cells. *Molecular and Cellular Biology, 29*(2), 559–569.

Gestl, E. E., & Anne Böttger, S. (2012). Cytoplasmic sequestration of the tumor suppressor p53 by a heat shock protein 70 family member, mortalin, in human colorectal adenocarcinoma cell lines. *Biochemical and Biophysical Research Communications, 423*(2), 411–416.

Gray, P. C., & Vale, W. (2012). Cripto/GRP78 modulation of the TGF-β pathway in development and oncogenesis. *FEBS Letters, 586*(4), 1836–1845.

Hanke, N. T., Garland, L. L., & Baker, A. F. (2016). Carfilzomib combined with suberanilohydroxamic acid (SAHA) synergistically promotes endoplasmic reticulum stress in non-small cell lung cancer cell lines. *Journal of Cancer Research and Clinical Oncology, 142*(3), 549–560.

Helmbrecht, K., & Rensing, L. (1999). Different constitutive heat shock protein 70 expression during proliferation and differentiation of rat C6 glioma cells. *Neurochemical Research, 24*(10), 1293–1299.

Howe, M. K., Bodoor, K., Carlson, D. A., et al. (2014). Identification of an allosteric small-molecule inhibitor selective for the inducible form of heat shock protein 70. *Chemistry & Biology, 21*(12), 1648–1659.

Hu, Y., Yang, L., Yang, Y., et al. (2016). Oncogenic role of mortalin contributes to ovarian tumorigenesis by activating the MAPK-ERK pathway. *Journal of Cellular and Molecular Medicine, 20*(11), 2111–2121.

Hung, C. M., Su, Y. H., Lin, J. N., et al. (2012). Demethoxycurcumin modulates prostate cancer cell proliferation via AMPK-induced down-regulation of HSP70 and EGFR. *Journal of Agricultural and Food Chemistry, 60*(34), 8427–8434.

Jäättelä, M., Wissing, D., Bauer, P. A., et al. (1992). Major heat shock protein hsp70 protects tumor cells from tumor necrosis factor cytotoxicity. *The EMBO Journal, 11*(10), 3507–3512.

Jakob, U., Scheibel, T., Bose, S., Reinstein, J., et al. (1996). Assessment of the ATP binding properties of Hsp90. *The Journal of Biological Chemistry, 271*(17), 10035–10041.

Jiang, C. C., Mao, Z. G., Avery-Kiejda, K. A., et al. (2009). Glucose-regulated protein 78 antagonizes cisplatin and adriamycin in human melanoma cells. *Carcinogenesis, 30*(2), 197–204.

Jin, H., Ji, M., Chen, L., et al. (2016). The clinicopathological significance of Mortalin overexpression in invasive ductal carcinoma of breast. *Journal of Experimental & Clinical Cancer Research, 35*, 42.

Jung, J. H., Lee, J. O., Kim, J. H., et al. (2010). Quercetin suppresses HeLa cell viability via AMPK-induced HSP70 and EGFR down-regulation. *Journal of Cellular Physiology, 223*(2), 408–414.

Kaiser, M., Lee, J. O., Kim, J. H., et al. (2011). Antileukemic activity of the HSP70 inhibitor pifithrin-μ in acute leukemia. *Blood Cancer Journal, 1*(7), e28.

Kampinga, H. H., Hageman, J., Vos, M. J., et al. (2009). Guidelines for the nomenclature of the human heat shock proteins. *Cell Stress & Chaperones, 14*(1), 105–111.

Khalouei, S., Chow, A. M., Brown, I. R. (2014). Localization of heat shock protein HSPA6 (HSP70B′) to sites of transcription in cultured differentiated human neuronal cells following thermal stress. *Journal of Neurochemistry*, *131*(6), 743–754.

Kim, J. A., Kim, Y., Kwon, B. M., et al. (2013). The natural compound cantharidin induces cancer cell death through inhibition of heat shock protein 70 (HSP70) and Bcl-2-associated athanogene domain 3 (BAG3) expression by blocking heat shock factor 1 (HSF1) binding to promoters. *The Journal of Biological Chemistry, 288*(40), 28713–28726.

Kim, J. A., Lee, S., Kim, D. E., et al. (2015). Fisetin, a dietary flavonoid, induces apoptosis of cancer cells by inhibiting HSF1 activity through blocking its binding to the hsp70 promoter. *Carcinogenesis, 36*(6), 696–706.

Ko, S. K., Kim, J., Na, D. C., et al. (2015). A small molecule inhibitor of ATPase activity of HSP70 induces apoptosis and has antitumor activities. *Chemistry & Biology, 22*(3), 391–403.

Kocsis, J., Mészáros, T., Madaras, B., et al. (2011). High levels of acute phase proteins and soluble 70 kDa heat shock proteins are independent and additive risk factors for mortality in colorectal cancer. *Cell Stress & Chaperones, 16*(1), 49–55.

Kosakowska-Cholody, T., Lin, J., Srideshikan, S. M., et al. (2014). HKH40A downregulates GRP78/BiP expression in cancer cells. *Cell Death & Disease, 5*, e1240.

Kuballa, P., Baumann, A. L., Mayer, K., et al. (2015). Induction of heat shock protein HSPA6 (HSP70B′) upon HSP90 inhibition in cancer cell lines. *FEBS Letters*, *589*(13):1450–1458.

Kubota, H., Yamamoto, S., Itoh, E., et al. (2010). Increased expression of co-chaperone HOP with HSP90 and HSC70 and complex formation in human colonic carcinoma. *Cell Stress & Chaperones, 15*(6), 1003–1011.

Lee, A. S. (2007). GRP78 induction in cancer: therapeutic and prognostic implications. *Cancer Research, 67*(8), 3496–3499.

Leung, T. K., Rajendran, M. Y., Monfries, C., et al. (1990). The human heat-shock protein family. Expression of a novel heat-inducible HSP70 (HSP70B′) and isolation of its cDNA and genomic DNA. *The Biochemical Journal*, *267*: 125–132.

Li, J., Ni, M., Lee, B., et al. (2008). The unfolded protein response regulator GRP78/BiP is required for endoplasmic reticulum integrity and stress-induced autophagy in mammalian cells. *Cell Death & Disease, 15*(9), 1460–1471.

Li, G., Xu, Y., Guan, D., et al. (2011). HSP70 protein promotes survival of C6 and U87 glioma cells by inhibition of ATF5 degradation. *The Journal of Biological Chemistry, 286*(23), 20251–20259.

Li, X., Srinivasan, S. R., Connarn, J., et al. (2013). Analogs of the allosteric heat shock protein 70 (Hsp70) inhibitor, MKT-077, as anti-cancer agents. *ACS Medicinal Chemistry Letters, 4*(11), 1042–1047.

Lin, Y., Wang, Z., Liu, L., & Chen, L. (2011). Akt is the downstream target of GRP78 in mediating cisplatin resistance in ER stress-tolerant human lung cancer cells. *Lung Cancer, 71*(3), 291–297.

Liu, T., Daniels, C. K., & Cao, S. (2012). Comprehensive review on the HSC70 functions, interactions with related molecules and involvement in clinical diseases and therapeutic potential. *Pharmacology & Therapeutics, 136*(3), 354–374.

Liu, T., Singh, R., Rios, Z., et al. (2015). Tyrosine phosphorylation of HSC70 and its interaction with RFC mediates methotrexate resistance in murine L1210 leukemia cells. *Cancer Letters, 357*(1), 231–241.

Liu, T., Krysiak, K., Shirai, C. L., et al. (2017). Knockdown of HSPA9 induces TP53-dependent apoptosis in human hematopoietic progenitor cells. *PLoS One, 12*(2), e0170470.

Lu, W. J., Lee, N. P., Kaul, S. C., et al. (2011). Induction of mutant p53-dependent apoptosis in human hepatocellular carcinoma by targeting stress protein mortalin. *International Journal of Cancer, 129*(8), 1806–1814.

Ma, Y., & Hendershot, L. M. (2004). The role of the unfolded protein response in tumor development: Friend or foe? *Nature Reviews. Cancer, 4*(12), 966–967.

Ma, X. H., Piao, S. F., Dey, S., et al. (2014). Targeting ER stress-induced autophagy overcomes BRAF inhibitor resistance in melanoma. *The Journal of Clinical Investigation, 124*(3), 1406–1417.

Macario, A. J., & Conway de Macario, E. (2007). Molecular chaperones: multiple functions, pathologies, and potential applications. *Frontiers in Bioscience, 1*(12), 2588–2600.

Misra, U. K., Mowery, Y., Kaczowka, S., et al. (2009). Ligation of cancer cell surface GRP78 with antibodies directed against its COOH-terminal domain up-regulates p53 activity and promotes apoptosis. *Molecular Cancer Therapeutics, 8*(5), 1350–1362.

Misra, U. K., Payne, S., & Pizzo, S. (2011). Ligation of prostate cancer cell surface GRP78 activates a proproliferative and antiapoptotic feedback loop: A role for secreted prostate-specific antigen. *The Journal of Biological Chemistry, 286*(2), 1248–1259.

Moghanibashi, M., Rastgar-Jazii, F., Soheili, Z. S., et al. (2013). Esophageal cancer alters the expression of nuclear pore complex binding protein Hsc70 and eIF5A-1. *Functional & Integrative Genomics, 13*(2), 253–260.

Moon, J. Y., & Cho, S. K. (2016). Nobiletin induces protective autophagy accompanied by ER-stress mediated apoptosis in human gastric cancer SNU-16 cells. *Molecules, 21*(7), 914.

Murphy, M. E. (2013). The HSP70 family and cancer. *Carcinogenesis, 34*(6), 1181–1188.

Na, Y., Kaul, S. C., Ryu, J., et al. (2016). Stress chaperone mortalin contributes to epithelial-mesenchymal transition and cancer metastasis. *Cancer Research, 76*(9), 2764–2765.

Nigam, N., Grover, A., Goyal, S., et al. (2015). Targeting mortalin by embelin causes activation of tumor suppressor p53 and deactivation of metastatic signaling in human breast cancer cells. *PLoS One, 10*(9), e0138192.

Noonan, E. J., Fournier, G., Hightower, L. E. (2008). Surface expression of Hsp70B′ in response to proteasome inhibition in human colon cells. *Cell Stress & Chaperones, 13*(1), 105–110.

Nylandsted, J., Gyrd-Hansen, M., Danielewicz, A., et al. (2004). Heat shock protein 70 promotes cell survival by inhibiting lysosomal membrane permeabilization. *The Journal of Experimental Medicine, 200*(4), 425–435.

Pilzer, D., Saar, M., Koya, K., et al. (2010). Mortalin inhibitors sensitize K562 leukemia cells to complement-dependent cytotoxicity. *International Journal of Cancer, 126*(6), 1428–1435.

Qi, W., White, M. C., Choi, W., et al. (2013). Inhibition of inducible heat shock protein-70 (hsp72) enhances bortezomib-induced cell death in human bladder cancer cells. *PLoS One, 8*(7), e69509.

Ramired, V. P., Krueger, W., & Aneskievich, B. J. (2015). TNIP1 reduction of HSPA6 gene expression occurs in promoter regions lacking binding sites for known TNIP1-repressed transcription factors. *Gene, 555*(2), 430–437.

Ramp, U., Mahotka, C., Heikaus, S., et al. (2007). Expression of heat shock protein 70 in renal cell carcinoma and its relation to tumor progression and prognosis. *Histology and Histopathology, 22*(10), 1099–1107.

Reddy, R. K., Mao, C., Baumeister, P., et al. (2003). Endoplasmic reticulum chaperone protein GRP78 protects cells from apoptosis induced by topoisomerase inhibitors: Role of ATP binding site in suppression of caspase-7 activation. *The Journal of Biological Chemistry, 278*(23), 20915–20924.

Regeling, A., Imhann, F., Volders, H. H., et al. (2016). HSPA6 is an ulcerative colitis susceptibility factor that is induced by cigarette smoke and protects intestinal epithelial cells by stabilizing anti-apoptotic Bcl-XL. *Biochimica et Biophysica Acta, 1862*, 788–796.

Rérole, A. L., Jego, G., & Garrido, C. (2011). Hsp70: Anti-apoptotic and tumorigenic protein. *Methods in Molecular Biology, 787*, 205–230.

Roberts, J. L., Tavallai, M., Nourbakhsh, A., et al. (2015). GRP78/Dna K is a target for nexavar/stivarga/votrient in the treatment of human malignancies, viral infections and bacterial diseases. *Journal of Cellular Physiology, 230*(10), 2552–2578.

Roller, C., & Maddalo, D. (2013). The Molecular chaperone GRP78/BiP in the development of chemoresistance: Mechanism and possible treatment. *Frontiers in Pharmacology, 4*, 10.

Rozenberg, P., Kocsis, J., Saar, M., et al. (2013). Elevated levels of mitochondrial mortalin and cytosolic HSP70 in blood as risk factors in patients with colorectal cancer. *International Journal of Cancer, 133*(2), 514–518.

Rusin, M., Zientek, H., Krześniak, M., et al. (2004). Intronic polymorphism (1541-1542delGT) of the constitutive heat shock protein 70 gene has functional significance and shows evidence of association with lung cancer risk. *Molecular Carcinogenesis, 39*(3), 155–163.

Sandoval, J. A., Hoelz, D. J., Woodruff, H. A., et al. (2006). Novel peptides secreted from human neuroblastoma: Useful clinical tools? *Journal of Pediatric Surgery, 41*(1), 245–251.

Sane, S., Abdullah, A., Nelson, M. E., et al. (2016). Structural studies of UBXN2A and mortalin interaction and the putative role of silenced UBXN2A in preventing response to chemotherapy. *Cell Stress & Chaperones, 21*(2), 313–326.

Sekihara, K., Harashima, N., Tongu, M., et al. (2013). Pifithrin-μ, an inhibitor of heat-shock protein 70, can increase the antitumor effects of hyperthermia against human prostate cancer cells. *PLoS One, 8*(11), e78772.

Smith, K. J. (2010). Heat shock protein 70B′ (HSP70B′) expression and release in response to human oxidized low density lipoprotein immune complexes in macrophages. *The Journal of Biological Chemistry, 285*(21), 15985–15993.

Sobolewski, C., Rhim, J., Legrand, N., et al. (2015). 2,5-Dimethyl-celecoxib inhibits cell cycle progression and induces apoptosis in human leukemia cells. *The Journal of Pharmacology and Experimental Therapeutics, 355*(2), 308–328.

Song, L., Liu, H., Ma, L., et al. (2013). Inhibition of autophagy by 3-MA enhances endoplasmic reticulum stress-induced apoptosis in human nasopharyngeal carcinoma cells. *Oncology Letters, 6*(4), 1031–1038.

Starenki, D., Hong, S. K., Lloyd, R. V., et al. (2015). Mortalin (GRP75/HSPA9) upregulation promotes survival and proliferation of medullary thyroid carcinoma cells. *Oncogene, 34*(35), 4624–4634.

Sugita, S., Ito, K., Yamashiro, Y., et al. (2015). EGFR-independent autophagy induction with gefitinib and enhancement of its cytotoxic effect by targeting autophagy with clarithromycin in non-small cell lung cancer cells. *Biochemical and Biophysical Research Communications, 461*(1), 28–34.

Sullivan, C. S., & Pipas, J. M. (2002). T antigens of simian virus 40: Molecular chaperones for viral replication and tumorigenesis. *Microbiology and Molecular Biology Reviews, 66*(2), 179–202.

Takahashi, K., Tanaka, M., Yashiro, M., et al. (2016). Protection of stromal cell-derived factor 2 by heat shock protein 72 prevents oxaliplatin-induced cell death in oxaliplatin-resistant human gastric cancer cells. *Cancer Letters, 378*(1), 8–15.

Tanaka, M., Mun, S., Harada, A., et al. (2014). Hsc70 contributes to cancer cell survival by preventing Rab1A degradation under stress conditions. *PLoS One, 9*(5), e96785.

Tavallai, M., Booth, L., Roberts, J. L., et al. (2016). Rationally repurposing Ruxolitinib (Jakafi (®)) as a solid tumor therapeutic. *Frontiers in Oncology, 6*, 142.

Trieb, K., Sulzbacher, I., & Kubista, B. (2016). Recurrence rate and progression of chondrosarcoma is correlated with heat shock protein expression. *Oncology Letters, 11*(1), 521–524.

Ulianich, L., & Insabato, L. (2014). Endoplasmic reticulum stress in endometrial cancer. *Frontiers of Medical (Lausanne), 1*, 55.

Wadhwa, R., Sugihara, T., Yoshida, A., et al. (2000). Selective toxicity of MKT-077 to cancer cells is mediated by its binding to the hsp70 family protein mot-2 and reactivation of p53 function. *Cancer Research, 60*(24), 6818–6821.

Wang, X., Wang, Q., & Lin, H. (2010). Correlation between clinicopathology and expression of heat shock protein 72 and glycoprotein 96 in human esophageal squamous cell carcinoma. *Clinical & Developmental Immunology, 2010*, 212537.

Wang, N., Wang, Z., Peng, C., et al. (2014a). Dietary compound isoliquiritigenin targets GRP78 to chemosensitize breast cancer stem cells via β-catenin/ABCG2 signaling. *Carcinogenesis, 35*(11), 2544–2554.

Wang, X., Chen, M., Zhou, J., et al. (2014b). HSP27, 70 and 90, anti-apoptotic proteins, in clinical cancer therapy (Review). *International Journal of Oncology, 45*(1), 18–30.

Wu, J., Liu, T., Rios, Z., et al. (2017). Heat shock proteins and cancer. *Trends in Pharmacological Sciences, 38*(3), 226–256.

Yang, L., Li, H., Jiang, Y., et al. (2013). Inhibition of mortalin expression reverses cisplatin resistance and attenuates growth of ovarian cancer cells. *Cancer Letters, 336*(1), 213–221.

Yang, Z., Zhuang, L., Szatmary, P., et al. (2015). Upregulation of heat shock proteins (HSPA12A, HSP90B1, HSPA4, HSPA5 and HSPA6) in tumour tissues is associated with poor outcomes from HBV-related early-stage hepatocellular carcinoma. *International Journal of Medical Sciences, 12*(3), 256–263.

Yerushalmi, R., Raiter, A., Nalbandvan, K., & Hardy, B. (2015). Cell surface GRP78: A potential marker of good prognosis and response to chemotherapy in breast cancer. *Oncology Letters, 10*(4), 2149–2155.

Yi, X., Luk, J. M., Lee, N. P., et al. (2008). Association of mortalin (HSPA9) with liver cancer metastasis and prediction for early tumor recurrence. *Molecular & Cellular Proteomics, 7*(2), 315–325.

Yoshidomi, K., Murakami, A., Yakabe, K., et al. (2014). Heat shock protein 70 is involved in malignant behaviors and chemosensitivities to cisplatin in cervical squamous cell carcinoma cells. *The Journal of Obstetrics and Gynaecology Research, 40*(5), 1188–1196.

Yuan, L., Zhang, L., Dong, X., et al. (2013). Apoptin selectively induces the apoptosis of tumor cells by suppressing the transcription of HSP70. *Tumour Biology, 34*(1), 577–585.

Zhang, H., Hu, H., Jiang, X., et al. (2005). Membrane HSP70: The molecule triggering gammadelta T cells in the early stage of tumorigenesis. *Immunological Investigations, 34*(4), 453–468.

Zhao, L., Li, H., Shi, Y., et al. (2014). Nanoparticles inhibit cancer cell invasion and enhance antitumor efficiency by targeted drug delivery via cell surface-related GRP78. *International Journal of Nanomedicine, 10*, 245–256.

Implication of HSP70 in the Pathogenesis of Gastric Cancer

Prakash Chand Sharma and Renu Verma

Abstract Gastric cancer is the third most common cause of cancer linked mortality in the world. Despite recent advances in treatment regimes, effective prognosis and diagnosis of GC is still a formidable task as the disease is asymptomatic in early stages. Heat Shock Proteins (HSP) constitute a class of ubiquitous and highly conserved proteins that show differential expression during different stresses including cancer. HSP70, an important member of the HSP family, regulate various biological processes like protein folding and degradation, apoptosis, and angiogenesis related with tumorigenesis. Differential expression of HSP70 has been explored in tumor versus normal tissues. Upregualtion of HSP70 has been validated in advanced stages making it a future potential prognostic marker in GC. Infection with *Helicobacter pylori*, a major causative agent of GC, is known to communicate with HSP70 and downregulate its expression. Genetic polymorphism has been correlated with susceptibility to many diseases and association of HSP70 polymorphism in GC is no exception. This chapter deals with various functional roles of HSP70 and its implication in tumorigenesis, particularly gastric cancer.

Keywords Angiogenesis · Apoptosis · Gastric cancer · *Helicobacter pylori* · HSP70 · Protein folding and degradation

Abbreviations

GC	gastric cancer
H. pylori	*Helicobacter pylori*
HSF	heat shock factor
HSP	heat shock proteins
IHC	immunohistochemistry

P. C. Sharma (✉) · R. Verma
University School of Biotechnology, Guru Gobind Singh Indraprastha University, New Delhi, India

A. A. A. Asea, P. Kaur (eds.), *HSP70 in Human Diseases and Disorders*, Heat Shock Proteins 14, https://doi.org/10.1007/978-3-319-89551-2_6

kDA	kilo dalton
RT-PCR	real time – polymerase chain reaction
sHSP	small heat shock proteins

Introduction

Cancer is a class of diseases characterized by the uncontrolled cell division that generally leads to the formation of a tumor, which tends to invade healthy tissues around and spread other regions of the body. Medical researchers have recognized many types of cancer and the classification has been expanding with time and investigations. Among the different type of cancers gastric cancer (GC) is the fourth most common cancer in men and the fifth most common cancer in women. The incidence and mortality rate of GC vary with respect to different factors like geographical location, age, sex and ethnicity of the subjects. Infection with *Helicobacter pylori* and Epstein Barr Virus (EBV), dietary habits, smoking, consumption of alcohol and red meat are various risk factors known to contribute towards occurrence of gastric cancer. The Cancer Genome Atlas (TCGA) Network (2014) has categorized GC into Chromosomal Instable (CIN; 50%), Microsatellite Instable (22%), genomically stable (20%) and EBV + (9%) GCs. WHO has classified GC on the basis of its histological patterns into tubular adenocarcinoma, papillary adenocarcinoma, mucinous adenocarcinoma, poorly cohesive carcinomas and mixed carcinomas.

Although a decline has been observed in the incidence of gastric cancer recently, it still occupies a higher rank in the cancer hierarchy in the world with huge number of cases reported every year (Stewart and Wild 2017). The mechanisms of initiation and proliferation of gastric cancer are not fully established and therefore comprehensive and effective GC treatment strategies are still obscure (Kim et al. 2011). Identification of the key players driving regulatory mechanisms of gastric tumorigenesis is an important and urgent goal of contemporary cancer research (Liu et al. 2015). Conventional treatment options for gastric cancer include surgery, chemotherapy, radiation therapy or multimodal therapy. Most of the symptoms of gastric cancer occur in the advanced stages of the disease and therefore early diagnosis of the disease is difficult.

Heat shock proteins (HSP) represent a large family of conserved proteins that are generally present in all cell types and primarily function as molecular chaperones involved in protein folding and refolding, and maintenance of cell integrity (Sessa et al. 2013). HSP were first discovered in Drosophila on the basis of their overexpression in response to heat stress and thus named as heat shock proteins (Seigneuric et al. 2011). They overexpress in response to many other physiological stresses also and hence they are often called as stress proteins. The stress inducing agents include radiation, magnetic fields, oxidative stress, cold, altered pH, nutrient depletion, presence of metal ions, alcohol, etc. (Macario et al. 2010). Some other pathophysiological factors like oncogenesis, neurodegenerative and autoimmune diseases, viral infections and aging also induce higher expression of HSP (Nollen and Morimoto 2002; Ranford and Henderson 2002; Pockley 2003; Takayama et al. 2003). HSP are designated on the basis of their molecular weight such as HSP40, HSP60, and HSP70. Mammalian HSP are classified according to their molecular weights into five families: large HSP, HSP90, HSP70, HSP60 and small HSP including HSP27. HSP are also required during normal cell growth in significant quantity as they channel the old proteins to proteasomes for proteolysis and recycling, and help in the proper folding of the newly synthesized proteins (Kapoor and Vaidya 2013). The transcription of the HSP is regulated by the Heat Shock Factor (HSF) family of the transcription factors. The HSF gene family includes hsf1 transcription factor which actively participates in the transcription for HSP and two other less studied factors, hsf2 and hsf4 (Ciocca and Calderwood 2005). HSP lack transmembrane domains and therefore they mostly exist as intracellular soluble proteins but certain HSP such as HSP27, HSP60, HSP70 and HSP90 exist as both intracellular and extracellular protein and are involved in various immunogenic events (Gallucci and Matzinger 2001; Schmitt et al. 2007). Intracellular heat shock proteins remain highly expressed in cancer cells and are essential to the survival of these cells. Interestingly, HSP have attained great significance in cancer research recently as they are found overexpressed in most cancers, particularly during tumor proliferation (Ciocca and Calderwood 2005). This chapter sheds light on the implication of HSP70 in gastric cancer and how the information is useful for the development of biomarkers for a better management of the disease.

Implication of HSP in Cancer

Overexpression of HSP has been linked to tumor growth and differentiation, and provides resistance to cancer treatment. HSP form multiprotein complexes that bind to their respective cofactors (Calderwood et al. 2006; Khalil et al. 2011). The implication of some important HSP in different cancers is briefly presented here.

HSP27 comes under small heat shock protein (sHSP) family located in the endothelial cells and helps in enhancing DNA repair mechanism. Increase in the expression level of HSP27 was observed during progression of prostate cancer and breast cancer. Furthermore, it has been reported as marker of squamous metaplasia in the uterine cervix (Ciocca and Calderwood 2005). Phosphorylation leads to the modification of the structure and function of HSP27. Moreover, HSP27 is also actively involved in protein refolding (Bruey et al. 2000). Overexpression of HSP27 has been associated with the good prognosis in esophageal cancer and endometrial adenocarcinomas (Seigneuric et al. 2011).

HSP60 is a highly conserved family of chaperones termed as chaperonin that help in the protein folding of mitochondrial proteins. HSP60 is present in the cytosol and enters into the mitochondria in response to the Mitochondrial Import Signal (MIS) (Vilasi et al. 2014). IHC expression of HSP60 has been reported to be associated with tumor progression of prostate cancer. In a clinical study, androgen

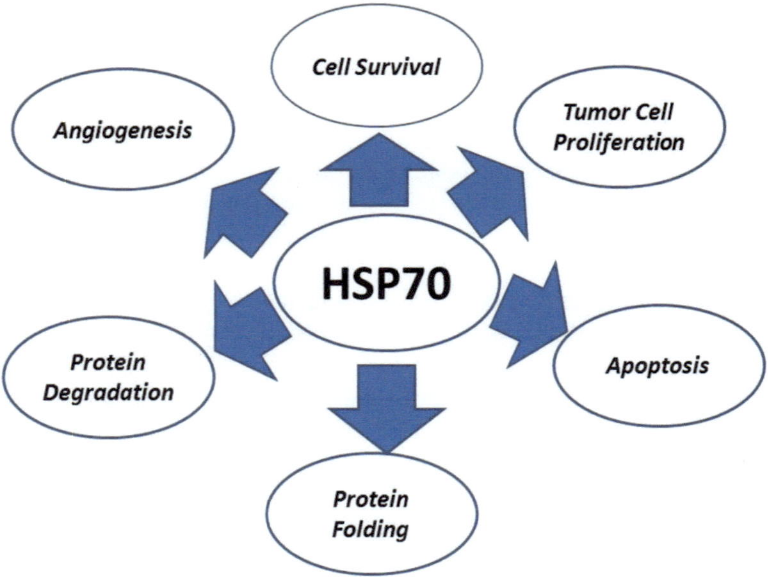

Fig. 1 Various roles of HSP70 in cancer related cellular processes

suppression treatment revealed that HSP60 is highly associated with androgen independence (Castilla et al. 2010). Higher expression of HSP60 is highly associated with prognosis of prostate cancer, cervical cancer and colorectal cancer (Glaessgen et al. 2008; Seigneuric et al. 2011).

HSP90, an ATP-dependent chaperone family protein, is involved in the regulation of a number of proteins such as focal adhesion kinase, epidermal growth factor receptor, hypoxia inducible factor-1, and vascular endothelial growth factor (Neckers and Workman 2012). It is also involved in the protein folding of myriad proteins. High expression of HSP90 has been observed in gastric and breast cancer indicating a positive association with vascular invasion and metastasis (Seigneuric et al. 2011). As most of the proteins regulated by HSP90 are associated with gastric cancer, expression of HSP90 has been used in the prognosis of gastric cancer (Liu et al. 2015).

HSP70 belongs to an ATP-dependent anti-apoptotic chaperone family of proteins. This protein is involved in the proper folding of the native proteins and also in refolding of the denatured proteins. Further, they also participate in the signaling pathways by translocating the membrane and secretory proteins. HSP70 is considered as the central core of the chaperone family as they perform a diverse range of functions in most of the cellular mechanisms (Didelot et al. 2007) (Fig. 1). These activities of HSP70 show that higher level of HSP70 expression is more crucial in cancer cells in order to cope up with the elevated level of metabolic rates and cellular mechanisms (Maehara et al. 2000). The homologs of HSP70 consist of an N-terminal ATPase domain and a C-terminal substrate binding domain of 45 kDa and 25 kDa, respectively. Sequence differences of HSP70s, as revealed by sequence alignment, have shown association with the variation in their function. Undulation between the ATP state with low affinity and fast exchange rates for substrates, and the ADP state with high affinity and low exchange rates for substrates takes place in the ATPase cycle of HSP70 (Mayer and Bukau 2005). A recent study has reported higher level of HSP70 in advanced gastric cancer cases concomitant with the reduction in the predominance of *Helicobacter pylori* suggesting thereby that HSP70 can be used as a prognostic marker for the efficient diagnosis of gastric cancer (Bodoor et al. 2016).

HSP70 Regulated Protein Folding and Degration

HSP70 plays a crucial role in the maintenance of the integrity of protein structures and their transport by facilitating precise protein folding and degradation of misfolded or old proteins across various cell organelles. A brief description of its role in protein folding and degradation is presented below.

Protein Folding

HSP70 facilitates the process of protein folding in cytoplasm, endoplasmic reticulum and mitochondria (Fig. 2). In cytoplasm, after synthesis of a protein on ribosome, the hydrophobic area of the protein is captured by HSP40 for further binding to substrate domain of HSP70 so that unwanted aggregation of proteins does not occur. Factors such as Hip further assist the binding of nascent protein to HSP70. The C-domain covers the bound protein forming a lid to the HSP70 molecule. ATP hydrolysis by HSP40 occurs within N-domain to promote the binding affinity of protein to the substrate domain. The conversion of ATP to ADP leads to decrease in the protein affinity towards substrate domain of HSP70 and finally the folded protein is released. In endoplasmic reticulum, the protein synthesized on ribosome gets transported to endoplasmic reticulum through a membrane translocon, which attracts HSP70 (BiP) and regulate its ATPase activity. BiP binds to hydrophobic area of the first protein with which it interacts and prevents the polypeptide chain to escape back into the cytoplasm. A continuous cycle of ATP and ADP conversion ensures the binding of BiP to the next protein and results in the release of polypeptide chain and folding of protein. In mitochondria, proteins follow the similar route for protein folding as they do in endoplasmic reticulum. The cytoplasmic HSP70 helps in maintaining the linear state of protein which further aids in the penetration of protein through translocons of the mitochondria. The polypeptide chain comes into mitochondrial matrix and formation of spatial protein structure occurs with ATP generated energy (Hartl et al. 1992).

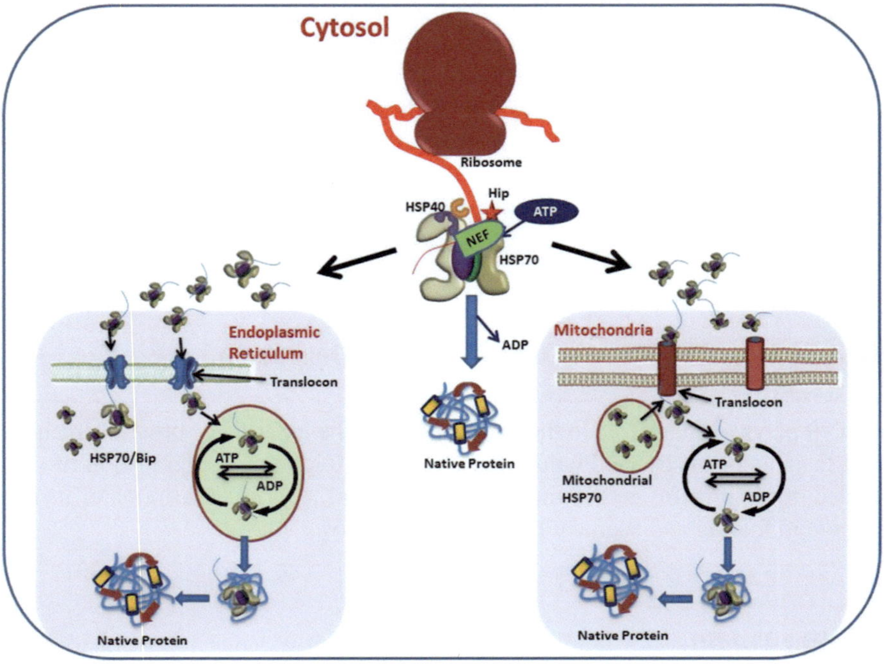

Fig. 2 Involvement of HSP70 in protein folding in cytosol, endoplasmic reticulum and mitochondria

Protein Degradation

HSP70 acts as a cleaning agent as it removes erroneous proteins that are harmful to the cell. HSP70 performs this job in two ways: proteasomal and lysosomal degradation. HSP70 along with HSP40 recognizes damaged proteins and transport them to proteasomes and lysosomes (Fig. 3). Protein ubiquitination occurs once the undesirable protein enters a proteasome. It includes binding of ubiquitin-activating enzyme El to ubiquitin, which further shifts ubiquitin to enzyme carrier E2. Ubiquitin is delivered by E2 to inaccurate protein and then E3 ligase transports ubiquitin from E2 to the incorrect protein. Finally, proteasomes identify and degrade the proteins attached with ubiquitin (Bercovich et al. 1997). Normal proteins start degrading when a cell undergoes starvation to complete the demand of essential proteins required by the cell for its survival. The normal proteins which need to be degraded are sent to lysosomes containing an amino acid sequence described as KFPRQ. This sequence is recognized by HSP70, HSP90, HSP40, Hip, Hop and Bag-1 forming a complex. The complex binds to lysosome membrane and sends the unwanted protein to the lysosome for degradation (Salvador et al. 2000).

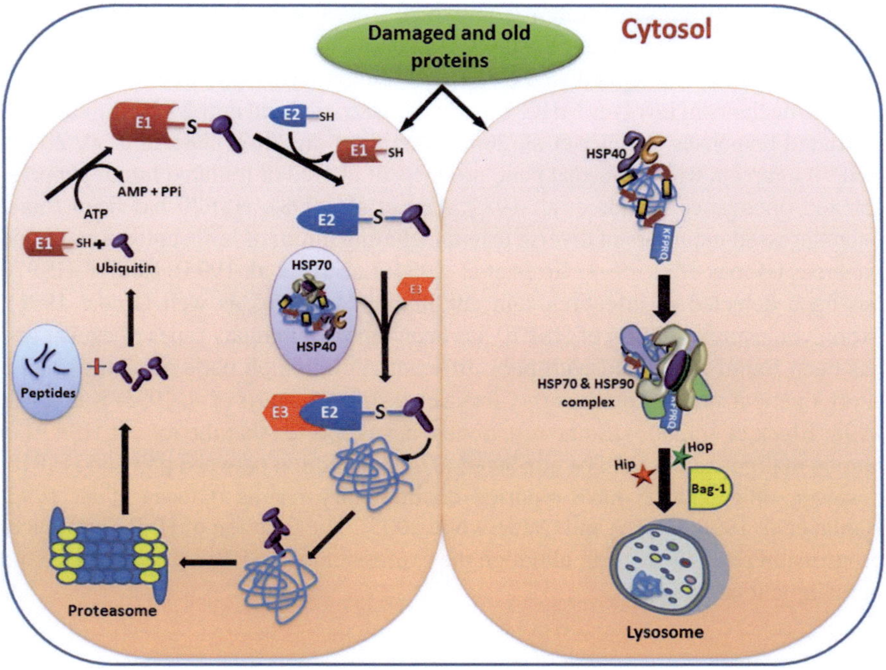

Fig. 3 Mechanism of HSP70 regulated protein degradation in lysosome and proteasome

HSP70 and Tumorigenesis

The HSP70 family is the most important and best characterized family of HSP. HSP70 along with HSP90 functions as potent regulator of the adaptive immune system as they work in the cross-presentation of tumor-specific peptides and therefore are potential biomarkers for cancer severity or mortality (Schild et al. 1999; Janetzki et al. 2000; Feng et al. 2001; Ciocca and Calderwood 2005). HSP70 can be considered as a tumor-specific marker since its unusual membrane expression occurs only on the surface of the tumor cells, but not on normal tissues (Multhoff et al. 1995; Gehrmann et al. 2003). HSP70 in conjunction with HSP60 and HSP90 also plays a significant role in folding of nascent polypeptide chains and translocation of precursor proteins across the membranes of cytoplasmic organelles (Hendrick and Hartl 1993; Pratt 1993; Hartl 1996). HSP70 is also required in protein maturation and regulation during cell growth under normal condition (Welch 1992; Kaur et al. 2000).

Various studies have implicated overexpression of HSP70 in tumorigenesis, development of aberrant phenotype, resistance towards cell death (Samali and Cotter 1996; Hoang et al. 2000; Kaur et al. 2000; Witkin 2001). Overexpression of HSP70 has been found to be a poor prognostic factor in colorectal cancer (Lazaris et al. 1995). Tumor derived HSP70 protein can evoke cancer specific immune response against the same tumor from which it was derived by binding to tumor-specific peptides (Graner et al. 2003). Some studies have suggested that tumor-specific response can be mediated by CD8+ T lymphocytes, γδT+ cells or NK cells, and the mechanism involves MHC-1 molecule-restricted and non-MHC-1 molecule-restricted responses (Graner et al. 2000; Manjili et al. 2002; Castelli et al. 2004). HSP70 interacts with abnormal gene products of altered or mutated tumor suppressor and oncogenes (Hinds et al. 1987; Abd et al. 1998). HSP70 has been found overexpressed in tumors of diverse origin including tumor of lung, colon, breast and pancreas (Bonay et al. 1994; Gress et al. 1994; Lazaris et al. 1995). Role of HSP70 has been detected in infectious and autoimmune diseases as well (Jindal 1996). Moreover, overexpression of HSP70 has been reported in many types of cancer and has been found associated with poor differentiation, lymph node metastasis and a shorter patient survival prognosis (Ciocca et al. 1990; Lazaris et al. 1995; Kaur et al. 1998; Elpek et al. 2003). Some researchers have reported that the role of HSP70 in human malignancies could be attributed to the interaction between p53 and HSP70, however, other studies have reported contradictory results (Ciocca et al. 1993; Nanbu et al. 1996; Ciocca and Calderwood 2005). The function of HSP70 in cancer progression remains unclear, although the expression of HSP70 is frequently found in malignant neoplasm.

Apoptosis

The programmed cell death or apoptosis is a highly regulated energy dependent metabolic pathway that performs the fundamental event of the developmental and physiological processes. Defects in the regulation of apoptotic process would lead to a wide range of disorders. HSP70 acts as a negative regulator of intrinsic and extrinsic apoptotic pathways. Overexpression of HSP70 leads to tumor formation by readily inhibiting the factors involved in apoptosis. In the intrinsic apoptotic pathway, HSP70 binds with Bax and Apaf1 and inhibits their function. During internal malfunction like DNA damage, the proapoptotic factors like Bax get activated and bind to the outer membrane of the mitochondria to help in the release of Cytochrome C. Along with Apaf1 and pro-caspase9, cytochrome-c binds to form apoptosome which leads to the activation of caspase-9. This caspase-9 acts on pro-caspase-3 (inactive caspase-3) and activates it to ensure subsequent apoptosis. As HSP70 has the capacity to inhibit Bax and Apaf1, activation of Caspase-9 gets inhibited and thereby further apoptotic processes are stopped leading to uncontrolled cell growth. In extrinsic pathway, HSP70 binds to death receptors, DR4 and DR5, and inhibits the assembly of death inducing signaling complex (DISC) (Murphy 2013) (Fig. 4).

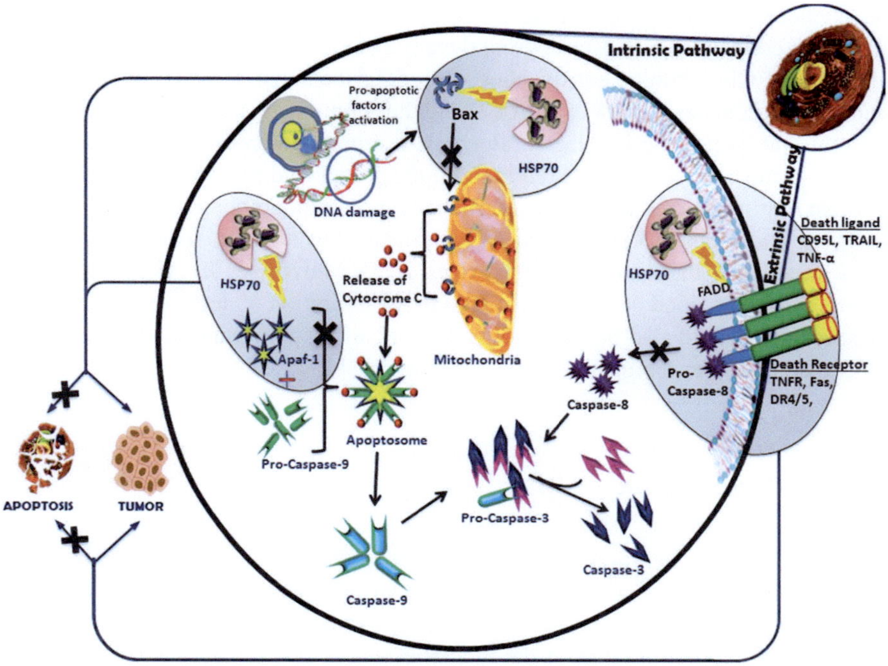

Fig. 4 Schematic representation of HSP70 regulated apoptotic pathways

Angiogenesis

Like cell proliferation and immune evasion, angiogenesis also plays a critical role in the development and progression of a tumor. Angiogenesis not only contributes to increased supply of nutrients to the tumor cells but also in the migration of these cells from the oncogenic tissue to different parts of the body. Like all other prominent tumors, in GC also, the major challenge remains to be either diagnosis of GC post metastasis or remission of tumor state with development of angiogenic vessels. Recent studies have shown that along with assisting in tumor cell proliferation and immune evasion, HSP70 also plays an important role in the formation of microvessels in tumor environment (Juhasz et al. 2014; Park et al. 2017). Role of HSP70 has been suggested to stimulate the human umbilical vein endothelial cell (HUVEC) migration and tube formation in vitro along with microvessel development *in vivo*, in a manner similar to as induced by recombinant vascular endothelial growth factor (VEGF). Multiple pathways have been explored and identified to explain the role of HSP70 family proteins in angiogenesis. HSP70-1A has been described to be overexpressed on the surface of HUVECs. This expression in turn increases the phosphorylation activity of mitogen-activated protein kinase (MAPK) proteins, which have been well established to play a crucial role in angiogenesis. However, it has also been suggested that HSP70 mediated activation of MAPK pathway is independent of VEGF and VEGF receptors (Kim et al. 2016). Another study by Park and colleagues has also reported the angiogenic role of HSP70 to be independent of VEGF. Simultaneously, they also described that HSP70 mediates IL-5 induced angiogenesis (Park et al. 2017) and thus connecting a link between immune system and angiogenesis. HSP70 along with other chaperonin TRiC (T-complex protein-1 ring complex) is also required for the proper folding of Von Hippel-Lindau protein that modulates the concentration of hypoxia-inducible transcription factor, HIF-1α, and other proteins, in an oxygen-dependent manner, required for tumor growth and vascularisation (Feldman et al. 2003; Melville et al. 2003). In conclusion, HSP70 is a good candidate involved in the inhibition of angiogenesis in tumor tissues and it could also be hypothesized as one of the important factors for VEGF tolerance in cancer therapies.

Tumor Proliferation

Cell Survival, proliferation and apoptosis processes ensure the regulation of life cycle of a normal tissue and the well-being of an organism. Activation of oncogenes, inactivation of tumor suppressor genes and aberrant mismatch repair machinery promote reprogramming of the metabolic pathways. Cancerous cells adapt to withstand deprivation of oxygen and nutrients so that they could proliferate indefinitely. A war is initiated between cancer and normal cells for the utilization of the limited resources and similar to Darwinian Theory, nature selects the ones more fit for the survival in the extreme environment. The stressed environment of a cell causes activation or deactivation of various proteins to combat the struggle of its survival. HSP70 is one such protein which modifies its own role under normal and stress conditions.

HSP70 is localized both as intracellular and extracellular. Intracellular HSP70 protects the cell and restricts cytokine production, whereas extracellular HSP70 stimulates cytokine production and labels cells for destruction. Intracellular HSP70 leads to tumor development by supporting homeostasis in tumor cells, stabilization of cyclin D1 and suppression of oncogene induced apoptosis. Extracellular and membrane bound HSP70 helps the immune system to destroy the tumor. Natural killer cells recognize HSP70 on tumor cell membrane and kill the cells. The second function depends on the type of receptors on the target cells that bind HSP70. Signaling receptors, such as the toll-like receptor (TLR), confer the ability to HSP70 to activate cytokine production and stimulate the innate response, whereas scavenger receptors help HSP70 to deliver antigens to antigen-presenting cells and therefore stimulate an adaptive response. The other function of HSP70 depends on the circumstances of synthesis and release of HSP70 from the cell. For example, in the case of microbial invasion, HSP70 is involved in the formation of antigen-dependent immune memory, and in the case of different stresses in the formation of antigen-independent immune memory (Guzhova and Margulis 2016).

HSP70 and Gastric Cancer

Helicobacter pylori *Associated GC*

Helicobacter pylori, is a Gram-negative, spiral-shaped microaerophilic bacterium, affecting 50% of the world population and has been classified under class I carcinogen by WHO (Brown 2000). *H. pylori* secretes urease enzyme which provides acid resistance and virulence, making the bacterium survival easy under highly acidic condition of stomach. The inflammatory response towards colonization of the bacteria can lead to gastritis, gastric ulcers, mucosa-associated lymphoid tissue lymphoma and gastric cancer (Kusters et al. 2006). Various studies have cited the interaction of *H. pylori* with HSP70 in the occurrence of gastric cancer. HSP70 facilitates attachment of *H. pylori* to gastric epithelia (Hoffman and Garduno 1999). Infection with *H. pylori* has been found to lower the expression of HSP70 in MKN7 cell lines. Furthermore, attenuation of this expression was observed when cagA and vacA positive *H. pylori* was cultured and incubated for 48 h. Likewise, *H. pylori* strain, negative for cagA and vacA, also showed a time dependent reduced expression, most prominent at 72 h of incubation. Moreover, complete disappearance of mRNA expression of HSP70 took place when cagA was added to cagA and vacA positive *H. pylori*, suggesting that cagA accelerate the inhibitory effect of *H. pylori* on HSP70. However, cagA alone failed to alter the expression of HSP70 (Targosz et al. 2012). Another study showed significant down regulation of the expression of HSP70 in response to *H. pylori* in gastric cell line RGM-1. On the contrary, the expression of HSP70 was upregulated in macrophages, dendritic cells and their exosomes. Furthermore, ELISA experiments corresponding to the above results suggested a potential role of HSP as a diagnostic marker. Upregulation of iNOS-2 and COX-2 in *H. pylori* infected RGM-1, macrophages and dendritic cells suppressed the HSP70 provoking *H. pylori* induced gastric damage and pointing towards the role of *H. pylori* infection in macrophage oxygen/nitrogen metabolism (Yao et al. 2016).

A recent study on association of HSP70 with *H. pylori* infection corroborates with other studies showing significantly downregulated HSP70 expression in GC patients. Patients harboring positive HSP70 expression in their tissues were shown to display worse disease free survival than those with negative HSP70 expression (Bodoor et al. 2016). Prevalence of *H. pylori* infection has been recorded to be higher in patients with gastritis in comparison of gastric cancer patients. Simultaneously, the proportion of cag positive strains is found to be higher in gastritis than gastric cancer (Ferrer-Ferrer et al. 2013). The effect of drug Geranylgeranylactone (GGA), used for the treatment of ulcers and gastritis, has been studied on *H. pylori* infection. It was observed that GGA binds to *H. pylori* HSP70 with 26-fold higher affinity as compared to human HSP70, causing conformational changes that leads to the suppressed activity of *H. pylori* (Grave et al. 2015). According to Yeo et al. (2004) *H. pylori* reduces the expression of HSP70 while exposure of GGA restores the expression of HSP70.

HSP70 Gene Polymorphism

Association of HSP70 gene polymorphism with gastric cancer has been an area of interest to many researchers. A study on Japanese patients including 159 men and 64 women diagnosed with gastric cancer reported association of HSP70-2 with the disease. Study of polymorphism was based on two alleles, P1 and P2, which correspond to the presence and absence of *Pst*I restriction site, respectively. P2/P2 polymorphism of HSP70-2 was proved to have a protective effect showing association with a significant lower risk of gastric cancer in females (Shibata et al. 2009). Polymorphism at allele C in HSP70-1 showed an increasingly significant association from non-atrophic gastritis to gastric cancer in Mexican patients as this polymorphism causes modifications in the expression and location of HSP70 that leads to increased risk of tissue damage and progress to more serious disease. In addition, there was a lack of association of HSP70-2 with GC (Partida-Rodrı'guez et al. 2010). Contrary to this study, Ferrer-Ferrer found association of HSP70-2 with GC. The possible reason for such a discrepancy may be the differences in genetic background of the cases studied. Polymorphism of HSP70-2 and HSP70-Hom were studied in Costa Rica, which is known for highest incidence and mortality rates for GC worldwide. Increased risk of GC was proposed for GA genotype in HSP70-2 and TT genotype in HSP70-Hom as compared to GG genotype and CT genotype (Ferrer-Ferrer et al. 2013).

HSP70 genes are linked to different pathological conditions other than GC also. One such condition, coal workers' pneumoconiosis caused probably by occupational exposure to coal mine dust, has shown susceptibility to polymorphism of HSP70 genes. The polymorphism in three members of HSP70 gene namely HSP70-1, HSP70-2 and HSP70-hom occur at nucleotide position 190, 1267 and 2437 displayed in amino acid substitutions Gla/Arg, Gln/Gln and Met/Thr, respectively. Association of these missense mutations and CWP implied a potential role of HSP70 genes in the disease process. Polymorphism in HSP70-2 and HSP70-hom suggest susceptibility towards development and progression of the disease (Zhang et al. 2011). Polymorphism of HSP70 with another disease, idiopathic pulmonary fibrosis (IPF) was observed and three out of the four SNPs detected were linked with decreased risk of IPF (Aquino-Gálvez et al. 2015).

Study of HSP70 Expression

Various techniques like immunohistochemistry, western blot, flow cytometry, etc. have been used to study the expression of HSP70. Using avidin-biotin method of IHC, HSP70 immunoreactivity was observed in gastric adenocarcinoma and described to be an important prognostic factor in GC in addition to invasion depth, lymph node metastasis, vascular invasion and tumor size (Canoz et al. 2002). Another study analysed the potential of HSP70 as a marker for GC through IHC using streptavidin biotin-peroxidase. Enhanced expression of HSP70 in cancerous tissues in comparison to the surrounding tissue was further validated by western blotting (Isomoto et al. 2003). Similarly, IHC was used to determine the expression level of HSP70 in archival tumor samples (Lee et al. 2013; Bodoor et al. 2016). Upregulation of HSP70 was analysed by immunocytochemistry, immunofluorescence cytochemistry and flow cytometry in GC cell line and was proposed to be related with the proliferation and tumor cell survival (Wang et al. 2005). The expression of HSP70 was observed by flow cytometry and it was concluded that membrane positive phenotype of HSP70 is associated with better prognosis and overall survival in 37% GC patients (Pfister et al. 2007). Suppression of HSP70 by RNA interference leads to the inhibition of tumor cells, cell cycle arrest and apoptosis in gastric cancer (Xiang et al. 2008). Inhibition of HSP70 expression by live strain of *H. pylori* was investigated by RT-PCR and immunoprecipitation (Targosz et al. 2012).

Clinical Studies Involving HSP70

HSP have garnered a great amount of interest in the recent past as a potential target for cancer treatment. Among these, HSP90 and HSP70 happen to be the major players for targeting tumor cells. Clinical trials conducted with HSP90 inhibitors are either in their primary stages or have been terminated, leaving HSP70 an interesting target for tumor therapy. HSP70 has been attributed to be a major cause of resistance of tumor cells against various drugs. It has been shown that salvage pathway occurring via HSP70 activation is one of the major reasons for tumor cells acquiring a resistance for HSP90 inhibitors. Due to overexpression of HSP70 on tumor cells, it has been used in clinical trials for identifying and specifically targeting them by immune system using vaccine based therapies for multiple tumors like NSCLC and cervical cancer (NCT00027144; NCT00121173; NCT00030303; NCT02118415). Direct downregulation of HSP70 is also under scrutiny by inhibiting its promoter, HSF1 (heat shock factor 1). Bioflavonoid quercetin has been found to be one such compound inhibiting the HSF1 activation leading to downregulation of HSP70.

Conclusions

HSP, particularly HSP70, have been implicated in various cancers and their role is being explored in tumor progression and response to therapy. HSP70 participates in different biological processes related with tumorigenesis providing important leads to understand pathogenesis of gastric cancer. The differential expression of HSP70 observed in tumor cells suggests it to be an important target for developing biomarkers and drug therapy. However, more efforts are required in this direction to explore the potential of HSP70 as biomarker useful in the management of gastric cancer.

Acknowledgements PCS acknowledges a faculty research grant and RV acknowledges short term research fellowship of Guru Gobind Singh Indraprastha University, Delhi.

References

Abd, E. A. H., Rey, A., & Duvillard, P. (1998). Expression of heat shock protein 70 and c-myc in cervical carcinoma. *Anticancer Research, 18*, 1533–1536.

Aquino-Gálvez, A., González-Ávila, G., Pérez-Rodríguez, M., Partida-Rodríguez, O., Nieves-Ramírez, M., Piña-Ramírez, I., Ramírez-Martínez, G., Castillejos-López, M., Checa, M., Ruiz, V., & Urrea, F. (2015). Analysis of heat shock protein 70 gene polymorphisms in Mexican patients with idiopathic pulmonary fibrosis. *BMC Pulmonary Medicine, 15*, 129–136.

Bercovich, B., Stancovski, I., Mayer, A., Blumenfeld, N., Laszlo, A., Schwartz, A. L., & Ciechanover, A. (1997). Ubiquitin-dependent degradation of certain protein substrates in vitro requires the molecular chaperone Hsc70. *The Journal of Biological Chemistry, 272*, 9002–9010.

Bodoor, K., Jalboush, S. A., Matalka, I., Abu-Sheikha, A., Al Waqfi, R., Ebwaini, H., Abu-Awad, A., Fayyad, L., Al-Arjat, J., & Haddad, Y. (2016). Heat shock protein association with clinico-pathological characteristics of gastric cancer in Jordan: HSP70 is predictive of poor prognosis. *Asian Pacific Journal of Cancer Prevention, 17*, 3927–3935.

Bonay, M., Soler, P., Riquet, M., Battesti, J. P., Hance, A. J., & Tazi, A. (1994). Expression of heat shock proteins in human lung and lung cancers. *American Journal of Respiratory Cell and Molecular Biology, 10*, 453–461.

Brown, L. M. (2000). *Helicobacter pylori*: Epidemiology and routes of transmission. *Epidemiologic Reviews, 22*, 283–297.

Bruey, J. M., Ducasse, C., Bonniaud, P., Ravagnan, L., Susin, S. A., Diaz-Latoud, C., Gurbuxani, S., Arrigo, A. P., Kroemer, G., Solary, E., & Garrido, C. (2000). Hsp27 negatively regulates cell death by interacting with cytochrome c. *Nature Cell Biology, 2*, 645–652.

Calderwood, S. K., Khaleque, M. A., Sawyer, D. B., & Ciocca, D. R. (2006). Heat shock proteins in cancer: Chaperones of tumorigenesis. *Trends in Biochemical Sciences, 31*, 164–172.

Cancer Genome Atlas Research Network. (2014). Comprehensive molecular characterization of gastric adenocarcinoma. *Nature, 513*, 202–209.

Canöz, Ö., Belenli, O., & Patiroglu, T. E. (2002). General features of gastric carcinomas and comparison of HSP70 and NK cell immunoreactivity with prognostic factors. *Pathology Oncology Research, 8*, 262–269.

Castelli, C., Rivoltini, L., Rini, F., Belli, F., Testori, A., Maio, M., Mazzaferro, V., Coppa, J., Srivastava, P. K., & Parmiani, G. (2004). Heat shock proteins: Biological functions and clinical application as personalized vaccines for human cancer. *Cancer Immunology, Immunotherapy, 53*, 227–233.

Castilla, C., Congregado, B., Conde, J. M., Medina, R., Torrubia, F. J., Japón, M. A., & Sáez, C. (2010). Immunohistochemical expression of Hsp60 correlates with tumor progression and hormone resistance in prostate cancer. *Urology, 76*, 1017e1–1017e6.

Ciocca, D. R., & Calderwood, S. K. (2005). Heat shock proteins in cancer: Diagnostic, prognostic, predictive, and treatment implications. *Cell Stress & Chaperones, 10*, 86–103.

Ciocca, D. R., Stati, A. O., & de Castro, M. M. A. (1990). Colocalization of estrogen and progesterone receptors with an estrogen regulated heat shock protein in paraffin sections of human breast and endometrial cancer tissue. *Breast Cancer Research and Treatment, 16*, 243–251.

Ciocca, D. R., Clark, G. M., Tandon, A. K., Fuqua, S. A., Welch, W. J., & McGuire, W. L. (1993). Heat shock protein hsp70 in patients with axillary lymph node-negative breast cancer: Prognostic implications. *Journal of the National Cancer Institute, 85*, 570–574.

Didelot, C., Lanneau, D., Brunet, M., Joly, A. L., Thonel, A. D., Chiosis, G., & Garrido, C. (2007). Anti-cancer therapeutic approaches based on intracellular and extracellular heat shock proteins. *Current Medicinal Chemistry, 14*, 2839–2847.

Elpek, G. O., Karaveli, S., Simsek, T., Keles, N., & Aksoy, N. H. (2003). Expression of heat-shock proteins hsp27, hsp70 and hsp90 in malignant epithelial tumour of the ovaries. *APMIS, 111*, 523–530.

Feldman, D. E., Spiess, C., Howard, D. E., & Frydman, J. (2003). Tumorigenic mutations in VHL disrupt folding in vivo by interfering with chaperonin binding. *Molecular Cell, 12*, 1213–1224.

Feng, H., Zeng, Y., Whitesell, L., & Katsanis, E. (2001). Stressed apoptotic tumor cells express heat shock proteins and elicit tumor-specific immunity. *Blood, 97*, 3505–3512.

Ferrer-Ferrer, M., Malespín-Bendaña, W., Ramírez, V., González, M. I., Carvajal, A., & Une, C. (2013). Polymorphisms in genes coding for HSP-70 are associated with gastric cancer and duodenal ulcer in a population at high risk of gastric cancer in Costa Rica. *Archives of Medical Research, 44*, 467–474.

Gallucci, S., & Matzinger, P. (2001). Danger signals: SOS to the immune system. *Current Opinion in Immunology, 13*, 114–119.

Gehrmann, M., Schmetzer, H., Eissner, G., Haferlach, T., Hiddemann, W., & Multhoff, G. (2003). Membrane-bound heat shock protein 70 in acute myeloid leukemia: A tumor-specific recognition structure for the cytolytic activity of autologous natural killer cells. *Haematologica, 88*, 474–476.

Glaessgen, A., Jonmarker, S., Lindberg, A., Nilsson, B. O., Lewensohn, R., Ekman, P., Valdman, A., & Egevad, L. (2008). Heat shock proteins 27, 60 and 70 as prognostic markers of prostate cancer. *APMIS, 116*, 888–895.

Graner, M., Raymond, A., Romney, D., He, L., Whitesell, L., & Katsanis, E. (2000). Immunoprotective activities of multiple chaperone proteins isolated from murine B-cell leukemia/lymphoma. *Clinical Cancer Research, 6*, 909–915.

Graner, M. W., Zeng, Y., Feng, H., & Katsanis, E. (2003). Tumor-derived chaperone-rich cell lysates are effective therapeutic vaccines against a variety of cancers. *Cancer Immunology, Immunotherapy, 52*, 226–234.

Grave, E., Yokota, S. I., Yamamoto, S., Tamura, A., Ohtaki-Mizoguchi, T., Yokota, K., Oguma, K., Fujiwara, K., Ogawa, N., Okamoto, T., & Otaka, M. (2015). Geranylgeranylacetone selectively binds to the HSP70 of *Helicobacter pylori* and alters its coccoid morphology. *Scientific Reports, 5*, 13738.

Gress, T. M., Müller-Pillasch, F., Weber, C., Lerch, M. M., Friess, H., Büchler, M., Beger, H. G., & Adler, G. (1994). Differential expression of heat shock proteins in pancreatic carcinoma. *Cancer Research, 54*, 547–551.

Guzhova, I. V., & Margulis, B. A. (2016). HSP70-based anti-cancer immunotherapy. *Human Vaccines & Immunotherapeutics, 12*, 2529–2535.

Hartl, F. U. (1996). Molecular chaperones in cellular protein folding. *Nature, 381*, 571.

Hartl, F. U., Martin, J., & Neupert, W. (1992). Protein folding in the cell: The role of molecular chaperones Hsp70 and Hsp60. *Annual Review of Biophysics and Biomolecular Structure, 21*, 293–322.

Hendrick, J. P., & Hartl, F. U. (1993). Molecular chaperone functions of heat-shock proteins. *Annual Review of Biochemistry, 62*, 349–384.

Hinds, P. W., Finlay, C. A., Frey, A. B., & Levine, A. J. (1987). Immunological evidence for the association of p53 with a heat shock protein, hsc70, in p53-plus-ras-transformed cell lines. *Molecular and Cellular Biology, 7*, 2863–2869.

Hoang, A. T., Huang, J., Rudra-Ganguly, N., Zheng, J., Powell, W. C., Rabindran, S. K., Wu, C., & Roy-Burman, P. (2000). A novel association between the human heat shock transcription factor 1 (HSF1) and prostate adenocarcinoma. *The American Journal of Pathology, 156*, 857–864.

Hoffman, P. S., & d Garduno, R. A. (1999). Surface-associated heat shock proteins of *Legionella pneumophila* and *Helicobacter pylori*: Roles in pathogenesis and immunity. *Infectious Diseases in Obstetrics and Gynecology, 7*, 58–63.

Isomoto, H., Oka, M., Yano, Y., Kanazawa, Y., Soda, H., Terada, R., Yasutake, T., Nakayama, T., Shikuwa, S., Takeshima, F., & Udono, H. (2003). Expression of heat shock protein (Hsp) 70 and Hsp 40 in gastric cancer. *Cancer Letters, 198*, 219–228.

Janetzki, S., Palla, D., Rosenhauer, V., Lochs, H., Lewis, J. J., & Srivastava, P. K. (2000). Immunization of cancer patients with autologous cancer-derived heat shock protein gp96 preparations: A pilot study. *International Journal of Cancer, 88*, 232–238.

Jindal, S. (1996). Heat shock proteins: Applications in health and disease. *Trends in Biotechnology, 14*, 17–20.

Juhasz, K., Lipp, A. M., Nimmervoll, B., Sonnleitner, A., Hesse, J., Haselgruebler, T., & Balogi, Z. (2014). The complex function of hsp70 in metastatic cancer. *Cancer, 6*, 42–66.

Kapoor, C., & Vaidya, S. (2013). Heat shock protein (HSP) and cancer: An overview. *American Journal of Medical Dental Sciences, 1*, 31–34.

Kaur, J., Das, S. N., Srivastava, A., & Ralhan, R. (1998). Cell surface expression of 70 kDa heat shock protein in human oral dysplasia and squamous cell carcinoma: Correlation with clinicopathological features. *Oral Oncology, 34*, 93–98.

Kaur, J., Kaur, J., & Ralhan, R. (2000). Induction of apoptosis by abrogation of HSP70 expression in human oral cancer cells. *International Journal of Cancer, 85*, 1–5.

Khalil, A. A., Kabapy, N. F., Deraz, S. F., & Smith, C. (2011). Heat shock proteins in oncology: Diagnostic biomarkers or therapeutic targets? *Biochimica et Biophysica Acta, 1816*, 89–104.

Kim, H. J., Eun, J. Y., Jeon, Y. W., Yun, J., Kim, K. H., Kim, S. H., Kim, H. J., Lee, S. C., Bae, S. B., Kim, C. K., & Lee, N. S. (2011). Efficacy and safety of oxaliplatin, 5-fluorouracil, and folinic acid combination chemotherapy as first-line treatment in metastatic or recurrent gastric cancer. *Cancer Research and Treatment, 43*, 154–159.

Kim, T. K., Na, H. J., Lee, W. R., Jeoung, M. H., & Lee, S. (2016). Heat shock protein 70-1A is a novel angiogenic regulator. *Biochemical and Biophysical Research Communications, 469*, 222–228.

Kusters, J. G., van Vliet, A. H., & Kuipers, E. J. (2006). Pathogenesis of *Helicobacter pylori* infection. *Clinical Microbiology Reviews, 19*, 449–490.

Lazaris, A. C., Theodoropoulos, G. E., Davaris, P. S., Panoussopoulos, D., Nakopoulou, L., Kittas, C., & Golematis, B. C. (1995). Heat shock protein 70 and HLA-DR molecules tissue expression. *Diseases of the Colon and Rectum, 38*, 739–745.

Lee, H. W., Lee, E. H., Kim, S. H., Roh, M. S., Jung, S. B., & Choi, Y. C. (2013). Heat shock protein 70 (HSP70) expression is associated with poor prognosis in intestinal type gastric cancer. *Virchows Archiv, 463*, 489–495.

Liu, H., Lu, J., Hua, Y., Zhang, P., Liang, Z., Ruan, L., Lian, C., Shi, H., Chen, K., & Tu, Z. (2015). Targeting heat-shock protein 90 with ganetespib for molecularly targeted therapy of gastric cancer. *Cell Death & Disease, 6*, e1595.

Macario, A. J., Cappello, F., Zummo, G., & Conway de Macario, E. (2010). Chaperonopathies of senescence and the scrambling of interactions between the chaperoning and the immune systems. *Annals of the New York Academy of Sciences, 1197*, 85–93.

Maehara, Y., Oki, E., Abe, T., Tokunaga, E., Shibahara, K., Kakeji, Y., & Sugimachi, K. (2000). Overexpression of the heat shock protein HSP70 family and p53 protein and prognosis for patients with gastric cancer. *Oncology, 58*, 144–151.

Manjili, M. H., Wang, X. Y., Park, J., Facciponte, J. G., Repasky, E. A., & Subjeck, J. R. (2002). Immunotherapy of cancer using heat shock proteins. *Frontiers in Bioscience, 7*, 43–52.

Mayer, M. P., & Bukau, B. (2005). Hsp70 chaperones: Cellular functions and molecular mechanism. *Cellular and Molecular Life Sciences, 62*, 670–684.

Melville, M. W., McClellan, A. J., Meyer, A. S., Darveau, A., & Frydman, J. (2003). The Hsp70 and TRiC/CCT chaperone systems cooperate in vivo to assemble the von Hippel-Lindau tumor suppressor complex. *Molecular and Cellular Biology, 23*, 3141–3151.

Multhoff, G., Botzler, C., Wiesnet, M., Müller, E., Meier, T., Wilmanns, W., & Issels, R. D. (1995). A stress-inducible 72-kDa heat-shock protein (HSP72) is expressed on the surface of human tumor cells, but not on normal cells. *International Journal of Cancer, 61*, 272–279.

Murphy, M. E. (2013). The HSP70 family and cancer. *Carcinogenesis, 34*, 1181–1188.

Nanbu, K., Konishi, I., Komatsu, T., Mandai, M., Yamamoto, S., Kuroda, H., Koshiyama, M., & Mori, T. (1996). Expression of heat shock proteins HSP70 and HSP90 in endometrial carcinomas: Correlation with clinicopathology, sex steroid receptor status, and p53 protein expression. *Cancer, 77*, 330–338.

Neckers, L., & Workman, P. (2012). Hsp90 molecular chaperone inhibitors: Are we there yet? *Clinical Cancer Research, 18*, 64–76.

Nollen, E. A., & Morimoto, R. I. (2002). Chaperoning signaling pathways: Molecular chaperones as stress-sensing heat shock proteins. *Journal of Cell Science, 115*, 2809–2816.

Park, S. L., Chung, T. W., Kim, S., Hwang, B., Kim, J. M., Lee, H. M., Cha, H. J., Seo, Y., Choe, S. Y., Ha, K. T., & Kim, G. (2017). HSP70-1 is required for interleukin-5-induced angiogenic responses through eNOS pathway. *Scientific Reports, 7*, 44687.

Partida-Rodríguez, O., Torres, J., Flores-Luna, L., Camorlinga, M., Nieves-Ramírez, M., Lazcano, E., & Perez-Rodríguez, M. (2010). Polymorphisms in TNF and HSP-70 show a significant association with gastric cancer and duodenal ulcer. *International Journal of Cancer, 126*, 1861–1868.

Pfister, K., Radons, J., Busch, R., Tidball, J. G., Pfeifer, M., Freitag, L., Feldmann, H. J., Milani, V., Issels, R., & Multhoff, G. (2007). Patient survival by Hsp70 membrane phenotype. *Cancer, 110*, 926–935.

Pockley, A. G. (2003). Heat shock proteins as regulators of the immune response. *Lancet, 362*, 469–476.

Pratt, W. B. (1993). The role of heat shock proteins in regulating the function, folding, and trafficking of the glucocorticoid receptor. *The Journal of Biological Chemistry, 268*, 21455–21455.

Ranford, J. C., & Henderson, B. (2002). Chaperonins in disease: Mechanisms, models, and treatments. *Molecular Pathology, 55*, 209–213.

Salvador, N., Aguado, C., Horst, M., & Knecht, E. (2000). Import of a cytosolic protein into lysosomes by chaperone-mediated autophagy depends on its folding state. *The Journal of Biological Chemistry, 275*, 27447–27456.

Samali, A., & Cotter, T. G. (1996). Heat shock proteins increase resistance to apoptosis. *Experimental Cell Research, 223*, 163–170.

Schild, H., Arnold-Schild, D., Lammert, E., & Rammensee, H. G. (1999). Stress proteins and immunity mediated by cytotoxic T lymphocytes. *Current Opinion in Immunology, 11*, 109–113.

Schmitt, E., Gehrmann, M., Brunet, M., Multhoff, G., & Garrido, C. (2007). Intracellular and extracellular functions of heat shock proteins: Repercussions in cancer therapy. *Journal of Leukocyte Biology, 81*, 15–27.

Seigneuric, R., Mjahed, H., Gobbo, J., Joly, A. L., Berthenet, K., Shirley, S., & Garrido, C. (2011). Heat shock proteins as danger signals for cancer detection. *Frontiers in Oncology, 1*, 37–46.

Sessa, C., Shapiro, G. I., Bhalla, K. N., Britten, C., Jacks, K. S., Mita, M., Papadimitrakopoulou, V., Pluard, T., Samuel, T. A., Akimov, M., & Quadt, C. (2013). First-in-human phase I dose-escalation study of the HSP90 inhibitor AUY922 in patients with advanced solid tumors. *Clinical Cancer Research, 19*, 3671–3680.

Shibata, T., Arisawa, T., Tahara, T., Yoshioka, D., Maruyama, N., Fujita, H., Kamiya, Y., Nakamura, M., Nagasaka, M., Iwata, M., & Takahama, K. (2009). Protective role of genetic polymorphism of heat shock protein 70-2 for gastric cancer risk. *Digestive Diseases and Sciences, 54*, 70–74.

Stewart, B. W., & Wild, C. P. (2017). *World Cancer Report 2014*. Lyon: WHO/IARC Publications.

Takayama, S., Reed, J. C., & Homma, S. (2003). Heat-shock proteins as regulators of apoptosis. *Oncogene, 22*, 9041–9047.

Targosz, A., Brzozowski, T., Pierzchalski, P., Szczyrk, U., Ptak-Belowska, A., Konturek, S. J., & Pawlik, W. (2012). *Helicobacter pylori* promotes apoptosis, activates cyclooxygenase (COX)-2 and inhibits heat shock protein HSP70 in gastric cancer epithelial cells. *Inflammation Research, 61*, 955–966.

Vilasi, S., Carrotta, R., Mangione, M. R., Campanella, C., Librizzi, F., Randazzo, L., Martorana, V., Gammazza, A. M., Ortore, M. G., Vilasi, A., & Pocsfalvi, G. (2014). Human Hsp60 with its mitochondrial import signal occurs in solution as heptamers and tetradecamers remarkably stable over a wide range of concentrations. *PLoS One, 9*, e97657.

Wang, X. P., Liao, J., Liu, G. Z., Wang, X. C., & Shang, H. W. (2005). Co-expression of heat shock protein 70 and glucose-regulated protein 94 in human gastric carcinoma cell line BGC-823. *World Journal of Gastroenterology, 11*, 3601–3604.

Welch, W. J. (1992). Mammalian stress response: Cell physiology, structure/function of stress proteins, and implications for medicine and disease. *Physiological Reviews, 72*, 1063–1081.

Witkin, S. S. (2001). Heat shock protein expression and immunity: Relevance to gynecologic oncology. *European Journal of Gynaecological Oncology, 22*, 249–256.

Xiang, T. X., Li, Y., Jiang, Z., Huang, A. L., Luo, C., Zhan, B., Wang, P. L., & Tao, X. H. (2008). RNA interference-mediated silencing of the Hsp70 gene inhibits human gastric cancer cell growth and induces apoptosis in vitro and in vivo. *Tumori, 94*, 539–550.

Yao, Y., Wu, J., Gu, T., Cheng, Y., & Li, G. (2016). Comparative analysis of the interaction of HSPs in dendritic cells, macrophages, RGM-1 cells infected by Helicobacter pylori. *American Journal of Translational Research, 8*, 4184–4194.

Yeo, M., Park, H. K., Kim, D. K., Cho, S. W., Kim, Y. S., Cho, S. Y., Paik, Y. K., & Hahm, K. B. (2004). Restoration of heat shock protein70 suppresses gastric mucosal inducible nitric oxide synthase expression induced by *Helicobacter pylori*. *Proteomics, 4*, 3335–3342.

Zhang, H., Jin, T., Zhang, G., Chen, L., Zou, W., & Li, Q. Q. (2011). Polymorphisms in heat-shock protein 70 genes are associated with coal workers' pneumoconiosis in southwestern China. *In Vivo, 25*, 251–257.

Part II
Hsp70 in Circulatory Disorders

Double Face of eHsp70 in Front of Different Situations

Multiple Role of eHsp70

Maria M. Barreca and Fabiana Geraci

Abstract The Hsp70 family is one of the best conserved and abundant member of the heat shock proteins (HSP). This family includes several members and in particular one constitutively expressed member (Hsc70) and another one inducibly expressed under several stress conditions (Hsp70). To date, the intracellular functions of Hsp70 are well defined, and increasing evidences establish its roles in the extracellular environment, such as cytoprotection and immunomodulation. Increasing evidences suggest that several cell types are able to release Hsp70 in the extracellular environment, both under physiological and stress conditions. At the same time many release mechanisms have been identified. This chapter briefly reviews recent advances in our understanding on extracellular Hsp70 role in both physiological and pathological conditions. A better comprehension will be useful to take advantage of its potential as a therapeutic target.

Keywords eHsp70 · Hsp70 export · Cellular receptors · Immune response · Cell migration

---

M. M. Barreca
Department of Biological, Chemical and Pharmaceutical Sciences and Technologies, University of Palermo, Palermo, Italy

F. Geraci (✉)
Department of Biological, Chemical and Pharmaceutical Sciences and Technologies, University of Palermo, Palermo, Italy

Euro-Mediterranean Institute of Science and Technology, Palermo, Italy
e-mail: fabiana.geraci@unipa.it

A. A. A. Asea, P. Kaur (eds.), *HSP70 in Human Diseases and Disorders*, Heat Shock Proteins 14, https://doi.org/10.1007/978-3-319-89551-2_7

Abbreviations

APC	antigen presenting cell
CNS	central nervous system
CSF	cerebrospinal fluid
CTL	cytotoxic T lymphocyte
DC	dendritic cells
EAE	experimental autoimmune encephalomyelitis
ECM	extracellular matrix
ER	endoplasmic reticulum
EV	extracellular vesicle
HSP	heat shock protein
HUVEC	human umbilical vein endothelial cell
LPS	lipopolysaccharide
MBP	myelin basic protein
MHC	major histocompatibility complex;
MS	multiple sclerosis
NF-kB	nuclear factor kB
NK	natural killer
PBL	peripheral blood lymphocyte
ROS	radical oxygen species
TLR	toll like receptor

Introduction

Heat shock proteins are highly conserved molecular chaperones involved in proper folding of newly synthesized proteins. They are also able to interact with naïve and denatured proteins to avoid inappropriate interactions and the formation of protein aggregates of aberrantly folded proteins (Liberek et al. 2008). Furthermore, HSP facilitate protein translocation, exhibit cytoprotective (Sharp et al. 1999; Giffard et al. 2004) and antiapoptosis functions (Martin et al. 1992; Aquino et al. 1993; Gao et al. 1995; Aquino et al. 1997), directly interacting with various components of the tightly regulated programmed cell death machinery, upstream and downstream of the mitochondrial events (Garrido et al. 2001; Beere 2004; Madden et al. 2008). More recently, it has been demonstrated that HSP are also involved in controlling cell signaling (Calderwood et al. 2007), in modulation of both immune response (Johnson and Fleshner 2006) and chronic disease conditions (Kampinga et al. 2007). Furthermore, HSP has cytostimolatory functions (Asea 2008). Almost all the HSP families consist of constitutively expressed members playing housekeeping roles and stress-induced members, which display a crucial role in recovery after different cellular stresses (e.g. heat shock, ultraviolet radiation, viral or bacterial infections, oxidative stress, ischemia, exercise, metabolic stress, heavy metals and so on) (Collins and Hightower 1982; Lindquist 1986; Patel et al. 1995; Richard et al. 1996; Yang et al. 1996; Feder and Hofmann 1999; Said Ali et al. 2010; Pierzchalski et al. 2014). For many years heat shock proteins were thought to be exclusive cytoplasmic proteins, acting only in intracellular compartments. The biology of both intracellular constitutive and inducible HSP has been extensively summarized by a large number of reviews and it will be not treated anymore (Lindquist and Craig 1988; Morimoto 1991; Bukau et al. 2006; Hartl and Hayer-Hartl 2009).

A new twist in the stress field was given by the finding that HSP could be also extracellular protein. In fact, it is now widely accepted that almost all the HSP are released in the extracellular environment, exerting multiple effects on different cell types (Tytell 2005). For example, in 1986 Tytell and coworkers identified an HSP-like protein as a glia-axon transfer protein in squid giant axon (Tytell et al. 1986). In addition, Hightower and Guidon found that Hsp70 is released extracellularly by a mechanism different from the classical secretion endoplasmic reticulum (ER)-Golgi pathway (Hightower and Guidon Jr 1989), as later confirmed by Hunter-Lavin et al. (Hunter-Lavin et al. 2004). Therefore, HSP could have a dual function depending on their intracellular or extracellular location. In the past years heat shock proteins were classified into different families depending on their molecular mass (i.e., Hsp110, Hsp90, Hsp70, Hsp60, Hsp40 and the small HSP). They comprise members with different inducibility and intracellular localization (Feder and Hofmann 1999). In 2009, Kampinga and coworkers proposed new guidelines for human HSP nomenclature (Kampinga et al. 2009). However, as most literature refers to HSP with their old name we decided to maintain the old nomenclature.

Hsp70 Family

One of the most conserved and abundant subsets of HSP is the Hsp70 family (Muchowski and Wacker 2005; Noble et al. 2008), consisting of at least four distinct proteins, two are cytosolic/nuclear, whereas the other two are localized within mitochondria and endoplasmic reticulum (ER) (Kregel 2002; Voos 2013). Furthermore, cytosolic Hsp70 includes the stress inducible form Hsp70 (Voellmy 2004) and the constitutively expressed Hsc70. Although Hsp70 synthesis is induced by several types of stressors (Wu et al. 1985; Milarski and Morimoto 1989), it is now well accepted that a basal concentration of Hsp70 is present in many tissues without any stress. Indeed, in addition to their cytoprotective properties member of the Hsp70 families are involved in physiological processes, such as cell differentiation, cell maturation and proliferation (Luft and Dix 1999; Lui and Kong 2007).

Extracellular Hsp70 and Its Release Mechanisms

As for other HSP (Chabas et al. 2001; Stadelmann et al. 2005), it has been demonstrated that Hsp70, in both basal and stress-induced conditions, is not only an intracellular (iHsp70) but also an extracellular protein (eHsp70), functioning as intercellular-signaling ligand (Asea et al. 2000; Asea 2003; Calderwood et al. 2007; Turturici et al. 2011). Indeed, it has been demonstrated that a variety of cell types, including neural cells (Guzhova et al. 2001; Taylor et al. 2007), epithelial cells (Broquet et al. 2003), embryo cells (Hightower and Guidon Jr 1989), B lymphocytes and dendritic cells (Théry et al. 1999, Clayton et al. 2005), maturing erythrocytes (Mathew et al. 1995) and tumor cells (Gastpar et al. 2005) is able to release Hsp70 in the culture medium.

Early studies on eHsp70 hypothesized that this protein was released by necrotic or dead cells, due to its aminoacidic sequence, and in particular to its lack of any exocytosis signals, (Gallucci et al. 1999, Basu et al. 2000; Saito et al. 2005). On the contrary, it is now well known that eHsp70 is also actively released by mammalian cultured cells through non classical secretory pathways, excluding the ER/Golgi compartment (Hightower and Guidon Jr 1989; Broquet et al. 2003), as already demonstrated for other proteins that lack secretion leader sequence. eHsp70 was also detected in the peripheral circulation of either healthy subjects or in several pathological states (Pockley et al. 1998, 2002; 2003; Asea 2007; Giraldo et al. 2008). All these data demonstrated that there are at least two different methods of Hsp70 release: one active due to an unconventional secretory pathway (Nickel and Seedorf 2008), and the other one passive, due to cell death and subsequent lysis. One of the possible active Hsp70 release pathway is similar to that identified for IL-1β (Eder et al. 2008). Specifically, Mambula and Calderwood demonstrated the involvement of a lysosome-endosome pathway, requiring the ABC-1 transporter (Mambula and Calderwood 2006). Hsp70 has also been proposed to be released by secretory-like granules (Evdonin et al. 2006), typical of specialized endocrine and exocrine cells, but already identified in non-secretory cells (Beuret et al. 2004). Another interesting idea images that Hsp70 release was dependent on its initial insertion into plasma membrane, depending on phosphatidylserine presence, membrane fluidity and, for certain cell types, lipid raft integrity (Triantafilou et al. 2002; Broquet et al. 2003; Arispe et al. 2004; Hunter-Lavin et al. 2004; Chen et al. 2005; Wang et al. 2006a; Horvath et al. 2008; Vega et al. 2008; Evdokimovskaya et al. 2010). The importance of Hsp70 insertion in cellular membrane is confirmed by the finding of Hsp70 containing extracellular vesicles (EVs) in cell culture medium, even in the absence of detectable cell death (Hunter-Lavin et al. 2004; Gastpar et al. 2005; Vega et al. 2008). Two principal EV types, i.e. exosomes and membrane vesicles, with different cellular origins have been demonstrated to be involved in Hsp70 release in the extracellular environment. Exosomes are nanovesicles derived from the multive-

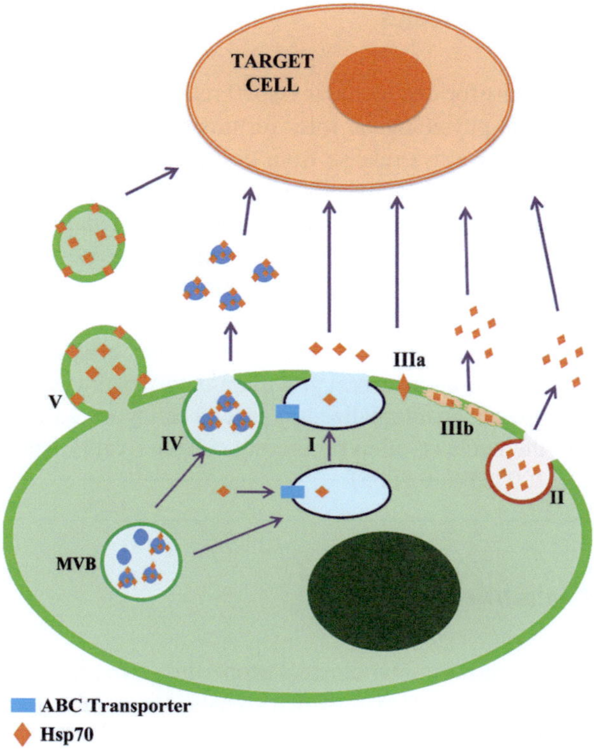

Fig. 1 Mechanism of Hsp70 release from cells. Hsp70 is able to leave cells using different way: *I* through endolysosome pathway; *II* by using secretory-like pathways; *IIIa* it can penetrate through the lipid bilayer via phospatidylserine *IIIb* or through of lipid raft interaction; *IV* it can be released inside exosomes; *V* or within membrane vesicles. In the extracellular milieu Hsp70 can interact with several cell types (target cells)

sicular endosomal cell compartment (Heijnen et al. 1999; Fevrier and Raposo 2004; Théry 2011). Several authors have demonstrated that *in vitro* different cell types are able to release Hsp70 in the extracellular environment by an unconventional pathway involving exosomes (Bausero et al. 2005; Gastpar et al. 2005; Lancaster and Febbraio 2005; Zhan et al. 2009). Indeed, for many years Hsp70/Hsc70 have been considered as exosome markers (Mathivanan and Simpson 2009). Contrary to this idea, for the first time Barreca et al. in 2017 demonstrated that mesoangioblasts, mouse vessel associated progenitor stem cells, are able to release Hsp70 in the extracellular milieu, even under basal growth conditions, by EVs originating directly from the plasmatic membrane (Barreca et al. 2017). Schematic model of different Hsp70 release mechanisms is represented in Fig. 1.

Extracellular Hsp70 Roles

As several mechanisms for extracellular Hsp70 release have been identified, there should be also several physiological roles for this protein in the external environment. In particular, its passive release from dying cells is considered as a danger signal. On the contrary, Hsp70 active release from living cells could indicate a successful stress response, and eHsp70 could have an active signaling role. An intriguing aspect of Hsp70 roles is its ability to induce antagonistic events, depending on its localization (Rodrigues-Krause et al. 2012). For example, iHsp70 exerts a strong anti-inflammatory effects through the interaction with the nuclear factor kB (NF-kB), blocking its activation (Jones et al. 2011), whereas eHsp70 has the opposite role, inducing the activation of several proinflammatory pathways. To date, there is evidence that eHsp70 can be internalized by several cell types, localizing both in the cytoplasm and in the nucleus, often promoting cell survival (Guzhova et al. 2001; Novoselova et al. 2005; Tytell 2005).

eHsp70 and Immune Responses

eHsp70 possesses powerful immunological properties. Two are the possible origins of eHsp70: pathogen derived Hsp70, which signals a local infection, and mammalian Hsp70, usually indicating an intracellular trauma. Hsp70, as other HSP associates itself with a broad variety of peptides, generated within the cells (Zhu et al. 1996). These peptides consist of normal self-peptides as well as antigenic peptides (i.e., tumor, bacterial and viral antigens) (Nieland et al. 1996; Castelli et al. 2001; Zugel et al. 2001). Several studies have demonstrated that eHsp70 is able to initiate both innate and adaptive immunity response as illustrated in Fig. 2 (Johnson and Fleshner 2006; Tsan and Gao 2009). For example, it is involved in the activation of cells of the innate immune pathway, such as macrophages, monocytes, which are directly induced in cytokine synthesis, neutrophils, dendritic cells (DCs), natural killer (NK) cells (Asea et al. 2000; Basu et al. 2001; Vabulas et al. 2002a; Gastpar et al. 2004; Aneja et al. 2006; Kovalchin et al. 2006; Wang et al. 2006a; Vega et al. 2008). In addition, eHsp70 also increases microbicidal capacity and chemiotaxis of neutrophils (Ortega et al. 2006, 2009) and phagocytosis (Wang et al. 2006a).

Phagocytosis characterizes the innate immune response and is initiated and executed by the antigen presenting cells (APCs). Some of these cells, including macrophages, are activated either by the presence of pathogens, or by physical, chemical stress, trauma and so on. All these conditions are responsible for HSP, and especially Hsp70, release in the extracellular environment. Immunologically, eHsp70 binds different receptors on macrophages membrane regulating several functions, such as cytokine release, phagocytosis, tumor-rejection and upregulation of co-stimulatory molecules (Basu et al. 2001; Delneste et al. 2002; Srivastava et al. 2002; Vabulas et al. 2002b; Asea 2005). Wang et al. demonstrated that eHsp70 is able to

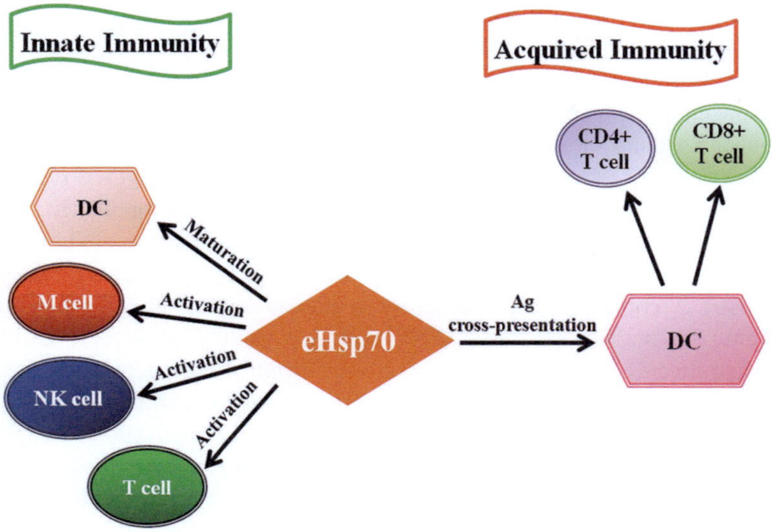

Fig. 2 Schematic model of eHsp70 immune response activation. eHsp70 is able to initiate both innate and adaptive immunity. On the one end, eHsp70 is involved in the activation of cells belonging to the innate immune pathway, such as macrophages, DCs and NK cells, directly induced in cytokine synthesizing (chaperokine role). On the other end, eHsp70 can acts as chaperone of antigens, which are transferred to APCs, processed, and cross-presented to T cells, inducing a strong T cell activation (chaperone role)

enhance murine macrophage-mediated antigen uptake. This process is concentration and time dependent. They also showed that eHsp70 mediated macrophage activation depends on actin cytoskeleton reorganization, and is not influenced by de novo protein synthesis. All these observations have suggested to the authors that the process of eHsp70 mediated phagocytosis occurs through the activation of a short signaling pathway, not involving gene upregulation. Moreover, they proved that eHsp70 and macrophages interaction occurs on lipid raft microdomains of macrophage plasmamembrane. Indeed, lipid raft disruption by cholesterol removal partially abrogates eHsp70 effects on phagocytosis. Finally, eHsp70 mediated phagocytosis enhances antigenic processing and presentation to $CD4^+$ T lymphocytes in a MHC II limited manner (Wang et al. 2006a). The involvement of eHsp70 in macrophage activation was also demonstrated by Vega and collaborators, confirming the paracrine role in cytokine expression and secretion by macrophages, already observed by Svensson and coworkers (Svensson et al. 2006). In particular they demonstrated that extracellular membrane bound Hsp70 activated macrophages that produced higher level of TNF-α in comparison with unstimulated cells. Furthermore, TNF-α production after stimulation with membrane positive eHsp70 was much higher than that observed with recombinant or purified Hsp70. According to these results the authors hypothesized that the membrane environment makes eHsp70 a better target to be recognized by macrophages that can engulf it via phagocytosis. On the other end, it is possible that eHsp70 insertion within extracellular membrane is responsible for its multimerization, making it more active in triggering a response by macrophages (Vega et al. 2008).

In vivo wound healing experiments carried out by injecting into naïve mice mimicking a wounding model, macrophages pretreated with Hsp70 showed an enhanced macrophage wound closure ability, compared to buffer treated cells (Kovalchin et al. 2006). It is well known that macrophage-mediated phagocytosis plays a fundamental role in wound healing by clearing debris and neutrophils. Kovalchin's group results indeed demonstrated that eHsp70 mediates some of its action on wound closure through the regulation of macrophage phagocytic activity (Kovalchin et al. 2006). According to HSP capability to interact with other cell types, such as neutrophils and platelets (Hilf et al. 2002; Radsak et al. 2003), it is also possible that eHsp70 stimulates wound closure not only by stimulating macrophage-mediated phagocytosis, but also altering the wound milieu via cytokine release.

The extracellular chaperone is also responsible for DC activation through their surface binding. In their immature form these cells are prevalently involved in antigen capture and processing, whereas mature DCs are potent APCs. Indeed, when their maturation is induced, they become potent stimulator of naïve T cells (Banchereau and Steinman 1998). According to this dual role, DCs represent a connection between the adaptive and the innate immune system. Therefore, in its function of DC activation, eHsp70 may stimulate a cross-talk between the two immune responses (Srivastava 2002). In particular, Kuppner et al. have observed that eHsp70 bound and stimulated maturation of immature differentiated DCs, but reduced their maturation from monocyte precursor cells, probably due to its incapacity to interact with these cells. In addition, DCs matured by eHsp70 increased cell proliferation level of PBL and showed an enlarged capability to stimulate IFN-γ synthesis from specifically stimulated CTL (Kuppner et al. 2001). DCs are also able to bind tumor eHsp70-peptide complex, which is endocyted through a receptor mediated process and cross-presented to MHC I (Noessner et al. 2002). Milani et al. demonstrated that eHsp70-peptide complex is not only involved in providing to DCs the peptide for presentation for T cell stimulation, but also stimulates DCs to release TNF-α inducing by an autocrine loop their maturation (Milani et al. 2002).

All the presented data indicate that eHsp70, acting as a danger signal to the innate immune system, on one hand induces: a cascade of proinflammatory cytokines (e.g. TNF-α, IL-1-IL-6, Il-12); NO and C-C chemokines release by monocytes, macrophages and DCs (Asea et al. 2000, 2002; Chen et al. 2002; Panjwani et al. 2002; Asea 2006); the stimulation of DCs and the activation of NK cells. Qiao et al. demonstrated a positive switch of the DC-NK cells cross activation. In particular, they found that soluble Hsp70, which *in vivo* is released as results of pathological cell stress and death (Basu et al. 2000), could amplify DC-NK cell interaction, leading to a higher IFN-γ response (Qiao et al. 2008). On the other hand, eHsp70 acts as chaperone of bound peptides, which are transferred to DCs, also induced to mature, and cross-presented to T cells, inducing stronger cell activation compared with antigenic peptide alone (Li et al. 2002; Bendz et al. 2007).

In the last years various papers have demonstrated that HSP-reactive T cells have an immunoregulatory phenotype, suggesting that these proteins, and in particular

Hsp60 and Hsp70, could activate immunoregulatory pathways, which can suppress the immune response occurring in human inflammatory diseases. In fact, several data highlighted that in various inflammatory diseases, eHsp70 exerts under certain circumstances also immunoregulatory and suppressive functions. (Wieten et al. 2007). In various autoimmune models, such as rheumatoid arthritis and diabetes, eHsp70 can also downregulate the immune response through its ability to stimulate T cells (Th2 and Tregs) to produce IL-10, the main anti-inflammatory and immunosuppressive cytokine (Moore et al. 2001; Borges et al. 2012; Stocki and Dickinson 2012). This process, depending on the form of eHsp70 (i.e., associated with peptide, with membranes, with nucleotides, peptide free) and the receptor activated (Li et al. 2012) shifts cell phenotype towards a tolerogenic one exhibiting anti-inflammatory properties. It has been demonstrated that several adaptive mechanisms could be involved in eHsp70 specific induced Tregs. In particular, peptides derived from eHsp70, either pathogen associated or secreted endogenous Hsp70, can be presented to MHC II on IFN stimulated APCs or non APCs. Most Tregs subset induce IL-10, responsible for inflammatory response suppression (Wendling et al. 2000; Prakken et al. 2001; Bluestone 2005). Another example of the anti-inflammatory eHsp70 role was observed in a mouse model of allergic bronchial asthma inflammatory process. eHsp70 treated mice showed a significant decrease in Th2 associated cytokines IL-4, IL-5 and IL-13 secretion compared to untreated mice. In addition, it has been observed that eHsp70 administration promoted survival of neutrophils at the site of inflammation, and reduced the influx of the eosinophils to the lungs (Shevchenko et al. 2016).

Under certain circumstances also crossreaction with self-Hsp70 induce T cells to synthesize of IL-10 (Tanaka et al. 1999; Wendling et al. 2000). Indeed, in animal models of experimental arthritis T cell reactivity to self Hsp70 downregulates inflammation developing Th2 $CD4^+$ T cells producing the regulatory cytokines IL-4 and IL-10 (Tanaka et al. 1999; Wendling et al. 2000; Wieten et al. 2007). Another anti-inflammatory role for eHsp70 was observed in rheumatoid arthritis. In particular, eHsp70 exerts an anti-inflammatory role on fibroblast-like synoviocytes inhibiting TNF-α induced IL-6, IL-8 and MCP-1 secretion. The silencing of this inflammatory response is due to the downregulation of NF-kB nuclear translocation (Luo et al. 2008). Immunization with the recognized bacterial Hsp70 peptide protects rats against adjuvant-induced arthritis (Tanaka et al. 1999). A protective effect from experimentally induced arthritis was also observed in rats pre-immunized with mycobacterial Hsp70 (Wendling et al. 2000; Prakken et al. 2001).

eHsp70 Role in Cancer Immunity

The eHsp70 or the one expressed on the tumor cell surface have been shown to elicit a strong anti-tumor immune response mediated by T cells, APCs and NK cells (Multhoff et al. 1997; Asea et al. 2000; Dressel et al. 2000; Clark and Ménoret 2001). It is by now demonstrated that many cancer cells contain a high level of chaperones, especially Hsp70 (Ciocca and Calderwood 2005). Indeed, contrary to normal cells, cancer cells overexpress Hsp70 in the cytoplasm, and also display the chaperone on their plasma membrane (mHsp70) (Multhoff et al. 1997; Gehrmann et al. 2003). This overexpressed Hsp70 has been found to stimulate an immune response specific to cancer cells (Suto and Srivastava 1995). Literature data highlighted that eHsp70 exerts a dual role in cancer, promoting survival of tumor cells, as well as contributing to anti-tumor immunity. In fact, in tumor environment eHsp70 can inhibit early-stage tumor growth by activating the immune response, whereas it stimulates late-stage tumor growth by inhibiting the immune system (Wu et al. 2012). Thus some eHsp70 features help cancer cells to escape from cytotoxic cells, while others facilitate an immune attack. According to the mechanism of Hsp70 release in the extracellular environment, i.e. as a free soluble molecule or in association with extracellular vesicles, it mediates distinct functions through the interaction with different target cells (e.g., immune cells, endothelial cells), and the activation of different intracellular pathways. Three different mechanisms could be involved in Hsp70 anti-tumor immunity:

(a) Hsp70, due to its chaperone role can act as carrier for tumor derived peptides which are re-presented by APCs, such as macrophages and DCs, the most powerful APC. This complex is recognized by specific receptors, internalized within the APCs, and cross-presented to $CD4^+$ T cells on MHC II (Tamura et al. 1997; Li et al. 2002), initiating an adaptive tumor-specific immune response (Arnold-Schild et al. 1999; Schild et al. 1999; Noessner et al. 2002; Enomoto et al. 2006). DCs are also able to efficiently transfer tumor and self-antigens on to MHC I (cross-presentation) (Heath and Carbone 2001). eHsp70 has also immunogenic functions related to its C-terminal substrate-binding domain, which is involved in interaction with APCs and consequently in antigen cross-presentation (MacAry et al. 2004).
(b) The interaction between eHsp70 and its receptors on APCs stimulates these cells to secrete immunostimulatory cytokines, playing a proinflammatory role, responsible for enhanced programming of cytotoxic lymphocytes (Calderwood et al. 2012). For this "chaperokine" effect immunogenic peptides are not required (Asea et al. 2000, 2002; Asea 2006).
(c) The last possible mechanism of action against tumor cells is NK dependent, and is correlated to plasmamembrane localization of Hsp70 (Multhoff et al. 1999). In fact, resting NK cell activation either by eHsp70 or by mHsp70 derived from tumor, even in the absence of immunogenic peptides (Multhoff et al. 1998), leads to an increased proliferative, migratory and cytotoxic activity against transmembrane Hsp70 positive tumor cells, whereas no killing activity was observed for Hsp70 negative cells (Multhoff et al. 1997; Gastpar et al. 2004, 2005).

Schematic model of different eHsp70 roles in cancer immunity is represented in Fig. 3.

Due to the observation that preactivated $CD94^+$ NK cells (Gross et al. 2003a, b), through an Hsp70 peptide (TDK) plus low doses of interleukin 2, were able to induce regression of Hsp70 membrane positive tumors in immunodeficient mice

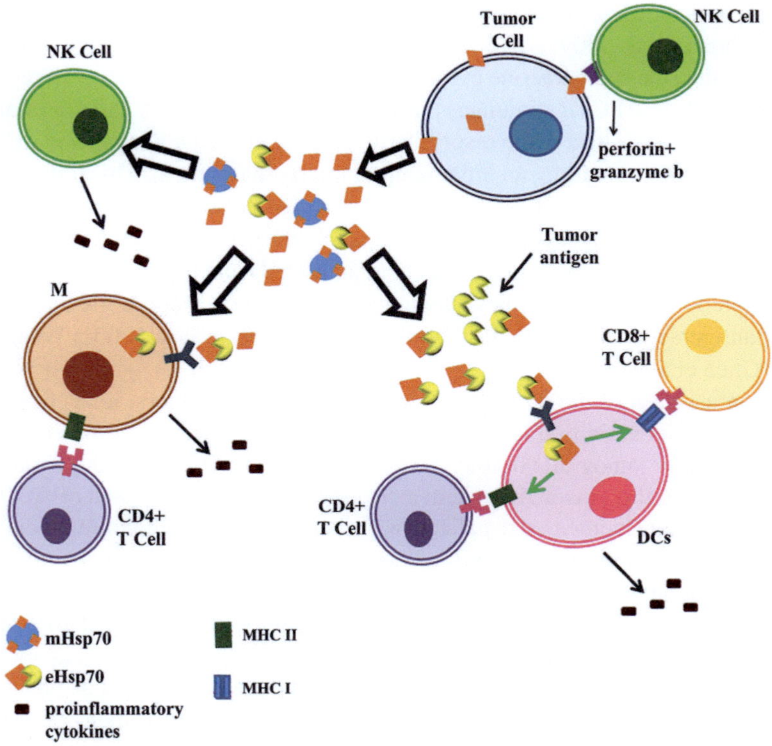

Fig. 3 Summary of Hsp70 activities in inducing anti-tumor immune responses. eHsp70 either alone or in combination with immunogenic peptides is able to induce an immune response. mHsp70 acts as a tumor-specific antigen, which is recognized by NK cells inducing their cytolytic, proliferative, and migratory capacity. eHsp70 also induces macrophages (M), NK and DCs to release proinflammatory cytokines. Moreover, eHsp70 could act as a carrier for tumor antigens and could support antigen uptake, processing, and presentation on MHC I to CD8$^+$ cytotoxic T lymphocytes, or on MHC II s to CD4$^+$ helper T cells

(Moser et al. 2002; Stangl et al. 2006), Hsp70 was used in clinical trials of patients with colorectal and non small lung cell cancer who failed standard therapies such as chemotherapy, radiotherapy or laser induced thermotherapy (Krause et al. 2004; Specht et al. 2015). The recognition by NK cells of fragment crystallizable region of antibodies that have bound to tumor specific antigens activate antibody-dependent cellular cytotoxicity, establishing a link between B cell and NK cell mediated immune responses.

It is now well demonstrated that increased membrane bound Hsp70 expression and its extracellular release can stimulate an anti-tumor immune response, rendering cancer cells more susceptible to immune cells. For this reason two possible methods have been tested to increase eHsp70 amount: pharmacological agents (Arispe et al. 2002), and physical factors (e.g., hyperthermia, photodynamic therapy and ionizing radiations) (Stangl et al. 2011). Some *in vivo* data indicate the prospect of therapeutic vaccines based on Hsp70 administration (Ito et al. 2004; Geng et al. 2006; Kumar et al. 2009; Abkin et al. 2013). eHsp70 released by tumor cells can serve as autocrine and paracrine cytokine and through the interaction with receptors on tumor cells or APCs it stimulates the secretion of chemokines, and proinflammatory cytokines and nitric oxide respectively (Asea et al. 2000; Panjwani et al. 2002; Vega et al. 2008), providing an inflammatory microenvironment. This inflammation exerts anti-tumor activity at early tumor stage, whereas it support tumor growth and metastasis formation at a chronic stage. Equally, eHsp70 has an effect also on dendritic cells, which were activated, and chemoattracted from tumor tissues to secondary lymphoid organs, where they showed an improved capacity in the induction of tumor antigen-specific cytotoxic T lymphocytes (Kuppner et al. 2001; Wang et al. 2005; Chen et al. 2009).

In confirmation of the protective role of the protein in tumors, it has been demonstrated that its inhibition is lethal, and its silencing induces *in vitro* cell death as well as in tumor xenografts in mice (Wei et al. 1995; Kaur et al. 2000; Nylandsted et al. 2000). For this reason, in several cancer patients a more elevated eHsp70 level was found in blood compared to patients with non-cancerous pathologies (Gehrmann et al. 2014) (e.g., chronic myeloid leukaemia, acute leukaemia, colorectal cancer, glioblastoma, pancreatic cancer) (Yeh et al. 2009, 2010; Kocsis et al. 2010; Elstner et al. 2011; Dutta et al. 2012). Wu and coworkers demonstrated that eHsp70 encourages hepatocarcinoma growth by promoting tumor cell proliferation and apoptosis resistance (Wu et al. 2012). Both the events were induced in a dose dependent manner through the activation of the NF-kB pathway. The involvement of eHsp70/Hsp70-peptide complex in hepatocellular carcinoma cell proliferation was also confirmed by Zhe et al. (2016). Protective effect of eHsp70 on induced apoptosis by H_2O_2 on lymphoma cells was also observed. Indeed, pretreatment of U937 cells with eHsp70 inhibited hydrogen peroxide apoptosis induction (Franco et al. 2016).

eHsp70 and the Nervous System

To date it has been extensively demonstrated that Hsp70 exerts a cellular protective role under stress conditions. However, some cell types are not able to express this protein, among them there are certain types of neurons (e.g., hippocampal neurons) (Sprang and Brown 1987; Guzhova et al. 2001; Robinson et al. 2008). These mature neurons are particularly susceptible to toxic conditions due to their inability to increase Hsp70 level after stress (Morimoto et al. 1997), even if they contain Hsc70 (Brown 1991). The inability to activate the stress response is responsible for a wide range of neurodegenerative diseases (Planas et al. 1997), such as amyotrophic lateral sclerosis, Parkinson's disease, Alzheimer's disease, and polyglutamine diseases (Adachi et al. 2009). In this context, it has been demonstrated that Hsp70 released by surrounding glial cells is able to be endocyted by them (Guzhova et al. 2001; Novoselova et al. 2005; Robinson et al. 2005). The internalization of eHsp70 makes the neuronal cells resistant to cytotoxic effects and thermal stresses (Guzhova et al.

2001). A protective effect of exogenous Hsp70, on moto or sensory neurons and glial cells has been demonstrated by several authors (Edbladh et al. 1994; Guzhova et al. 2001; Tidwell et al. 2004; Robinson et al. 2005; Tytell 2005). The eHsp70 neuroprotective effect was also observed both under stress growth conditions (e.g., oxidative stress common in several neurodegenerative disease) and for those neurons undergoing natural cell death (Robinson et al. 2005).

Luo and colleagues demonstrated that eHsp70 in a dose dependent fashion is responsible for Schwann cell protection against apoptosis induced by H_2O_2. In particular, this protective effect was exerted through failure of caspase 3 and 9 activation and Bcl2 upregulation (Luo et al. 2012). Oxidative stress is the cause of several neurodegenerative diseases, as ROS induces membrane lipid peroxidation, nitration of proteins, and degradation of DNA, all inducing cell apoptosis (Fiskum 2004; Park et al. 2005). Because Schwann cells can regulate neuron survival and nerve regeneration, their preservation by eHsp70 is fundamental for peripheral nerve protection. Oxidative stress also impairs motoneurons during exercise. In fact, exercise induces ROS production, depending on the intensity, the type and the duration of the exercise. Although motoneurons during exercise are submitted to stress they still continue to function without any apparent damage, even if they are not able to increase iHsp70 synthesis (Robinson et al. 2005). This resistance can be explained by the release of eHsp70 by the liver, or by in *vitro* external addiction, and its internalization in motoneurons. eHsp70 could also act as extracellular signaling molecule, and this ability, together with its chaperone role after endocytosis, helps the cells against stress and damage (Robinson et al. 2005; Krause and Rodrigues-Krause 2011).

On the one end, many papers have demonstrated the involvement of HSP in central nervous system (CNS) diseases, such as those characterized by the presence and accumulation of misfolded proteins. Because HSP are found in protein aggregates, along with disease proteins, ubiquitin or other cellular molecules, we can hypothesize that both iHSP and eHSP have a role in refolding the misfolded proteins. In the other end, diseases in which CNS immune activation is a prominent feature, such as ischemia, immune-mediated disorders, infections, and trauma, may involve eHSP, because they are able to induce the innate immune response and to enhance the adaptive immunity. The most important HSP implicated in the immune response is eHsp70 (for a review see Turturici et al. 2011). eHsp70, through the interaction with specific cell surface receptors, are responsible for the expression of proinflammatory cytokines, chemokines and for DC activation, which are involved in antigen presentation to B and T lymphocytes.

High levels of anti-Hsp70 autoantibodies were found in the cerebrospinal fluid (CSF) of patients affected by multiple sclerosis (MS), than in that of healthy controls. In particular, the highest levels were detected in patients with progressive MS, in contrast to those with a stable disease (Chiba et al. 2006). This increase induces a higher production of IL-8 in THP-1 monocytes with consequent higher inflammatory levels (Yokota et al. 2010). eHsp70 was also found in and around MS lesions, and it may be involved in the induction or exacerbation of the immunologic response, because of its ability to act as a proinflammatory cytokine (for a review see Turturici et al. 2014). In addition, Lund and collaborators demonstrated that eHsp70 was associated with myelin basic protein (MBP)-derived peptides in normal appearing white matter of both MS ad normal human brain. They hypothesize that part of eHsp70-MBP peptide is secreted by stressed oligodendrocytes and it can act as an adjuvant molecule that stimulates an adaptive immune response against specific autoantigens. This event could be involved in the initiation of MS, and in the subsequent immune mediated destruction of myelin (Lund et al. 2006).

Galazka et al., obtained conflicting results which demonstrated that mouse immunization with eHsp70 fraction associated with peptide complexes isolated from animals with experimental autoimmune encephalomyelitis (EAE) reduced its subsequent induction (Galazka et al. 2006). On the contrary, the disease was not induced using eHsp70-complexes isolated from healthy donors. These results suggest substantial differences in the peptide that bind Hsp70 in normal versus pathological CNS. In contrast, in EAE, Hsp70 pharmacologically induced (e.g., with geldamycin) has a protective role because it suppresses the glial inflammatory response and ameliorates clinical signs (Dello Russo et al. 2006). All the evidences according to which Hsp70 is not only an intracellular chaperone but has also extracellular functions, such as neuroprotective effects in many brain diseases, could open a new scenario in CNS disease therapy.

Other Functions of eHsp70

In recent years new roles for eHsp70 have been found. For example, several studies have demonstrated that circulating eHsp70 is a marker for cardiovascular disease (Pockley et al. 2003). In addition to its marker role, eHsp70 actively participate in the inflammatory damage typical of the cardiovascular damage through its binding to TLR4 (Yang et al. 2005). Furthermore, in 2013 Gonzáles-Ramos et al. demonstrated that eHsp70 acts as a profibrotic regulator of the ECM protein synthesis (fibronectin and type I collagen) by the vascular smooth muscle cells. Also this stimulation depends on eHsp70 binding to the membrane receptor TLR4 and the subsequent activation of the ERK and JNK pathways. These two kinases determine the activation of the transcription factor AP-1, which increased the transcriptional capacity of the TGF-β1 promoter involved in increasing fibronectin and collagen I amount (Gonzáles-Ramos et al. 2013).

In vitro and *in vivo* assays showed that eHsp70 could be also an angiogenic regulator. Indeed, experiments on HUVEC cells demonstrated that eHsp70 activates the ERK signaling pathway involved in cell migration, tube formation and microvessel formation (Kim et al. 2016). Furthermore, eHsp70 promotes osteogenic differentiation of mesenchymal stem cells. Chen et al. showed that the addition of Hsp70 to the culture medium of the stem cells stimulate the expression of the alkaline phosphatase, an early marker of osteoblastic differentiation. eHsp70 also enhanced calcium deposits in both dose and time dependent manners, via the activation of the ERK signaling pathway (Chen et al. 2015).

Additionally, eHsp70 is involved in stem cell migration as demonstrated by Barreca et al. (Barreca et al. 2017). One of the fundamental problems in stem cells based therapies is their homing to the site of injury. It is well known that MMPs play a fundamental role in mesenchymal stem cell migration (Ries et al. 2007). Barreca and coworkers demonstrated that mouse mesoangioblast release Hsp70 in membrane derived vesicles as a transmembrane protein. mHsp70 acts via an autocrine signaling pathway by increasing mesoangioblast MMP2 release and accentuating their migration and invasion capability. A predominant role in the activation of these processes was carried out by the eHsp70 receptors TLR4 and CD91 and the consequent NF-kB pathway activation (Barreca et al. 2017).

eHsp70 Receptors

As described above Hsp70 may be released in the extracellular environment by both active and passive mechanisms, it can travel through the circulation to reach cellular target cells, such as immune effector cells. HSP have two essential properties that allow them to activate the immune response. First they are chaperones of antigenic peptides produced by specific cell types and presented via MHC I and MHC II. Second HSP are capable of binding to membrane receptors on APCs. Indeed, for eHsp70 activating innate and adaptive immune responses, it must bind to specific cell receptors. In the last years many cell-surface receptors involved in eHsp70 mediated signaling have been identified, including TLR2 and TLR4 with their cofactor CD14 (Asea et al. 2000, 2002), the co-stimulatory molecule CD40 (Becker et al. 2002; Wang et al. 2001), the chemokine receptor CCR5, the scavenger receptors (SR) LOX-1 (Delneste et al. 2002; Thériault et al. 2006), SCREC-1 (Thériault et al. 2006), FEEL-1 (Thériault et al. 2006), CD94 (C-type lectin) (Gross et al. 2003a, b), CD91 (Basu et al. 2001), CD36 (Delneste et al. 2002). All these immune receptors are differentially expressed on various cell types and have different roles in eHsp70 signaling (Fig. 4). Indeed, the receptors can be divided in three groups: pattern recognition receptors (e.g., TLR2, TLR4), uptake of antigens by APCs (e.g., LOX-1, CD91), mediating co-stimulatory signals (e.g., CCR5, CD40).

The TLR family consists of several members recognizing specific ligands, e.g., TLR2 (triacyled lipoproteins) and TLR4 (LPS). Activation of TLR pathway by ligand binding is responsible for activation of the transcription factors NF-kB and AP1 (Takeda and Akira 2004). NF-kB is central to the inflammatory response by innate cells (neutrophils, macrophages and DCs). For the first time Asea et al. in 2002 demonstrated that eHsp70 induced proinflammatory cytokine by binding monocyte TLR2 and TLR4 with their cofactor CD14 and the adaptor molecule MyD88, and by activating the NF-kB pathway involved in IL-1β, IL-6 and TNF-α (Asea et al. 2000, 2002). Furthermore, LRs located on macrophages have been demonstrated to be involved in antigen uptake and phagocytosis induction (Blander and Medzhitov 2004 Science). In particular, another member of TLR family which is able to interact with eHsp70 is the lipid raft associated TLR7, generally involved in

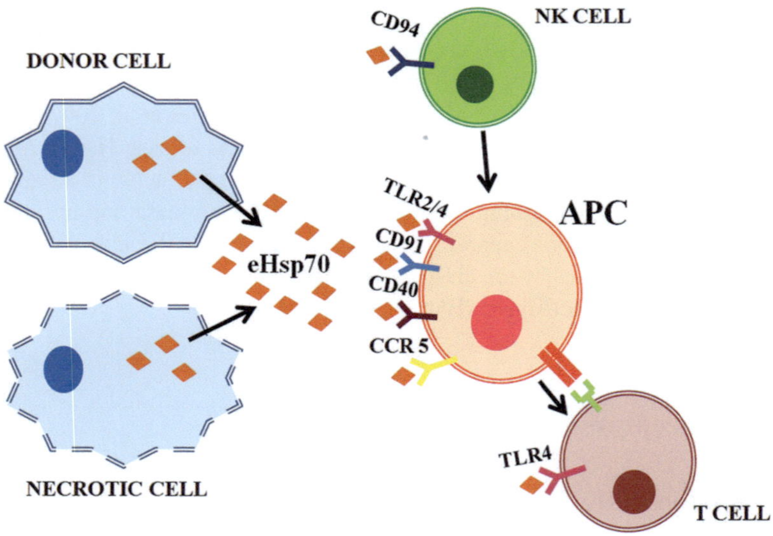

Fig. 4 Hsp70 release from the donor cells and its binding to cell surface receptors on the target cells. Several cell types release the Hsp70 into the extracellular microenvironment through different active secretion mechanism or through a passive release after necrosis. eHsp70 can be recognized by several receptors on the target cells. Among them there are TLR2, TLR4, LOX-1, CD91, CD94, CCR5 and CD40 expressed on APCs, NK cells, and T cells Their integrated function is discussed in the text

viral RNA recognition. Wang et al. demonstrated that exogenous Hsp70 binds TLR7 on the lipid raft, which is essential for this binding. The interaction between eHsp70 and this member of the TLR family stimulates phagocytosis and antigen presentation through PI3K and the p38 MAPK pathway (Wang et al. 2006b).

TLR2 and TLR4 are not only expressed on immune cells, but also on endothelial cells, and airway epithelial cells (Greene et al. 2005). In heart eHsp70 directly decreased cardiomyocyte contractility and increased cell death through the activation of the inflammatory response by the signaling pathway TLR2-MyD88-NF-kB (Mathur et al. 2011). On the other hand, eHsp70 induced inflammation in the lung is TLR4-NF-kB dependent (Chase et al. 2007).

Another receptor involved in eHsp70 binding is CD40, a member of the TNF family. This binding was competed by the cochaperone Bip, whereas it is enhanced by the antigenic peptide. As observed for other receptor, binding of eHsp70-peptide complex to CD40 increases peptide uptake and activate the p38 intracellular pathway (Becker et al. 2002). Moreover, CD40^{+} cells were stimulated to produce chemokines (Lehner et al. 2003). As TLRs, CD40 is not only expressed on immune cells, but also on other cell types, such as HUVEC endothelial cells, where it functions by competing CD40 ligand and decreasing tubular formation process (Futagami et al. 2008). By binding assays Multhoff's group demonstrates the involvement of the C-type lectin receptor CD94 in the interaction of NK cells with eHsp70/mHsp70 (Gross et al. 2003b). This interaction induce both their proliferation and cytolytic activity (Multhoff et al. 1999, 2001).

Another receptor implicated in eHsp70 binding is CD91, a transmembrane scavenger receptor ubiquitously expressed in multiple tissues where it plays a key role in intracellular signaling and endocytosis. Basu et al. identified CD91 as a receptor for several HSP including eHsp70 (Basu et al. 2001). Its role in binding and internalization of HSP by APCs, necessary for HSP chaperoned peptides cross-presentation, was reported in several papers (Delneste et al. 2002; Martin et al. 2003; Tobian et al. 2004). CD91-HSP was recently also be shown to be a signaling complex regulating multiple pathways in APCs (Pawaria and Binder 2011). The signaling pathway started up by CD91 upon HSP stimulation includes the activation of NF-kB and p38 MAPK (Pawaria and Binder 2011). As a consequence of this intracellular signaling several cytokines are released by HSP stimulated APCs (such as TNF-α, IL-1β, IL-6, IL-12 and GM-CSF) (Pawaria and Binder 2011). APCs also upregulate the expression of costimulatory molecules, including CD80, CD86, CD40 and MHC II (Basu et al. 2000). The costimulatory molecules and cytokines expressed depends on both the APC stimulated and the HSP involved in stimulation. According to these data CD91 has a role in cross-presentation and in costimulation given by the APCs to T cells in response to eHSP. The receptor CD91 expressed on monocytes, after eHsp70-peptide binding stimulates antigen specific $CD4^+$ T cell proliferation (Fischer et al. 2010). Although CD91 is highly expressed on macrophages, it is very low on immature DCs. Delneste and coworkers demonstrated that on these cells the scavenger receptor LOX-1 is the main eHsp70 binding receptor and is involved in antigen cross-presentation to MHC I *in vitro* and *in vivo* (Delneste et al. 2002).

One of the last eHsp70 receptor identified is the receptor for advanced glycation endproducts (RAGE), a receptor that interacts with several inflammation and stress mediators to activate the inflammatory pathway (Xie et al. 2013). The activation of RAGE by eHsp70 evokes strong proinflammatory changes in human lung carcinoma cell line. In fact, interaction between eHsp70 and RAGE trigger the typical cellular effects of this receptor activation (i.e., ERK phosphorylation and activation of NF-kB, increased expression of RAGE itself and secretion of proinflammatory cytokines) (Somensi et al. 2017).

Conclusions

Over the years several authors have detected Hsp70 in the extracellular environment. Early studies hypothesized that this protein was released consequently to cell death, as it lacks any exocytosis signals. Nowadays, it is well demonstrated that Hsp70 is released through active mechanism, involving non classical secretory pathways (e.g., extracellular vesicles, lysosome-endosome pathway and so on). Several roles have been demonstrated for eHsp70, especially at cell signaling and cell communication level, depending on its state (i.e. membrane-bound or membrane-free). An intriguing aspect of eHsp70 biology is its ability to induce antagonist responses, such as inflammatory and anti-inflammatory events. A particularly abundant field on eHsp70 functions regards its ability to modulate the immune system. Indeed, it has been displayed that eHsp70 possesses powerful immunological properties, acting both on innate and adaptive immune system. eHsp70 activates innate immune system cells by inducing cytokine release, or stimulate the adaptive immune response via antigen cross presentation. In addition to its immunological role, eHsp70 has protective functions in nervous cells, such as motoneurons, sensory neurons, glial cells. Finally, eHsp70 has been demonstrated to be involved in cell migration and angiogenesis, and it is a marker for several diseases. A better comprehension of eHsp70 multiple roles will be useful for therapeutic applications in inflammatory diseases, cancer and autoimmunity.

Acknowledgements This research was supported by grants from the University of Palermo.

References

Abkin, S. V., Pankratova, K. M., Komarova, E. Y., Guzhova, I. V., & Margulis, B. A. (2013). Hsp70 chaperone-based gel composition as a novel immunotherapeutic anti-tumor tool. *Cell Stress & Chaperones, 18*, 391–396.

Adachi, H., Katsuno, M., Waza, M., Minamiyama, M., Tanaka, F., & Sobue, G. (2009). Heat shock proteins in neurodegenerative diseases: Pathogenic roles and therapeutic implications. *International Journal of Hyperthermia, 25*, 647–654.

Aneja, R., Odoms, K., Dunsmore, K., Shanley, T. P., & Wong, H. R. (2006). Extracellular heat shock protein-70 induces endotoxin tolerance in THP-1 cells. *Journal of Immunology, 177*, 7184–7192.

Aquino, D. A., Klipfel, A. A., Brosnan, C. F., & Norton, W. T. (1993). The 70-kDa heat shock cognate protein (HSC70) is a major constituent of the central nervous system and is up-regulated only at the mRNA level in acute experimental autoimmune encephalomyelitis. *Journal of Neurochemistry, 61*, 1340–1348.

Aquino, D. A., Capello, E., Weisstein, J., Sanders, V., Lopez, C., Tourtellotte, W. W., Brosnan, C. F., Raine, C. S., & Norton, W. T. (1997). Multiple sclerosis: Altered expression of 70- and 27-kDa heat shock proteins in lesions and myelin. *Journal of Neuropathology and Experimental Neurology, 56*, 664–672.

Arispe, N., Doh, M., & De Maio, A. (2002). Lipid interaction differentiates the constitutive and stress-induced heat shock proteins Hsc70 and Hsp70. *Cell Stress & Chaperones, 7*, 330–338.

Arispe, N., Doh, M., Simakova, O., Kurganov, B., & De Maio, A. (2004). Hsc70 and Hsp70 interact with phosphatideylserine on the surface of PC12 cells resulting in a decrease of viability. *The FASEB Journal, 18*, 1636–1645.

Arnold-Schild, D., Hanau, D., Spehner, D., Schmid, C., Rammensee, H. G., de la Salle, H., & Schild, H. (1999). Cutting edge: Receptor-mediated endocytosis of heat shock proteins by professional antigen-presenting cells. *Journal of Immunology, 162*, 3757–3760.

Asea, A. (2003). Chaperokine-induced signal transduction pathways. *Exercise Immunology, 9*, 25–33.

Asea, A. (2005). Stress proteins and initiation of immune response: Chaperokine activity of hsp72. *Exercise Immunology Review, 11*, 34–45.

Asea, A. (2006). Initiation of the immune response by extracellular Hsp72: Chaperokine activity of Hsp72. *Current Immunology Reviews, 2*, 209–215.

Asea, A. (2007). Hsp72 release: Mechanisms and methodologies. *Methods, 43*, 194–198.

Asea, A. (2008). Hsp70: A chaperokine. *Novartis Foundation Symposium, 291*, 173–179. discussion 179-183:221-224.

Asea, A., Kraeft, S. K., Kurt-Jones, E. A., Stevenson, M. A., Chen, L. B., Finberg, R. W., Koo, G. C., & Calderwood, S. K. (2000). HSP70 stimulates cytokine production through a CD14-dependant pathway, demonstrating its dual role as a chaperone and cytokine. *Nature Medicine, 6*, 435–442.

Asea, A., Rehli, M., Kabingu, E., Boch, J. A., Bare, O., Auron, P. E., Stevenson, M. A., & Calderwood, S. K. (2002). Novel signal transduction pathway utilized by extracellular HSP70: Role of toll-like receptor (TLR) 2 and TLR4. *The Journal of Biological Chemistry, 277*, 15028–15034.

Banchereau, J., & Steinman, R. M. (1998). Dendritic cells and the control of immunity. *Nature, 392*, 245–252.

Barreca, M. M., Spinello, W., Cavalieri, V., Turturici, G., Sconzo, G., Kaur, P., Tinnirello, R., Asea, A. A., & Geraci, F. (2017). Extracellular Hsp70 enhances mesoangioblast migration via an autocrine signaling pathway. *Journal of Cellular Physiology, 232*, 1845–1861.

Basu, S., Binder, R. J., Suto, R., Anderson, K. M., & Srivastava, P. K. (2000). Necrotic but not apoptotic cell death releases heat shock proteins, which deliver a partial maturation signal to dendritic cells and activate the NF-kappa B pathway. *International Immunology, 12*, 1539–1546.

Basu, S., Binder, R. J., Ramalingam, T., & Srivastava, P. K. (2001). CD91 is a common receptor for heat shock proteins gp96, hsp90, hsp70, and calreticulin. *Immunity, 14*, 303–313.

Bausero, M. A., Gastpar, R., Multhoff, G., & Asea, A. (2005). Alternative mechanism by which IFN-gamma enhances tumor recognition: Active release of heat shock protein 72. *Journal of Immunology, 175*, 2900–2912.

Becker, T., Hartl, F. U., & Wieland, F. (2002). CD40, an extracellular receptor for binding and uptake of Hsp70-peptide complexes. *The Journal of Cell Biology, 158*, 1277–1285.

Beere, H. M. (2004). "The stress of dying": The role of heat shock proteins in the regulation of apoptosis. *Journal of Cell Science, 117*, 2641–2651.

Bendz, H., Ruhland, S. C., Pandya, M. J., Hainzl, O., Riegelsberger, S., Braüchle, C., Mayer, M. P., Buchner, J., Issels, R. D., & Noessner, E. (2007). Human heat shock protein 70 enhances tumor antigen presentation through complex formation and intracellular antigen delivery without innate immune signaling. *The Journal of Biological Chemistry, 282*, 31688–31702.

Beuret, N., Stettler, H., Renold, A., Rutishauser, J., & Spiess, M. (2004). Expression of regulated secretory proteins is sufficient to generate granule-like structures in constitutively secreting cells. *The Journal of Biological Chemistry, 279*, 20242–20249.

Blander, J. M., & Medzhitov, R. (2004). Regulation of phagosome maturation by signals from toll-like receptors. *Science, 304*, 1014–1018.

Bluestone, J. A. (2005). Regulatory T-cell therapy: Is it ready for the clinic? *Nature Reviews. Immunology, 5*, 343–349.

Borges, T. J., Wieten, L., van Herwijnen, M. J., Broere, F., van der Zee, R., Bonorino, C., & van Eden, W. (2012). The anti-inflammatory mechanisms of Hsp70. *Frontiers in Immunology, 3*, 1–12.

Broquet, A. H., Thomas, G., Masliah, J., Trugnan, G., & Bachelet, M. (2003). Expression of the molecular chaperone Hsp70 in detergent-resistant microdomains correlates with its membrane delivery and release. *The Journal of Biological Chemistry, 278*, 21601–21606.

Brown, I. R. (1991). Expression of heat shock genes (hsp70) in the mammalian nervous system. *Results and Problems in Cell Differentiation, 17*, 217–229.

Bukau, B., Weissman, J., & Horwich, A. (2006). Molecular chaperones and protein quality control. *Cell, 125*, 443–451.

Calderwood, S. K., Mambula, S. S., Gray, P. J., Jr., & Theriault, J. R. (2007). Extracellular heat shock proteins in cell signalling. *FEBS Letters, 581*, 3689–3694.

Calderwood, S. K., Murshid, A., & Gong, J. (2012). Heat shock proteins: Conditional mediators of inflammation in tumor immunity. *Frontiers in Immunology, 3*, 1–10.

Castelli, C., Ciupitu, A. M., Rini, F., Rivoltini, L., Mazzocchi, A., Kiessling, R., & Parmiani, G. (2001). Human heat shock protein 70 peptide complexes specifically activate antimelanoma T cells. *Cancer Research, 61*, 222–227.

Chabas, D., Baranzini, S. E., Mitchell, D., Bernard, C. C., Rittling, S. R., Denhardt, D. T., Sobel, R. A., Lock, C., Karpuj, M., Pedotti, R., Heller, R., Oksenberg, J. R., & Steinman, L. (2001). The influence of the pro-inflammatory cytokine, osteopontin, on autoimmune demyelinating disease. *Science, 294*, 1731–1735.

Chase, M. A., Wheeler, D. S., Lierl, K. M., Hughes, V. S., Wong, H. R., & Page, K. (2007). Hsp72 induces inflammation and regulates cytokine production in airway epithelium through a TLR4- and NF-kappaB-dependent mechanism. *Journal of Immunology, 179*, 6318–6324.

Chen, X., Tao, Q., Yu, H., Zhang, L., & Cao, X. (2002). Tumor cell membrane-bound heat shock protein 70 elicits antitumor immunity. *Immunology Letters, 84*, 81–87.

Chen, S., Bawa, D., Besshoh, S., Gurd, J. W., & Brown, I. R. (2005). Association of heat shock proteins and neuronal membrane components with lipid rafts from the rat brain. *Journal of Neuroscience Research, 81*(4), 522–529.

Chen, T., Guo, J., Han, C., Yang, M., & Cao, X. (2009). Heat shock protein 70, released from heat-stressed tumor cells, initiates antitumor immunity by inducing tumor cell chemokine production and activating dendritic cells via TLR4 pathway. *Journal of Immunology, 182*, 1449–1459.

Chen, E., Xue, D., Zhang, W., Lin, F., & Pan, Z. (2015). Extracellular heat shock protein 70 promotes osteogenesis of human mesenchymal stem cells through activation of the ERK signaling pathway. *FEBS Letters, 589*, 4088–4096.

Chiba, S., Yokota, S., Yonekura, K., Tanaka, S., Furuyama, H., Kubota, H., Fujii, N., & Matsumoto, H. (2006). Autoantibodies against HSP70 family proteins were detected in the cerebrospinal fluid from patients with multiple sclerosis. *Journal of the Neurological Sciences, 241*, 39–43.

Ciocca, D. R., & Calderwood, S. K. (2005). Heat shock proteins in cancer: Diagnostic, prognostic, predictive, and treatment implications. *Cell Stress & Chaperones, 10*, 86–103.

Clark, P. R., & Ménoret, A. (2001). The inducible Hsp70 as a marker of tumor immunogenicity. *Cell Stress & Chaperones, 6*, 121–125.

Clayton, A., Turkes, A., Navabi, H., Mason, M. D., & Tabi, Z. (2005). Induction of heat shock proteins in B-cell exosomes. *Journal of Cell Science, 118*, 3631–3638.

Collins, P. L., & Hightower, L. E. (1982). Newcastle disease virus stimulates the cellular accumulation of stress (heat shock) mRNAs and proteins. *Journal of Virology, 44*, 703–707.

Dello Russo, C., Polak, P. E., Mercado, P. R., Spagnolo, A., Sharp, A., Murphy, P., Kamal, A., Burrows, F. J., Fritz, L. C., & Feinstein, D. L. (2006). The heat-shock protein 90 inhibitor17-allylamino-17-demethoxygeldanamycin suppresses glial inflammatory responses and ameliorates experimental autoimmune encephalomyelitis. *Journal of Neurochemistry, 99*, 1351–1362.

Delneste, Y., Magistrelli, G., Gauchat, J., Haeuw, J., Aubry, J., Nakamura, K., Kawakami-Honda, N., Goetsch, L., Sawamura, T., Bonnefoy, J., & Jeannin, P. (2002). Involvement of LOX-1 in dendritic cell-mediated antigen cross-presentation. *Immunity, 17*, 353–362.

Dressel, R., Elsner, L., Quentin, T., Walter, L., & Günther, E. (2000). Heat shock protein 70 is able to prevent heat shock-induced resistance of target cells to CTL. *Journal of Immunology, 164*, 2362–2371.

Dutta, S. K., Girotra, M., Singla, M., Dutta, A., Otis Stephen, F., Nair, P. P., & Merchant, N. B. (2012). Serum HSP70: A novel biomarker for early detection of pancreatic cancer. *Pancreas, 41*, 530–534.

Edbladh, M., Ekstrom, P. A., & Edstrom, A. (1994). Retrograde axonal transport of locally synthesized proteins, e.g., actin and heat shock protein 70, in regenerating adult frog sciatic sensory axons. *Journal of Neuroscience Research, 38*, 424–432.

Eder, K., Guan, H., Sung, H. Y., Ward, J., Angyal, A., Janas, M., Sarmay, G., Duda, E., Turner, M., Dower, S. K., Francis, S. E., Crossman, D. C., & Kiss-Toth, E. (2008). Tribbles-2 is a novel regulator of inflammatory activation of monocytes. *International Immunology, 20*, 1543–1550.

Elstner, A., Stockhammer, F., Nguyen-Dobinsky, T. N., Nguyen, Q. L., Pilgermann, I., Gill, A., Guhr, A., Zhang, T., von Eckardstein, K., Picht, T., Veelken, J., Martuza, R. L., von Deimling,

A., & Kurtz, A. (2011). Identification of diagnostic serum protein profiles of glioblastoma patients. *Journal of Neuro-Oncology, 102*, 71–80.

Enomoto, Y., Bharti, A., Khaleque, A. A., Song, B., Liu, C., Apostolopoulos, V., Xing, P. X., Calderwood, S. K., & Gong, J. (2006). Enhanced immunogenicity of heat shock protein 70 peptide complexes from dendritic cell-tumor fusion cells. *Journal of Immunology, 177*, 5946–5955.

Evdokimovskaya, Y., Skarga, Y., Vrublevskaya, V., & Morenkov, O. (2010). Secretion of the heat shock proteins HSP70 and HSC70 by baby hamster kidney (BHK-21) cells. *Cell Biology International, 34*, 985–990.

Evdonin, A. L., Martynova, M. G., Bystrova, O. A., Guzhova, I. V., Margulis, B. A., & Medvedeva, N. D. (2006). The release of Hsp70 from A431 carcinoma cells is mediated by secretory-like granules. *European Journal of Cell Biology, 85*, 443–455.

Feder, M. E., & Hofmann, G. E. (1999). Heat-shock proteins, molecular chaperones, and the stress response: Evolutionary and ecological physiology. *Annual Review of Physiology, 61*, 243–282.

Fevrier, B., & Raposo, G. (2004). Exosomes: Endosomal-derived vesicles shipping extra- cellular messages. *Current Opinion in Cell Biology, 16*, 415–421.

Fischer, N., Haug, M., Kwok, W. W., Kalbacher, H., Wernet, D., Dannecker, G. E., & Holzer, U. (2010). Involvement of CD91 and scavenger receptors in Hsp70-facilitated activation of human antigen-specific CD4+ memory T cells. *European Journal of Immunology, 40*, 986–997.

Fiskum, G. (2004). Mechanisms of neuronal death and neuroprotection. *Journal of Neurosurgical Anesthesiology, 16*, 108–110.

Franco, L., Terrinca, J., Rodríguez, A. B., Espino, J., & Pariente, J. A. (2016). Extracellular heat shock proteins protect U937 cells from H2O2-induced apoptotic cell death. *Molecular and Cellular Biochemistry, 412*, 19–26.

Futagami, S., Hiratsuka, T., Shindo, T., Hamamoto, T., Horie, A., Ueki, N., Kusunoki, M., Gudis, K., Miyake, K., Tsukui, T., & Sakamoto, C. (2008). Extracellular HSP70 blocks CD40L-induced apoptosis and tubular formation in endothelial cells. *Journal of Gastroenterology and Hepatology, 23*, S222–S228.

Galazka, G., Stasiolek, M., Walczak, A., Jurewicz, A., Zylicz, A., Brosnan, C. F., Raine, C. S., & Selmaj, K. W. (2006). Brain-derived heat shock protein 70-peptide complexes induce NK cell-dependent tolerance to experimental autoimmune encephalomyelitis. *Journal of Immunology, 176*, 1588–1599.

Gallucci, S., Lolkema, M., & Matzinger, P. (1999). Natural adjuvants: Endogenous activators of dendritic cells. *Nature Medicine, 5*, 1249–1255.

Gao, Y. L., Brosnan, C. F., & Raine, C. S. (1995). Experimental autoimmune encephalomyelitis. Qualitative and semiquantitative differences in heat shock protein 60 expression in the central nervous system. *Journal of Immunology, 154*, 3548–3556.

Garrido, C., Gurbuxani, S., Ravagnan, L., & Kroemer, G. (2001). Heat shock proteins: Endogenous modulators of apoptotic cell death. *Biochemical and Biophysical Research Communications, 286*, 433–442.

Gastpar, R., Gross, C., Rossbacher, L., Ellwart, J., Riegger, J., & Multhoff, G. (2004). The cell surface-localized heat shock protein 70 epitope TKD induces migration and cytolytic activity selectively in human NK cells. *Journal of Immunology, 172*, 972–980.

Gastpar, R., Gehrmann, M., Bausero, M. A., Asea, A., Gross, C., Schroeder, J. A., & Multhoff, G. (2005). Heat shock protein 70 surface-positive tumor exosomes stimulate migratory and cytolytic activity of natural killer cells. *Cancer Research, 65*, 5238–5247.

Gehrmann, M., Schmetzer, H., Eissner, G., Haferlach, T., Hiddemann, W., & Multhoff, G. (2003). Membrane-bound heat shock protein 70 (Hsp70) in acute myeloid leukemia: A tumor specific recognition structure for the cytolytic activity of autologous NK cells. *Haematologica, 88*, 474–476.

Gehrmann, M., Cervello, M., Montalto, G., Cappello, F., Gulino, A., Knape, C., Specht, H. M., & Multhoff, G. (2014). Heat shock protein 70 serum levels differ significantly in patients with chronic hepatitis, liver cirrhosis, and hepatocellular carcinoma. *Frontiers in Immunology, 5*, 1–7.

Geng, H., Zhang, G. M., Xiao, H., Yuan, Y., Li, D., Zhang, H., Qiu, H., He, Y. F., & Feng, Z. H. (2006). HSP70 vaccine in combination with gene therapy with plasmid DNA encoding sPD-1 overcomes immune resistance and suppresses the progression of pulmonary metastatic melanoma. *International Journal of Cancer, 118*, 2657–2664.

Giffard, R. G., Xu, L., Zhao, H., Carrico, W., Ouyang, Y., Qiao, Y., Sapolsky, R., Steinberg, G., Hu, B., & Yenari, M. A. (2004). Chaperones, protein aggregation, and brain protection from hypoxic/ischemic injury. *The Journal of Experimental Biology, 207*, 3213–3220.

Giraldo, E., Hinchado, M. D., Garcia, J. J., & Ortega, E. (2008). Influence of gender and oral contraceptives intake on innate and inflammatory response. Role of neuroendocrine factors. *Molecular and Cellular Biochemistry, 313*, 147–153.

González-Ramos, M., Calleros, L., López-Ongil, S., Raoch, V., Griera, M., Rodríguez-Puyol, M., de Frutos, S., & Rodríguez-Puyol, D. (2013). HSP70 increases extracellular matrix production by human vascular smooth muscle through TGF-β1 upregulation. *The International Journal of Biochemistry & Cell Biology, 45*, 232–242.

Greene, C. M., Carroll, T. P., Smith, S. G., Taggart, C. C., Devaney, J., Griffin, S., O'neill, S. J., & McElvaney, N. G. (2005). TLR-induced inflammation in cystic fibrosis and non-cystic fibrosis airway epithelial cells. *Journal of Immunology, 174*, 1638–1646.

Gross, C., Schmidt-Wolf, I. G., Nagaraj, S., Gastpar, R., Ellwart, J., Kunz-Schughart, L. A., & Multhoff, G. (2003a). Heat shock protein 70-reactivity is associated with increased cell surface density of CD94/CD56 on primary natural killer cells. *Cell Stress & Chaperones, 8*, 348–360.

Gross, C., Hansch, D., Gastpar, R., & Multhoff, G. (2003b). Interaction of heat shock protein 70 peptide with NK cells involves the NK receptor CD94. *Biological Chemistry, 384*, 267–279.

Guzhova, I., Kislyakova, K., Moskaliova, O., Fridlanskaya, I., Tytell, M., Cheetham, M., & Margulis, B. (2001). In vitro studies show that Hsp70 can be released by glia and that exogenous Hsp70 can enhance neuronal stress tolerance. *Brain Research, 914*, 66–73.

Hartl, F. U., & Hayer-Hartl, M. (2009). Converging concepts of protein folding in vitro and in vivo. *Nature Structural & Molecular Biology, 16*, 574–581.

Heath, W. R., & Carbone, F. R. (2001). Cross-presentation, dendritic cells, tolerance and immunity. *Annual Review of Immunology, 19*, 47–64.

Heijnen, H. F., Schiel, A. E., Fijnheer, R., Geuze, H. J., & Sixma, J. J. (1999). Activated platelets release two types of membrane vesicles: Microvesicles by surface shedding and exosomes derived from exocytosis of multivesicular bodies and alpha-granules. *Blood, 94*, 3791–3799.

Hightower, L. E., & Guidon, P. T., Jr. (1989). Selective release from cultured mammalian cells of heat-shock (stress) proteins that resemble glia-axon transfer proteins. *Journal of Cellular Physiology, 138*, 257–266.

Hilf, N., Singh-Jasuja, H., Schwarzmaier, P., Gouttefangeas, C., Rammensee, H. G., & Schild, H. (2002). Human platelets express heat shock protein receptors and regulate dendritic cell maturation. *Blood, 99*, 3676–3682.

Horváth, I., Multhoff, G., Sonnleitner, A., & Vígh, L. (2008). Membrane-associated stress proteins: More than simply chaperones. *Biochimica et Biophysica Acta, 1778*, 1653–1664.

Hunter-Lavin, C., Davies, E. L., Bacelar, M. M., Marshall, M. J., Andrew, S. M., & Williams, J. H. (2004). Hsp70 release from peripheral blood mononuclear cells. *Biochemical and Biophysical Research Communications, 324*, 511–517.

Ito, A., Matsuoka, F., Honda, H., & Kobayashi, T. (2004). Atitumor effects of combined therapy of recombinant heat shock protein 70 and hyperthermia using magnetic nanoparticles in an experimental subcutaneous murine melanoma. *Cancer Immunology, Immunotherapy, 53*, 26–32.

Johnson, J. D., & Fleshner, M. (2006). Releasing signals, secretory pathways, and immune function of endogenous extracellular heat shock protein 72. *Journal of Leukocyte Biology, 79*, 425–434.

Jones, Q., Voegeli, T. S., Li, G., Chen, Y., & Currie, R. W. (2011). Heat shock proteins protect against ischemia and inflammation through multiple mechanisms. *Inflammation Allergy Drug Targets, 10*, 247–259.

Kampinga, H. H., Henning, R. H., van Gelder, I. C., & Brundel, B. J. (2007). Beat shock proteins and atrial fibrillation. *Cell Stress & Chaperones, 12*, 97–100.

Kampinga, H. H., Hageman, J., Vos, M. J., Kubota, H., Tanguay, R. M., Bruford, E. A., Cheetham, M. E., Chen, B., & Hightower, L. E. (2009). Guidelines for the nomenclature of the human heat shock proteins. *Cell Stress & Chaperones, 14*, 105–111.

Kaur, J., Kaur, J., & Ralhan, R. (2000). Induction of apoptosis by abrogation of HSP70 expression in human oral cancer cells. *International Journal of Cancer, 85*, 1–5.

Kim, T. K., Na, H. J., Lee, W. R., Jeoung, M. H., & Lee, S. (2016). Heat shock protein 70-1A is a novel angiogenic regulator. *Biochemical and Biophysical Research Communications, 469*, 222–228.

Kocsis, J., Madaras, B., Tóth, E. K., Füst, G., & Prohászka, Z. (2010). Serum level of soluble 70-kD heat shock protein is associated with high mortality in patients with colorectal cancer without distant metastasis. *Cell Stress & Chaperones, 15*, 143–151.

Kovalchin, J. T., Wang, R., Wagh, M. S., Azoulay, J., Sanders, M., & Chandawarkar, R. Y. (2006). In vivo delivery of heat shock protein 70 accelerates wound healing by up-regulating macrophage-mediated phagocytosis. *Wound Repair and Regeneration, 14*, 129–137.

Krause, M., & Rodrigues-Krause, J. C. (2011). Extracellular heat shock proteins (eHSP70) in exercise: Possible targets outside the immune system and their role for neurodegenerative disorders treatment. *Medical Hypotheses, 76*, 286–290.

Krause, S. W., Gastpar, R., Andreesen, R., Gross, C., Ullrich, H., Thonigs, G., Pfister, K., & Multhoff, G. (2004). Treatment of colon and lung cancer patients with ex vivo heat shock protein 70-peptide-activated, autologous natural killer cells: A clinical phase I trial. *Clinical Cancer Research, 10*, 3699–36707.

Kregel, K. C. (2002). Heat shock proteins: Modifying factors in physiological stress responses and acquired thermotolerance. *Journal of Applied Physics, 92*, 2177–2186.

Kumar, S., Deepak, P., & Acharya, A. (2009). Autologous Hsp70 immunization induces anti-tumor immunity and increases longevity and survival of tumor-bearing mice. *Neoplasma, 56*, 259–268.

Kuppner, M. C., Gastpar, R., Gelwer, S., Nössner, E., Ochmann, O., Scharner, A., & Issels, R. D. (2001). The role of heat shock protein (hsp70) in dendritic cell maturation: Hsp70 induces the maturation of immature dendritic cells but reduces DC differentiation from monocyte precursors. *European Journal of Immunology, 31*, 1602–1609.

Lancaster, G. I., & Febbraio, M. A. (2005). Exosome-dependent trafficking of HSP70: A novel secretory pathway for cellular stress proteins. *The Journal of Biological Chemistry, 280*, 23349–23355.

Lehner, T., Wang, Y., & Kelly, C. (2003). Heat shock protein receptors, functions and their effect on monocytes and dendritic cells. In W. van Eden & D. B. Young (Eds.), *Heat shock proteins and inflammation* (pp. 193–198). Basel: Dekker.

Li, Z., Menoret, A., & Srivastava, P. (2002). Roles of heat-shock proteins in antigen presentation and cross-presentation. *Current Opinion in Immunology, 14*, 45–51.

Li, D., Romain, G., Flamar, A. L., Duluc, D., Dullaers, M., Li, X. H., Zurawski, S., Bosquet, N., Palucka, A. K., Le Grand, R., O'Garra, A., Zurawski, G., Banchereau, J., & Oh, S. (2012). Targeting self- and foreign antigens to dendritic cells via DC-ASGPR generates IL-10-producing suppressive CD4+ T cells. *The Journal of Experimental Medicine, 209*, 109–121.

Liberek, K., Lewandowska, A., & Zietkiewicz, S. (2008). Chaperones in control of protein disaggregation. *The EMBO Journal, 23*, 27328–27335.

Lindquist, S. (1986). The heat-shock response. *Annual Review of Biochemistry, 55*, 1151–1191.

Lindquist, S., & Craig, E. A. (1988). The heat-shock proteins. *Annual Review of Genetics, 22*, 631–677.

Luft, J. C., & Dix, D. J. (1999). Hsp70 expression and function during embryogenesis. *Cell Stress & Chaperones, 4*, 162–170.

Lui, J. C., & Kong, S. K. (2007). Heat shock protein 70 inhibits the nuclear import of apoptosis-inducing factor to avoid DNA fragmentation in TF-1 cells during erythropoiesis. *FEBS Letters, 581*, 109–117.

Lund, B. T., Chakryan, Y., Ashikian, N., Mnatsakanyan, L., Bevan, C. J., Aguilera, R., Gallaher, T., & Jakowec, M. W. (2006). Association of MBP peptides with Hsp70 in normal appearing human white matter. *Journal Neurology Science Novel, 249*, 122–134.

Luo, X., Zuo, X., Zhou, Y., Zhang, B., Shi, Y., Liu, M., Wang, K., McMillian, D. R., & Xiao, X. (2008). Extracellular heat shock protein 70 inhibits tumour necrosis factor-alpha induced proinflammatory mediator production in fibroblast-like synoviocytes. *Arthritis Research & Therapy, 10*, R41.

Luo, X., Tao, L., Lin, P., Mo, X., & Chen, H. (2012). Extracellular heat shock protein 72 protects schwann cells from hydrogen peroxide-induced apoptosis. *Journal of Neuroscience Research, 90*, 1261–1269.

MacAry, P. A., Javid, B., Floto, R. A., Smith, K. G., Oehlmann, W., Singh, M., & Lehner, P. J. (2004). HSP70 peptide binding mutants separate antigen delivery from dendritic cell stimulation. *Immunity, 20*, 95–106.

Madden, L. A., Sandström, M. E., Lovell, R. J., & McNaughton, L. (2008). Inducible heat shock protein 70 and its role in preconditioning and exercise. *Amino Acids, 34*, 511–516.

Mambula, S. S., & Calderwood, S. K. (2006). Heat shock protein 70 is secreted from tumor cells by a nonclassical pathway involving lysosomal endosomes. *Journal of Immunology, 177*, 7849–7857.

Martin, R., McFarland, H. F., & McFarlin, D. E. (1992). Immunological aspects of demyelinating diseases. *Annual Review of Immunology, 10*, 153–187.

Martin, C. A., Carsons, S. E., Kowalewski, R., Bernstein, D., Valentino, M., & Santiago-Schwarz, F. (2003). Aberrant extracellular and dendritic cell (DC) surface expression of heat shock protein (hsp)70 in the rheumatoid joint: Possible mechanisms of hsp/DC-mediated cross-priming. *Journal of Immunology, 171*, 5736–5742.

Mathew, A., Bell, A., & Johnstone, R. M. (1995). Hsp-70 is closely associated with the transferrin receptor in exosomes from maturing reticulocytes. *The Biochemical Journal, 308*, 823–830.

Mathivanan, S., & Simpson, R. J. (2009). ExoCarta: A compendium of exosomal proteins and RNA. *Proteomics, 9*, 4997–5000.

Mathur, S., Walley, K. R., Wang, Y., Indrambarya, T., & Boyd, J. H. (2011). Extracellular heat shock protein 70 induces cardiomyocyte inflammation and contractile dysfunction via TLR2. *Circulation Journal, 75*, 2445–2452.

Milani, V., Noessner, E., Ghose, S., Kuppner, M., Ahrens, B., Scharner, A., Gastpar, R., & Issels, R. D. (2002). Heat shock protein 70: Role in antigen presentation and immune stimulation. *International Journal of Hyperthermia, 18*, 563–575.

Milarski, K. L., & Morimoto, R. I. (1989). Mutational analysis of the human HSP70 protein: Distinct domains for nucleolar localization and adenosine triphosphate binding. *The Journal of Cell Biology, 109*, 1947–1962.

Moore, K. W., de Waal Malefyt, R., Coffman, R. L., & O'Garra, A. (2001). Interleukin-10 and the interleukin-10 receptor. *Annual Review of Immunology, 19*, 683–765.

Morimoto, R. I. (1991). Heat shock: The role of transient inducible responses in cell damage, transformation, and differentiation. *Cancer Cells, 3*, 295–301.

Morimoto, R. I., Kline, M. P., Bimston, D. N., & Cotto, J. J. (1997). The heat-shock response: Regulation and function of heat-shock proteins and molecular chaperones. *Essays in Biochemistry, 32*, 17–29.

Moser, C., Schmidbauer, C., Gürtler, U., Gross, C., Gehrmann, M., Thonigs, G., Pfister, K., & Multhoff, G. (2002). Inhibition of tumor growth in mice with severe combined immunodeficiency is mediated by heat shock protein 70 (Hsp70)-peptide-activated, CD94 positive natural killer cells. *Cell Stress & Chaperones, 7*, 365–373.

Muchowski, P. J., & Wacker, J. L. (2005). Modulation of neurodegeneration by molecular chaperones. *Nature Reviews. Neuroscience, 6*, 11–22.

Multhoff, G., Botzler, C., Jennen, L., Schmidt, J., Ellwart, J., & Issels, R. (1997). Heat shock protein 72 on tumor cells: A recognition structure for natural killer cells. *Journal of Immunology, 158*, 4341–4350.

Multhoff, G., Botzler, C., & Issels, R. (1998). The role of heat shock proteins in the stimulation of an immune response. *Biological Chemistry, 379*, 295–300.

Multhoff, G., Mizzen, L., Winchester, C. C., Milner, C. M., Wenk, S., Eissner, G., Kampinga, H. H., Laumbacher, B., & Johnson, J. (1999). Heat shock protein 70 (Hsp70) stimulates proliferation and cytolytic activity of natural killer cells. *Experimental Hematology, 27*, 1627–1636.

Multhoff, G., Pfister, K., Gehrmann, M., Hantschel, M., Gross, C., Hafner, M., & Hiddemann, W. (2001). A 14-mer Hsp70 peptide stimulates natural killer (NK) cell activity. *Cell Stress & Chaperones, 6*, 337–344.

Nickel, W., & Seedorf, M. (2008). Unconventional mechanisms of protein transport to the cell surface of eukaryotic cells. *Annual Review of Cell and Developmental Biology, 24*, 287–308.

Nieland, T. J., Tan, M. C., Monne-van Muijen, M., Koning, F., Kruisbeek, A. M., & van Bleek, G. M. (1996). Isolation of an immunodominant viral peptide that is endogenously bound to the stress protein GP96/GRP94. *Proceedings of the National Academy of Sciences of the United States of America, 93*, 6135–6139.

Noble, E. G., Milne, K. J., & Melling, C. W. (2008). Heat shock proteins and exercise: A primer. *Applied Physiology, Nutrition, and Metabolism, 33*, 1050–1065.

Noessner, E., Gastpar, R., Milani, V., Brandl, A., Hutzler, P. J., Kuppner, M. C., Roos, M., Kremmer, E., Asea, A., Calderwood, S. K., & Issels, R. D. (2002). Tumor-derived heat shock protein 70 peptide complexes are cross-presented by human dendritic cells. *Journal of Immunology, 169*, 5424–5432.

Novoselova, T. V., Margulis, B. A., Novoselov, S. S., Sapozhnikov, A. M., van der Spuy, J., Cheetham, M. E., & Guzhova, I. V. (2005). Treatment with extracellular HSP70/HSC70 protein can reduce polyglutamine toxicity and aggregation. *Journal of Neurochemistry, 94*, 597–606.

Nylandsted, J., Rohde, M., Brand, K., Bastholm, L., Elling, F., & Jäättelä, M. (2000). Selective depletion of heat shock protein 70 (Hsp70) activates a tumor-specific death program that is independent of caspases and bypasses Bcl-2. *Proceedings of the National Academy of Sciences of the United States of America, 97*, 7871–7876.

Ortega, E., Giraldo, E., Hinchado, M. D., Martinez, M., Ibanez, S., Cidoncha, A., Collazos, M. E., & Garcia, J. J. (2006). Role of Hsp72 and norepinephrine in the moderate exercise-induced stimulation of neutrophils' microbicide capacity. *European Journal of Applied Physiology, 98*, 250–255.

Ortega, E., Hinchado, M. D., Martin-Cordero, L., & Asea, A. (2009). The effect of stress-inducible extracellular Hsp72 on human neutrophil chemotaxis: A role during acute intense exercise. *Stress, 12*, 240–249.

Panjwani, N. N., Popova, L., & Srivastava, P. K. (2002). Heat shock proteins gp96 and hsp70 activate the release of nitric oxide by APCs. *Journal of Immunology, 168*, 2997–3003.

Park, C. J., Park, S. A., Yoon, T. G., Lee, S. J., Yum, K. W., & Kim, H. J. (2005). Bupivacaine induces apoptosis via ROS in the Schwann cell line. *Journal of Dental Research, 84*, 852–857.

Patel, B., Khaliq, A., Jarvis-Evans, J., Boulton, M., Arrol, S., Mackness, M., & McLeod, D. (1995). Hypoxia induces HSP 70 gene expression in human hepatoma (HEP G2) cells. *Biochemistry and Molecular Biology International, 36*, 907–912.

Pawaria, S., & Binder, R. J. (2011). CD91-dependent programming of T-helper cell responses following heat shock protein immunization. *Nature Communications, 2*, 1–11.

Pierzchalski, P., Jastrzebska, M., Link-Lenczowski, P., Leja-Szpak, A., Bonior, J., Jaworek, J., Okon, K., & Wojcik, P. (2014). The dynamics of heat shock system activation in Monomac-6 cells upon Helicobacter pylori infection. *Journal of Physiology and Pharmacology, 65*, 791–800.

Planas, A. M., Soriano, M. A., Estrada, A., Sanz, O., Martin, F., & Ferrer, I. (1997). The heat shock stress response after brain lesions: Induction of 72 kDa heat shock protein (cell types involved, axonal transport, transcriptional regulation) and protein synthesis inhibition. *Progress in Neurobiology, 51*, 607–636.

Pockley, A. G., Shepherd, J., & Corton, J. M. (1998). Detection of heat shock protein 70 (Hsp70) and anti-Hsp70 antibodies in the serum of normal individuals. *Immunological Investigations, 27*, 367–377.

Pockley, A. G., De Faire, U., Kiessling, R., Lemne, C., Thulin, T., & Frostegård, J. (2002). Circulating heat shock protein and heat shock antibody levels in established hypertension. *Journal of Hypertension, 20*, 1815–1820.

Pockley, A. G., Georgiades, A., Thulin, T., de Faire, U., & Frostegård, J. (2003). Serum heat shock protein 70 levels predict the development of atherosclerosis in subjects with established hypertension. *Hypertension, 42*, 235–238.

Prakken, B. J., Wendling, U., van der Zee, R., Rutten, V. P., Kuis, W., & van Eden, W. (2001). Induction of IL-10 and inhibition of experimental arthritis are specific features of microbial heat shock proteins that are absent for other evolutionarily conserved immunodominant proteins. *Journal of Immunology, 167*, 4147–4153.

Qiao, Y., Liu, B., & Li, Z. (2008). Activation of NK cells by extracellular heat shock protein 70 through induction of NKG2D ligands on dendritic cells. *Cancer Immunity, 10*(8), 12.

Radsak, M. P., Hilf, N., Singh-Jasuja, H., Braedel, S., Brossart, P., Rammensee, H. G., & Schild, H. (2003). The heat shock protein gp96 binds to human neutrophils and monocytes and stimulates effector functions. *Blood, 101*, 2810–2281.

Richard, V., Kaeffer, N., & Thuillez, C. (1996). Delayed protection of the ischemic heart from pathophysiology to therapeutic applications. *Fundamental & Clinical Pharmacology, 10*, 409–415.

Ries, C., Egea, V., Karow, M., Kolb, H., Jochum, M., & Neth, P. (2007). MMP-2, MT1-MMP, and TIMP-2 are essential for the invasive capacity of human mesenchymal stem cells: Differential regulation by inflammatory cytokines. *Blood, 109*, 4055–4063.

Robinson, M. B., Tidwell, J. L., Gould, T., Taylor, A. R., Newbern, J. M., Graves, J., Tytell, M., & Milligan, C. E. (2005). Extracellular heat shock protein 70: A critical component for motoneuron survival. *The Journal of Neuroscience, 25*, 9735–9745.

Robinson, M. B., Taylor, A. R., Gifondorwa, D. J., Tytell, M., & Milligan, C. E. (2008). Exogenous Hsc70, but not thermal preconditioning, confers protection to motoneurons subjected to oxidative stress. *Developmental Neurobiology, 68*, 1–17.

Rodrigues-Krause, J., Krause, M., O'Hagan, C., De Vito, G., Boreham, C., Murphy, C., Newsholme, P., & Colleran, G. (2012). Divergence of intracellular and extracellular HSP72 in type 2 diabetes: Does fat matter? *Cell Stress & Chaperones, 17*, 293–302.

Said Ali, K., Ferencz, A., Nemcsok, J., & Hermesz, E. (2010). Expressions of heat shock and metallothionein genes in the heart of common carp (Cyprinus carpio): Effects of temperature shock and heavy metal exposure. *Acta Biologica Hungarica, 61*, 10–23.

Saito, K., Dai, Y., & Ohtsuka, K. (2005). Enhanced expression of heat shock proteins in gradually dying cells and their release from necrotically dead cells. *Experimental Cell Research, 310*, 229–236.

Schild, H., Arnold-Schild, D., Lammert, E., & Rammensee, H. G. (1999). Stress proteins and immunity mediated by cytotoxic T lymphocytes. *Current Opinion in Immunology, 11*, 109–113.

Sharp, F. R., Massa, S. M., & Swanson, R. A. (1999). Heat-shock protein protection. *Trends in Neurosciences, 22*, 97–99.

Shevchenko, M. A., Troyanova, N. I., Servuli, E. A., Bolkhovitina, E. L., Fedorina, A. S., & Sapozhnikov, A. M. (2016). Study of immunomodulatory effects of extracellular HSP70 in a mouse model of allergic airway inflammation. *Biochemistry (Mosc), 81*, 1384–1395.

Somensi, N., Brum, P. O., de Miranda Ramos, V., Gasparotto, J., Zanotto-Filho, A., Rostirolla, D. C., da Silva Morrone, M., Moreira, J. C. F., & Pens Gelain, D. (2017). Extracellular HSP70 activates ERK1/2, NF-kB and pro-inflammatory gene transcription through binding with RAGE in A549 human lung cancer cells. *Cellular Physiology and Biochemistry, 42*, 2507–2522.

Specht, H. M., Ahrens, N., Blankenstein, C., Duell, T., Fietkau, R., Gaipl, U. S., Günther, C., Gunther, S., Habl, G., Hautmann, H., Hautmann, M., Huber, R. M., Molls, M., Offner, R., Rödel, C., Rödel, F., Schütz, M., Combs, S. E., & Multhoff, G. (2015). Heat shock protein 70 (Hsp70) peptide activated natural killer (NK) cells for the treatment of patients with non-small cell lung cancer (NSCLC) after radiochemotherapy (RCTx) – from preclinical studies to a clinical phase II trial. *Frontiers in Immunology, 6*, 1–9.

Sprang, G. K., & Brown, I. R. (1987). Selective induction of a heat shock gene in fibre tracts and cerebellar neurons of the rabbit brain detected by in situ hybridization. *Brain Research, 427*, 89–93.

Srivastava, P. (2002). Roles of heat-shock proteins in innate and adaptive immunity. *Nature Reviews. Immunology, 2*, 185–194.

Stadelmann, C., Ludwin, S., Tabira, T., Guseo, A., Lucchinetti, C. F., Leel-Ossy, L., Ordinario, A. T., Brück, W., & Lassmann, H. (2005). Tissue preconditioning may explain concentric lesions in Balo's type of multiple sclerosis. *Brain, 128*, 979–987.

Stangl, S., Wortmann, A., Guertler, U., & Multhoff, G. (2006). Control of metastasized pancreatic carcinomas in SCID/beige mice with human IL-2/TKD-activated NK cells. *Journal of Immunology, 176*, 6270–6276.

Stangl, S., Themelis, G., Friedrich, L., Ntziachristos, V., Sarantopoulos, A., Molls, M., Skerra, A., & Multhoff, G. (2011). Detection of irradiation-induced, membrane heat shock protein 70 (Hsp70) in mouse tumors using Hsp70 Fab fragment. *Radiotherapy and Oncology, 99*, 313–316.

Stocki, P., & Dickinson, A. M. (2012). The immunosuppressive activity of heat shock protein70. *Autoimmune Disease, 2012*, 1–6.

Suto, R., & Srivastava, P. K. (1995). A mechanism for the specific immunogenicity of heat shock protein-chaperoned peptides. *Science, 269*, 1585–1588.

Svensson, P. A., Asea, A., Englund, M. C., Bausero, M. A., Jernås, M., Wiklund, O., Ohlsson, B. G., Carlsson, L. M., & Carlsson, B. (2006). Major role of HSP70 as a paracrine inducer of cytokine production in human oxidized LDL treated macrophages. *Atherosclerosis, 185*, 32–38.

Takeda, K., & Akira, S. (2004). TLR signaling pathways. *Seminars in Immunology, 16*, 3–9.

Tamura, Y., Peng, P., Liu, K., Daou, M., & Srivastava, P. K. (1997). Immunotherapy of tumors with autologous tumor-derived heat shock protein preparations. *Science, 278*, 117–120.

Tanaka, S., Kimura, Y., Mitani, A., Yamamoto, G., Nishimura, H., Spallek, R., Singh, M., Noguchi, T., & Yoshikai, Y. (1999). Activation of T cells recognizing an epitope of heat-shock protein 70 can protect against rat adjuvant arthritis. *Journal of Immunology, 163*, 5560–55605.

Taylor, A. R., Robinson, M. B., Gifondorwa, D. J., Tytell, M., & Milligan, C. E. (2007). Regulation of heat shock protein 70 release in astrocytes: Role of signaling kinases. *Developmental Neurobiology, 67*, 1815–1829.

Thériault, J. R., Adachi, H., & Calderwood, S. K. (2006). Role of scavenger receptors in the binding and internalization of heat shock protein 70. *Journal of Immunology, 177*, 8604–8611.

Théry, C. (2011). Exosomes: Secreted vesicles and intercellular communications. *F1000 Biology Reports, 3*, 1–8.

Théry, C., Regnault, A., Garin, J., Wolfers, J., Zitvogel, L., Ricciardi-Castagnoli, P., Raposo, G., & Amigorena, S. (1999). Molecular characterization of dendritic cell-derived exosomes. Selective accumulation of the heat shock protein hsc73. *The Journal of Cell Biology, 147*, 599–610.

Tidwell, J. L., Houenou, L. J., & Tytell, M. (2004). Administration of Hsp70 in vivo inhibits motor and sensory neuron de generation. *Cell Stress and Chaperones, 9*, 88–98.

Tobian, A. A., Canaday, D. H., Boom, W. H., & Harding, C. V. (2004). Bacterial heat shock proteins promote CD91-dependent class I MHC cross-presentation of chaperoned peptide to CD8+ T cells by cytosolic mechanisms in dendritic cells versus vacuolar mechanisms in macrophages. *Journal of Immunology, 172*, 5277–5286.

Triantafilou, M., Miyake, K., Golenbock, D. T., & Triantafilou, K. (2002). Mediators of innate immune recognition of bacteria concentrate in lipids rafts and facilitate lipopolysaccharide-induced cell activation. *Journal of Cell Science, 115*, 2603–2611.

Tsan, M. F., & Gao, B. (2009). Heat shock proteins and immune system. *Journal of Leukocyte Biology, 85*, 905–910.

Turturici, G., Sconzo, G., & Geraci, F. (2011). Hsp70 and its molecular role in nervous system diseases. *Biochemistry Research International, 2011*, 1–18.

Turturici, G., Tinnirello, R., Sconzo, G., Asea, A., Savettieri, G., Ragonese, P., & Geraci, F. (2014). Positive or negative involvement of heat shock proteins in multiple sclerosis pathogenesis: An overview. *Journal of Neuropathology and Experimental Neurology, 73*, 1092–1106.

Tytell, M. (2005). Release of heat shock proteins (HSP) and the effects of extracellular HSP on neural cells and tissues. *International Journal of Hyperthermia, 21*, 445–455.

Tytell, M., Greenberg, S. G., & Lasek, R. J. (1986). Heat shock-like protein is transferred from glia to axon. *Brain Research, 363*, 161–164.

Vabulas, R. M., Ahmad-Nejad, P., Ghose, S., Kirschning, C., Issels, R., & Wagner, H. (2002a). HSP70 as endogenous stimulus of the Toll/interleukin-1 receptor signal pathway. *The Journal of Biological Chemistry, 277*, 15107–15112.

Vabulas, R. M., Wagner, H., & Schild, H. (2002b). Heat shock proteins as ligands of toll-like receptors. *Current Topics in Microbiology and Immunology, 270*, 169–184.

Vega, V. L., Rodriguez-Silva, M., Frey, T., Gehrmann, M., Diaz, J. C., Steinem, C., Multhoff, G., Arispe, N., & De Maio, A. (2008). Hsp70 translocates into the plasma membrane after stress and is released into the extracellular environment in a membrane-associated form that activates macrophages. *Journal of Immunology, 180*, 4299–4307.

Voellmy, R. (2004). On mechanisms that control heat shock transcription factor activity in metazoan cells. *Cell Stress & Chaperones, 9*, 122–133.

Voos, W. (2013). Chaperone-protease networks in mitochondrial protein homeostasis. *Biochimica et Biophysica Acta, 1833*, 388–399.

Wang, Y., Kelly, C. G., Karttunen, J. T., Whittall, T., Lehner, P. J., Duncan, L., MacAry, P., Younson, J. S., Singh, M., Oehlmann, W., Cheng, G., Bergmeier, L., & Lehner, T. (2001). CD40 is a cellular receptor mediating mycobacterial heat shock protein 70 stimulation of CC-chemokines. *Immunity, 15*, 971–983.

Wang, Y., Whittall, T., McGowan, E., Younson, J., Kelly, C., Bergmeier, L. A., Singh, M., & Lehner, T. (2005). Identification of stimulating and inhibitory epitopes within the heat shock protein 70 molecule that modulate cytokine production and maturation of dendritic cells. *Journal of Immunology, 174*, 3306–3316.

Wang, R., Kovalchin, J. T., Muhlenkamp, P., & Chandawarkar, R. Y. (2006a). Exogenous heat shock protein 70 binds macrophage lipid raft microdomain and stimulates phagocytosis, processing, and MHC-II presentation of antigens. *Blood, 107*, 1636–1642.

Wang, R., Town, T., Gokarn, V., Flavell, R. A., & Chandawarkar, R. Y. (2006b). HSP70 enhances macrophage phagocytosis by interaction with lipid raft-associated TLR-7 and upregulating p38 MAPK and PI3K pathways. *The Journal of Surgical Research, 136*, 58–69.

Wei, Y. Q., Zhao, X., Kariya, Y., Teshigawara, K., & Uchida, A. (1995). Inhibition of proliferation and induction of apoptosis by abrogation of heat-shock protein (HSP) 70 expression in tumor cells. *Cancer Immunology, Immunotherapy, 40*, 73–78.

Wendling, U., Paul, L., van der Zee, R., Prakken, B., Singh, M., & van Eden, W. (2000). A conserved mycobacterial heat shock protein (hsp) 70 sequence prevents adjuvant arthritis upon nasal administration and induces IL-10-producing T cells that cross-react with the mammalian self-hsp70 homologue. *Journal of Immunology, 164*, 2711–2717.

Wieten, L., Broere, F., van der Zee, R., Koerkamp, E. K., Wagenaar, J., & van Eden, W. (2007). Cell stress induced HSP are targets of regulatory T cells: A role for HSP inducing compounds as anti-inflammatory immuno-modulators? *FEBS Letters, 581*, 3716–3722.

Wu, B., Hunt, C., & Morimoto, R. (1985). Structure and expression of the human gene encoding major heat shock protein HSP70. *Molecular and Cell Biology, 5*, 330–341.

Wu, F. H., Yuan, Y., Li, D., Liao, S. J., Yan, B., Wei, J. J., Zhou, Y. H., Zhu, J. H., Zhang, G. M., & Feng, Z. H. (2012). Extracellular HSPA1A promotes the growth of hepatocarcinoma by augmenting tumor cell proliferation and apoptosis-resistance. *Cancer Letters, 317*, 157–164.

Xie, J., Méndez, J. D., Méndez-Valenzuela, V., & Aguilar-Hernández, M. M. (2013). Cellular signalling of the receptor for advanced glycation end products (RAGE). *Cellular Signalling, 25*, 2185–2197.

Yang, X. M., Baxter, G. F., Heads, R. J., Yellon, D. M., Downey, J. M., & Cohen, M. V. (1996). Infarct limitation of the second window of protection in a conscious rabbit model. *Cardiovascular Research, 31*, 777–783.

Yang, X., Coriolan, D., Murthy, V., Schultz, K., Golenbock, D. T., & Beasley, D. (2005). Proinflammatory phenotype of vascular smooth muscle cells: Role of efficient Toll-like receptor 4 signaling. *American Journal of Physiology. Heart and Circulatory Physiology, 289*, H1069–H1076.

Yeh, C. H., Tseng, R., Zhang, Z., Cortes, J., O'Brien, S., Giles, F., Hannah, A., Estrov, Z., Keating, M., Kantarjian, H., & Albitar, M. (2009). Circulating heat shock protein 70 and progression in patients with chronic myeloid leukemia. *Leukemia Research, 33*, 212–217.

Yeh, C. H., Tseng, R., Hannah, A., Estrov, Z., Estey, E., Kantarjian, H., & Albitar, M. (2010). Clinical correlation of circulating heat shock protein 70 in acute leukemia. *Leukemia Research, 34*, 605–609.

Yokota, S., Chiba, S., Furuyama, H., & Fujii, N. (2010). Cerebrospinal fluids containing anti-HSP70 autoantibodies from multiple sclerosis patients augment HSP70-induced proinflammatory cytokine production in monocytic cells. *Journal of Neuroimmunology, 218*, 129–133.

Zhan, R., Leng, X., Liu, X., Wang, X., Gong, J., Yan, L., Wang, L., Wang, Y., Wang, X., & Qian, L. J. (2009). Heat shock protein 70 is secreted from endothelial cells by a non-classical pathway involving exosomes. *Biochemical and Biophysical Research Communications, 387*, 229–233.

Zhe, Y., Li, Y., Liu, D., Su, D. M., Liu, J. G., & Li, H. Y. (2016). Extracellular HSP70-peptide complexes promote the proliferation of hepatocellular carcinoma cells via TLR2/4/JNK1/2MAPK pathway. *Tumour Biology, 37*, 13951–13959.

Zhu, X., Zhao, X., Burkholder, W. F., Gragerov, A., Ogata, C. M., Gottesman, M. E., & Hendrickson, W. A. (1996). Structural analysis of substrate binding by the molecular chaperone DnaK. *Science, 272*, 1606–1614.

Zügel, U., Sponaas, A. M., Neckermann, J., Schoel, B., & Kaufmann, S. H. (2001). Gp96-peptide vaccination of mice against intracellular bacteria. *Infection and Immunity, 69*, 4164–4167.

HSP70 Is a Major Contributor to the MHCII Ligandome and Inducer of Regulatory T Cells

Willem van Eden, Femke Broere, and Ruurd van der Zee

Abstract Experimental models of autoimmunity have revealed anti-inflammatory effects of immunization with HSP70 or its derivative peptides. In depth cellular analysis of the effects of HSP70 immunization has shown the capacity of HSP70 to induce and expand self-tolerance promoting regulatory T cells (Tregs). In other words, in the models tolerance was re-established by the action of HSP70 specific Tregs. For the inflammation suppressive activity of antigen specific Tregs it is essential that the targeted antigen is ubiquitously expressed in the tissues. HSP70 family members, especially those that are stress-inducible, are widely expressed by stressed cells in the inflamed tissue due to the local presence of inflammatory mediators. In addition, cell stress is known to lead to autophagy, which in the case of chaperone mediated autophagy does lead to the preferential loading of HSP70 in MHC class II molecules. MHCII peptide elution profiles obtained from cells in a steady state have also revealed the dominating presence of HSP70 derived peptides in MHC class II molecules. For these reasons HSP70 is one of the most frequent cytosolic/nuclear MHCII natural ligand sources. HSP70, when presented by tolerizing antigen presenting cells in tissues, does induce Tregs, which seem to contribute to the tolerance promoting default setting of the healthy immune system.

Keywords Autoimmunity · Hsp70 · MHC · Peptide · Tolerance · Treg

Abbreviations

BMDC	bone marrow derived dendritic cells
CMA	chaperone mediated autophagy
DC	dendritic cells
ER	endoplasmic reticulum
ERK	extracellular signal regulated kinase
GAPDH	glyceraldehyde-3-phosphate dehydrogenase

W. van Eden (✉) · F. Broere · R. van der Zee
Division of Immunology, Faculty of Veterinary Medicine, Utrecht University, Utrecht, The Netherlands
e-mail: W.vaneden@uu.nl

A. A. A. Asea, P. Kaur (eds.), *HSP70 in Human Diseases and Disorders*, Heat Shock Proteins 14, https://doi.org/10.1007/978-3-319-89551-2_8

IL-10	interleukin 10
JNK	c-jun N-terminal kinase
LAG-3	lymphocyte activating gene-3
MAPK	mitogen activated protein kinase
MHC	major histocompatibility complex
MS	multiple sclerosis
NFkB	nuclear factor kappa beta
OVA	ovalbumin
PBMC	peripheral blood mononuclear cells
PBS	phosphate buffered saline
PGIA	proteoglycan induced arthritis
RA	rheumatoid arthritis
TcR	T cell receptor
TNF	tumor necrosis factor
tolDC	tolerized dendritic cells
Treg	regulatory T cells

Introduction

The first observations showing the disease suppressive effects of HSP in autoimmune disease models were made in the model of adjuvant arthritis in rats (van Eden et al. 1988). Since then the effects of HSP immunizations were analyzed in many different disease models, amongst which were arthritis, diabetes, EAE and allergies. Almost without exception HSP was found to prevent or diminish the expression disease (reviewed by (van Eden et al. 2005). In a study by Anderton et al. (1995) it was shown that mycobacterial HSP60 was protective in adjuvant arthritis due to the induction of a regulatory response of T cells that cross-recognized the mammalian HSP60, as expressed in the tissues. By this and subsequent studies it was shown that due to their evolutionary conservation, microbial HSP were triggering self-HSP recognizing T cells with IL-10 mediated anti-inflammatory activities. In addition to this, when it was shown that other conserved and immunogenic microbial proteins had no disease inhibitory effect in the models, it became apparent that HSP were exceptional in this quality of modulating inflammatory diseases (Prakken et al. 2001). It is possible that besides conservation, also the stress inducibility in tissue adds to the unique features of HSP as disease modulators.

HSP70 and Its Anti-inflammatory Effects in Experimental Models

For HSP70, disease suppressive effects in adjuvant arthritis were seen with *M. tuberculosis* derived recombinant HSP70. Again, a conserved mycobacterial HSP70 sequence was found to be immunogenic and to induce T cells that cross-reacted with the rat homologue sequence. In this case parenteral immunization with the peptide containing the critical cross-reactive T cell epitope did not suppress disease. Upon analysis of cytokines produced by the peptide-specific T cells, IL-10 production was found, as was the case with T cells responsive to whole HSP70 protein. Nasal administration of this peptide led to inhibition of subsequent adjuvant arthritis induction (Wendling et al. 2000). Another HSP70 family member, ER chaperone BiP, a 78-kDa glucose regulated protein, was proposed as an autoantigen in RA, as it stimulated proliferative responses of synovium derived T cells from RA patients and not from patients with other joint diseases. Attempts to induce arthritis with BiP in CFA failed, and in line with earlier observations, also this immunization led to resistance against induction of arthritis (Corrigall et al. 2001). Interestingly, more recently, BiP was studied in a first phase clinical trial in RA patients (see below).

A recent analysis of mycobacterial HSP70 was performed in the model of proteoglycan induced arthritis (PGIA) in Balb/c mice. First of all, a mapping exercise was done with mycobacterial HSP70, which revealed the presence of a very conserved T cell epitope at positions 141–155, which we termed the "B29" epitope. When this epitope, in the form of the B29 synthetic peptide, was administered intranasally in PBS (100 ugr) at days -7, -5 and -3 prior to PG immunizations at day 0, a reduced arthritis was noted. And this was also seen in control groups that received whole mycobacterial HSP70 in PBS (30 ugr) and not with control peptide pOVA (100 ugr). Follow-up experiments were then performed with B29 immunization in the presence of DDA as adjuvant, in naïve recipient mice. Splenocytes were sampled 10 days later and from these cells CD4+ CD25+ T cells were collected by cell-sorting and found to be suppressive in a standard suppression assay, using CD3 antibody stimulation. Phenotypic analysis showed upregulated neuropilin (Nrp-1) and LAG-3, both known as markers of regulatory T cells. In addition the percentage of IL-10 positive T cells was also increased. Not surprisingly, a similar phenotypic and functional behavior was noted in these assays with splenocytes from pOVA immunized mice. Interestingly however, when CD4+ CD25+ T cells were transferred into naïve recipient mice prior to induction of PGIA, only the B29 primed T cells protected from disease, whereas the pOVA primed T cells did not. Upon interpretation of all available findings it was concluded that the B29 peptide was inducing a T cell response that cross-reacted with the self-homologs present in the mouse. And indeed, the B29 specific T cell were also showing *in vitro* proliferative responses in the presence of mammalian B29a (mB29a) and B29b (mB29b). Therefore, on the basis of the cross-recognition of endogenously expressed HSP70 self-homologs, the B29 induced Tregs suppressed disease whereas the pOVA induced Tregs did not.

Subsequent studies showed an exceptional tolerance promoting impact of the B29 induced Tregs in the PGIA model. Making use of a congenic marker (CD90.1) present on the transferred Tregs, it was shown that the transferred T cells survived for a long period of time. Even 50 days after transfer, the Tregs were still present in spleen, draining lymph-nodes and the joints. And they were still having the phenotypic qualities of Tregs. Interestingly, when 5 weeks after transfer a depleting CD90.1 monoclonal antibody was infused, the disease suppressive activity of the Tregs was halted. Disease returned in the anti-CD90.1 infused animals and not in the controls, where apparently the Tregs were still actively engaged in suppression of arthritis. The prevention of disease by early transfer of B29 Tregs was possible with as few as 3.10E5 transferred CD4+ CD25+ T cells. When the T cells were selected on the basis of LAG-3 expression, even the minimal number of 4000 T cells was sufficient to suppress disease (van Herwijnen et al. 2012). Altogether, the B29 mouse experiments have indicated that the targeting of endogenous HSP70 for Treg recognition is a potentially powerful intervention to restore tolerance in inflammatory conditions.

HSP70 in Human Autoimmune Diseases

Analysis of T cell responses to mycobacterial HSP60 in juvenile idiopathic arthritis has shown that such responses were present mainly in cases with self-remitting forms of the disease, namely the subgroup of patients with so-called oligo-articular arthritis and not in the systemic or poly-articular patients. In addition, it was seen that HSP60 specific T cell responses peaked just prior to disease remission. These findings were suggestive of a protective effect of HSP60 specific T cell responses (De Graeff-Meeder et al. 1991, 1995; Prakken et al. 1996). Possibly, similar protective effects can be seen for HSP70. In the case of multiple sclerosis (MS), the presence of HSP70 specific auto-antibodies were found to be a characteristic of patients with relapsing-remitting MS. In patients with primary or secondary progressive MS such antibodies were not seen (Quintana et al. 2008). For BiP comprehensive mapping exercises have been performed with human peripheral blood mononuclear cells (PBMC). In a study by Shoda et al. both effector and regulatory T cell epitopes were identified to be present in BiP (Shoda et al. 2015). The BiP epitope 336–355 was found to induce IL-10 secretion in CD25+ PBMCs obtained from RA patients. In addition, suppression of proliferative T cell responses and pro-inflammatory cytokine production was noted. In mice suppression of disease and induction of Foxp3+ Tregs was obtained following immunization with this epitope.

A recent phase 1 clinical trial with BiP in RA patients showed, besides safety of the approach, some positive clinical effects at the higher doses of peptide intravenously administered (5 and 15 mgr) and not a the 1 mgr dosis. In addition to slight clinical effects, C-reactive protein (CRP), an acute phase protein that acts as a marker of inflammation, was significantly suppressed (Kirkham et al. 2016). Besides induction of Tregs, HSP70 seems to mediate other potentially anti-inflammatory effects. In RA fibroblast-like synoviocytes, the TNF induced production of pro-inflammatory cytokines was inhibited by addition of human HSP70. This effect was claimed to be caused by inhibition of ERK, JNK and p38 MAPK and by inhibitory effects on NF-κB (Luo et al. 2008). Others have shown the HSP70 mediated degradation of the p65 subunit of NF-κB (Tanaka et al. 2014).

Presence of HSP70 Fragments in MHCII in Various Tissues

Elution studies have shown the frequent presence of HSP70 derived protein fragments (peptides) in the groove of MHCII molecules. This can be learned, amongst others, from the SYFPEITHY database of peptides known to bind MHC molecules. In the supplementary data of the Paludan paper (Paludan et al. 2005) it is argued that the cytosol or nucleus derived natural peptides in MHCII are mostly originating from long-lived proteins and their paper shows that among them HSP70 and HSC70 next to GAPDH stand out. And this can be most relevant for the recognition of the HSP70 peptides by Tregs. Tregs are CD4+ T cells that recognize antigens in the context of MHCII. And, as indicated by Shevach, "*it is unlikely that suppression is secondary to the presence of the very small number of autoantigen-specific Treg cells present in the polyclonal population. A more likely scenario is that polyclonal Treg cells are able to control various responses because they are continuously being activated via their TcR by complexes of MHC class II and ubiquitous self-peptides*" (Shevach 2009). It seems that HSP70 is a significant provider of such ubiquitous self-peptides. And as already shown in the case of the B29 peptide of HSP70, the recognizing Tregs can be demonstrated to exist.

A relevant question in this regard is of course how central tolerance with respect to these peptides is organized in the thymus. Following our common understanding of the thymic selection process, it is possible that high affinity T cells will be negatively selected through deletion, whereas intermediate and still relatively high affinities will be leading to the generation of Tregs. A study that explored the MHC-peptide matrix in the human thymus has already revealed the presence of HSP70 fragments in the MHCII molecules of positively selecting thymic epithelial cells (Adamopoulou et al. 2013). And interestingly, a relatively high number of stress response related protein fragments was detected in the thymic DC depleted MHCII positive antigen presenting cell preparations. Also a larger protein fragment was found in the MHC cleft of these cells that included the HSP70-B29 mammalian homolog.

Dengjel et al. have analyzed the MHCII ligandome obtained from nutrient deprived human B cells (Dengjel et al. 2005). The stress caused by nutrient deprivation had led to autophagy, which had influenced the loading of the MHCII compartments of the cell. Possibly through the mechanism of so-called chaperone mediated autophagy (CMA), a chaperone dependent targeting of cytosolic proteins to lysosomes, a preferential loading of MHCII with HSP70 fragments was seen. In the cleft of the HLA-DR4 molecules in this case also our HSP70-B29 was present. Given the known association of HLA-DR4 with RA, this finding is of interest. Apparently, also RA patients with disease predisposing HLA molecules have in principle the genetic capacity to present a proposed disease protective peptide to their T cells. And indeed, with the use of HLA-DR-B29 tetramers the presence of B29 specific Tregs has now been shown in the human T cell repertoire (de Wolf et al. 2016). And very similar to what was seen in mice, the B29 specific T cells were again cross-reactive with the mammalian B29 homologs mB29a and mB29b.

HSP70 Loaded MHCII Is a Treg Inducer

The studies with HSP70-B29 have shown that HSP70 harbors epitopes that can be used to induce disease suppressive anti-inflammatory Tregs (van Herwijnen et al. 2012). And also that these Tregs are functionally active through their cross-recognition of the endogenously expressed and presented mammalian homologs of peptide B29. However, whether or not the endogenous HSP by itself can function as an inducer of Tregs remained unclear. And certainly one may ask the question, whether upregulated HSP at sites of tissue inflammation would be capable of doing so.

In studies by Wieten (Wieten et al. 2010) we have exploited the capacity of a so-called HSP-co-inducer to increase the expression of HSP in the context of cell-stress. This co-inducer was carvacrol, an essential oil obtained from Oregano plant species. When cells, mouse splenocytes or peripheral blood lymphocytes, were exposed to carvacrol for one hour and after that heated (42.5 °C) or exposed to low dose arsenite for two hours and rested in overnight culture, there was a firmly upregulated expression of HSP70 visible with intracellular staining's. When administered intragastrically in mice, carvacrol treated mice showed an upregulated HSP70 in the Peyer's patches, the lymphoid organs lining the gut. Upon analysis, the mesenteric lymphnodes and speen cells appeared to be enriched for Foxp3 positive T cells in carvacrol treated mice and transfer of these cells (selected on the basis of CD3 expression) led to suppression of PGIA in recipient animals. Herewith it seemed that the *in vivo* upregulated HSP70 expression had indeed induced a regulatory T cell population with disease suppressive functionality. When T cell responses against HSP70 were analyzed in carvacrol treated mice, it was found that HSP70 specific T cells had expanded as a consequence of the intragastric carvacrol administration. In more general terms, we have shown herewith that food components can boost the protective cellular response to stress and that the immune system may transduce such information into regulation and suppression of inflammation. What we eat can up-regulate immune regulatory T cells and down-regulate disease.

The induction of Tregs depends most likely on the presence of so-called tolerant DCs (tolDC). Generally, tolDC are characterized by low expression of T cell co-stimulatory molecules, low production of proinflammatory cytokines and high production of immunoregulatory cytokines compared with immunogenic DC (Stoop et al. 2011). It seems that the endogenous upregulation of HSP70 in DCs can turn immunogenic DC into tolerogenic DC. When mouse bone marrow derived DCs (BMDC) was exposed to carvacrol in combination with thermal stress, it turned out that these treated DCs had a reduced capacity to activate pro-inflammatory T cells (Spiering et al. 2012). When such treated DCs were transferred into naïve recipient mice together with CFSE labelled OVA specific T cells, it was found that, of the transferred T cells, especially the Foxp3+ Treg population had expanded. In addition, injection of the carvacrol-thermal stress treated BMDC caused a prophylactic suppression of subsequently induced PGIA.

Altogether, experiments with HSP co-induction have shown that upregulated endogenous HSP70 does trigger an anti-inflammatory Treg cell population. It is possible that the above mentioned effect of HSP70 on NFκB does contribute to this through the production of tolDC. A specific and irreversible alternative NFκB inhibitor, Bay11–7082 was already tested in a clinical trial using the autologous tolDC as a treatment for RA (Benham et al. 2015). In the same vein, we are preparing grounds for such an autologous tolDC trial, with B29 loaded tolDC. In the latter case, we will follow a recently published protocol of tolDC prepared with exposure of DC to corticosteroids in combination with vitamin D3 (Bell et al. 2017).

Conclusions

Regulatory T cells (Treg) are effector T cells that have the capacity to control inflammation and to maintain herewith the state of self-tolerance. Effective control of inflammation will depend on the cognate interactions with abundantly expressed self-antigens at the site of inflammation. Due to stress imposed on cells exposed to inflammatory mediators HSP are abundantly expressed at sites of inflammation. In the absence of inflammation molecules such as HSP70 can function as default peptide donors for MHCII molecules, which always depend on peptide loading for their cell-surface expression. In this manner, HSP may have a function in the maintenance of self-tolerance.

Acknowledgements We thank the Dutch Arthritis Foundation for its financial support.

References

Adamopoulou, E., Tenzer, S., Hillen, N., Klug, P., Rota, I. A., Tietz, S., Gebhardt, M., Stevanovic, S., Schild, H., Tolosa, E., Melms, A., & Stoeckle, C. (2013). Exploring the MHC-peptide matrix of central tolerance in the human thymus. *Nature Communications, 4*, 2039.

Anderton, S. M., van der Zee, R., Prakken, B., Noordzij, A., & van Eden, W. (1995). Activation of T cells recognizing self 60-kD heat shock protein can protect against experimental arthritis. *The Journal of Experimental Medicine, 181*, 943–952.

Bell, G. M., Anderson, A. E., Diboll, J., Reece, R., Eltherington, O., Harry, R. A., Fouweather, T., MacDonald, C., Chadwick, T., McColl, E., Dunn, J., Dickinson, A. M., Hilkens, C. M., & Isaacs, J. D. (2017). Autologous tolerogenic dendritic cells for rheumatoid and inflammatory arthritis. *Annals of the Rheumatic Diseases, 76*, 227–234.

Benham, H., Nel, H. J., Law, S. C., Mehdi, A. M., Street, S., Ramnoruth, N., Pahau, H., Lee, B. T., Ng, J., Brunck, M. E., Hyde, C., Trouw, L. A., Dudek, N. L., Purcell, A. W., O'Sullivan, B. J., Connolly, J. E., Paul, S. K., Le-Cao, K. A., & Thomas, R. (2015). Citrullinated peptide dendritic cell immunotherapy in HLA risk genotype-positive rheumatoid arthritis patients. *Science Translational Medicine, 7*, 290ra87.

Corrigall, V. M., Bodman-Smith, M. D., Fife, M. S., Canas, B., Myers, L. K., Wooley, P., Soh, C., Staines, N. A., Pappin, D. J., Berlo, S. E., van Eden, W., van Der Zee, R., Lanchbury, J. S., & Panayi, G. S. (2001). The human endoplasmic reticulum molecular chaperone BiP is an autoantigen for rheumatoid arthritis and prevents the induction of experimental arthritis. *Journal of Immunology, 166*, 1492–1498.

De Graeff-Meeder, E. R., van der Zee, R., Rijkers, G. T., Schuurman, H. J., Kuis, W., Bijlsma, J. W., Zegers, B. J., & van Eden, W. (1991). Recognition of human 60 kD heat shock protein by mononuclear cells from patients with juvenile chronic arthritis. *Lancet, 337*, 1368–1372.

de Graeff-Meeder, E. R., van Eden, W., Rijkers, G. T., Prakken, B. J., Kuis, W., Voorhorst-Ogink, M. M., van der Zee, R., Schuurman, H. J., Helders, P. J., & Zegers, B. J. (1995). Juvenile chronic arthritis: T cell reactivity to human HSP60 in patients with a favorable course of arthritis. *The Journal of Clinical Investigation, 95*, 934–940.

de Wolf, C., van der Zee, R., den Braber, I., Glant, T., Maillere, B., Favry, E., van Lummel, M., Koning, F., Hoek, A., Ludwig, I., van Eden, W., & Broere, F. (2016). An arthritis-suppressive and treg cell-inducing CD4+ T cell epitope is functional in the context of HLA-restricted T cell responses. *Arthritis & Rhematology, 68*, 639–647.

Dengjel, J., Schoor, O., Fischer, R., Reich, M., Kraus, M., Muller, M., Kreymborg, K., Altenberend, F., Brandenburg, J., Kalbacher, H., Brock, R., Driessen, C., Rammensee, H. G., & Stevanovic, S. (2005). Autophagy promotes MHC class II presentation of peptides from intracellular source proteins. *Proceedings of the National Academy of Sciences of the United States of America, 102*, 7922–7927.

Kirkham, B., Chaabo, K., Hall, C., Garrood, T., Mant, T., Allen, E., Vincent, A., Vasconcelos, J. C., Prevost, A. T., Panayi, G. S., & Corrigall, V. M. (2016). Safety and patient response as indicated by biomarker changes to binding immunoglobulin protein in the phase I/IIA RAGULA clinical trial in rheumatoid arthritis. *Rheumatology (Oxford), 55*, 1993–2000.

Luo, X., Zuo, X., Zhou, Y., Zhang, B., Shi, Y., Liu, M., Wang, K., McMillian, D. R., & Xiao, X. (2008). Extracellular heat shock protein 70 inhibits tumour necrosis factor-alpha induced proinflammatory mediator production in fibroblast-like synoviocytes. *Arthritis Research & Therapy, 10*, R41.

Paludan, C., Schmid, D., Landthaler, M., Vockerodt, M., Kube, D., Tuschl, T., & Munz, C. (2005). Endogenous MHC class II processing of a viral nuclear antigen after autophagy. *Science, 307*, 593–596.

Prakken, A. B., van Eden, W., Rijkers, G. T., Kuis, W., Toebes, E. A., de Graeff-Meeder, E. R., van der Zee, R., & Zegers, B. J. (1996). Autoreactivity to human heat-shock protein 60 predicts disease remission in oligoarticular juvenile rheumatoid arthritis. *Arthritis and Rheumatism, 39*, 1826–1832.

Prakken, B. J., Wendling, U., van der Zee, R., Rutten, V. P., Kuis, W., & van Eden, W. (2001). Induction of IL-10 and inhibition of experimental arthritis are specific features of microbial heat shock proteins that are absent for other evolutionarily conserved immunodominant proteins. *Journal of Immunology, 167*, 4147–4153.

Quintana, F. J., Basso, A. S., Iglesias, A. H., Korn, T., Farez, M. F., Bettelli, E., Caccamo, M., Oukka, M., & Weiner, H. L. (2008). Control of T(reg) and T(H)17 cell differentiation by the aryl hydrocarbon receptor. *Nature, 453*, 65–71.

Shevach, E. M. (2009). Mechanisms of foxp3+ T regulatory cell-mediated suppression. *Immunity, 30*, 636–645.

Shoda, H., Fujio, K., Sakurai, K., Ishigaki, K., Nagafuchi, Y., Shibuya, M., Sumitomo, S., Okamura, T., & Yamamoto, K. (2015). Autoantigen BiP-derived HLA-DR4 epitopes differentially recognized by effector and regulatory T cells in rheumatoid arthritis. *Arthritis & Rhematology, 67*, 1171–1181.

Spiering, R., van der Zee, R., Wagenaar, J., Kapetis, D., Zolezzi, F., van Eden, W., & Broere, F. (2012). Tolerogenic dendritic cells that inhibit autoimmune arthritis can be induced by a combination of carvacrol and thermal stress. *PLoS One, 7*, e46336.

Stoop, J. N., Robinson, J. H., & Hilkens, C. M. (2011). Developing tolerogenic dendritic cell therapy for rheumatoid arthritis: What can we learn from mouse models? *Annals of the Rheumatic Diseases, 70*, 1526–1533.

Tanaka, T., Shibazaki, A., Ono, R., & Kaisho, T. (2014). HSP70 mediates degradation of the p65 subunit of nuclear factor kappaB to inhibit inflammatory signaling. *Science Signaling, 7*, ra119.

van Eden, W., Thole, J. E., van der Zee, R., Noordzij, A., van Embden, J. D., Hensen, E. J., & Cohen, I. R. (1988). Cloning of the mycobacterial epitope recognized by T lymphocytes in adjuvant arthritis. *Nature, 331*, 171–173.

van Eden, W., van der Zee, R., & Prakken, B. (2005). Heat-shock proteins induce T-cell regulation of chronic inflammation. *Nature Reviews. Immunology, 5*, 318–330.

van Herwijnen, M. J., Wieten, L., van der Zee, R., van Kooten, P. J., Wagenaar-Hilbers, J. P., Hoek, A., den Braber, I., Anderton, S. M., Singh, M., Meiring, H. D., van Els, C. A., van Eden, W., & Broere, F. (2012). Regulatory T cells that recognize a ubiquitous stress-inducible self-antigen are long-lived suppressors of autoimmune arthritis. *Proceedings of the National Academy of Sciences of the United States of America, 109*, 14134–14139.

Wendling, U., Paul, L., van der Zee, R., Prakken, B., Singh, M., & van Eden, W. (2000). A conserved mycobacterial heat shock protein (hsp) 70 sequence prevents adjuvant arthritis upon nasal administration and induces IL-10-producing T cells that cross-react with the mammalian self-hsp70 homologue. *Journal of Immunology, 164*, 2711–2717.

Wieten, L., van der Zee, R., Spiering, R., Wagenaar-Hilbers, J., van Kooten, P., Broere, F., & van Eden, W. (2010). A novel HSP co-inducer boosts stress protein HSP70 to activate T cell regulation of inflammation in autoimmune arthritis. *Arthritis and Rheumatism, 62*, 1026–1035.

Immune Properties of HSP70

Yves Delneste, Vincent Larochette, and Pascale Jeannin

Abstract In addition to their conventional chaperon activity, numerous studies have reported that heat shock protein 70 (HSP0) exhibit immune properties and especially the capacity (i) to induce the presentation and cross-presentation of associated or client proteins and, (ii) to control myeloid cell activation. Several studies were focused on the identification of HSP70-binding elements that contribute to their immune properties. A general consensus was reached on the nature of the endocytic receptors involved in the internalization of extracellular HSP70 with belong, for most of them, to the innate immunity receptor family. However, the nature of signaling receptors recruited by HSP70 remains unclear, because the stimulatory versus regulatory properties of HSP70 remains a subject of debate. Nevertheless, these unique immune properties allowed developing innovative prophylactic and therapeutic vaccines, especially in the treatment of cancers and chronic viral infections. Although HSP70 constitute potent vaccine vehicles in different preclinical models, clinical studies remain disappointing. The fact that the immune properties of HSP70 have not been totally clarified may explain their relative efficacy in human. In this review are presented the main immune properties of HSP70 related to the HSP70-binding elements identified to date, and discuss our current knowledge on their intrinsic immune properties.

Keywords Adaptive immunity · Hsp70 · Immune regulation · Innate immune receptors · Innate immunity · Myeloid cell · Vaccine

Y. Delneste (✉) · P. Jeannin
Laboratory of Immunology and Allergology, University Hospital of Angers, Angers, France

CRCINA, INSERM, Université de Nantes, Université d'Angers, Angers, France

LabEx ImmunoGraftOnco, Angers, France
e-mail: yves.delneste@inserm.fr

V. Larochette
CRCINA, INSERM, Université de Nantes, Université d'Angers, Angers, France

LabEx ImmunoGraftOnco, Angers, France

A. A. A. Asea, P. Kaur (eds.), *HSP70 in Human Diseases and Disorders*, Heat Shock Proteins 14, https://doi.org/10.1007/978-3-319-89551-2_9

Abbreviations

A2M	alpha2 macroglobulin
Ac-LDL	acetylated low-density lipoprotein
ADP	adenosine dipohosphate
AGE	advanced glycation end product
AIF	apoptosis inducing factor
APAf-1	apoptotic peptidase activating factor 1
APC	antigen-presenting cell
APOER	apolipoprotein E receptor
ATP	adenosine triphosphate
Bax	Bcl-2-associated X protein
BCR	B cell receptor
BiP	binding immunoglobulin protein
CCL	C-C motif ligand
CCR	C-C chemokine receptor
CD	cluster of differentiation
CLEC8A	C-type lectin domain family 8 member A
CLEVER-1	common lymphatic endothelial and vascular endothelial receptor-1
CTL	cytotoxic T cell
DAMP	danger-associated molecular pattern
DC-SIGN	dendritic cell-specific ICAM-grabbing non-integrin
EBV	Epstein-Barr virus
EGF-like	and link domain-containing scavenger receptor-1
ER	endoplasmic reticulum
ERK	Extracellular signal-regulated kinases
FAT	fatty acid translocase
FEEL-1	fasciclin EGF-like laminin-type
HBV	hepatitis B virus
HCV	hepatitis C virus
Her2/Neu	human epidermal growth factor receptor 2/proto-oncogene Neu
HLA	human leukocyte antigen
HMGB1	high–mobility group box 1
HPV	human papilloma virus
HSP	heat shock protein
IFN	interferon
IL	interleukin
IRAK	IL-1 receptor-associated kinase
IRE1α	inositol requiring enzyme 1α
JAK	Janus kinase
LBP	LPS-binding protein
LDL	low-density lipoprotein
LRP1	low density lipoprotein receptor-related protein 1
MAGE-1	melanoma-associated antigen 1

Mart-1	melanoma antigen recognized by T-cells 1
MD2	myeloid differentiation factor 2
MDSC	myeloid-derived suppressive cells
MHC	major histocompatibility complex
MyD88	Myeloid differentiation primary response 88
NBD	nucleotide-binding domain
NF-κB	nuclear factor-kappa B
NK	natural killer
Ox-LDL	oxidized low-density lipoprotein
PAMP	pathogen-associated molecular pattern
PDZK	PDZ domain-containing protein 1
PRM	pattern recognition molecule
PRR	pattern recognition receptor
PSA	prostate-specific antigen
PTX3	pentraxin 3
RAP	receptor-associated protein
SBD	substrate-binding domain
SIGLEC	sialic-acid-binding immunoglobulin-like lectins
SP-D	surfactant protein D
SREC	scavenger receptor expressed by endothelial cells
STAT	signal transducers and activators of transcription
TAB1	TAK1-binding protein 1
TAK1	TGFβ-activated kinase
TAM	tumor-associated macrophages
TCR	T cell receptor
Th	helper T cell
TLR	toll-like receptor
TNF	tumor necrosis factor
TNFSF	TNF superfamily
TRAF	TNF receptor-associated factor
Trp2	tyrosinase-related protein 2
TSP-1	thrombospondin 1

Introduction

Heat shock proteins (HSP) are involved in 3D-folding of newly synthesized proteins and protect them against endogenous and exogenous assaults. Besides their classical biochemical properties, several studies have demonstrated that HSP70, one of the largest and highly conserved family of HSP, exhibit intrinsic immune properties. The potential immune properties of HSP70 were first hypothesized in the 1970s by showing that proteins with an apparent molecular weight of 70 kDa and isolated from tumor cells may induce a protective immune response in vivo, in the absence of adjuvant. Thereafter, a huge quantity of studies aimed at deciphering, in vivo and in vitro, the biological mechanisms involved in the immune properties of HSP. Most of the immune properties of HSP were first elucidated for gp96 (HSP90B1), a member of the HSP90 family, and thereafter for HSP70. One of the most remarkable immune properties of HSP70 is their ability to mediate the cross-presentation of exogenous antigens and to initiate protective antitumor immune responses. In agreement with the cross-presentation process, HSP70 have been shown to interact with different immune receptors, especially innate receptors. However, the exact nature of the endocytic and signaling receptors engaged by HSP70, as well as whether HSP70 exhibit stimulatory or regulatory immune properties, remain a subject of debate. Nevertheless, and whatever the mechanism involved, HSP70 constitute interesting vehicles to induce, in vivo, antigen-specific cytotoxic responses. This review thus addresses the immune properties of HSP70, with a focus on the innate immune receptors engaged, and the consequences on vaccine strategies.

HSP70 Family of Chaperones

Common Features of HSP70

The 70 kDa heat shock proteins (HSP70s) constitute a family of highly conserved proteins that are ubiquitously expressed in prokaryotes and eukaryotes. Prokaryotes express three HSP70 proteins (DnaK, HscA (Hsc66), and HscC (Hsc62)) whereas eukaryotes express several HSP70s. As an example, the human family comprises 13 proteins, that differ from each other by their amino acid sequence, their levels of expression and their localization (Radons 2016). The mostly expressed inducible human HSP70s are Hsp70–1 (encoded by the *HSPA1A* gene) and Hsp70–2 (*HSPA1B*); these two genes are closely located in the genome (Brocchieri et al. 2008). In line with the topic of this review, it is important to mention human HSP70–14 (also referred to as Hsp70L1), a stress-induced HSP identified in dendritic cells, exhibits potent immunostimulatory properties by favoring Th1 responses (Wan et al. 2004). Interestingly, the levels of HSP0L1 are elevated in some tumors (Yang et al. 2015).

HSP70s display a common functional domain structure: (i) a 44 kDa N-terminal nucleotide binding domain (NBD) that binds and hydrolyzes ATP, (ii) a middle protease-sensitive domain and, (iii) a 28 kDa C-terminal substrate-binding domain (SBD) that can interact with 6–9 amino acid-long peptides; SBD has preferential affinity for neutral and hydrophobic amino acids. HSP70s are monomeric chaperons that participate, in physiological situations, to the folding of neo-synthesized proteins, to their transport and to the assembly of multi-protein complexes. The binding/release of polypeptides is dependent on the ATPase activity: the ADP- and

ATP-bound states favor peptide binding and release, respectively. HSP70s also regulate the activity of proteins and prevent their aggregation. Indeed, HSP70s exhibit an « unfoldase » activity, meaning that they recognize unfolded or aggregated proteins and unfold them before native refolding (Radons 2016). The expression of HSP70s is strongly upregulated by cellular stress, such as heat, toxic chemicals (such as heavy metals), ischemia, irradiations, infection, inflammation and nutrient deprivation. In stressed cells (induced by endogenous or exogenous challenges), HSP70s promote cell survival and allow cells to restore cellular homeostasis.

HSP70s act as anti-apoptoti proteins via their capacity to prevent mitochondrial translocation and activation of Bax (Yang et al. 2012) and to inhibit assembly of the death-inducing signaling complexes (Guo et al. 2005). HSP70s also inhibit the activity of different pro-apoptotic molecules (such as Apaf-1, AIF and caspase 3) (Beere et al. 2000; Ravagnan et al. 2001). HSP70s can protect cells against apoptosis induced by endoplasmic reticulum stress via their interaction with the ER stress sensor protein IRE1α (Wei et al. 2013).

Intracellular, Membrane and Extracellular HSP70

As the other HSP, members of the HSP70 family are mainly expressed in intracellular compartments, including cytosol, endoplasmic reticulum (such as binding immunoglobulin protein (BiP) or Grp78) and mitochondria (mtHsp70 or Grp75), to exert their chaperon activities. Nevertheless, HSP70s can be expressed at the membrane surface or present, in a soluble form, in different biological fluids. Membrane HSP70s (mHSP70) accumulate at the surface of infected and tumor cells but not living cells (Multhoff et al. 1995; Multhoff and Hightower 1996; Poccia et al. 1996). Different mechanisms have been proposed to explain the membrane relocalization of proteins devoid of secretion signals, such as the release of secretory granules (Evdonin et al. 2004; Mambula and Calderwood 2006a, b) and of exosomes (Gastpar et al. 2005; Lancaster and Febbraio 2005). The apparently « passive » accumulation of HSP70 at the cell membrane appears similar to the one reported for other intracellular molecules, such as PTX3 (Jaillon et al. 2009), or danger molecules, such as histones and nucleic acids (Cunin et al. 2016) that have also been reported to accumulate at the surface of stressed cells. Accordingly, mHSP70 may have an important role in the recognition of stressed and tumor cells by innate immune cells (Multhoff 2007; Radons and Multhoff 2005). Based on their biochemical properties, one can suspect that mHSP70 are associated to membrane proteins; however, to date, the nature of these putative mHSP70-associated molecules remains unknown.

Another explanation should be that extracellular HSP70s may bind to altered molecules, such as oxidized proteins, that are present at the surface of stressed cells, in a way similar to the one of bridging molecules (Henson 2017). HSP70s can be released by numerous cell types in response to various stimuli (Campisi and Fleshner 2003). Different ways of secretion of HSP70 have been described:

- A passive release by necrotic cells (Fleshner and Johnson 2005),
- The disruption of HSP70-containing secretory vesicles/exosomes released via a non-classical secretion pathway (MacKenzie et al. 2001) and,
- A release via the secretion of lysosomal endosomes (Baraldi et al. 2004; Mambula and Calderwood 2006b).

HSP70s can be used as biomarkers. As an example, mHSP70s are considered as selective markers of aggressive tumors. Circulating HSP70 have been proposed as biomarkers of inflammation in healthy subjects (Gehrmann et al. 2014a; Marotta et al. 2007). HSP70s have been detected in the serum of patients suffering from cancer or chronic infection (Pockley et al. 2014) and are proposed as biomarkers of tumor outcome after treatment (Gehrmann et al. 2014a, b). These data suggest that the immune properties of HSP70s may depend not only on the nature of the chaperoned peptides but also on their localization. Indeed, membrane and extracellular HSP70 may influence the nature and amplitude of signals delivered to immune cells present in the close vicinity or at distance.

HSP70 Induce Protective Immune Responses Against Tumors and Microbial Pathogens

Most of the studies on the use of HSP70s as vaccine vehicles have been performed in preclinical models of tumor growth. Nevertheless, and based on their vaccine potential, HSP70 have been also used for the induction of protective immune responses against microbes. In this paragraph, we focus on the antitumor immune responses induced by HSP70s.

HSP70s Mediate Antigen Presentation and Cross Presentation

The pioneering studies by the group of PK Srivastava demonstrated that tumor-derived gp96 initiate protective tumor-specific CTL responses, suggesting that HSP may act as major tumor-rejection antigens (Suto and Srivastava 1995; Tamura et al. 1997). This capacity to induce protective antitumor immune responses has been extended to other HSP families, especially HSP60 and HSP70 families (Castellino et al. 2000). The capacity of HSP70s to induce protective antitumor immune responses was mainly reported in both prophylactic and therapeutic murine models of tumor development (reviewed in (Srivastava 2002). Moreover, HSP70s also induce cross-presentation of human tumor antigens (Castelli et al. 2001; Milani et al. 2002; Noessner 2006; Noessner et al. 2002), the rationale for their use in tumor vaccines in humans. In fact, numerous studies have clearly demonstrated that the specificity of the anti-tumor immune response was determined by the chaperoned peptides (Binder and Srivastava 2005; Ishii et al. 1999; Suto and Srivastava 1995). Biochemical analysis have shown that peptides associated to HSP70 are very diverse, deriving from self-proteins, tumor, microbial or minor histocompatibility antigens (Srivastava 2002). As an example, HSP70s isolated from melanoma cell lines can chaperon peptides derived from the tumor antigens Mart-1, gp100, Trp2 and gp100 (Castelli et al. 2001; Noessner et al. 2002).

According to the classical view of antigen presentation, antigen-derived peptides are presented into the MHC class I (MHC-I) and MHC class II (MHC-II) molecules of antigen-presenting cells (APC) to epitope-specific T cell receptors (TCR) expressed by CD8+ and CD4+ T cells, respectively. This interaction is the basis of the antigen-specificity of the adaptive immunity. Although different myeloid cells are able to present antigens to T cells (such as macrophages and neutrophils), dendritic cells are the only APC able to prime naive T cells and to initiate immune responses (Banchereau et al. 2000; Cella et al. 1997).

The process by which some exogenous antigens are endocytosed by APCs, gain access to the MHC class I pathway, and stimulate CD8+ T cells is called cross-presentation (Heath and Carbone 2001; Yewdell et al. 1999). Indeed, the classical view of antigen presentation claimed that exogenous antigens endocytosed by APC are mainly loaded into the MHC-II molecules for recognition by CD4+ T cells, while, in contrast, endogenous antigens (self and viral proteins) are loaded in the MHC-I molecules for recognition by CD8+ T cells. Antigen cross-presentation has revolutionized our view of the induction of antigen-specific immune responses and allowed to propose « conventional » vaccine approaches for the treatment of cancers. The HSP-mediated antigen cross-presentation is dependent on three essential mechanisms:

- A receptor-mediated internalization of the antigens by professional APCs, especially dendritic cells,
- The functional maturation and activation of professional APC, rendering them able to prime naive T cells and,
- An intracellular trafficking allowing exogenous antigens/peptides to get access to the MHC-I presentation pathway.

One of the most important, and also most debated immune property of HSP, is their capacity to activate APCs. Indeed, the priming of CD8+ T cells by exogenous antigens requires that APC are fully activated (Banchereau et al. 2000), a status that may require CD4+ T cell help (Heath and Carbone 1999). The cross-presentation of an antigen by non-activated dendritic cells maintains or induces antigen-specific tolerance; this process is referred to as cross-tolerance. Ideally, the induction of functional CD8+ T cell responses against exogenous antigens may require both antigen cross-presentation in MHC-I molecules to CD8+ T cells and presentation in MHC-II molecules to CD4+ T cells. In agreement with the classical view of antigen presentation, HSP70 also induce antigen-specific CD4+ T cell activation, as evidenced by the induction of humoral immune responses and the interaction of microbial HSP70 with HLA-DR and their peptide fragments (Haug et al. 2007). Moreover, HSP70 chaperon both MHC-I and MHC-II epitopes (Stocki et al. 2010; 2011).

Potential Immunomodulatory Roles of Extracellular HSP70

The fact that (i) HSP70s mediate the cross-presentation of chaperoned peptides and (ii) that the cross-presentation requires receptor-mediated internalization of the peptide/chaperone complex, suggest that HSP70s have to be released in the extracellular milieu (Stocki and Dickinson 2012). According to their capacity to induce effective immune response, several studies reported that HSP70s are immunostimulatory molecules and, more interestingly, can induce the maturation and activation of dendritic cells, rendering them fully functional (reviewed in (Kuppner et al. 2001; Milani et al. 2002; Srivastava 2002). The term chaperokine was attributed to HSP to define this unique capacity of chaperons to activate immune cells (Asea et al. 2000). However, the intrinsic potential of HSP70s to activate APC remains a subject of debate (Borges et al. 2012) (see paragraph 5).

HSP70-Binding Elements

Antigen cross-presentation requires that exogenous antigens are internalized via endocytic receptors [3, 4]. Accordingly, HSP70s bind to dendritic cells and macrophages (Arnold-Schild et al. 1999; Todryk et al. 1999; Wassenberg et al. 1999) before being internalized in a receptor-dependent manner (Arnold-Schild et al. 1999; Basu et al. 2001; Binder et al. 2000; Castellino et al. 2000; Singh-Jasuja et al. 2000; Sondermann et al. 2000; Wassenberg et al. 1999). Most if not all HSP70-binding elements identified are innate immune receptors. These results are in agreement with the fact that, in an immunological point of view, extracellular HSP70 can be viewed (or detected) as a danger signal (modified self) released by altered/dying cells. These motifs are detected by the innate immune system.

By opposition to the adaptive immunity, innate immunity is defined as a non-antigen specific system. It is involved in numerous processes, such as antimicrobial activity, induction and resolution of inflammation, maintenance of tissue homeostasis and wound healing. The innate immune system includes a large variety of molecular and cellular actors, such as epithelial barriers, numerous soluble molecules (including the complement system) and innate lymphoid and myeloid cells. The most remarkable characteristic of innate immune cells is their capacity to discriminate self from non self (microbes) and altered or modified self (such as the detection of biochemical modification of cell surface molecules). The recognition of non self and modified self is mediated by a restricted number of molecules (compared to the TCR and BCR repertoires) called pattern recognition molecules (PRM); this term is now preferred to the ancient nomenclature pattern-recognition receptor (PRR). PRM recognize microbial moieties called pathogen-associated molecular patterns (PAMPs) and motifs expressed by altered self and called danger-associated molecular patterns (DAMPs). Remarkably, a same PRM can detect different PAMPs and DAMPs and exhibiting diverse biochemical characteristics (such as nucleic acids, lipids, proteins or glucids). Innate immune cells also orchestrate the adaptive immune response via the production of soluble immune mediators (cytokines and chemokines) and the priming/activation of antigen-specific lymphocytes, thanks to the antigen-presenting functions of myeloid cells.

PRM can be classified into three families, based on their functions: (i) endocytic receptors, involved in ligand recognition and internalization, (ii) signaling receptors, involved in ligand-induced cell activation and (iii) bridging molecules (also called opsonins), that bind to and favor the recognition of extracellular ligands by innate cells. To date, and except some individual cases, most of the HSP70-binding elements are endocytic and/or signaling PRM.

Endocytic Receptors

CD91

CD91 was the first HSP-binding element identified, initially as a gp96 receptor (Binder et al. 2000). CD91 was then reported as a receptor for human HSP70s on macrophages (Basu et al. 2001). CD91, also known as the α2 macroglobulin (A2M) receptor, low density lipoprotein receptor-related protein 1 (LRP1) or apolipoprotein E receptor (APOER), is an endocytic and signaling receptor belonging to the lipoprotein receptor family. CD91 is expressed by numerous cell types, including hepatocytes, fibroblasts, keratinocytes, smooth muscle cells and myeloid cells (Herz and Strickland 2001). CD91 is a multimeric receptor consisting of a 420 kDa α subunit, a 85 kDa β subunit and a 39-kDa associated molecule. CD91 binds to the activated form of α2M, a soluble molecule that binds to and inhibits a wide variety of proteinases and growth factors. CD91 is also suspected involved in lipid metabolism and can bind, in addition to activated A2M, tissue-specific plasminogen-activator–inhibitor complex and urokinase-PAI1 complex.

The identification of CD91 as an HSP-binding structure was mainly based on competitive binding assays with the CD91 ligand A2M and with a neutralizing anti-CD91 mAb. The role of CD91 in HSP70-mediated antigen presentation to CD4+ and CD8+ T cells was confirmed by several independent studies (Fischer et al. 2010; Salimu et al. 2015; Tobian et al. 2004a).

Scavenger Receptors

Scavenger receptors represent a family of non-related cell-surface glycoproteins that recognize a large repertoire of ligands, ranging from bacteria and yeast to self (native proteins) and modified-self such as oxidized LDL (Ox-LDL) and apoptotic cells (reviewed in (Yamada et al. 1998; Yu et al. 2015)). Some scavenger receptors can also bind chemically modified LDL (acetylated LDL) that constitutes reliable tools in identifying HSP-binding elements, especially as binding competitors. Scavenger receptors thus represent an important family of endocytic PRM through their ability to bind endogenous and exogenous danger molecules (Jeannin et al. 2008). Different studies reported that scavenger receptor-binding molecules, such as modified LDL (OxLDL or AcLDL), apoliprotein B, fucoidan and poly[IC], inhibited the binding of HSP70 to human APCs, suggesting that these PRM are the main cell surface HSP70 binding elements on human APCs (Delneste et al. 2002; Facciponte et al. 2007; Theriault et al. 2006; Theriault et al. 2005).

The first identified HSP70-binding scavenger receptors was SR-E1. Initially identified as an Ox-LDL receptor expressed by endothelial cells; this molecule, also known as LOX-1 or CLEC8A (C-type lectin domain family 8 member A), binds multiple ligands, including advanced glycation end products (AGE), activated platelets, and apoptotic cells. SR-E1 also binds exogenous ligands, such as virus and bacteria. In addition to endothelial cells, SR-E1 is constitutively expressed by macrophages and dendritic cells (Delneste et al. 2002). By using a collection of scavenger receptor-expressing CHO cells, we have identified SR-E1 as one of the main HSP70-binding element on human macrophages and dendritic cells. Interestingly, the in vitro and in vivo cross-presentation of an antigen coupled to HSP70 is dependent on SR-E1 internalization (Delneste et al. 2002). Finally, the in vivo targeting of an exogenous antigen to SR-E1, by coupling to an anti-SR-E1 antibody, induced an antigen-specific CD8+ T cell response (Delneste et al. 2002).

SR-E1 also exhibits features of C-type lectins (Sawamura et al. 1997) which represent another important family of highly conserved PRM. Several studies have underlined the important role played by C-type lectins in mediating antigen cross-presentation, both in vitro and in vivo. Among these PRM, one can mention CD205 (DEC205) (Bozzacco et al. 2007) and CD209 (DC-SIGN) (Garcia-Vallejo et al. 2013). However, no binding of HSP70s to some other C-type lectins, such as CD209 (Theriault et al. 2005) and CLEC7A (dectin-1), was reported so far. A binding of HSP70 family members was also reported to other scavenger receptors:

- SR-A1 (Facciponte et al. 2007); SR-A1 (CD204) is a 220–250 kDa trimeric molecule that shares a collagen-like domain, essential for ligand binding. SR-A1 binds numerous endogenous and exogenous ligands, such as AGE products, as well as microbes and microbial moieties.
- SR-B1 (Fischer et al. 2010). Also known as CD36, platelet glycoprotein 4, fatty acid translocase (FAT) or glycoproteins 88 (GP88), IIIb (GPIIIB), or IV (GPIV), SR-B1 preferentially accumulates in caveolae. SR-B1 bind modified LDL (OxLDL and AcLDL), unmodified low density lipoproteins (LDL), very low

density lipoproteins (VLDL) and apoptotic cells. SR-B1 is a heavily N-glycosylated protein with the C-terminal cytoplasmic tail that interacts with the multisubunit adaptor protein PDZK.
- SR-F1 (Facciponte et al. 2007; Gong et al. 2010; Theriault et al. 2006). Also known as SREC-I, SR-F1 mediates the internalization of OxLDL (as well as AcLDL). SR-F1 can also interact with SREC-2, a homologous of SR-F1, via its extracellular domain; although the role played by this heterophilic interaction remains unclear, it is suppressed by SR-F1 ligands.
- SR-H1 (Theriault et al. 2006). Also known as FEEL-1, CLEVER-1 or stabilin-1 (STAB1), SR-H1 binds AcLDL, AGEs as well as Gram-negative and Gram-positive bacteria (Adachi and Tsujimoto 2002; Tamura et al. 2003).

These endocytic receptors are involved in the internalization and presentation of associated antigens to CD4+ and CD8+ T cells (Facciponte et al. 2007; Gong et al. 2010).

CD40

In 2001, the group of T Lehner reported a very elegant study showing that CD40 is a binding and uptake receptor for *Mycobacterium tuberculosis* HSP70 (MtbHSP70), but not human HSP70 (Binder 2009; Wang et al. 2001). CD40, a membrane molecule belonging to the TNF receptor superfamily (TNFRSF5), is constitutively expressed by APCs, including macrophages, dendritic cells and B lymphocytes. Triggering CD40 on myeloid cells induces the production of inflammatory cytokines. The engagement of CD40 on B cells induces their maturation, antibody isotype switching, and their differentiation into plasma cells. The ligand of CD40 is a member of the TNF superfamily, TNFSF5, also called CD40 ligand (CD40L) or CD154. CD154 is mainly expressed by activated CD4+ T cells and acts as a costimulatory molecule for B cells and myeloid cells; the engagement of CD154 participates to the T cell priming process.

A subsequent study by the same group showed that the binding of MtbHSP70 to the extracellular domain of CD40 was localized in the N-terminal nucleotide-binding domain in its ADP (peptide-binding) state (Becker et al. 2002). To date, CD40 has been mainly reported as a signaling receptor (reviewed in (Banchereau et al. 1994); its role in myeloid cell activation by HSP70 is discussed in the paragraph “Signaling receptors”.

C-type Lectins

As mentioned above, HSP70 can bind to the C-type lectin/scavenger receptor SR-E1 which is constitutively expressed on myeloid cells. Studies have also reported that HSP70 can bind to the C-type lectin CD94 (Gross et al. 2003a; Moser et al. 2002). CD94/NKG2 is a family of receptors mainly expressed on natural killer (NK) cells and a subset of CD8+ T cells. The consequences of HSP70 binding to CD94 on the biology of NK cells is detailed in the paragraph 4.

Signaling Receptors

HSP70 can efficiently cross-prime naive T cells, a process that required an optimal activation of APCs. However, some of the endocytic receptors identified are not signaling molecules, such as SR-E1, suggesting that signaling receptors are recruited by HSP70s. Numerous studies were thus focused on identifying HSP70 signaling receptors.

TLRs

Through their pivotal role in the activation of myeloid APCs, members of the Toll-like receptor (TLR) family are crucial in the initiation of innate and adaptive immune responses. The induction of in vivo immune responses by HSP70 suggested that they can activate myeloid cells and members of the TLR family rapidly emerged as candidate molecules. TLRs are type I integral membrane glycoproteins belonging to the IL-1 receptor (IL-1R) superfamily. TLR and IL-1R have a conserved region of ≈ 200 amino acids in their cytoplasmic domain, known as Toll/IL-1R (TIR) domain. The TIR domain is required for the intracellular signaling induced by TLR ligands. TLR can sense a large variety of microbes and microbial moieties, as well as host motifs. They are localized either at the cell surface or in endosomes where there are specialized in detecting microbes or nucleic acids, respectively. After ligand binding, TLRs dimerize and undergo a conformational change required for the recruitment of downstream signaling molecules that include the adaptor molecule myeloid differentiation primary-response protein 88 (MyD88), IL-1R-associated kinases (IRAKs), TGFβ-activated kinase (TAK1), TAK1-binding protein 1 (TAB1), TAB2 and TNF-receptor-associated factor 6 (TRAF6). The engagement of TLRs generates potent activation signals for myeloid cells, as evidenced by the production of numerous pro-inflammatory cytokines, chemokines and interferons (IFNs). TLR agonists also induce the maturation of dendritic cells, a process required for naïve T cell priming. TLR agonists may also participate to the activation of innate and conventional lymphoid cells.

Asea et al. reported that TLR2/TLR4 are involved in the activation of myeloid cells by HSP70 (Asea et al. 2000, 2002), as evidenced by the production of the inflammatory cytokines IL-1β, TNFα and IL-12, and an elevated expression of the costimulatory molecule CD86. The HSP70-induced signaling was mediated via the MyD88/IRAK/NF-kB signaling pathway. The role played by TLR4 in the activation of dendritic cells and in the induction of a Th1 response by HSP70s was confirmed by others (Fang et al. 2011).

TLR4 was initially described as an element of the multimeric LPS receptor which includes, in addition to TLR4, the binding elements CD14 and LBP and the accessory molecule MD2. Accordingly, the activation of APCs induced by HSP70 requires CD14, in addition to TLR4; however, no data supported a direct binding of HSP70 to CD14 (Asea et al. 2000; Moroi et al. 2000). In line with the capacity of HSP70 to signal via TLR2/TLR4, several endogenous ligands have reported to bind to these molecules, including, in addition to other HSP (HSP60, HSP96), HMGB1, surfactant protein D (SP-D), fibrinogen, fibronectin, and hyaluronic acid (Bryant et al. 2015).

CD91

CD91 is phosphorylated in response to HSP, triggering signaling cascades that ultimately lead to the activation of NF-κB (Pawaria and Binder 2011). The stimulatory function of CD91 was confirmed in another study showing that the proliferation and cytokine production by CD4+ T cells in response to APC pulsed with complexes between HSP70 and antigenic peptides was inhibited by CD91 siRNA (Fischer et al. 2010). Accordingly, the CD91 ligand A2M acts as an adjuvant to prime CD8+ T cells in vivo (Kropp et al. 2010).

CD40

MtbHSP70 induced a CD40-mediated production of numerous cytokines and chemokines, such as CCL3 (MIP1α) involved in the recruitment and activation of polymorphonuclear cells, CCL4 (MIP1β) chemotactic for numerous immune cells and, CCL5 (RANTES), chemotactic for T cells. This property can be explaind by the adjuvanticity of MtbHSP70 (Wang et al. 2001). The binding of MtbHSP70 to CD40 induces an intracellular signaling via p38, associated with the internalization of the complex HSP70-CD40. P38 is a transduction molecule involved in the signaling cascade downstream CD40 that is involved in the production of pro-inflammatory cytokines, such as TNFα and IFNγ (Pullen et al. 1999).

CCR5

C-C chemokine receptor type 5 (CCR5), also known as CD195, acts as a receptor for the C-C chemokines CCL3, CCL4, and CCL5, three chemokines induced by MtbHSP70. CCR5 is mainly involved in the attraction of T cells in specific tissues and organs. The signaling via CCR5 induces dendritic cell activation and aggregation and participates to the formation of the immune synapse between dendritic cells and T lymphocytes (Floto et al. 2006). The binding of MtbHSP70 to CCR5 induces a Ca2+ signaling and the engagement of CCR5 participates in the generation effector immune responses (MacAry et al. 2004). In this study, the authors showed that the activation by MtbHSP70 was not dependent on TLR signaling.

Cooperation Between Endocytic and Signaling Receptors

Except some HSP70-binding elements, such as CD91 and CD40 which can act as both endocytic and signaling receptors, several studies suggest that the activation of innate immune cells by HSP70 requires cooperation between an endocytic receptor and a signaling receptor. This mechanism of interaction is observed in several examples of innate immune cell activation by non self and altered self. Nevertheless, in most cases, this model is supported by indirect in vivo experiments using HSP70 as a vaccine carrier molecule. As an example, Gong et al have demonstrated that the induction of antitumor immunity by HSP70 isolated from tumor-dendritic cell fusions is dependent on functional SR-F1 and TLR2/TLR4 expression by dendritic cells (Gong et al. 2009). In a similar manner, we have reported that targeting in vivo a vaccine antigen to SR-E1 is not sufficient to induce a protective antitumor response and that protection was only observed when antigen targeting was associated with the use of a TLR-activating adjuvant (Delneste et al. 2002). Interestingly, Mizukami et al showed that the cross-priming capacity of HSP70s was mediated by a TLR-independent mechanism, while the MyD88/IRAK signaling was required to induce tumor rejection (Mizukami et al. 2012).

Hsp70 in the Regulation of Innate Immune Cell Activation

HSP70s, as other members of the HSP superfamily, induce antigen-specific immune responses, in vitro and in vivo, a process that requires activation of APCs. As a consequence, several studies have reported that HSP70 induce the activation of innate immune cells, and especially myeloid cells.

Activation of Myeloid Cells

In addition to their role as sentinels of the innate immune system, macrophages and dendritic cells also act as professional antigen-presenting cells. Although macrophages are only able to present antigens to memory T cells, dendritic cells have the unique capacity to prime naive T cells. HSP70s can induce not only the cross-presentation but also the presentation of chaperoned peptides and coupled antigens to CD4+ and CD8+ T cells. The fact that HSP70s induce protective immune CD4+ and CD8+ T cell responses suggests that they are able to activate APCs. It is important to mention that antigen presentation and cross-presentation induced by HSP70s involve the same endocytic and signaling receptors. As an example, CD91 and scavenger receptors are involved in the activation of antigen-specific memory CD4+ T cells (Fischer et al. 2010). The role of HSP70 in the induction of CD4+ T cells has been also reported by other studies (Mycko et al. 2004; Tobian et al. 2004a, b; Wang et al. 2006).

As mentioned above, extracellular HSP70s induce the production of pro-inflammatory cytokines (TNFα, IL-1β, IL-6) and C-C chemokines by myeloid cells (Asea et al. 2000, 2002; Redzovic et al. 2015; Wang et al. 2001). HSP70s also induce the production of IL-12 (Vabulas et al. 2002), a pro-Th1 cytokine, and of IL-15 (Redzovic et al. 2015), a cytokine involved in the activation of innate and adaptive cytotoxic lymphoid cells, as well as type I IFNs by plasmacytoid cells (Jacquemin et al. 2017). The activation of myeloid cells is also associated with an increase of expression of MHC-II and of the costimulatory molecules CD40, CD80 and CD86 (Asea et al. 2002; Fang et al. 2011; Wang et al. 2002) which provide stimulatory signals required for optimal T cell activation.

Interestingly, HSP70s were reported to induce dendritic cell maturation (Kuppner et al. 2001; Vabulas et al. 2002; Wang et al. 2002), a process required to prime naive T cells against neo-antigens. Importantly, the capacity of HSP70 to activate and induce the maturation of dendritic cells supports their ability to generate immune responses in vivo, in the absence of adjuvant (Srivastava et al. 1998). As a consequence of their dendritic cell-stimulatory activity, HSP70s have been reported able to convert T cell tolerance to autoimmunity in vivo, in a murine model of model of autoimmune diabetes (Millar et al. 2003). Studies also reported that an intratumoral infusion of HSP70 increase the infiltration of cytotoxic lymphoid and the production of IFNγ (Shevtsov et al. 2014).

Activation of Innate Lymphoid Cells

NK cells are innate lymphoid cells that play a pivotal role in the destruction of virus-infected cells and tumoral cells. Their activation is dictated by a delicate balance between stimulatory and regulatory signals (Narni-Mancinelli et al. 2013). The overexpression of HSP70 by tumor cells has been reported as a marker of tumor immunogenicity (Clark and Menoret 2001) and in vivo studies showed that tumors secreting HSP70s display increased immunogenicity, with induction of strong and specific CTL responses (Massa et al. 2004). Interestingly, CD94+ CD3- NK cells can recognize and kill membrane HSP70-positive tumor cells in a granzyme-dependent manner (Moser et al. 2002; Multhoff et al. 1995). The interaction of HSP70 with CD94 was demonstrated by competitive binding experiments (Gross et al. 2003a, b).

Moreover, in the absence of HLA-E expression, the activation of NK cells with the HSP70 peptide TKD render them cytotoxic against target cells (Bottger et al. 2012; Gross et al. 2008). More intriguingly, Massa et al reported that the interaction of HSP70 with NK cells may also participate in the initiation of antigen-specific T cell responses. More precisely, the authors demonstrated that the adjuvant activity of HSP70 requires NK cells at the site of dendritic cell-HSP70 interaction whereas its ability to induce antigen cross-presentation of chaperoned peptides is independent of NK cells (Massa et al. 2005). A recent study reported that HSP70-positive exosomes derived from genotoxic drug-treated tumor cells induce the production of cytokines by CD56^{high} NK cells in a TLR2-dependent manner (Vulpis et al. 2017).

The Immune Properties of HSP70: An Unsolved Mystery

Innate immunity receptors act as immune sensors able to discriminate self from non-self and modified self. In most cases, the elimination of modified self does not activate immune cells, maintaining tissue homeostasis and preventing the initiation of a potentially harmful autoimmune response. Internalization without recruitment of a signaling receptor may lead to tolerance, as reported for the internalization of apoptotic cells. As an example, the interaction of TSP-1 with CD91 and CD36, two HSP70-binding elements, may participate to the tolerogenic status of dendritic cells after apoptotic cell phagocytosis (Poon et al. 2014). In agreement with these studies, the presentation of apoptotic-cell-derived antigens requires the presence of a TLR agonist in the same cargo as that of apoptotic cells (Blander and Medzhitov 2006).

However, most of innate endocytic receptors also recognize non self (microbes and microbial moieties) that induce a huge activation of immune cells. The concept of cooperation between endocytic and signaling receptors has thus emerged to explain this dichotomy. Innate cell activation induces the recruitment of signaling PRM after the binding of ligands to the endocytic receptor. In a similar manner, HSP70 endocytosis precedes signaling in myeloid cells (Kuppner et al. 2001; Moroi et al. 2000). In contrast to early apoptotic cells, necrotic cells induce a pro-inflammatory response, thanks to the release of stimulatory endogenous molecules called alarmins. These alarmins, such as HMGB1, IL-33 and ATP, activate immune cells and participate in the initiation of antigen-specific immune responses. Based on the studies showing that HSP70s activate APCs, it has been proposed that HSP70 may be considered as a danger signal. However, other authors refute to classify HSP70 as a DAMP (van Eden et al. 2012). Indeed, circulating HSP70 are detected in the serum of healthy subjects, without signs of inflammation. Moreover, different studies argue that the stimulatory activities of HSP70s are associated to the presence of contaminating molecules in HSP70 preparation.

Do HSP70 Interact with a Limited Number of Receptors?

To date, HSP70s have been reported to interact with a large variety of immune receptors. However, and even though most studies concur to recognize that HSP70 are able to induce antigen (cross) presentation, different studies refuted the HSP70-binding capacity to some receptors. Although this initial study was confirmed by subsequent studies (Binder and Srivastava 2004; Salimu et al. 2015), authors contested the fact that CD91 may be considered as an HSP-binding structure, either by using CD91^{null} cells or by demonstrating that the binding of HSP to CD91 was not altered by the conventional CD91 ligand activated α2 macroglobulin (A2M*) or the CD91 antagonist molecule, receptor-associated protein (RAP) (Berwin et al. 2002). Moreover, although highly expressed on macrophages, the expression of CD91 is very low on dendritic cells, suggesting that its role in in vivo T cell priming should be marginal. Moreover, Theriault et al. reported that scavenger receptors, but not CD40 and CD91, are the main HSP70-binding elements (Theriault et al. 2005). In a similar manner, Bendz et al have shown that the cross-presentation capacity of different dendritic cell subsets was equivalent, irrespective of the level of CCR5 expression (Bendz et al. 2008).

HSP70 are endogenous molecules that, theoretically, cannot activate APCs. Nevertheless, several studies reported that HSP70s can directly activate myeloid cells (and, at a lower extent, lymphoid cells) via innate receptors, such as TLR4. All the studies reporting a direct stimulatory activity of HSP70 claimed that this process was independent of contaminating molecules, especially endotoxins (Wang et al. 2010). However, independent studies reported that contaminating endotoxins were responsible for the activation of myeloid cells (Bausinger et al. 2002; Gao and Tsan 2003, 2004). Another study reported that calcium signaling induced by human HSP70 and MtbHSP70 may be caused by contaminating nucleotides (Bendz et al. 2008).

The Immunoregulatory Properties of HSP70

HSP70s are highly conserved molecules which can be released in the extracellular milieu. Consequently, the initiation of HSP70-specific immune responses remains exceptional. Contrary to the vast majority of studies, some authors have reported that HSP70 exhibit regulatory properties (reviewed in (Borges et al. 2012; van Eden et al. 2005). The initial study on a potential regulatory role for HSP70s was the demonstration that MtbHSP70 has an anti-inflammatory role in an in vivo model of autoimmune arthritis model (van Eden et al. 1998). The anti-inflammatory and protective roles of HSP70s and of selected HSP70 peptides have been thereafter reported in different models of severe or chronic inflammation (Vinokurov et al. 2012; Yurinskaya et al. 2009), infection (Kimura et al. 1998) and skin allografts (Borges et al. 2010). Extracellular HSP70s also induce endotoxin tolerance in macrophages (Aneja et al. 2006).

At the cellular level, HSP70 have been also reported to inhibit the maturation of dendritic cells, to induce the differentiation of monocyte-derived dendritic cells into tolerogenic cells (Motta et al. 2007; Stocki and Dickinson 2012) and to potentiate the suppressive activity of myeloid derived suppressive cells (MDSC) (Diao et al. 2015). By inducing suppressive myeloid cells, HSP70s increase the immunosuppressive activity of CD4+ CD25+ FoxP3+ regulatory T cells (Wachstein et al. 2012); regulatory T cells (Treg) have the unique capacity to dampen inflammation and to maintain an immunoregulatory environment (Josefowicz et al. 2012). HSP70s also favor the production of immunosuppressive Th2 cytokines by CD4+ T cells (Tsan and Gao 2004).

Macrophages are involved in numerous processes, such as immune surveillance and wound healing. A binary classification has thus been proposed to define their functional polarization with M1 and M2 cells representing the extremes of a continuum of polarization profiles. M1 cells exhibit antimicrobial and antitumor properties while M2 cells, mainly involved in tissue homeostasis and repair, exhibit immunoregulatory and protumoral properties. In established solid tumors, tumor-associated macrophages (TAM) exhibit a M2 phenotype (Mantovani et al. 2017). As examples, SR-A1 promotes tumor progression in murine models of ovarian and pancreatic cancer (Neyen et al. 2013a, b). HSP70s favor the polarization of macrophages into regulatory M2 cells (Lopes et al. 2014). An intriguing study recently reported that HSP70 regulates the M2-like polarization of tumor-associated macrophages in a SR-A1-dependent manner, favoring in vivo glioma regression (Zhang et al. 2016). However, whether this mechanism can be extended to other HSP70-binding elements, and especially scavenger receptors and C-type lectins, remains unknown.

The immunoregulatory activity of HSP70 was demonstrated to be dependent on the induction of the suppressive cytokine IL-10 by myeloid cells ((Detanico et al. 2004; Kimura et al. 1998; van Eden et al. 2005; Wendling et al. 2000). HSP70-induced IL-10 may then favor the generation of regulatory cells (Treg) as Treg cell depletion completely abolished this effect (Hauet-Broere et al. 2006). The capacity of HSP70 to favor a M2 polarization was also suspected dependent on the induction of IL-10 (Lopes et al. 2016). More recent studies have shown that the capacity of HSP70 to inhibit the production of IL-10 is driven by a down-regulation of the transcription factors C/EBPβ and C/EBPδ (Borges et al. 2013); this inhibition was correlated with a decreased production of pro-inflammatory cytokines and abrogated upon pretreatment of cells with ERK and JAK2/STAT3 inhibitors. These results are in agreement with studies reporting the pivotal role played by STAT3 in the establishment of an immunoregulatory environment (Yu et al. 2007). More intriguingly, Chandarwakar et al reported that the regulatory versus stimulatory activity of HSP was dependent on their concentrations with low doses being efficient to initiate antitumor immune responses and high doses being inefficient or even immunosuppressive (Chandawarkar et al. 1999).

Can SIGLEC4/15 Reconciliate Stimulatory and Regulatory Immune Properties of HSP70?

In 2015, Fong et al reported that the immune properties of extracellular HSP70s are mediated via the receptors Siglec-5 and Siglec-14 (Fong et al. 2015). Siglecs are transmembrane sialic acid-binding immunoglobulin-like lectins mainly expressed on leukocytes (Macauley and Paulson 2014; Schwarz et al. 2015). Siglec-5 and Siglec-14 belong to the rapidly evolving CD33-related Siglecs (CD33rSiglecs) family (Angata 2006). Siglec-5 and Siglec-14 are immune-suppressive and immune-activating paired receptors. This study reported that HSP70s bind Siglec-5 and Siglec-14 in a sialic-acid independent manner and that Hsp70 suppresses inflammation through Siglec-5 while, in contrast, its augments inflammation through Siglec-14. Interestingly, Siglec-5 and Siglec-14 can interact with other HSP70-binding elements, such as TLR and scavenger receptors/C-type lectins which have been defined as HSP70-binding elements. In addition to shed new light on the complexity of the biology of extracellular HSP70, this study provides an elegant molecular demonstration of the dichotomous immune properties of HSP70s and may help clarifying contradictory studies on the immune properties of HSP70s.

HSP70-Based Strategies to Induce Protective Immune Responses

Even though the intrinsic immune properties of HSP70 remain unclear, most authors agree on the fact that HSP70s represent unique vehicles to induce protective antitumoral and antimicrobial CD8+ and CD4+ T cell responses in vivo (Blachere et al. 1997; SenGupta et al. 2004; Udono et al. 1994; Udono and Srivastava 1993), thanks to their capacity to carry exogenous antigens into MHC-I and MHC-II antigen presentation pathways in professional APCs (Castellino et al. 2000; Srivastava 2002). Accordingly, different vaccine strategies have been proposed based on these unique immune properties. A recent review summarized some ongoing antitumoral vaccine clinical trials using HSP (Shevtsov and Multhoff 2016). In this paragraph are only mentioned the different HSP70-based vaccine strategies to induce protective antitumor and antiviral immune responses.

The initial HSP70-based vaccine strategies relied on the peptide-biding capacity of HSP. Vaccines contained HSP70 isolated from tumor (or virus infected) cells (Noessner et al. 2002). In this approach, the antigen specificity is determined by the chaperone-assisted peptides, allowing proposing individual vaccines irrespective of the nature of the vaccine antigens and of the MHC restriction (Suto and Srivastava 1995). HSP70s can be also isolated from the supernatants of apoptotic or stressed tumor cells (Chen et al. 2009; Masse et al. 2004) or from HSP70-transfected tumor cells (Massa et al. 2004). An original approach was the use of HSP70 isolated from dendritic/tumor cell fusion which induce potent antigen-specific antitumor immune responses, superior to the one of HSP70 isolated from tumor cells. (Enomoto et al. 2006). Another peptide-based approach is to reconstitute HSP-peptides complexes with immunodominant tumor antigen peptides (Blachere et al. 1997). Vaccinations with dendritic cells pulsed with tumor-derived HSP70 can also induce protective immune responses (Toomey et al. 2008).

Other strategies are based on the capacity of HSP70 to target dendritic cells in vivo, allowing the antigen to get access to the antigen presentation pathways. In this case, HSP70, coupled to the vaccine antigen, is used as a vaccine vehicle. Vaccine antigens can be chemically coupled to HSP70 (Delneste et al. 2002) or produced as a recombinant fusion molecule (Zhang and Huang 2006). Fusion proteins, consisting in HSP70 coupled to a vaccine antigen, have been validated using several different tumor antigens, such mesothelin, MAGE-1, PSA, carcinoembyonic antigen and Her2/neu (Ge et al. 2009; Yuan et al. 2014; Dong et al. 2013; Jiang et al. 2013; Pakravan et al. 2010; Wu et al. 2005) as well as viral antigens, such as the HPV16 E7 antigen and a dominant epitope of the EBV latent protein 2A (Zong et al. 2009, 2013).

Based on its immunostimulatory properties, other strategies use HSP70s as an adjuvant molecule for vaccines using tumor cell lysates (Wang et al. 2010; Li et al. 2010) or DNA vaccines as a source of antigens (Li et al. 2007; Zhang et al. 2007; Farzanehpour et al. 2013; Garrod et al. 2014) or in classical subunit vaccines (Lewis et al. 2014; Li et al. 2010; Shevtsov et al. 2014) to initiate effective immune responses. Nevertheless, in order to ameliorate the efficacy of the vaccine or to overcome the tumor immunosuppressive environment, several studies associated HSP70 vaccines with an adjuvant. In most cases, combining HSP70 vaccine with adjuvant (Delneste et al. 2002) or CD40L (Gao et al. 2012) induced more potent protective immune responses.

Conclusions

The identification of tumor-associated antigens and the presence of circulating tumor specific cytotoxic T lymphocytes (CTLs) in tumor-bearing patients suggested that initiating a protective antitumor immune response is feasible. A lot of work was thus done to propose antitumor vaccine strategies, such as the use of recombinant virus encoding human tumor antigens, tumor cell-derived exosomes or nucleic acid-based vaccines. Following the identification of the antigen cross-presentation process, numerous studies aimed at identifying carrier proteins that may selectively target and activate immature dendritic cells in vivo. HSP70s have thus emerged in the 1990s as powerful vaccine vehicles to initiate potent and protective antitumoral (and antiviral) CD8+ T cell immune responses. This remarkable property is related to the ability of extracellular HSP70s to interact with innate immune receptors involved in antigen cross-presentation. The initiation of an antigen-specific CD8+ T cells response against a neo-antigen requires that dendritic cells are fully activated. In agreement with their vaccine potentials, the initial studies suggested that HSP70s are intrinsically able to activate APCs. However, the exact nature of PRM involved in the immune properties of HSP70s remains, to date, controversed. Members of the scavenger receptor family appear as the less discussed endocytic receptors. Contrastingly, whether HSP70s are suppressive or stimulatory molecules remain debated. Based on the current view of the capacity of the innate immune system to maintain immune homeostasis in response to modified self, one could hypothesize that HSP70s are not stimulating molecules. Nevertheless, to precisely determine the immune status of HSP70s remains a big challenge that would allow reevaluating their therapeutic use.

Acknowledgements Studies in our laboratory are supported by institutional grants from Inserm and the University of Angers and by grants from the Ligue contre le Cancer and the Cancéropole Grand Ouest. Vincent Larochette is supported by a grant fellowship from the French Ministry of Research and Higher Education. This manuscript was prepared in the context of the LabEX IGO program supported by the National Research Agency via the investment of the future program (ANR-11-LABEX-0016-01).

References

Adachi, H., & Tsujimoto, M. (2002). FEEL-1, a novel scavenger receptor with in vitro bacteria-binding and angiogenesis-modulating activities. *The Journal of Biological Chemistry, 277*, 34264–34270.

Aneja, R., Odoms, K., Dunsmore, K., Shanley, T. P., & Wong, H. R. (2006). Extracellular heat shock protein-70 induces endotoxin tolerance in THP-1 cells. *Journal of Immunology, 177*, 7184–7192.

Angata, T. (2006). Molecular diversity and evolution of the Siglec family of cell-surface lectins. *Molecular Diversity, 10*, 555–566.

Arnold-Schild, D., Hanau, D., Spehner, D., et al. (1999). Cutting edge: Receptor-mediated endocytosis of heat shock proteins by professional antigen-presenting cells. *Journal of Immunology, 162*, 3757–3760.

Asea, A., Kraeft, S. K., Kurt-Jones, E. A., et al. (2000). HSP70 stimulates cytokine production through a CD14-dependant pathway, demonstrating its dual role as a chaperone and cytokine. *Nature Medicine, 6*, 435–442.

Asea, A., Rehli, M., Kabingu, E., et al. (2002). Novel signal transduction pathway utilized by extracellular HSP70: Role of toll-like receptor (TLR) 2 and TLR4. *The Journal of Biological Chemistry, 277*, 15028–15034.

Banchereau, J., Bazan, F., Blanchard, D., et al. (1994). The CD40 antigen and its ligand. *Annual Review of Immunology, 12*, 881–922.

Banchereau, J., Briere, F., Caux, C., et al. (2000). Immunobiology of dendritic cells. *Annual Review of Immunology, 18*, 767–811.

Baraldi, P. G., Di Virgilio, F., & Romagnoli, R. (2004). Agonists and antagonists acting at P2X7 receptor. *Current Topics in Medicinal Chemistry, 4*, 1707–1717.

Basu, S., Binder, R. J., Ramalingam, T., & Srivastava, P. K. (2001). CD91 is a common receptor for heat shock proteins gp96, hsp90, hsp70, and calreticulin. *Immunity, 14*, 303–313.

Batra, L., Verma, S. K., Nagar, D. P., et al. (2014). HSP70 domain II of Mycobacterium tuberculosis modulates immune response and protective potential of F1 and LcrV antigens of Yersinia pestis in a mouse model. *PLoS Neglected Tropical Diseases, 8*, e3322.

Bausinger, H., Lipsker, D., Ziylan, U., et al. (2002). Endotoxin-free heat-shock protein 70 fails to induce APC activation. *European Journal of Immunology, 32*, 3708–3713.

Becker, T., Hartl, F. U., & Wieland, F. (2002). CD40, an extracellular receptor for binding and uptake of Hsp70-peptide complexes. *The Journal of Cell Biology, 158*, 1277–1285.

Beere, H. M., Wolf, B. B., Cain, K., et al. (2000). Heat-shock protein 70 inhibits apoptosis by preventing recruitment of procaspase-9 to the Apaf-1 apoptosome. *Nature Cell Biology, 2*, 469–475.

Bendz, H., Marincek, B. C., Momburg, F., et al. (2008). Calcium signaling in dendritic cells by human or mycobacterial Hsp70 is caused by contamination and is not required for Hsp70-mediated enhancement of cross-presentation. *The Journal of Biological Chemistry, 283*, 26477–26483.

Berwin, B., Hart, J. P., Pizzo, S. V., & Nicchitta, C. V. (2002). Cutting edge: CD91-independent cross-presentation of GRP94(gp96)-associated peptides. *Journal of Immunology, 168*, 4282–4286.

Binder, R. J. (2009). CD40-independent engagement of mammalian hsp70 by antigen-presenting cells. *Journal of Immunology, 182*, 6844–6850.

Binder, R. J., & Srivastava, P. K. (2004). Essential role of CD91 in re-presentation of gp96-chaperoned peptides. *Proceedings of the National Academy of Sciences of the United States of America, 101*, 6128–6133.

Binder, R. J., & Srivastava, P. K. (2005). Peptides chaperoned by heat-shock proteins are a necessary and sufficient source of antigen in the cross-priming of CD8+ T cells. *Nature Immunology, 6*, 593–599.

Binder, R. J., Harris, M. L., Menoret, A., & Srivastava, P. K. (2000). Saturation, competition, and specificity in interaction of heat shock proteins (hsp) gp96, hsp90, and hsp70 with CD11b+ cells. *Journal of Immunology, 165*, 2582–2587.

Blachere, N. E., Li, Z., Chandawarkar, R. Y., et al. (1997). Heat shock protein-peptide complexes, reconstituted in vitro, elicit peptide-specific cytotoxic T lymphocyte response and tumor immunity. *The Journal of Experimental Medicine, 186*, 1315–1322.

Blander, J. M., & Medzhitov, R. (2006). Toll-dependent selection of microbial antigens for presentation by dendritic cells. *Nature, 440*, 808–812.

Borges, T. J., Porto, B. N., Teixeira, C. A., et al. (2010). Prolonged survival of allografts induced by mycobacterial Hsp70 is dependent on CD4+CD25+ regulatory T cells. *PLoS One, 5*, e14264.

Borges, T. J., Wieten, L., van Herwijnen, M. J., et al. (2012). The anti-inflammatory mechanisms of Hsp70. *Frontiers in Immunology, 3*, 95.

Borges, T. J., Lopes, R. L., Pinho, N. G., Machado, F. D., Souza, A. P., & Bonorino, C. (2013). Extracellular Hsp70 inhibits pro-inflammatory cytokine production by IL-10 driven down-regulation of C/EBPbeta and C/EBPdelta. *International Journal of Hyperthermia, 29*, 455–463.

Bottger, E., Multhoff, G., Kun, J. F., & Esen, M. (2012). Plasmodium falciparum-infected erythrocytes induce granzyme B by NK cells through expression of host-Hsp70. *PLoS One, 7*, e33774.

Bozzacco, L., Trumpfheller, C., Siegal, F. P., et al. (2007). DEC-205 receptor on dendritic cells mediates presentation of HIV gag protein to CD8+ T cells in a spectrum of human MHC I haplotypes. *Proceedings of the National Academy of Sciences of the United States of America, 104*, 1289–1294.

Brocchieri, L., Conway de Macario, E., & Macario, A. J. (2008). hsp70 genes in the human genome: Conservation and differentiation patterns predict a wide array of overlapping and specialized functions. *BMC Evolutionary Biology, 8*, 19.

Bryant, C. E., Gay, N. J., Heymans, S., Sacre, S., Schaefer, L., & Midwood, K. S. (2015). Advances in toll-like receptor biology: Modes of activation by diverse stimuli. *Critical Reviews in Biochemistry and Molecular Biology, 50*, 359–379.

Campisi, J., & Fleshner, M. (2003). Role of extracellular HSP72 in acute stress-induced potentiation of innate immunity in active rats. *Journal of Applied Physiology (1985), 94*, 43–52.

Castelli, C., Ciupitu, A. M., Rini, F., et al. (2001). Human heat shock protein 70 peptide complexes specifically activate antimelanoma T cells. *Cancer Research, 61*, 222–227.

Castellino, F., Boucher, P. E., Eichelberg, K., et al. (2000). Receptor-mediated uptake of antigen/heat shock protein complexes results in major histocompatibility complex class I antigen presentation via two distinct processing pathways. *The Journal of Experimental Medicine, 191*, 1957–1964.

Cella, M., Sallusto, F., & Lanzavecchia, A. (1997). Origin, maturation and antigen presenting function of dendritic cells. *Current Opinion in Immunology, 9*, 10–16.

Chandawarkar, R. Y., Wagh, M. S., & Srivastava, P. K. (1999). The dual nature of specific immunological activity of tumor-derived gp96 preparations. *The Journal of Experimental Medicine, 189*, 1437–1442.

Chen, T., Guo, J., Han, C., Yang, M., & Cao, X. (2009). Heat shock protein 70, released from heat-stressed tumor cells, initiates antitumor immunity by inducing tumor cell chemokine production and activating dendritic cells via TLR4 pathway. *Journal of Immunology, 182*, 1449–1459.

Clark, P. R., & Menoret, A. (2001). The inducible Hsp70 as a marker of tumor immunogenicity. *Cell Stress & Chaperones, 6*, 121–125.

Cunin, P., Beauvillain, C., Miot, C., et al. (2016). Clusterin facilitates apoptotic cell clearance and prevents apoptotic cell-induced autoimmune responses. *Cell Death & Disease, 7*, e2215.

Dabaghian, M., Latifi, A. M., Tebianian, M., Dabaghian, F., & Ebrahimi, S. M. (2015). A truncated C-terminal fragment of Mycobacterium tuberculosis HSP70 enhances cell-mediated immune response and longevity of the total IgG to influenza A virus M2e protein in mice. *Antiviral Research, 120*, 23–31.

Delneste, Y., Magistrelli, G., Gauchat, J., et al. (2002). Involvement of LOX-1 in dendritic cell-mediated antigen cross-presentation. *Immunity, 17*, 353–362.

Detanico, T., Rodrigues, L., Sabritto, A. C., et al. (2004). Mycobacterial heat shock protein 70 induces interleukin-10 production: Immunomodulation of synovial cell cytokine profile and dendritic cell maturation. *Clinical and Experimental Immunology, 135*, 336–342.

Diao, J., Yang, X., Song, X., et al. (2015). Exosomal Hsp70 mediates immunosuppressive activity of the myeloid-derived suppressor cells via phosphorylation of Stat3. *Medical Oncology, 32*, 453.

Dong, L., Zhang, X., Ren, J., et al. (2013). Human prostate stem cell antigen and HSP70 fusion protein vaccine inhibits prostate stem cell antigen-expressing tumor growth in mice. *Cancer Biotherapy & Radiopharmaceuticals, 28*, 391–397.

Enomoto, Y., Bharti, A., Khaleque, A. A., et al. (2006). Enhanced immunogenicity of heat shock protein 70 peptide complexes from dendritic cell-tumor fusion cells. *Journal of Immunology, 177*, 5946–5955.

Evdonin, A. L., Guzhova, I. V., Margulis, B. A., & Medvedeva, N. D. (2004). Phospholipse c inhibitor, u73122, stimulates release of hsp-70 stress protein from A431 human carcinoma cells. *Cancer Cell International, 4*, 2.

Facciponte, J. G., Wang, X. Y., & Subjeck, J. R. (2007). Hsp110 and Grp170, members of the Hsp70 superfamily, bind to scavenger receptor-A and scavenger receptor expressed by endothelial cells-I. *European Journal of Immunology, 37*, 2268–2279.

Fang, H., Wu, Y., Huang, X., et al. (2011). Toll-like receptor 4 (TLR4) is essential for Hsp70-like protein 1 (HSP70L1) to activate dendritic cells and induce Th1 response. *The Journal of Biological Chemistry, 286*, 30393–30400.

Farzanehpour, M., Soleimanjahi, H., Hassan, Z. M., Amanzadeh, A., Ghaemi, A., & Fazeli, M. (2013). HSP70 modified response against HPV based tumor. *European Review for Medical and Pharmacological Sciences, 17*, 228–234.

Fischer, N., Haug, M., Kwok, W. W., et al. (2010). Involvement of CD91 and scavenger receptors in Hsp70-facilitated activation of human antigen-specific CD4+ memory T cells. *European Journal of Immunology, 40*, 986–997.

Fleshner, M., & Johnson, J. D. (2005). Endogenous extra-cellular heat shock protein 72: Releasing signal(s) and function. *International Journal of Hyperthermia, 21*, 457–471.

Floto, R. A., MacAry, P. A., Boname, J. M., et al. (2006). Dendritic cell stimulation by mycobacterial Hsp70 is mediated through CCR5. *Science, 314*, 454–458.

Fong, J. J., Sreedhara, K., Deng, L., et al. (2015). Immunomodulatory activity of extracellular Hsp70 mediated via paired receptors Siglec-5 and Siglec-14. *The EMBO Journal, 34*, 2775–2788.

Gao, B., & Tsan, M. F. (2003). Endotoxin contamination in recombinant human heat shock protein 70 (Hsp70) preparation is responsible for the induction of tumor necrosis factor alpha release by murine macrophages. *The Journal of Biological Chemistry, 278*, 174–179.

Gao, B., & Tsan, M. F. (2004). Induction of cytokines by heat shock proteins and endotoxin in murine macrophages. *Biochemical and Biophysical Research Communications, 317*, 1149–1154.

Gao, J., Luo, S. M., Peng, M. L., & Deng, T. (2012). Enhanced immunity against hepatoma induced by dendritic cells pulsed with Hsp70-H22 peptide complexes and CD40L. *Journal of Cancer Research and Clinical Oncology, 138*, 917–926.

Garcia-Vallejo, J. J., Unger, W. W., Kalay, H., & van Kooyk, Y. (2013). Glycan-based DC-SIGN targeting to enhance antigen cross-presentation in anticancer vaccines. *Oncoimmunology, 2*, e23040.

Garrod, T., Grubor-Bauk, B., Yu, S., Gargett, T., & Gowans, E. J. (2014). Encoded novel forms of HSP70 or a cytolytic protein increase DNA vaccine potency. *Human Vaccines & Immunotherapeutics, 10*, 2679–2683.

Gastpar, R., Gehrmann, M., Bausero, M. A., et al. (2005). Heat shock protein 70 surface-positive tumor exosomes stimulate migratory and cytolytic activity of natural killer cells. *Cancer Research, 65*, 5238–5247.

Ge, W., Hu, P. Z., Huang, Y., et al. (2009). The antitumor immune responses induced by nanoemulsion-encapsulated MAGE1-HSP70/SEA complex protein vaccine following different administration routes. *Oncology Reports, 22*, 915–920.

Gehrmann, M., Cervello, M., Montalto, G., et al. (2014a). Heat shock protein 70 serum levels differ significantly in patients with chronic hepatitis, liver cirrhosis, and hepatocellular carcinoma. *Frontiers in Immunology, 5*, 307.

Gehrmann, M., Specht, H. M., Bayer, C., et al. (2014b). Hsp70 – A biomarker for tumor detection and monitoring of outcome of radiation therapy in patients with squamous cell carcinoma of the head and neck. *Radiation Oncology, 9*, 131.

Gong, J., Zhu, B., Murshid, A., et al. (2009). T cell activation by heat shock protein 70 vaccine requires TLR signaling and scavenger receptor expressed by endothelial cells-1. *Journal of Immunology, 183*, 3092–3098.

Gong, J., Zhang, Y., Durfee, J., et al. (2010). A heat shock protein 70-based vaccine with enhanced immunogenicity for clinical use. *Journal of Immunology, 184*, 488–496.

Gross, C., Koelch, W., DeMaio, A., Arispe, N., & Multhoff, G. (2003a). Cell surface-bound heat shock protein 70 (Hsp70) mediates perforin-independent apoptosis by specific binding and uptake of granzyme B. *The Journal of Biological Chemistry, 278*, 41173–41181.

Gross, C., Schmidt-Wolf, I. G., Nagaraj, S., et al. (2003b). Heat shock protein 70-reactivity is associated with increased cell surface density of CD94/CD56 on primary natural killer cells. *Cell Stress & Chaperones, 8*, 348–360.

Gross, C., Holler, E., Stangl, S., et al. (2008). An Hsp70 peptide initiates NK cell killing of leukemic blasts after stem cell transplantation. *Leukemia Research, 32*, 527–534.

Guo, F., Sigua, C., Bali, P., et al. (2005). Mechanistic role of heat shock protein 70 in Bcr-Abl-mediated resistance to apoptosis in human acute leukemia cells. *Blood, 105*, 1246–1255.

Hauet-Broere, F., Wieten, L., Guichelaar, T., Berlo, S., van der Zee, R., & Van Eden, W. (2006). Heat shock proteins induce T cell regulation of chronic inflammation. *Annals of the Rheumatic Diseases, 65*(Suppl 3), iii65–iii68.

Haug, M., Schepp, C. P., Kalbacher, H., Dannecker, G. E., & Holzer, U. (2007). 70-kDa heat shock proteins: Specific interactions with HLA-DR molecules and their peptide fragments. *European Journal of Immunology, 37*, 1053–1063.

Heath, W. R., & Carbone, F. R. (1999). Cytotoxic T lymphocyte activation by cross-priming. *Current Opinion in Immunology, 11*, 314–318.

Heath, W. R., & Carbone, F. R. (2001). Cross-presentation in viral immunity and self-tolerance. *Nature Reviews. Immunology, 1*, 126–134.

Henson, P. M. (2017). Cell removal: Efferocytosis. *Annual Review of Cell and Developmental Biology, 33*, 127–144.

Herz, J., & Strickland, D. K. (2001). LRP: A multifunctional scavenger and signaling receptor. *The Journal of Clinical Investigation, 108*, 779–784.

Ishii, T., Udono, H., Yamano, T., et al. (1999). Isolation of MHC class I-restricted tumor antigen peptide and its precursors associated with heat shock proteins hsp70, hsp90, and gp96. *Journal of Immunology, 162*, 1303–1309.

Jacquemin, C., Rambert, J., Guillet, S., et al. (2017). HSP70 potentiates interferon-alpha production by plasmacytoid dendritic cells: Relevance for cutaneous lupus and vitiligo pathogenesis. *British Journal of Dermatology, 177*(5), 1367–1375.

Jaillon, S., Jeannin, P., Hamon, Y., et al. (2009). Endogenous PTX3 translocates at the membrane of late apoptotic human neutrophils and is involved in their engulfment by macrophages. *Cell Death and Differentiation, 16*, 465–474.

Jeannin, P., Jaillon, S., & Delneste, Y. (2008). Pattern recognition receptors in the immune response against dying cells. *Current Opinion in Immunology, 20*, 530–537.

Jiang, J., Xie, D., Zhang, W., Xiao, G., & Wen, J. (2013). Fusion of Hsp70 to Mage-a1 enhances the potency of vaccine-specific immune responses. *Journal of Translational Medicine, 11*, 300.

Josefowicz, S. Z., Lu, L. F., & Rudensky, A. Y. (2012). Regulatory T cells: Mechanisms of differentiation and function. *Annual Review of Immunology, 30*, 531–564.

Karyampudi, L., & Ghosh, S. K. (2008). Mycobacterial HSP70 as an adjuvant in the design of an idiotype vaccine against a murine lymphoma. *Cellular Immunology, 254*, 74–80.

Kimura, Y., Yamada, K., Sakai, T., et al. (1998). The regulatory role of heat shock protein 70-reactive CD4+ T cells during rat listeriosis. *International Immunology, 10*, 117–130.

Kropp, L. E., Garg, M., & Binder, R. J. (2010). Ovalbumin-derived precursor peptides are transferred sequentially from gp96 and calreticulin to MHC class I in the endoplasmic reticulum. *Journal of Immunology, 184*, 5619–5627.

Krupka, M., Zachova, K., Cahlikova, R., et al. (2015). Endotoxin-minimized HIV-1 p24 fused to murine hsp70 activates dendritic cells, facilitates endocytosis and p24-specific Th1 response in mice. *Immunology Letters, 166*, 36–44.

Kuppner, M. C., Gastpar, R., Gelwer, S., et al. (2001). The role of heat shock protein (hsp70) in dendritic cell maturation: hsp70 induces the maturation of immature dendritic cells but reduces DC differentiation from monocyte precursors. *European Journal of Immunology, 31*, 1602–1609.

Lancaster, G. I., & Febbraio, M. A. (2005). Exosome-dependent trafficking of HSP70: A novel secretory pathway for cellular stress proteins. *The Journal of Biological Chemistry, 280*, 23349–23355.

Lewis, D. J., Wang, Y., Huo, Z., et al. (2014). Effect of vaginal immunization with HIVgp140 and HSP70 on HIV-1 replication and innate and T cell adaptive immunity in women. *Journal of Virology, 88*, 11648–11657.

Li, H., Ou, X., & Xiong, J. (2007). Modified HPV16 E7/HSP70 DNA vaccine with high safety and enhanced cellular immunity represses murine lung metastatic tumors with downregulated expression of MHC class I molecules. *Gynecologic Oncology, 104*, 564–571.

Li, H., Yu, Y., Sun, L., et al. (2010). Vaccination with B16 tumor cell lysate plus recombinant Mycobacterium tuberculosis Hsp70 induces antimelanoma effect in mice. *Cancer Biotherapy & Radiopharmaceuticals, 25*, 185–191.

Li, J., Xing, Y., Zhou, Z., et al. (2016). Microbial HSP70 peptide epitope 407-426 as adjuvant in tumor-derived autophagosome vaccine therapy of mouse lung cancer. *Tumour Biology, 37*, 15097–15105.

Liu, G., Yao, K., Wang, B., et al. (2009). Immunotherapy of Epstein-Barr virus associated malignancies using mycobacterial HSP70 and LMP2A356-364 epitope fusion protein. *Cellular & Molecular Immunology, 6*, 423–431.

Liu, G., Yao, K., Wang, B., et al. (2011). Reconstituted complexes of mycobacterial HSP70 and EBV LMP2A-derived peptides elicit peptide-specific cytotoxic T lymphocyte responses and anti-tumor immunity. *Vaccine, 29*, 7414–7423.

Lopes, R. L., Borges, T. J., Araujo, J. F., et al. (2014). Extracellular mycobacterial DnaK polarizes macrophages to the M2-like phenotype. *PLoS One, 9*, e113441.

Lopes, R. L., Borges, T. J., Zanin, R. F., & Bonorino, C. (2016). IL-10 is required for polarization of macrophages to M2-like phenotype by mycobacterial DnaK (heat shock protein 70). *Cytokine, 85*, 123–129.

MacAry, P. A., Javid, B., Floto, R. A., et al. (2004). HSP70 peptide binding mutants separate antigen delivery from dendritic cell stimulation. *Immunity, 20*, 95–106.

Macauley, M. S., & Paulson, J. C. (2014). Siglecs induce tolerance to cell surface antigens by BIM-dependent deletion of the antigen-reactive B cells. *Journal of Immunology, 193*, 4312–4321.

MacKenzie, A., Wilson, H. L., Kiss-Toth, E., Dower, S. K., North, R. A., & Surprenant, A. (2001). Rapid secretion of interleukin-1beta by microvesicle shedding. *Immunity, 15*, 825–835.

Mambula, S. S., & Calderwood, S. K. (2006a). Heat induced release of Hsp70 from prostate carcinoma cells involves both active secretion and passive release from necrotic cells. *International Journal of Hyperthermia, 22*, 575–585.

Mambula, S. S., & Calderwood, S. K. (2006b). Heat shock protein 70 is secreted from tumor cells by a nonclassical pathway involving lysosomal endosomes. *Journal of Immunology, 177*, 7849–7857.

Mantovani, A., Marchesi, F., Malesci, A., Laghi, L., & Allavena, P. (2017). Tumour-associated macrophages as treatment targets in oncology. *Nature Reviews. Clinical Oncology, 14*, 399–416.

Marotta, F., Koike, K., Lorenzetti, A., et al. (2007). Nutraceutical strategy in aging: Targeting heat shock protein and inflammatory profile through understanding interleukin-6 polymorphism. *Annals of the New York Academy of Sciences, 1119*, 196–202.

Massa, C., Guiducci, C., Arioli, I., Parenza, M., Colombo, M. P., & Melani, C. (2004). Enhanced efficacy of tumor cell vaccines transfected with secretable hsp70. *Cancer Research, 64*, 1502–1508.

Massa, C., Melani, C., & Colombo, M. P. (2005). Chaperon and adjuvant activity of hsp70: Different natural killer requirement for cross-priming of chaperoned and bystander antigens. *Cancer Research, 65*, 7942–7949.

Masse, D., Ebstein, F., Bougras, G., Harb, J., Meflah, K., & Gregoire, M. (2004). Increased expression of inducible HSP70 in apoptotic cells is correlated with their efficacy for antitumor vaccine therapy. *International Journal of Cancer, 111*, 575–583.

Milani, V., Noessner, E., Ghose, S., et al. (2002). Heat shock protein 70: Role in antigen presentation and immune stimulation. *International Journal of Hyperthermia, 18*, 563–575.

Millar, D. G., Garza, K. M., Odermatt, B., et al. (2003). Hsp70 promotes antigen-presenting cell function and converts T-cell tolerance to autoimmunity in vivo. *Nature Medicine, 9*, 1469–1476.

Mizukami, S., Kajiwara, C., Tanaka, M., Kaisho, T., & Udono, H. (2012). Differential MyD88/IRAK4 requirements for cross-priming and tumor rejection induced by heat shock protein 70-model antigen fusion protein. *Cancer Science, 103*, 851–859.

Moroi, Y., Mayhew, M., Trcka, J., et al. (2000). Induction of cellular immunity by immunization with novel hybrid peptides complexed to heat shock protein 70. *Proceedings of the National Academy of Sciences of the United States of America, 97*, 3485–3490.

Moser, C., Schmidbauer, C., Gurtler, U., et al. (2002). Inhibition of tumor growth in mice with severe combined immunodeficiency is mediated by heat shock protein 70 (Hsp70)-peptide-activated, CD94 positive natural killer cells. *Cell Stress & Chaperones, 7*, 365–373.

Motta, A., Schmitz, C., Rodrigues, L., et al. (2007). Mycobacterium tuberculosis heat-shock protein 70 impairs maturation of dendritic cells from bone marrow precursors, induces interleukin-10 production and inhibits T-cell proliferation in vitro. *Immunology, 121*, 462–472.

Multhoff, G. (2007). Heat shock protein 70 (Hsp70): Membrane location, export and immunological relevance. *Methods, 43*, 229–237.

Multhoff, G., & Hightower, L. E. (1996). Cell surface expression of heat shock proteins and the immune response. *Cell Stress & Chaperones, 1*, 167–176.

Multhoff, G., Botzler, C., Wiesnet, M., et al. (1995). A stress-inducible 72-kDa heat-shock protein (HSP72) is expressed on the surface of human tumor cells, but not on normal cells. *International Journal of Cancer, 61*, 272–279.

Mycko, M. P., Cwiklinska, H., Szymanski, J., et al. (2004). Inducible heat shock protein 70 promotes myelin autoantigen presentation by the HLA class II. *Journal of Immunology, 172*, 202–213.

Narni-Mancinelli, E., Ugolini, S., & Vivier, E. (2013). Tuning the threshold of natural killer cell responses. *Current Opinion in Immunology, 25*, 53–58.

Neyen, C., Mukhopadhyay, S., Gordon, S., & Hagemann, T. (2013a). An apolipoprotein A-I mimetic targets scavenger receptor A on tumor-associated macrophages: A prospective anticancer treatment? *Oncoimmunology, 2*, e24461.

Neyen, C., Pluddemann, A., Mukhopadhyay, S., et al. (2013b). Macrophage scavenger receptor a promotes tumor progression in murine models of ovarian and pancreatic cancer. *Journal of Immunology, 190*, 3798–3805.

Noessner, E. (2006). Thermal stress-related modulation of tumor cell physiology and immune responses. *Cancer Immunology, Immunotherapy, 55*, 289–291.

Noessner, E., Gastpar, R., Milani, V., et al. (2002). Tumor-derived heat shock protein 70 peptide complexes are cross-presented by human dendritic cells. *Journal of Immunology, 169*, 5424–5432.

Pakravan, N., Langroudi, L., Hajimoradi, M., & Hassan, Z. M. (2010). Co-administration of GP96 and Her2/neu DNA vaccine in a Her2 breast cancer model. *Cell Stress & Chaperones, 15*, 977–984.

Paliwal, P. K., Bansal, A., Sagi, S. S., & Sairam, M. (2011). Intraperitoneal immunization of recombinant HSP70 (DnaK) of Salmonella Typhi induces a predominant Th2 response and protective immunity in mice against lethal Salmonella infection. *Vaccine, 29*, 6532–6539.

Pawaria, S., & Binder, R. J. (2011). CD91-dependent programming of T-helper cell responses following heat shock protein immunization. *Nature Communications, 2*, 521.

Poccia, F., Piselli, P., Vendetti, S., et al. (1996). Heat-shock protein expression on the membrane of T cells undergoing apoptosis. *Immunology, 88*, 6–12.

Pockley, A. G., Henderson, B., & Multhoff, G. (2014). Extracellular cell stress proteins as biomarkers of human disease. *Biochemical Society Transactions, 42*, 1744–1751.

Poon, I. K., Lucas, C. D., Rossi, A. G., & Ravichandran, K. S. (2014). Apoptotic cell clearance: Basic biology and therapeutic potential. *Nature Reviews. Immunology, 14*, 166–180.

Pullen, S. S., Dang, T. T., Crute, J. J., & Kehry, M. R. (1999). CD40 signaling through tumor necrosis factor receptor-associated factors (TRAFs). Binding site specificity and activation of downstream pathways by distinct TRAFs. *The Journal of Biological Chemistry, 274*, 14246–14254.

Radons, J. (2016). The human HSP70 family of chaperones: Where do we stand? *Cell Stress & Chaperones, 21*, 379–404.

Radons, J., & Multhoff, G. (2005). Immunostimulatory functions of membrane-bound and exported heat shock protein 70. *Exercise Immunology Review, 11*, 17–33.

Ravagnan, L., Gurbuxani, S., Susin, S. A., et al. (2001). Heat-shock protein 70 antagonizes apoptosis-inducing factor. *Nature Cell Biology, 3*, 839–843.

Redzovic, A., Gulic, T., Laskarin, G., Eminovic, S., Haller, H., & Rukavina, D. (2015). Heat-shock proteins 70 induce pro-inflammatory maturation program in decidual CD1a(+) dendritic cells. *American Journal of Reproductive Immunology, 74*, 38–53.

Salimu, J., Spary, L. K., Al-Taei, S., et al. (2015). Cross-presentation of the oncofetal tumor antigen 5T4 from irradiated prostate cancer cells – A key role for heat-shock protein 70 and receptor CD91. *Cancer Immunology Research, 3*, 678–688.

Sawamura, T., Kume, N., Aoyama, T., et al. (1997). An endothelial receptor for oxidized low-density lipoprotein. *Nature, 386*, 73–77.

Schwarz, F., Pearce, O. M., Wang, X., et al. (2015). Siglec receptors impact mammalian lifespan by modulating oxidative stress. *eLife, 4*, e06184.

SenGupta, D., Norris, P. J., Suscovich, T. J., et al. (2004). Heat shock protein-mediated cross-presentation of exogenous HIV antigen on HLA class I and class II. *Journal of Immunology, 173*, 1987–1993.

Shevtsov, M., & Multhoff, G. (2016). Heat shock protein-peptide and HSP-based immunotherapies for the treatment of cancer. *Frontiers in Immunology, 7*, 171.

Shevtsov, M. A., Pozdnyakov, A. V., Mikhrina, A. L., et al. (2014). Effective immunotherapy of rat glioblastoma with prolonged intratumoral delivery of exogenous heat shock protein Hsp70. *International Journal of Cancer, 135*, 2118–2128.

Singh-Jasuja, H., Toes, R. E., Spee, P., et al. (2000). Cross-presentation of glycoprotein 96-associated antigens on major histocompatibility complex class I molecules requires receptor-mediated endocytosis. *The Journal of Experimental Medicine, 191*, 1965–1974.

Sondermann, H., Becker, T., Mayhew, M., Wieland, F., & Hartl, F. U. (2000). Characterization of a receptor for heat shock protein 70 on macrophages and monocytes. *Biological Chemistry, 381*, 1165–1174.

Srivastava, P. (2002). Interaction of heat shock proteins with peptides and antigen presenting cells: Chaperoning of the innate and adaptive immune responses. *Annual Review of Immunology, 20*, 395–425.

Srivastava, P. K., Menoret, A., Basu, S., Binder, R. J., & McQuade, K. L. (1998). Heat shock proteins come of age: Primitive functions acquire new roles in an adaptive world. *Immunity, 8*, 657–665.

Stocki, P., & Dickinson, A. M. (2012). The immunosuppressive activity of heat shock protein 70. *Autoimmune Diseases, 2012*, 617213.

Stocki, P., Morris, N. J., Preisinger, C., et al. (2010). Identification of potential HLA class I and class II epitope precursors associated with heat shock protein 70 (HSPA). *Cell Stress & Chaperones, 15*, 729–741.

Stocki, P., Wang, X. N., Morris, N. J., & Dickinson, A. M. (2011). HSP70 natively and specifically associates with an N-terminal dermcidin-derived peptide that contains an HLA-A*03 antigenic epitope. *The Journal of Biological Chemistry, 286*, 12803–12811.

Suto, R., & Srivastava, P. K. (1995). A mechanism for the specific immunogenicity of heat shock protein-chaperoned peptides. *Science, 269*, 1585–1588.

Suzue, K., & Young, R. A. (1996). Adjuvant-free hsp70 fusion protein system elicits humoral and cellular immune responses to HIV-1 p24. *Journal of Immunology, 156*, 873–879.

Tamura, Y., Peng, P., Liu, K., Daou, M., & Srivastava, P. K. (1997). Immunotherapy of tumors with autologous tumor-derived heat shock protein preparations. *Science, 278*, 117–120.

Tamura, Y., Adachi, H., Osuga, J., et al. (2003). FEEL-1 and FEEL-2 are endocytic receptors for advanced glycation end products. *The Journal of Biological Chemistry, 278*, 12613–12617.

Theriault, J. R., Mambula, S. S., Sawamura, T., Stevenson, M. A., & Calderwood, S. K. (2005). Extracellular HSP70 binding to surface receptors present on antigen presenting cells and endothelial/epithelial cells. *FEBS Letters, 579*, 1951–1960.

Theriault, J. R., Adachi, H., & Calderwood, S. K. (2006). Role of scavenger receptors in the binding and internalization of heat shock protein 70. *Journal of Immunology, 177*, 8604–8611.

Tobian, A. A., Canaday, D. H., Boom, W. H., & Harding, C. V. (2004a). Bacterial heat shock proteins promote CD91-dependent class I MHC cross-presentation of chaperoned peptide to

CD8+ T cells by cytosolic mechanisms in dendritic cells versus vacuolar mechanisms in macrophages. *Journal of Immunology, 172*, 5277–5286.

Tobian, A. A., Canaday, D. H., & Harding, C. V. (2004b). Bacterial heat shock proteins enhance class II MHC antigen processing and presentation of chaperoned peptides to CD4+ T cells. *Journal of Immunology, 173*, 5130–5137.

Todryk, S., Melcher, A. A., Hardwick, N., et al. (1999). Heat shock protein 70 induced during tumor cell killing induces Th1 cytokines and targets immature dendritic cell precursors to enhance antigen uptake. *Journal of Immunology, 163*, 1398–1408.

Toomey, D., Conroy, H., Jarnicki, A. G., Higgins, S. C., Sutton, C., & Mills, K. H. (2008). Therapeutic vaccination with dendritic cells pulsed with tumor-derived Hsp70 and a COX-2 inhibitor induces protective immunity against B16 melanoma. *Vaccine, 26*, 3540–3549.

Tsan, M. F., & Gao, B. (2004). Heat shock protein and innate immunity. *Cellular & Molecular Immunology, 1*, 274–279.

Udono, H., & Srivastava, P. K. (1993). Heat shock protein 70-associated peptides elicit specific cancer immunity. *The Journal of Experimental Medicine, 178*, 1391–1396.

Udono, H., Levey, D. L., & Srivastava, P. K. (1994). Cellular requirements for tumor-specific immunity elicited by heat shock proteins: Tumor rejection antigen gp96 primes CD8+ T cells in vivo. *Proceedings of the National Academy of Sciences of the United States of America, 91*, 3077–3081.

Vabulas, R. M., Ahmad-Nejad, P., Ghose, S., Kirschning, C. J., Issels, R. D., & Wagner, H. (2002). HSP70 as endogenous stimulus of the Toll/interleukin-1 receptor signal pathway. *The Journal of Biological Chemistry, 277*, 15107–15112.

van Eden, W., van der Zee, R., Taams, L. S., Prakken, A. B., van Roon, J., & Wauben, M. H. (1998). Heat-shock protein T-cell epitopes trigger a spreading regulatory control in a diversified arthritogenic T-cell response. *Immunological Reviews, 164*, 169–174.

van Eden, W., van der Zee, R., & Prakken, B. (2005). Heat-shock proteins induce T-cell regulation of chronic inflammation. *Nature Reviews. Immunology, 5*, 318–330.

van Eden, W., Spiering, R., Broere, F., & van der Zee, R. (2012). A case of mistaken identity: HSPs are no DAMPs but DAMPERs. *Cell Stress & Chaperones, 17*, 281–292.

Verma, S. K., Batra, L., & Tuteja, U. (2016). A recombinant trivalent fusion protein F1-LcrV-HSP70(II) augments humoral and cellular immune responses and imparts full protection against Yersinia pestis. *Frontiers in Microbiology, 7*, 1053.

Vinokurov, M., Ostrov, V., Yurinskaya, M., et al. (2012). Recombinant human Hsp70 protects against lipoteichoic acid-induced inflammation manifestations at the cellular and organismal levels. *Cell Stress & Chaperones, 17*, 89–101.

Vulpis, E., Cecere, F., Molfetta, R., et al. (2017). Genotoxic stress modulates the release of exosomes from multiple myeloma cells capable of activating NK cell cytokine production: Role of HSP70/TLR2/NF-kB axis. *Oncoimmunology, 6*, e1279372.

Wachstein, J., Tischer, S., Figueiredo, C., et al. (2012). HSP70 enhances immunosuppressive function of CD4(+)CD25(+)FoxP3(+) T regulatory cells and cytotoxicity in CD4(+)CD25(−) T cells. *PLoS One, 7*, e51747.

Wan, T., Zhou, X., Chen, G., et al. (2004). Novel heat shock protein Hsp70L1 activates dendritic cells and acts as a Th1 polarizing adjuvant. *Blood, 103*, 1747–1754.

Wang, Y., Kelly, C. G., Karttunen, J. T., et al. (2001). CD40 is a cellular receptor mediating mycobacterial heat shock protein 70 stimulation of CC-chemokines. *Immunity, 15*, 971–983.

Wang, Y., Kelly, C. G., Singh, M., et al. (2002). Stimulation of Th1-polarizing cytokines, C-C chemokines, maturation of dendritic cells, and adjuvant function by the peptide binding fragment of heat shock protein 70. *Journal of Immunology, 169*, 2422–2429.

Wang, R., Kovalchin, J. T., Muhlenkamp, P., & Chandawarkar, R. Y. (2006). Exogenous heat shock protein 70 binds macrophage lipid raft microdomain and stimulates phagocytosis, processing, and MHC-II presentation of antigens. *Blood, 107*, 1636–1642.

Wang, Y., Seidl, T., Whittall, T., Babaahmady, K., & Lehner, T. (2010). Stress-activated dendritic cells interact with CD4+ T cells to elicit homeostatic memory. *European Journal of Immunology, 40*, 1628–1638.

Wang, H., Feng, F., Wang, X. P., et al. (2016). Dendritic cells pulsed with Hsp70 and HBxAg induce specific antitumor immune responses in hepatitis B virus-associated hepatocellular carcinoma. *Molecular Medicine Reports, 13*, 1077–1082.

Wassenberg, J. J., Dezfulian, C., & Nicchitta, C. V. (1999). Receptor mediated and fluid phase pathways for internalization of the ER Hsp90 chaperone GRP94 in murine macrophages. *Journal of Cell Science, 112*(Pt 13), 2167–2175.

Wei, Y., Xu, Y., Han, X., et al. (2013). Anti-cancer effects of dioscin on three kinds of human lung cancer cell lines through inducing DNA damage and activating mitochondrial signal pathway. *Food and Chemical Toxicology, 59*, 118–128.

Wendling, U., Paul, L., van der Zee, R., Prakken, B., Singh, M., & van Eden, W. (2000). A conserved mycobacterial heat shock protein (hsp) 70 sequence prevents adjuvant arthritis upon nasal administration and induces IL-10-producing T cells that cross-react with the mammalian self-hsp70 homologue. *Journal of Immunology, 164*, 2711–2717.

Wu, Y., Wan, T., Zhou, X., et al. (2005). Hsp70-like protein 1 fusion protein enhances induction of carcinoembryonic antigen-specific CD8+ CTL response by dendritic cell vaccine. *Cancer Research, 65*, 4947–4954.

Yamada, Y., Doi, T., Hamakubo, T., & Kodama, T. (1998). Scavenger receptor family proteins: Roles for atherosclerosis, host defence and disorders of the central nervous system. *Cellular and Molecular Life Sciences, 54*, 628–640.

Yang, X., Wang, J., Zhou, Y., Wang, Y., Wang, S., & Zhang, W. (2012). Hsp70 promotes chemoresistance by blocking Bax mitochondrial translocation in ovarian cancer cells. *Cancer Letters, 321*, 137–143.

Yang, Z., Zhuang, L., Szatmary, P., et al. (2015). Upregulation of heat shock proteins (HSPA12A, HSP90B1, HSPA4, HSPA5 and HSPA6) in tumour tissues is associated with poor outcomes from HBV-related early-stage hepatocellular carcinoma. *International Journal of Medical Sciences, 12*, 256–263.

Yewdell, J. W., Norbury, C. C., & Bennink, J. R. (1999). Mechanisms of exogenous antigen presentation by MHC class I molecules in vitro and in vivo: Implications for generating CD8+ T cell responses to infectious agents, tumors, transplants, and vaccines. *Advances in Immunology, 73*, 1–77.

Yu, H., Kortylewski, M., & Pardoll, D. (2007). Crosstalk between cancer and immune cells: Role of STAT3 in the tumour microenvironment. *Nature Reviews. Immunology, 7*, 41–51.

Yu, X., Guo, C., Fisher, P. B., Subjeck, J. R., & Wang, X. Y. (2015). Scavenger receptors: Emerging roles in cancer biology and immunology. *Advances in Cancer Research, 128*, 309–364.

Yuan, J., Kashiwagi, S., Reeves, P., et al. (2014). A novel mycobacterial Hsp70-containing fusion protein targeting mesothelin augments antitumor immunity and prolongs survival in murine models of ovarian cancer and mesothelioma. *Journal of Hematology & Oncology, 7*, 15.

Yurinskaya, M. M., Vinokurov, M. G., Zatsepina, O. G., et al. (2009). Exogenous heat shock proteins (HSP70) significantly inhibit endotoxin-induced activation of human neutrophils. *Doklady Biological Sciences, 426*, 298–301.

Zhang, H., & Huang, W. (2006). Fusion proteins of Hsp70 with tumor-associated antigen acting as a potent tumor vaccine and the C-terminal peptide-binding domain of Hsp70 being essential in inducing antigen-independent anti-tumor response in vivo. *Cell Stress & Chaperones, 11*, 216–226.

Zhang, X., Yu, C., Zhao, J., et al. (2007). Vaccination with a DNA vaccine based on human PSCA and HSP70 adjuvant enhances the antigen-specific CD8+ T-cell response and inhibits the PSCA+ tumors growth in mice. *The Journal of Gene Medicine, 9*, 715–726.

Zhang, H., Zhang, W., Sun, X., et al. (2016). Class A1 scavenger receptor modulates glioma progression by regulating M2-like tumor-associated macrophage polarization. *Oncotarget, 7*, 50099–50116.

Zong, J., Peng, Q., Wang, Q., Zhang, T., Fan, D., & Xu, X. (2009). Human HSP70 and modified HPV16 E7 fusion DNA vaccine induces enhanced specific CD8+ T cell responses and anti-tumor effects. *Oncology Reports, 22*, 953–961.

Zong, J., Wang, C., Wang, Q., et al. (2013). HSP70 and modified HPV 16 E7 fusion gene without the addition of a signal peptide gene sequence as a candidate therapeutic tumor vaccine. *Oncology Reports, 30*, 3020–3026.

Molecular Chaperones Regulating the Dynamics, Composition and Functionality of RNP Granules: Implications for Age-Related Diseases

Daniel Mateju, Laura Mediani, Federica F. Morelli, Simon Alberti, and Serena Carra

Abstract The maturation, storage and degradation of RNAs occur in RNA-protein membrane-less assemblies that have properties of liquid droplets and arise from the surrounding aqueous cytoplasm or nucleoplasm through a process known as liquid-liquid phase separation. In healthy cells, ribonucleoprotein (RNP) granules are highly dynamic compartments. In contrast, in aging cells or due to environmental stresses or genetic mutations, RNP granules, in particular stress granules (SGs), convert into solid, aggregate-like inclusions. The accumulation of these RNA-protein inclusions is linked to an increasing number of age-related neurodegenerative diseases, such as amyotrophic lateral sclerosis and frontotemporal dementia. Thus, a detailed understanding of the molecular causes underlying the conversion of liquid-like RNPs into aggregates and the identification of the cellular players that can prevent this conversion may represent a valid approach to combat these diseases.

In this book chapter, we summarize the current knowledge about stress granule formation. We focus on recent findings demonstrating that liquid-like SGs can sequester aggregation-prone misfolded proteins with detrimental consequences for SG dynamics and functionality. We further discuss a specific protein quality control process, referred to as granulostasis, which prevents the accumulation of misfolding-prone proteins in SGs, thereby maintaining the physiological state of SGs and ensuring timely SG disassembly.

Simon Alberti and Serena Carra contributed equally to this work.

D. Mateju · S. Alberti (✉)
Max Planck Institute of Molecular Cell Biology and Genetics, Dresden, Germany
e-mail: alberti@mpi-cbg.de

L. Mediani · F. F. Morelli · S. Carra (✉)
Department of Biomedical, Metabolic and Neural Sciences, Center for Neuroscience and Neurotechnology, University of Modena and Reggio Emilia, Modena, Italy
e-mail: serena.carra@unimore.it

A. A. A. Asea, P. Kaur (eds.), *HSP70 in Human Diseases and Disorders*, Heat Shock Proteins 14, https://doi.org/10.1007/978-3-319-89551-2_10

Keywords Membrane-less compartments · Molecular chaperones and co-chaperones · Neurodegenerative diseases · Phase transition · Protein aggregation · Stress granules

Abbreviations

AD	Alzheimer's disease
ALS	amyotrophic lateral sclerosis
BAG	Bcl-2 associated athanogenes
CCT/TRiC	chaperonin-containing t-complex polypeptide 1/TCP-1 ring complex
CHIP	carboxyl terminus of Hsc70-interacting protein
FTD	frontotemporal dementia
FUS	fused in sarcoma
HD	Huntington's disease
hnRNPA1	heterogeneous nuclear ribonucleoprotein A1
HSP	heat shock proteins
IBM	inclusion body myopathy
IDP	intrinsically disordered protein
IDR	intrinsically disordered region
LC3	microtubule-associated protein 1A/1B-light chain 3
LCS	low-complexity sequences
mTOR	mammalian target of rapamycin
NEFs	nucleotide-exchange factors
PD	Parkinson's disease
PML	promyelocytic leukemia protein
PQC	protein quality control
RACK1	receptor for activated C kinase 1
RAN	repeat-associated non-ATG
RBP	RNA-binding protein
RNP	ribonucleoprotein particle
SG	stress granule
SOD1	superoxide dismutase 1
SQSTM1	sequestosome 1
TDP-43	TAR DNA-binding protein 43
TIA-1	T-cell intracellular antigen 1
TRAF	TNF receptor associated factor

Introduction

Protein homeostasis refers to the ability of cells to maintain a healthy equilibrium between protein synthesis, folding, assembly and protein clearance (Morimoto and Cuervo 2014). Protein homeostasis is maintained by the protein quality control (PQC) system and is essential for the function and viability of eukaryotic cells. However, cells are constantly challenged by changes in their environment, such as temperature upshift, changes in salinity and osmolarity, exposure to environmental toxic agents and oxidative stress. These environmental changes cause protein unfolding and/or misfolding, thereby perturbing protein homeostasis and increasing the risk of protein aggregation. But even in absence of external stressors, a large fraction of proteins is constantly at risk for aggregation, because it is present at barely soluble, supersaturated concentrations (Ciryam et al. 2015). This is particularly true for proteins that contain regions of low sequence complexity. Under physiological conditions, these low complexity domains promote protein assembly into higher order macromolecular complexes, or membrane-less compartments, which are thought to exert specific cellular functions. Examples of membrane-less compartments are cytoplasmic ribonucleoprotein (RNP) particles, such as stress granules, and nuclear bodies including speckles, paraspeckles, nucleoli, PML (Banani et al. 2017). However, the presence of low complexity domains also makes these proteins prone to misfold and irreversibly aggregate. Thus, to maintain an intact and functional proteome, cells depend on the optimal functionality of the PQC system and imbalances in that system will inevitably lead to loss of protein homeostasis and accumulation of protein aggregates, with severe consequences on cell viability. In agreement, there are many neurodegenerative and neuromuscular diseases, referred to as protein conformational diseases, which are characterized by the accumulation of proteinaceous aggregates (Chiti and Dobson 2006). Not surprisingly, recent genetic and experimental evidence has demonstrated the accumulation of low complexity proteins in pathological proteinaceous aggregates in e.g. amyotrophic lateral sclerosis (ALS), Alzheimer's disease (AD), frontotemporal dementia (FTD), Parkinson's disease (PD), Huntington's disease (HD) and inclusion body myopathies (IBM). Most importantly, genetic mutations in these proteins that enhance protein misfolding and aggregation have been documented and are causative for protein conformation diseases (Banani et al. 2017; Knowles et al. 2014).

In this chapter, we will provide a general introduction to the PQC. In particular, we will focus on the HSP70 chaperone machine and the BAG family of co-chaperones and how they regulate protein homeostasis. We will then summarize recent findings demonstrating that pathological aggregates may arise from the deregulated assembly of low complexity proteins into membrane-less compartments, focussing on stress granules. Finally, we will examine emerging evidence showing that specific components of the PQC act in concert to prevent the pathological conversion of stress granules into irreversible aggregates.

Protein Quality Control: Protein Folding and Degradation in Balance

To preserve the proteome, prokaryotic and eukaryotic cells evolved a sophisticated PQC system that assists proteins in their folding, assembly and translocation from the moment of their synthesis and throughout their life, until their destruction (Hartl et al. 2011). The PQC system includes molecular chaperones, co-chaperones and degradation systems, namely ubiquitin-proteasome system and the autophagolysosomal system. Molecular chaperones recognize and transiently bind to hydrophobic regions of unfolded or partially folded proteins, preventing their premature or aberrant folding and avoiding their aggregation (Hartl et al. 2011). Many heat shock protein (HSP) family members function as molecular chaperones (Ellis and Hartl 1999). The human genome encodes for more than 100 different HSP that are classified in 7 families based on their molecular weight: HSPH (HSP110), HSPC (HSP90), HSPA (HSP70), DNAJ (HSP40), HSPB (small HSP), the human chaperonin families HSPD/E (HSP60/HSP10) and CCT/TRiC (Kampinga et al. 2009). While some HSP are constitutively expressed and not induced by stress (e.g. HSPA8), others are both constitutively expressed and upregulated upon stress (e.g. HSPA1A) or only expressed upon exposure of the cells to specific stressors (e.g. HSPA6) (Morimoto and Cuervo 2014). Together, these families of HSP constitute the cellular chaperone network, of which HSPAs are a central and crucial component. Specialized types of HSPAs assist the cotranslational folding and interact with the nascent polypeptide as it emerges from the ribosome tunnel and in the cytoplasm. Acting in concert with the HSP40/DNAJs co-chaperones, these specialized HSPAs detect the unfolding states of the nascent polypeptides during their elongation and recruit specific degradation factors to destroy defective nascent chains, which may exert potentially toxic effects (Beckmann et al. 1990; Gloge et al. 2014). When nascent chains are terminated, they are released in the cytoplasm to exert their functions or they translocate, with the assistance of specific chaperones to their final destination. For example, HSPA8 facilitates the transport of proteins across the mitochondrial membrane (Stojanovski et al. 2012). Alternatively, when proteins cannot be completely folded by the HSPAs, they are transferred to the chaperonins or to the HSP90/HSPCs, which can also handle specific client proteins independently of HSPAs (Buchner 1999).

HSPA and the Bag Family of Co-chaperones

In contrast to HSPBs, which possess no ATPase activity and are generally considered as "holdases", HSPAs hydrolyze ATP to cyclically release the unfolded substrate, which will spontaneously fold until it will reach its native state. HSPAs are thus considered as "foldases". When bound to ADP, HSPAs have a high binding affinity for the substrate, but substrate binding and release rate is low. Instead, binding of HSPAs to ATP increases the association and dissociation rates for the substrate, substantially decreasing the affinity for the substrate (10- to 50-fold) (Mayer and Bukau 2005; Schmid et al. 1994). Although essential for HSPA chaperone function, their intrinsic ATPase activity is very low and requires the stimulation by specific co-chaperones. These include the HSP40s/DNAJs and nucleotide-exchange factors (NEFs), such as the BAG (Bcl-2 associated athanogenes) family of proteins (BAG1-6), HSPBP1 and HSPHs (Andreasson et al. 2008; Laufen et al. 1999; Mayer et al. 2000; Takayama and Reed 2001). Co-chaperones with J domains stimulate the ATPase activity of HSPAs, whereas the NEFs assist protein folding by accelerating ADP-ATP exchange on HSPAs (Laufen et al. 1999; Mayer et al. 2000; Xu et al. 2008). As a consequence, DNAJs mainly confer client specificity to the HSPAs, while NEFs are thought to be mainly involved in client fate, including client degradation (Kampinga and Craig 2010). This is exemplified by BAG1 and BAG3, which target HSPA-bound substrates to the proteasome and macroautophagy system, respectively, for degradation (Alberti et al. 2002; Carra et al. 2008; Demand et al. 2001; Qian et al. 2006).

BAG1 possesses an ubiquitin-like domain and interacts with CHIP, an E3 ligase that ubiquitinates the HSPA-bound substrates and negatively regulates the folding and ATPase activities of HSPAs. Consequently, the BAG1-HSPA-CHIP complex promotes the transfer of HSPA-bound ubiquitinated clients to the proteasome for disposal (Alberti et al. 2002; Demand et al. 2001). In contrast to BAG1, BAG3 lacks the ubiquitin-like domain, but rather it acts as a scaffold that simultaneously recruits different players. BAG3 possesses two IPV motifs required for binding to HSPBs. Although BAG3 can bind to several members of the HSPB family (HSPB1-HSPB10), experimental evidence obtained in test tube, cell cultures and tissue extracts demonstrates that BAG3 has the highest binding affinity for HSPB8 (Carra et al. 2008; Fuchs et al. 2009; Rauch et al. 2017). BAG3 also contains a PXXP region that allows it to interact with the retrograde motor protein dynein (Carra et al. 2008; Gamerdinger et al. 2011). Moreover, BAG3 also interacts with the autophagy receptor p62/SQSTM1, a protein capable of simultaneously binding to ubiquitinated proteins and the autophagosome anchored protein LC3-II (Bjorkoy et al. 2005; Gamerdinger et al. 2011). Therefore, it has been proposed that the HSPB8-BAG3-HSPA complex promotes the retrograde transport of HSPB8-HSPA-bound clients to the microtubule organization center (MTOC). Here, at the MTOC, the exceeding amounts of misfolded proteins are accumulated and stored in form of aggresome. The latter can be efficiently engulfed, and subsequently cleared, by the autophagic vacuoles, which are also retrogradely transported and concentrated at the MTOC (Carra et al. 2008; Gamerdinger et al. 2011; Minoia et al. 2014). Thus, BAG3 regulates the processing of both HSPA- and HSPB-bound clients; in particular, HSPAs will promote the release of the HSPB-bound substrate for subsequent refolding or targeting to aggresome or autophagosomes. Interestingly, while the BAG domain is highly conserved in all the six BAG proteins and represents the signature of the BAG family of proteins, only BAG3 possesses the IPV motifs that are required for binding to HSPBs. The ability of BAG3 to simultaneously bind to HSPBs and HSPAs (Rauch et al. 2017) and modulate the fate of their clients might explain why, among the six BAG proteins, it is the only one that is induced upon proteotoxic stress conditions (Gamerdinger et al. 2009; Minoia et al. 2014). Moreover, BAG3 has the highest binding affinity for HSPAs compared to the other BAG proteins (Rauch and Gestwicki 2014). These findings suggest that, under proteotoxic stress conditions or during aging, when BAG3 levels are induced, the fate of the majority of the HSPA-bound clients will be dictated by BAG3, and to a lesser extent by BAG1, whose expression levels remain unchanged or even decrease (Gamerdinger et al. 2009).

Combined these data support the interpretation that the HSPB8-BAG3-HSPA complex plays a crucial role in the maintenance of protein homeostasis, especially under stressful conditions and during aging (Gamerdinger et al. 2009). This interpretation is further supported by genetic evidence linking mutations of the HSPB8-BAG3-HSPA complex to inherited diseases, including motor neuropathies, myopathies and cardiomyopathies. In particular, the K141E and K141N mutations of HSPB8 linked to motor neuropathies decrease the ability of HSPB8 to bind to BAG3 and negatively impair its chaperone-like activity (Carra et al. 2010; Irobi et al. 2004; Shemetov and Gusev 2011). Moreover, the P209L mutation of BAG3, which is located within the HSPB-binding motif, leads to a severe form of myopathy with yet unknown mechanisms (Selcen et al. 2009).

Protein/RNA Aggregates in Neurodegenerative Diseases

Protein misfolding frequently occurs in cells. Under normal growth conditions, the capacity of the PQC system is sufficient to deal with misfolded proteins either by refolding or degradation. However, the functionality of the PQC machinery can be compromised by chronic stress or genetic mutations. This can cause disruptions in the balance between protein folding, misfolding, and degradation. In such cases, we observe that misfolded proteins accumulate in cells and form aggregates of tightly interacting proteins. When aggregates and misfolded proteins persist for extended times, this can lead to a pathological condition. Indeed, protein aggregates are a defining feature of various types of age-related disorders, including Alzheimer's, Parkinson's and Huntington's disease as well as fronto-temporal dementia (FTD) and amyotrophic lateral sclerosis (ALS). Why the accumulation of protein aggregates causes a pathological state is not completely understood. There is evidence for both gain-of-function and loss-of-function effects upon protein misfolding, but the specific molecular effects may differ from protein to protein (Hipp et al. 2014). One striking feature of protein misfolding diseases is that they most frequently affect the nervous system. The reasons for this tissue specificity are still debated. Possible explanations are the long life-time of neurons, the complex neuronal morphology or the specific metabolic needs of neurons.

One protein folding disorder, ALS, has received a lot of attention in recent years. ALS is characterized by the degeneration of motor neurons, which is followed by progressive muscle atrophy, usually leading to death within 3–5 years of disease onset (Morgan and Orrell 2016). ALS is the most common motor neuron disorder with a life time risk of 1 in 400. The disease is rare in young individuals, but the risk increases exponentially with age. ALS cases are divided into familial and sporadic forms with unknown aetiology. Although there is evidence that environmental factors increase the risk of ALS, the disease appears to have a very strong genetic component. Regardless of whether ALS is due to genetic or environmental causes, the affected motor neurons and their neighbouring cells share one characteristic feature; they contain cytosolic protein aggregates, which are composed of specific proteins such as SOD1, TDP-43, FUS, and/or hnRNPA1 (Ruegsegger and Saxena 2016; Scotter et al. 2015).

Genetic studies of families with recurrent cases of ALS and FTD have given us important insight into the molecular underpinnings of ALS. ALS-causing mutations have been identified in genes coding for SOD1, FUS and TDP-43, the very same proteins that are frequently aggregated in patient tissues (Kwiatkowski et al. 2009; Rosen et al. 1993; Sreedharan et al. 2008; Vance et al. 2009). Typically, these ALS-associated mutations increase the propensity of these proteins to misfold and aggregate (Ruegsegger and Saxena 2016). More recent findings have implicated repeat expansion mutations in the *C9orf72* gene as the most frequent cause of familial ALS. In these cases, the toxicity might arise from the repeat-containing RNA as well as from the encoded proteins; *C9orf72* repeat RNAs have been shown to deplete cells of important factors that are required for nucleocytoplasmic shuttling (Freibaum et al. 2015; Jovicic et al. 2015; Zhang et al. 2015); in addition, there is evidence that expansions in *C9orf72* promote spurious translation events, called RAN (repeat-associated non-ATG) translation, which leads to the production of aggregation-prone and highly toxic dipeptide repeat proteins (Ash et al. 2013; Mori et al. 2013). Importantly, there are additional, often rare mutations that do not increase the aggregation propensity of proteins. These genetic mutations have been identified in PQC factors, such as the autophagy receptors p62/SQSTM1 or optineurin (Taylor et al. 2016b). Together, these findings suggest that ALS arises due to a disruption of protein homeostasis and formation of protein aggregates that cannot be cleared.

Findings in recent years have provided evidence for another important disease mechanism: ALS-associated proteins often have functional roles in RNA metabolism (Li et al. 2013). The ALS proteins TDP-43 and FUS for example bind RNA and regulate various steps in transcription and RNA processing. This suggests that misregulation of RNA homeostasis could be an important disease mechanism. This idea has further been bolstered by the finding that *C9orf72* patients contain aberrant repeat RNAs that sequester diverse RNA-binding proteins (Barker et al. 2017). Recently, the focus has moved to the role of these proteins and RNAs in forming large RNA/protein assemblies or RNP granules. One particular type of RNP granule, called stress granules (SGs), has emerged as a nexus of ALS and other protein misfolding diseases. This idea has been further supported by the recent finding that mutations in the gene encoding for TIA-1, an RNA-binding protein that participates in RNA metabolism and drives the assembly of SGs, is causative for ALS and FTD (Mackenzie et al. 2017).

SGs form in response to stress conditions such as oxidative stress or heat stress (Li et al. 2013). The assembly of SGs is thought to be a protective response, which contributes to silencing of mRNA translation and storage of non-translating mRNA during cellular stress (Buchan 2014; Li et al. 2013). Moreover, SGs are also involved in the regulation of signaling pathways, such as mTOR, TRAF2 and RACK1, by sequestering key signaling proteins (Arimoto et al. 2008; Kim et al. 2005; Takahara and Maeda 2012). In agreement with these important functional roles, SGs are increasingly implicated in age-related diseases such as neurodegeneration and cancer (Li et al. 2013; Protter and Parker 2016). SGs are membrane-less organelles with material properties of dynamic viscous liquids. For a long time, it was mysterious how these compartments form in the absence of membranes. However, it is now becoming clear that SG assembly relies on a special class of compartment-forming proteins. Importantly, these compartment-forming proteins often misfold and aggregate in ALS and FTD patients and frequently carry mutations that affect the dynamics and functionality of SGs.

Intrinsically Disordered Proteins, Membrane-Less Compartments and Protein Conformational Age-Related Diseases

FUS and TDP-43 are two ALS- and FTD-associated proteins that have been implicated in the formation of SGs. The primary sequences of these two proteins are very unusual because they contain so-called low-complexity sequences (LCS). The amino acid composition of such LCS is biased towards only a small subset of the 20 natural amino acids. Another characteristic is that LCS contain repeats of very short amino acid motifs. LCS impart unusual biochemical properties on proteins, because they do not fold into a three-dimensional structure but adopt a range of different conformational states. This ability to adopt different conformations is critical for the functionality of these proteins.

FUS and TDP-43 are a subclass of LCS proteins called prion-like proteins, because the amino acid composition of their LCS resembles that of yeast prion domains (Alberti et al. 2009; King et al. 2012). Prion domains are characteristically enriched for asparagines, glutamines, tyrosines and serines and they have a propensity to form self-propagating amyloid fibrils (March et al. 2016). These amyloid fibrils can promote heritable phenotypic variation in fungi and have been associated with diseases such as prion disease in humans. However, it is now emerging that amyloid or prion formation is often not the function for which these proteins have been selected by evolution. Rather amyloid and prion formation often seems to be a structural and functional aberration of these domains, which manifests when an organism loses control over the conformational dynamics of these proteins.

Recent biochemical studies with prion-like proteins have given us important insight into the molecular function of prion-like domains. The current model is that these domains promote the formation of membrane-less compartments such as RNP granules through a process called liquid-liquid phase separation (LLPS) (Molliex et al. 2015; Murakami et al. 2015; Patel et al. 2015). In LLPS an initially homogenous solution of molecules separates ("demixes") into two phases, one that is enriched for the molecule and one that is depleted. When the conditions are stable, these two phases can co-exist indefinitely, while the molecules in the dense and depleted phase are in equilibrium with each other and undergo dynamic interactions. This has for example been demonstrated for the protein FUS (Patel et al. 2015). When purified FUS protein is exposed to low salt or low temperature conditions, it demixes from solution into a protein-dense and protein-depleted phase. The dense phase has properties of a liquid, as demonstrated by the fact that the FUS molecules undergo rapid rearrangements and movement within the compartment. Moreover, when two dense liquid FUS droplets touch, they fuse rapidly and relax into a larger droplet. Despite the dynamic nature of these assemblies, they can concentrate proteins against steep concentration gradients. In addition, they are surrounded by an interface that acts as a boundary and a selectivity filter, thus allowing these droplets to function as compartments with distinct composition. This suggests that prion-like proteins have been shaped by evolution to promote the formation of membrane-less compartments with properties of dynamic liquids. Importantly, prion-like domains plays a key role in this process, because their removal prevents phase separation and compartment formation (Molliex et al. 2015; Patel et al. 2015). The current thinking is that prion-like domains provide the molecular platform for the weak multivalent interactions that are required for the formation of liquid droplet compartments.

Although prion-like proteins initially form liquid droplet compartments, these droplets are metastable and mature into a more solid form with time. For example, the liquid droplets formed from purified FUS protein harden when incubated for several hours in the test tube (Patel et al. 2015). Initially, these hardened FUS droplets have properties of gels, but later these gels convert into solid-like fibrils. These fibrils are reminiscent of inclusion structures seen in ALS and FTD patients. This has led to the hypothesis that this conversion from liquid to solid is an aberrant phase transition that is associated with disease (Patel et al. 2015). This is strongly supported by experiments with mutant versions of FUS that have been identified in ALS patients. For example, the presence of a point mutation in the prion-like domain of FUS strongly accelerates the rate with which the FUS protein converts into a solid state (Patel et al. 2015). Similar findings have been reported for ALS-associated mutations in the prion-like proteins hnRNPA1 and TIA-1 (Mackenzie et al. 2017; Molliex et al. 2015). This suggests that the inherent propensity of proteins to form liquid droplet compartments has to be tightly controlled, because otherwise it will lead to a disease state.

In agreement with this model, RNP granules such as SGs initially have properties of dynamic liquid droplets in living cells, but they can turn into a more solid-like state, for example when the inducing stress conditions are too severe (Mateju et al. 2017). Once formed, such SGs persist, meaning that they do not dissolve anymore when the stress conditions subside. It has been hypothesized that these persisting "aberrant" SGs could slowly convert into cytotoxic inclusions that cause disease (Alberti et al. 2017). In agreement with this idea, the nervous tissue of ALS and FTD patients often contains inclusions that are enriched for various SGs components, such as the proteins TIA-1, FUS and TDP-43. This raises important questions: How does the cell normally regulate and prevent the formation of aberrant SGs? Why do aberrant forms of SGs and inclusions only appear with increasing age?

Granulostasis: A Specific PQC System that Surveys and Maintains Stress Granule Dynamics

SGs are highly dynamic compartments that disassemble within a few hours after their formation, thereby allowing the cells to restore proper translation and ensuring cellular response or adaptation to stress (Mazroui et al. 2007). During their formation, SGs can entrap misfolded proteins, an event that may have functional consequences for their dynamic behaviour and functionality. The co-aggregation of SG components with misfolded aggregation-prone proteins has been demonstrated for both yeast and mammalian SGs; however, profound mechanistic differences exist. In fact, yeast SGs can be nucleated by misfolded proteins and are more solid-like compared to mammalian SGs (Cherkasov et al. 2013; Kroschwald et al. 2015). The disassembly of yeast SGs depends on the activity of the yeast chaperone Hsp104, which acts as a disaggregase. Hsp104-assisted SG disaggregation also ensures the restoration of mRNA translation during the recovery phase (Cherkasov et al. 2013; Kroschwald et al. 2015). Similarly, also in *Drosophila melanogaster* cells SGs co-aggregate with misfolding proteins, such as firefly luciferase and VHL (von Hippel-Lindau tumor suppressor) (Cherkasov et al. 2013; Kroschwald et al. 2015; Walters and Parker 2015).

In contrast to yeast and *Drosophila melanogaster* cells, in mammalian cells, misfolding proteins are not required for SG nucleation (Cherkasov et al. 2013). Among the misfolded aggregation-prone species that accumulate in mammalian cells prior to SG assembly are the so called defective ribosomal products (DRiPs) (Schubert et al. 2000; Yewdell 2011). DRiPs are nascent chains whose translation has been prematurely terminated and that are released by disassembling polyribosome immediately prior to SG formation. Because DRiPs are truncated proteins they are highly aggregation prone. Immediately after their release, DRiPs are recognized and bound by chaperones such as HSPAs and VCP/p97, which target them to degradation by the proteasome and autophagy systems (Verma et al. 2013; Wang et al. 2013). We found that, although occurring with a low frequency, DRiPs and other misfolded proteins can accumulate inside SGs in healthy mammalian cells. Importantly, this event has profound consequences on SG behavior. DRiPs and misfolded proteins strongly delay the dynamic dissolution of mammalian SGs. They also change the biochemical properties of SGs: DRiP-enriched SGs become resistant to RNase digestion. Thus, when misfolded proteins are entrapped inside SGs, they convert into an aberrant state where RNA is no longer required for their structural integrity and thus resemble protein aggregates. In agreement, key SG proteins, such as G3BP1, become less dynamic in SGs that contain high levels of DRiPs and other misfolded proteins (Ganassi et al. 2016; Mateju et al. 2017). This observation is in line with recent findings demonstrating that SGs are composed of a protein-rich, RNase-resistant core and a dynamic, RNase-sensitive shell (Jain et al. 2016). Whether the core proteinaceous structures are solid RNPs that nucleate the formation of mammalian SGs is still unknown.

Interestingly, we found that "aberrant" SGs recruit high amount of chaperones, including e.g. the small heat shock proteins HSPB1, HSPB8, the HSPA co-chaperone BAG3 and VCP, likely in attempt to assist in misfolded protein processing/clearance and avoid their irreversible aggregation with SG components (Ganassi et al. 2016; Jain et al. 2016; Mateju et al. 2017). This interpretation is supported by several findings. First, the concentration of HSPB8 and, to a lesser extent, HSPB1 inside SGs is proportional to the amount of misfolded proteins: the higher the accumulation of DRiPs and aberrantly folded proteins inside SGs, the higher the recruitment of HSPB8 and HSPB1 inside SGs. This suggests that misfolded proteins recruit the chaperones. Second, depletion of HSPB8 and, to a lesser extent of HSPB1, leads to a significant accumulation of DRiPs inside SGs and delays their disassembly. HSPB1 and HSPB8 are ATP-independent chaperones that bind to and hold unfolded/misfolded clients maintaining them in a competent state for processing by the HSPA ATP-dependent chaperones (Haslbeck et al. 2005); HSPAs, in turn, will either refold or target to degradation the bound clients (Hartl et al. 2011). Combined these results support the hypothesis that HSPB1 and HSPB8 are recruited inside SGs to neutralize misfolded proteins, including DRiPs.

This interpretation is reinforced by our finding that the HSPA co-chaperone BAG3, which forms a complex with HSPB8 and HSPAs (Carra et al. 2008), is also recruited inside SGs that become enriched for misfolded proteins. Moreover, BAG3 depletion, which also destabilizes and compromises the function of HSPB8 (Carra et al. 2008), also promotes the conversion of physiological SGs into "aberrant" RNAse resistant SGs (Ganassi et al. 2016). A similar, and even stronger, effect is observed upon inhibition of the ATPase activity of HSPAs (Ganassi et al. 2016; Mateju et al. 2017). Finally, as previously discussed, HSPB8, BAG3 and the stress-inducible form HSPA1A are all upregulated by many of the stressors that elicit SG formation, including proteasome inhibition and temperature upshift (Minoia et al. 2014). Altogether these data support the interpretation that the HSPB8-BAG3-HSPA1A complex is one of the key players that surveys SG composition and avoids the irreversible co-aggregation of SGs with misfolded proteins. Because this specific type of proteostasis specifically affects RNP granules, we named the surveillance function exerted by the HSPB8-BAG3-HSPA1A complex granulostasis (Ganassi et al. 2016; Mateju et al. 2017).

Besides HSPB8-BAG3-HSPA1A, other chaperones seem to participate in granulostasis. Our data suggest that HSPB1 would act as a second player, which is recruited at later time points inside SGs that become enriched for additional misfolded proteins. Also VCP participates in the maintenance of SG composition and dynamics. Besides targeting DRiPs and other misfolded proteins to degradation, thereby maintaining SG composition and dynamics (Seguin et al. 2014; Verma et al. 2013), VCP seems to be also involved in the targeting of irreversibly aggregated SGs to autophagy for disposal (Buchan et al. 2013). Future studies will unravel if other chaperones and co-chaperones also participate with similar (or distinct) mechanisms to survey SG composition and disaggregation/dissolution, or in their digestion.

Concerning SG degradation by autophagy, our data suggest that only a minor fraction of irreversibly damaged SGs is targeted to autophagosomes for destruction. Instead, in mammalian cells, the majority of SGs are preferentially disassembled with the assistance of dedicated chaperones, especially HSPAs and co-factors. On the one hand, the disassembly of SGs allows the cells to recycle essential components, such as proteins, signalling molecules, mRNAs and 40S ribosomal subunits. From an energetic point of view, the storage and chaperone-assisted release of SG components represents a more attractive solution than protein degradation and de novo synthesis. On the other hand, the deposition of misfolded proteins in membrane-less compartments seems to be a more general phenomenon that occurs also in other membrane-less compartments. For example, upon heat or acid stress conditions, misfolded proteins accumulate in the nucleolus, where they form the so-called amyloid or A body (Audas et al. 2016). The aggregation of proteins inside the nucleoli is not irreversible and A body dissolution requires the action of chaperones such as HSPAs, similarly to what we observed for the dissolution of SGs that become enriched for misfolded proteins.

As mentioned previously, increasing experimental and genetic evidence points to deregulated SG dynamics and RNA metabolism as key pathomechanisms in age-related neurodegenerative diseases, including AD, ALS, FTD, inclusion body myopathy (IBM) and multisystem proteinopathy (Taylor et al. 2016a). Of note, the protein deposits that accumulate in disease-affected cells are enriched for TDP-43, a nuclear RNA-binding protein that is recruited into SGs upon stress and regulates RNA processing at every stage of its life cycle; however, together with TDP-43, and other SG components such as TIA-1, markers of the PQC system such as ubiquitin and p62/SQSTM1 also accumulate in the inclusion bodies; this observation suggests an intimate connection between SGs and the PQC pathways (Lagier-Tourenne et al. 2010; Prudlo et al. 2016).

This interpretation is further supported by our finding that mutated disease-linked RBPs that accumulate inside SGs have a higher propensity to co-aggregate with misfolding-prone proteins; this, in turn, may further accelerate the conversion of SGs into dysfunctional aggregates (Ganassi et al. 2016). Similarly, pathogenic *C9orf72*-DPRs, in particular the highly toxic arginine-containing DPR proteins GR and PR, could interact with RBPs and decrease the dynamics of SGs; in parallel they could also negatively affect the material properties and functions of other membrane-less organelles such as nucleoli (Lee et al. 2016; Taylor et al. 2016b). Altogether these findings point to aberrant RNP granules as crucibles of age-related disease.

Conclusions

Recent findings discussed here highlight the importance of maintaining protein and RNA homeostasis. Failure to maintain homeostasis has severe consequences for normal cellular physiology and leads to many age-related diseases. In agreement with this, increasing genetic evidence links PQC factors to neurological diseases. For example, mutations in the genes coding for the chaperones VCP/p97, HSPB1, HSPB8 and BAG3 as well as the autophagy receptor p62/SQSTM1 cause ALS, motor neuropathies and IBM (Evgrafov et al. 2004; Ghaoui et al. 2016; Irobi et al. 2004; Johnson et al. 2010; Mizuno et al. 2006; Rea et al. 2014; Selcen et al. 2009; Teyssou et al. 2013). In addition, there is increasing experimental data emphasizing the importance of a well-balanced interplay between PQC and RNP homeostasis. VCP as well as the granulostasis chaperone complex HSPB8-BAG3-HSP70 are emerging as key regulators of this interplay and thus prevent steps that lead to disease. This suggests that boosting specific chaperones and degradative systems could rescue protein and RNP homeostasis, and thus could represent an effective strategy to combat ALS, FTD, as well as other protein conformation age-related diseases.

Acknowledgments S.C. is grateful to AriSLA, The Italian Ministry of Foreign Affair and International Cooperation and the Cariplo Foundation for financial support. S.C. and S.A. acknowledge EU Joint Programme – Neurodegenerative Disease Research (JPND) project for financial support. The project is supported through funding organisations under the aegis of JPND (http://www.neurodegenerationresearch.eu/). This project has received funding from the European Union's Horizon 2020 Research and Innovation Programme under grant agreement No 643417. S.A. acknowledges funding from the Max Planck Society.

References

Alberti, S., Demand, J., Esser, C., Emmerich, N., Schild, H., & Hohfeld, J. (2002). Ubiquitylation of BAG-1 suggests a novel regulatory mechanism during the sorting of chaperone substrates to the proteasome. *The Journal of Biological Chemistry, 277*, 45920–45927.

Alberti, S., Halfmann, R., King, O., Kapila, A., & Lindquist, S. (2009). A systematic survey identifies prions and illuminates sequence features of prionogenic proteins. *Cell, 137*, 146–158.

Alberti, S., Mateju, D., Mediani, L., & Carra, S. (2017). Granulostasis: Protein quality control of RNP granules. *Frontiers in Molecular Neuroscience, 10*, 84.

Andreasson, C., Fiaux, J., Rampelt, H., Mayer, M. P., & Bukau, B. (2008). Hsp110 is a nucleotide-activated exchange factor for Hsp70. *The Journal of Biological Chemistry, 283*, 8877–8884.

Arimoto, K., Fukuda, H., Imajoh-Ohmi, S., Saito, H., & Takekawa, M. (2008). Formation of stress granules inhibits apoptosis by suppressing stress-responsive MAPK pathways. *Nature Cell Biology, 10*, 1324–1332.

Ash, P. E., Bieniek, K. F., Gendron, T. F., Caulfield, T., Lin, W. L., Dejesus-Hernandez, M., van Blitterswijk, M. M., Jansen-West, K., Paul, J. W., 3rd, Rademakers, R., Boylan, K. B., Dickson, D. W., & Petrucelli, L. (2013). Unconventional translation of C9ORF72 GGGGCC expansion generates insoluble polypeptides specific to c9FTD/ALS. *Neuron, 77*, 639–646.

Audas, T. E., Audas, D. E., Jacob, M. D., Ho, J. J., Khacho, M., Wang, M., Perera, J. K., Gardiner, C., Bennett, C. A., Head, T., Kryvenko, O. N., Jorda, M., Daunert, S., Malhotra, A., Trinkle-Mulcahy, L., Gonzalgo, M. L., & Lee, S. (2016). Adaptation to stressors by systemic protein amyloidogenesis. *Developmental Cell, 39*, 155–168.

Banani, S. F., Lee, H. O., Hyman, A. A., & Rosen, M. K. (2017). Biomolecular condensates: Organizers of cellular biochemistry. *Nature Reviews. Molecular Cell Biology, 18*, 285–298.

Barker, H. V., Niblock, M., Lee, Y. B., Shaw, C. E., & Gallo, J. M. (2017). RNA misprocessing in C9orf72-linked neurodegeneration. *Frontiers in Cellular Neuroscience, 11*, 195.

Beckmann, R. P., Mizzen, L. E., & Welch, W. J. (1990). Interaction of Hsp 70 with newly synthesized proteins: Implications for protein folding and assembly. *Science, 248*, 850–854.

Bjorkoy, G., Lamark, T., Brech, A., Outzen, H., Perander, M., Overvatn, A., Stenmark, H., & Johansen, T. (2005). p62/SQSTM1 forms protein aggregates degraded by autophagy and has a protective effect on huntingtin-induced cell death. *The Journal of Cell Biology, 171*, 603–614.

Buchan, J. R. (2014). mRNP granules. Assembly, function, and connections with disease. *RNA Biology, 11*, 1019–1030.

Buchan, J. R., Kolaitis, R. M., Taylor, J. P., & Parker, R. (2013). Eukaryotic stress granules are cleared by autophagy and Cdc48/VCP function. *Cell, 153*, 1461–1474.

Buchner, J. (1999). Hsp90 & Co. – a holding for folding. *Trends in Biochemical Sciences, 24*, 136–141.

Carra, S., Seguin, S. J., Lambert, H., & Landry, J. (2008). HspB8 chaperone activity toward poly(Q)-containing proteins depends on its association with Bag3, a stimulator of macroautophagy. *The Journal of Biological Chemistry, 283*, 1437–1444.

Carra, S., Boncoraglio, A., Kanon, B., Brunsting, J. F., Minoia, M., Rana, A., Vos, M. J., Seidel, K., Sibon, O. C., & Kampinga, H. H. (2010). Identification of the Drosophila ortholog of HSPB8: Implication of HSPB8 loss of function in protein folding diseases. *The Journal of Biological Chemistry, 285*, 37811–37822.

Cherkasov, V., Hofmann, S., Druffel-Augustin, S., Mogk, A., Tyedmers, J., Stoecklin, G., & Bukau, B. (2013). Coordination of translational control and protein homeostasis during severe heat stress. *Current Biology, 23*, 2452–2462.

Chiti, F., & Dobson, C. M. (2006). Protein misfolding, functional amyloid, and human disease. *Annual Review of Biochemistry, 75*, 333–366.

Ciryam, P., Kundra, R., Morimoto, R. I., Dobson, C. M., & Vendruscolo, M. (2015). Supersaturation is a major driving force for protein aggregation in neurodegenerative diseases. *Trends in Pharmacological Sciences, 36*, 72–77.

Demand, J., Alberti, S., Patterson, C., & Hohfeld, J. (2001). Cooperation of a ubiquitin domain protein and an E3 ubiquitin ligase during chaperone/proteasome coupling. *Current Biology, 11*, 1569–1577.

Ellis, R. J., & Hartl, F. U. (1999). Principles of protein folding in the cellular environment. *Current Opinion in Structural Biology, 9*, 102–110.

Evgrafov, O. V., Mersiyanova, I., Irobi, J., Van Den Bosch, L., Dierick, I., Leung, C. L., Schagina, O., Verpoorten, N., Van Impe, K., Fedotov, V., Dadali, E., Auer-Grumbach, M., Windpassinger, C., Wagner, K., Mitrovic, Z., Hilton-Jones, D., Talbot, K., Martin, J. J., Vasserman, N., Tverskaya, S., Polyakov, A., Liem, R. K., Gettemans, J., Robberecht, W., De Jonghe, P., & Timmerman, V. (2004). Mutant small heat-shock protein 27 causes axonal Charcot-Marie-Tooth disease and distal hereditary motor neuropathy. *Nature Genetics, 36*, 602–606.

Freibaum, B. D., Lu, Y., Lopez-Gonzalez, R., Kim, N. C., Almeida, S., Lee, K. H., Badders, N., Valentine, M., Miller, B. L., Wong, P. C., Petrucelli, L., Kim, H. J., Gao, F. B., & Taylor, J. P. (2015). GGGGCC repeat expansion in C9orf72 compromises nucleocytoplasmic transport. *Nature, 525*, 129–133.

Fuchs, M., Poirier, D. J., Seguin, S. J., Lambert, H., Carra, S., Charette, S. J., & Landry, J. (2009). Identification of the key structural motifs involved in HspB8/HspB6-Bag3 interaction. *The Biochemical Journal, 425*, 245–255.

Gamerdinger, M., Hajieva, P., Kaya, A. M., Wolfrum, U., Hartl, F. U., & Behl, C. (2009). Protein quality control during aging involves recruitment of the macroautophagy pathway by BAG3. *The EMBO Journal, 28*, 889–901.

Gamerdinger, M., Kaya, A. M., Wolfrum, U., Clement, A. M., & Behl, C. (2011). BAG3 mediates chaperone-based aggresome-targeting and selective autophagy of misfolded proteins. *EMBO Reports, 12*, 149–156.

Ganassi, M., Mateju, D., Bigi, I., Mediani, L., Poser, I., Lee, H. O., Seguin, S. J., Morelli, F. F., Vinet, J., Leo, G., Pansarasa, O., Cereda, C., Poletti, A., Alberti, S., & Carra, S. (2016). A surveillance function of the HSPB8-BAG3-HSP70 chaperone complex ensures stress granule integrity and dynamism. *Molecular Cell, 63*, 796–810.

Ghaoui, R., Palmio, J., Brewer, J., Lek, M., Needham, M., Evila, A., Hackman, P., Jonson, P. H., Penttila, S., Vihola, A., Huovinen, S., Lindfors, M., Davis, R. L., Waddell, L., Kaur, S., Yiannikas, C., North, K., Clarke, N., MacArthur, D. G., Sue, C. M., & Udd, B. (2016). Mutations in HSPB8 causing a new phenotype of distal myopathy and motor neuropathy. *Neurology, 86*, 391–398.

Gloge, F., Becker, A. H., Kramer, G., & Bukau, B. (2014). Co-translational mechanisms of protein maturation. *Current Opinion in Structural Biology, 24*, 24–33.

Hartl, F. U., Bracher, A., & Hayer-Hartl, M. (2011). Molecular chaperones in protein folding and proteostasis. *Nature, 475*, 324–332.

Haslbeck, M., Franzmann, T., Weinfurtner, D., & Buchner, J. (2005). Some like it hot: The structure and function of small heat-shock proteins. *Nature Structural & Molecular Biology, 12*, 842–846.

Hipp, M. S., Park, S. H., & Hartl, F. U. (2014). Proteostasis impairment in protein-misfolding and –aggregation diseases. *Trends in Cell Biology, 24*, 506–514.

Irobi, J., Van Impe, K., Seeman, P., Jordanova, A., Dierick, I., Verpoorten, N., Michalik, A., De Vriendt, E., Jacobs, A., Van Gerwen, V., Vennekens, K., Mazanec, R., Tournev, I., Hilton-Jones, D., Talbot, K., Kremensky, I., Van Den Bosch, L., Robberecht, W., Van Vandekerckhove, J., Van Broeckhoven, C., Gettemans, J., De Jonghe, P., & Timmerman, V. (2004). Hot-spot residue in small heat-shock protein 22 causes distal motor neuropathy. *Nature Genetics, 36*, 597–601.

Jain, S., Wheeler, J. R., Walters, R. W., Agrawal, A., Barsic, A., & Parker, R. (2016). ATPase-modulated stress granules contain a diverse proteome and substructure. *Cell, 164*, 487–498.

Johnson, J. O., Mandrioli, J., Benatar, M., Abramzon, Y., Van Deerlin, V. M., Trojanowski, J. Q., Gibbs, J. R., Brunetti, M., Gronka, S., Wuu, J., Ding, J., McCluskey, L., Martinez-Lage, M., Falcone, D., Hernandez, D. G., Arepalli, S., Chong, S., Schymick, J. C., Rothstein, J., Landi, F., Wang, Y. D., Calvo, A., Mora, G., Sabatelli, M., Monsurro, M. R., Battistini, S., Salvi, F., Spataro, R., Sola, P., Borghero, G., Galassi, G., Scholz, S. W., Taylor, J. P., Restagno, G., Chio, A., & Traynor, B. J. (2010). Exome sequencing reveals VCP mutations as a cause of familial ALS. *Neuron, 68*, 857–864.

Jovicic, A., Mertens, J., Boeynaems, S., Bogaert, E., Chai, N., Yamada, S. B., Paul, J. W., 3rd, Sun, S., Herdy, J. R., Bieri, G., Kramer, N. J., Gage, F. H., Van Den Bosch, L., Robberecht, W., & Gitler, A. D. (2015). Modifiers of C9orf72 dipeptide repeat toxicity connect nucleocytoplasmic transport defects to FTD/ALS. *Nature Neuroscience, 18*, 1226–1229.

Kampinga, H. H., & Craig, E. A. (2010). The HSP70 chaperone machinery: J proteins as drivers of functional specificity. *Nature Reviews Molecular Cell Biology, 11*, 579–592.

Kampinga, H. H., Hageman, J., Vos, M. J., Kubota, H., Tanguay, R. M., Bruford, E. A., Cheetham, M. E., Chen, B., & Hightower, L. E. (2009). Guidelines for the nomenclature of the human heat shock proteins. *Cell Stress & Chaperones, 14*, 105–111.

Kim, W. J., Back, S. H., Kim, V., Ryu, I., & Jang, S. K. (2005). Sequestration of TRAF2 into stress granules interrupts tumor necrosis factor signaling under stress conditions. *Molecular and Cellular Biology, 25*, 2450–2462.

King, O. D., Gitler, A. D., & Shorter, J. (2012). The tip of the iceberg: RNA-binding proteins with prion-like domains in neurodegenerative disease. *Brain Research, 1462*, 61–80.

Knowles, T. P., Vendruscolo, M., & Dobson, C. M. (2014). The amyloid state and its association with protein misfolding diseases. *Nature Reviews Molecular Cell Biology, 15*, 384–396.
Kroschwald, S., Maharana, S., Mateju, D., Malinovska, L., Nuske, E., Poser, I., Richter, D., & Alberti, S. (2015). Promiscuous interactions and protein disaggregases determine the material state of stress-inducible RNP granules. *eLife, 4*, e06807.
Kwiatkowski, T. J., Jr., Bosco, D. A., Leclerc, A. L., Tamrazian, E., Vanderburg, C. R., Russ, C., Davis, A., Gilchrist, J., Kasarskis, E. J., Munsat, T., Valdmanis, P., Rouleau, G. A., Hosler, B. A., Cortelli, P., de Jong, P. J., Yoshinaga, Y., Haines, J. L., Pericak-Vance, M. A., Yan, J., Ticozzi, N., Siddique, T., McKenna-Yasek, D., Sapp, P. C., Horvitz, H. R., Landers, J. E., & Brown, R. H., Jr. (2009). Mutations in the FUS/TLS gene on chromosome 16 cause familial amyotrophic lateral sclerosis. *Science, 323*, 1205–1208.
Lagier-Tourenne, C., Polymenidou, M., & Cleveland, D. W. (2010). TDP-43 and FUS/TLS: Emerging roles in RNA processing and neurodegeneration. *Human Molecular Genetics, 19*, R46–R64.
Laufen, T., Mayer, M. P., Beisel, C., Klostermeier, D., Mogk, A., Reinstein, J., & Bukau, B. (1999). Mechanism of regulation of hsp70 chaperones by DnaJ cochaperones. *Proceedings of the National Academy of Sciences of the United States of America, 96*, 5452–5457.
Lee, K. H., Zhang, P., Kim, H. J., Mitrea, D. M., Sarkar, M., Freibaum, B. D., Cika, J., Coughlin, M., Messing, J., Molliex, A., Maxwell, B. A., Kim, N. C., Temirov, J., Moore, J., Kolaitis, R. M., Shaw, T. I., Bai, B., Peng, J., Kriwacki, R. W., & Taylor, J. P. (2016). C9orf72 dipeptide repeats impair the assembly, dynamics, and function of membrane-less organelles. *Cell, 167*, 774–788. e17.
Li, Y. R., King, O. D., Shorter, J., & Gitler, A. D. (2013). Stress granules as crucibles of ALS pathogenesis. *The Journal of Cell Biology, 201*, 361–372.
Mackenzie, I. R., Nicholson, A. M., Sarkar, M., Messing, J., Purice, M. D., Pottier, C., Annu, K., Baker, M., Perkerson, R. B., Kurti, A., Matchett, B. J., Mittag, T., Temirov, J., Hsiung, G. R., Krieger, C., Murray, M. E., Kato, M., Fryer, J. D., Petrucelli, L., Zinman, L., Weintraub, S., Mesulam, M., Keith, J., Zivkovic, S. A., Hirsch-Reinshagen, V., Roos, R. P., Zuchner, S., Graff-Radford, N. R., Petersen, R. C., Caselli, R. J., Wszolek, Z. K., Finger, E., Lippa, C., Lacomis, D., Stewart, H., Dickson, D. W., Kim, H. J., Rogaeva, E., Bigio, E., Boylan, K. B., Taylor, J. P., & Rademakers, R. (2017). TIA1 mutations in amyotrophic lateral sclerosis and frontotemporal dementia promote phase separation and alter stress granule dynamics. *Neuron, 95*, 808–816. e9.
March, Z. M., King, O. D., & Shorter, J. (2016). Prion-like domains as epigenetic regulators, scaffolds for subcellular organization, and drivers of neurodegenerative disease. *Brain Research, 1647*, 9–18.
Mateju, D., Franzmann, T. M., Patel, A., Kopach, A., Boczek, E. E., Maharana, S., Lee, H. O., Carra, S., Hyman, A. A., & Alberti, S. (2017). An aberrant phase transition of stress granules triggered by misfolded protein and prevented by chaperone function. *The EMBO Journal, 36*, 1669–1687.
Mayer, M. P., & Bukau, B. (2005). Hsp70 chaperones: Cellular functions and molecular mechanism. *Cellular and Molecular Life Sciences, 62*, 670–684.
Mayer, M. P., Schroder, H., Rudiger, S., Paal, K., Laufen, T., & Bukau, B. (2000). Multistep mechanism of substrate binding determines chaperone activity of Hsp70. *Nature Structural Biology, 7*, 586–593.
Mazroui, R., Di Marco, S., Kaufman, R. J., & Gallouzi, I. E. (2007). Inhibition of the ubiquitin-proteasome system induces stress granule formation. *Molecular Biology of the Cell, 18*, 2603–2618.
Minoia, M., Boncoraglio, A., Vinet, J., Morelli, F. F., Brunsting, J. F., Poletti, A., Krom, S., Reits, E., Kampinga, H. H., & Carra, S. (2014). BAG3 induces the sequestration of proteasomal clients into cytoplasmic puncta: Implications for a proteasome-to-autophagy switch. *Autophagy, 10*, 1603–1621.

Mizuno, Y., Amari, M., Takatama, M., Aizawa, H., Mihara, B., & Okamoto, K. (2006). Immunoreactivities of p62, an ubiqutin-binding protein, in the spinal anterior horn cells of patients with amyotrophic lateral sclerosis. *Journal of the Neurological Sciences, 249*, 13–18.

Molliex, A., Temirov, J., Lee, J., Coughlin, M., Kanagaraj, A. P., Kim, H. J., Mittag, T., & Taylor, J. P. (2015). Phase separation by low complexity domains promotes stress granule assembly and drives pathological fibrillization. *Cell, 163*, 123–133.

Morgan, S., & Orrell, R. W. (2016). Pathogenesis of amyotrophic lateral sclerosis. *British Medical Bulletin, 119*, 87–98.

Mori, K., Weng, S. M., Arzberger, T., May, S., Rentzsch, K., Kremmer, E., Schmid, B., Kretzschmar, H. A., Cruts, M., Van Broeckhoven, C., Haass, C., & Edbauer, D. (2013). The C9orf72 GGGGCC repeat is translated into aggregating dipeptide-repeat proteins in FTLD/ALS. *Science, 339*, 1335–1338.

Morimoto, R. I., & Cuervo, A. M. (2014). Proteostasis and the aging proteome in health and disease. *The Journals of Gerontology. Series A, Biological Sciences and Medical Sciences, 69* (Suppl 1), S33–S38.

Murakami, T., Qamar, S., Lin, J. Q., Schierle, G. S., Rees, E., Miyashita, A., Costa, A. R., Dodd, R. B., Chan, F. T., Michel, C. H., Kronenberg-Versteeg, D., Li, Y., Yang, S. P., Wakutani, Y., Meadows, W., Ferry, R. R., Dong, L., Tartaglia, G. G., Favrin, G., Lin, W. L., Dickson, D. W., Zhen, M., Ron, D., Schmitt-Ulms, G., Fraser, P. E., Shneider, N. A., Holt, C., Vendruscolo, M., Kaminski, C. F., & St George-Hyslop, P. (2015). ALS/FTD mutation-induced phase transition of FUS liquid droplets and reversible hydrogels into irreversible hydrogels impairs RNP granule function. *Neuron, 88*, 678–690.

Patel, A., Lee, H. O., Jawerth, L., Maharana, S., Jahnel, M., Hein, M. Y., Stoynov, S., Mahamid, J., Saha, S., Franzmann, T. M., Pozniakovski, A., Poser, I., Maghelli, N., Royer, L. A., Weigert, M., Myers, E. W., Grill, S., Drechsel, D., Hyman, A. A., & Alberti, S. (2015). A liquid-to-solid phase transition of the ALS protein FUS accelerated by disease mutation. *Cell, 162*, 1066–1077.

Protter, D. S., & Parker, R. (2016). Principles and properties of stress granules. *Trends in Cell Biology, 26*, 668–679.

Prudlo, J., Konig, J., Schuster, C., Kasper, E., Buttner, A., Teipel, S., & Neumann, M. (2016). TDP-43 pathology and cognition in ALS: A prospective clinicopathologic correlation study. *Neurology, 87*, 1019–1023.

Qian, S. B., McDonough, H., Boellmann, F., Cyr, D. M., & Patterson, C. (2006). CHIP-mediated stress recovery by sequential ubiquitination of substrates and Hsp70. *Nature, 440*, 551–555.

Rauch, J. N., & Gestwicki, J. E. (2014). Binding of human nucleotide exchange factors to heat shock protein 70 (Hsp70) generates functionally distinct complexes in vitro. *The Journal of Biological Chemistry, 289*, 1402–1414.

Rauch, J. N., Tse, E., Freilich, R., Mok, S. A., Makley, L. N., Southworth, D. R., & Gestwicki, J. E. (2017). BAG3 is a modular, scaffolding protein that physically links heat shock protein 70 (Hsp70) to the small heat shock proteins. *Journal of Molecular Biology, 429*, 128–141.

Rea, S. L., Majcher, V., Searle, M. S., & Layfield, R. (2014). SQSTM1 mutations – bridging Paget disease of bone and ALS/FTLD. *Experimental Cell Research, 325*, 27–37.

Rosen, D. R., Siddique, T., Patterson, D., Figlewicz, D. A., Sapp, P., Hentati, A., Donaldson, D., Goto, J., O'Regan, J. P., Deng, H. X., et al. (1993). Mutations in Cu/Zn superoxide dismutase gene are associated with familial amyotrophic lateral sclerosis. *Nature, 362*, 59–62.

Ruegsegger, C., & Saxena, S. (2016). Proteostasis impairment in ALS. *Brain Research, 1648*, 571–579.

Schmid, D., Baici, A., Gehring, H., & Christen, P. (1994). Kinetics of molecular chaperone action. *Science, 263*, 971–973.

Schubert, U., Anton, L. C., Gibbs, J., Norbury, C. C., Yewdell, J. W., & Bennink, J. R. (2000). Rapid degradation of a large fraction of newly synthesized proteins by proteasomes. *Nature, 404*, 770–774.

Scotter, E. L., Chen, H. J., & Shaw, C. E. (2015). TDP-43 proteinopathy and ALS: Insights into disease mechanisms and therapeutic targets. *Neurotherapeutics, 12*, 352–363.

Seguin, S. J., Morelli, F. F., Vinet, J., Amore, D., De Biasi, S., Poletti, A., Rubinsztein, D. C., & Carra, S. (2014). Inhibition of autophagy, lysosome and VCP function impairs stress granule assembly. *Cell Death and Differentiation, 21*, 1838–1851.

Selcen, D., Muntoni, F., Burton, B. K., Pegoraro, E., Sewry, C., Bite, A. V., & Engel, A. G. (2009). Mutation in BAG3 causes severe dominant childhood muscular dystrophy. *Annals of Neurology, 65*, 83–89.

Shemetov, A. A., & Gusev, N. B. (2011). Biochemical characterization of small heat shock protein HspB8 (Hsp22)-Bag3 interaction. *Archives of Biochemistry and Biophysics, 513*, 1–9.

Sreedharan, J., Blair, I. P., Tripathi, V. B., Hu, X., Vance, C., Rogelj, B., Ackerley, S., Durnall, J. C., Williams, K. L., Buratti, E., Baralle, F., de Belleroche, J., Mitchell, J. D., Leigh, P. N., Al-Chalabi, A., Miller, C. C., Nicholson, G., & Shaw, C. E. (2008). TDP-43 mutations in familial and sporadic amyotrophic lateral sclerosis. *Science, 319*, 1668–1672.

Stojanovski, D., Bohnert, M., Pfanner, N., & van der Laan, M. (2012). Mechanisms of protein sorting in mitochondria. *Cold Spring Harbor Perspective Biology, 4*, a011320.

Takahara, T., & Maeda, T. (2012). Transient sequestration of TORC1 into stress granules during heat stress. *Molecular Cell, 47*, 242–252.

Takayama, S., & Reed, J. C. (2001). Molecular chaperone targeting and regulation by BAG family proteins. *Nature Cell Biology, 3*, E237–E241.

Taylor, A. A., Fournier, C., Polak, M., Wang, L., Zach, N., Keymer, M., Glass, J. D., & Ennist, D. L. (2016a). Predicting disease progression in amyotrophic lateral sclerosis. *Annals of Clinical Translational Neurology, 3*, 866–875.

Taylor, J. P., Brown, R. H., Jr., & Cleveland, D. W. (2016b). Decoding ALS: From genes to mechanism. *Nature, 539*, 197–206.

Teyssou, E., Takeda, T., Lebon, V., Boillee, S., Doukoure, B., Bataillon, G., Sazdovitch, V., Cazeneuve, C., Meininger, V., LeGuern, E., Salachas, F., Seilhean, D., & Millecamps, S. (2013). Mutations in SQSTM1 encoding p62 in amyotrophic lateral sclerosis: Genetics and neuropathology. *Acta Neuropathologica, 125*, 511–522.

Vance, C., Rogelj, B., Hortobagyi, T., De Vos, K. J., Nishimura, A. L., Sreedharan, J., Hu, X., Smith, B., Ruddy, D., Wright, P., Ganesalingam, J., Williams, K. L., Tripathi, V., Al-Saraj, S., Al-Chalabi, A., Leigh, P. N., Blair, I. P., Nicholson, G., de Belleroche, J., Gallo, J. M., Miller, C. C., & Shaw, C. E. (2009). Mutations in FUS, an RNA processing protein, cause familial amyotrophic lateral sclerosis type 6. *Science, 323*, 1208–1211.

Verma, R., Oania, R. S., Kolawa, N. J., & Deshaies, R. J. (2013). Cdc48/p97 promotes degradation of aberrant nascent polypeptides bound to the ribosome. *eLife, 2*, e00308.

Walters, R. W., & Parker, R. (2015). Coupling of ribostasis and proteostasis: Hsp70 proteins in mRNA metabolism. *Trends in Biochemical Sciences, 40*, 552–559.

Wang, F., Durfee, L. A., & Huibregtse, J. M. (2013). A cotranslational ubiquitination pathway for quality control of misfolded proteins. *Molecular Cell, 50*, 368–378.

Xu, Z., Page, R. C., Gomes, M. M., Kohli, E., Nix, J. C., Herr, A. B., Patterson, C., & Misra, S. (2008). Structural basis of nucleotide exchange and client binding by the Hsp70 cochaperone Bag2. *Nature Structural & Molecular Biology, 15*, 1309–1317.

Yewdell, J. W. (2011). DRiPs solidify: Progress in understanding endogenous MHC class I antigen processing. *Trends in Immunology, 32*, 548–558.

Zhang, K., Donnelly, C. J., Haeusler, A. R., Grima, J. C., Machamer, J. B., Steinwald, P., Daley, E. L., Miller, S. J., Cunningham, K. M., Vidensky, S., Gupta, S., Thomas, M. A., Hong, I., Chiu, S. L., Huganir, R. L., Ostrow, L. W., Matunis, M. J., Wang, J., Sattler, R., Lloyd, T. E., & Rothstein, J. D. (2015). The C9orf72 repeat expansion disrupts nucleocytoplasmic transport. *Nature, 525*, 56–61.

Heat Shock Proteins in Cardiovascular Diseases: From Bench to Bedside

Francesca Bonomini, Gaia Favero, Valentina Trapletti, and Rita Rezzani

Abstract Heat shock proteins (HSP) are stress proteins induced in response to a wide variety of physiological and environmental insults. HSP function as molecular chaperones and they are required to maintain the proteome in a folded and functional state, allowing the cells to survive stress conditions. These key proteins, which may be located intracellularly or extracellularly, have multiple functions that range from the regulation of essential cells function to the renaturation of misfolded proteins. In the last decades, the HSP involvement in both normal cell function and disease pathogenesis is widely studied, especially in the context of cardiovascular diseases (CVDs). This chapter covers the current knowledge on the function HSP in the cardiovascular system and particular in the relationship between these proteins and CVDs. Initially, the roles of HSP in cardiovascular health are outlined, followed by an evaluation of the role of HSP in CVDs key processes, such as atherosclerosis, vascular hypertrophy and heart failure. Finally, the therapeutic potential of roles HSP are examined in a CVDs context, considering how the knowledge actually gained may be capitalized in future clinical studies.

Keywords Cardiovascular diseases · Heat shock proteins · Hsp40 · Hsp60 · Hsp70 · Hsp90 · Small heat shock proteins · Therapeutics

F. Bonomini · R. Rezzani (✉)
Anatomy and Physiopathology Division, Department of Clinical and Experimental Sciences, University of Brescia, Brescia, Italy

Interdipartimental University Center of Research "Adaption and Regeneration of Tissues and Organs- (ARTO)", University of Brescia, Brescia, Italy
e-mail: rita.rezzani@unibs.it

G. Favero · V. Trapletti
Anatomy and Physiopathology Division, Department of Clinical and Experimental Sciences, University of Brescia, Brescia, Italy

A. A. A. Asea, P. Kaur (eds.), *HSP70 in Human Diseases and Disorders*, Heat Shock Proteins 14, https://doi.org/10.1007/978-3-319-89551-2_11

Abbreviations

AIF	apoptosis inducing factor
$ApoE^{-/-}$	apolipoproteinE knock out mice
Ca^{2+}	calcium
CEL	celastrol
CVDs	cardiovascular diseases
GGA	geranylgeranylacetone
HSEs	heat shock elements
Hsf1	heat transcription factor 1
HSP	heat shock protein
LDL-C	low-density lipoprotein-cholesterol
RIPC	remote ischemic preconditioning
ROS	reactive oxygen species
sHSP	small heat shock protein
sHSP60	soluble heat shock protein60
SMCs	smooth muscle cells
TLR4s	toll-like receptors 4
VSMCs	vascular smooth muscle cells

Introduction

Nowadays, despite a variety of therapeutic advances, cardiovascular diseases (CVDs) remain a leading cause of mortality worldwide (Kessing et al. 2016). CVDs include various diseases that affect heart and/or blood vessels, such as coronary artery diseases like angina and myocardial infarction. Heat shock proteins (HSP) are stress proteins, also called chaperones that are massively produced by almost all prokaryotic and eukaryotic species when exposed to elevated temperatures. These proteins were discovered in 1962 by Ritossa in the *Drosophila melanogaster* as a set of highly conserved proteins whose expression was induced in salivary gland chromosome in response to transient exposures to elevated temperatures (Kliková et al. 2016; Ritossa 1962). The increased expression of HSP in cells during the heat shock response was demonstrated to inhibit stress-mediated cellular death as shown also by recent experiments indicating a highly versatile role for these proteins (Garrido et al. 2006). These functions are summarized in Fig. 1. HSP play an important role in cell-cycle control, folding, defolding and assembling of protein complexes (Nakai and Ishikawa 2001) and they have been classified into six families on the basis of their approximate molecular weight: small HSP, HSP40, HSP60, HSP70, HSP90 and HSP100 (Khalil et al. 2011). These proteins constitute the 5–10% of the total protein content in cells in physiological conditions and play important roles in cellular homeostasis. They control maturation and turnover of intracellular proteins and play significant roles in the maintenance of cellular integrity.

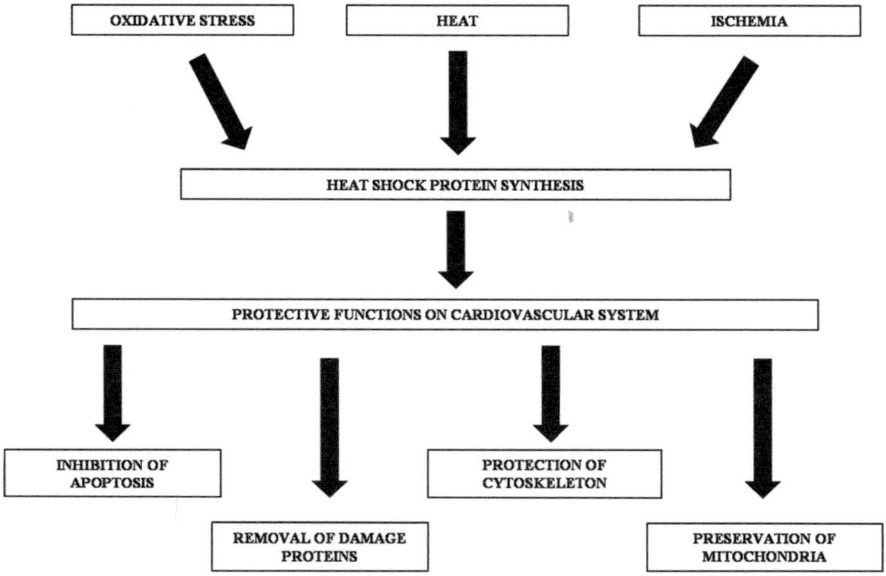

Fig. 1 Main heat shock proteins protective functions at cardiovascular level

Interestingly, the synthesis of these proteins can be markedly induced by various cellular insults ranging from elevation in temperature to exposure to different stimuli which are potentially damaging to the cell (Pockley 2002). These latter conditions include several known risk factors for CVDs onset and development. It is clear, however, from the extensive studies which have been carried out in a variety of cell types that the HSP are induced in response to cellular stress and they confer protection against nonthermal cytotoxic stimuli. A recent in vitro experiment demonstrated that induction of the stress response protects endothelial cells against superoxide generation (Chen et al. 2017). Other examples in vivo include protection of cardiomyocytes against apoptosis, induced by ischemia/reperfusion injury or by prolonged beta-agonist stimulation (Fan et al. 2004, 2005). Even if new functions continue to be discovered, the stress proteins are generally thought to maintain cellular homeostasis by acting as molecular chaperones, facilitating the proper folding and assembly of nascent polypeptides, as well as assisting in the refolding and stabilization of damaged peptides (Bond and Schlesinger 1985). HSP act as intracellular chaperones, protecting protein structure and folding under stress condition, but are also able to assist with numerous reparative processes including the refolding of denatured proteins and removal of irreparably damaged proteins (Kampinga and Craig 2010). Under particular conditions HSP could be released from cells into the extracellular space, where they exhibit a range of immunoregulatory activities (Chebotareva et al. 2017).

To date numerous studies suggest that the upregulation of HSP could be associated with cardioprotection (Jimenez et al. 2014). In fact, before myocardial infarction, an increase in HSP70 and HSPB1 levels induced through an infection with

Table 1 Heat shock proteins involvement in cardiovascular diseases

Superfamily	Protein names	Alternative names	Low protein expression levels or mutations
sHSPs	HSPB1	HSP27,HSP25, HMN2B	Coronary artery disease
	HSPB2	MKBP	Altered mitochondria functions
	HSPB3	HSPL27	Deregulation of myoblast viability
	HSPB5	alphaB-crystallin	Desmin-related myopathy
	HSPB6	HSP20	Decreased contractile function
	HSPB7	CvHSP	No protection against age-related dysfuction
	HSPB8	H11, HSP22	Lack of protection after myocardial ischemia
HSP40	DNAJA3	Tid-1; Tid11	Respiratory chain deficiency
HSP60	HSP60	Cpn60, Hsp60	Protection from atherosclerosis and proliferation of vascular smooth muscle cells
HSP70	HSPA1A	HSP72; HSPA1	Plaque rupture and thrombotic complications
	HSPA8	HSC70; HSC71	
HSP90	HSPC4	GP96, endoplasmin	Increased in plaque stability
			Cardiomyocyte apoptosis

replication deficient adenovirus encoding for these two HSP has been shown to protect heart (Wei et al. 2006). Moreover, Chen et al. (2005) showed that along with the rise in reactive oxygen species (ROS) levels, there was a significant increase in the heart levels of HSP (HSP70, HSPB1) in response to remote ischemic preconditioning (RIPC). Moreover, treatment with antioxidants reduced the levels of HSP and also attenuated cardioprotection (Chen et al. 2005). These results suggest that HSP production may be induced in condition characterized by an increase in free radical production and this enhancement, in turn, may confer resistance to myocardium against sustained ischemia reperfusion injury. Furthermore, Madamanchi et al. (2001) in an in vitro study showed that upregulation of HSP may be secondary to increase in ROS in vascular smooth muscle cells. In contrast, other studies showed that HSP may stimulate the production of ROS (Baruah et al. 2014; Multhoff et al. 2015). So, more studies are required to better clarify the possible interaction between heat shock proteins HSP in cardioprotection (Singh et al. 2017). Table 1 summarizes the main involvement of H HSP in CVDs. In the following sections, we present an overview of the HSP roles and functions in cardiovascular health conditions and their implication in the onset and development of cardiovascular diseases, firstly we presented the small HSP, characterised by a low molecular weight (15–43 kDa), and then the HSP with a higher molecular weight (HSP40, HSP60, HSP70, HSP90). Finally, we provided a brief summary about potentially preventive and therapeutic approaches that involved HSP modulation.

Emerging Roles of Heat Shock Proteins in Cardiovascular Diseases

Small Heat Shock Proteins

Small HSP (sHSP), also called HSPβ (HSPBs), have molecular weight which varied from 15 to 43 kDa and are mainly known for their chaperoning function in the process of embryonic development (Jee 2016). This family of HSP consists of ten members (HSPB1-HSPB10), which are characterized by different functions and display different expression profiles. Some members of this family are widely expressed, like HSPB1, HSPB5 and HSPB8, while some other members show exclusively a restricted expression pattern. In particular, HSPB2 and HSPB3 are only found in cardiac and muscle cells, HSPB4 is lens specific, while HSPB9 and HSPB10 are specifically expressed in testis (Clark et al. 2012; Fontaine et al. 2003; Kappe et al. 2003; Sugiyama et al. 2000; Suzuki et al. 1998; Verschuure et al. 2003). Some members of this family, such as HSPB1 and HSPB4, exert mainly refolding activities, while other members display only anti-aggregation function (HSPB6, HSPB7 and HSPB8). Furthermore, other members exert very specialized activities, for example HSPB2 and HSPB3 seem to have an essential role in muscle differentiation, while HSPB1, HSPB5 and HSPB7 show very specific chaperone functions at the levels of nuclear structures forming during stress (Boncoraglio et al. 2012).

sHSP present a highly conserved sequence of 80–100 amino acids called the α-crystallin domain. This structural domain is responsible for many intra- and intermolecular interactions leading to the formation of dimers, which are considered as sHSP basic unit (Kim et al. 1998; Van Montfort et al. 2001a, b;). These dimers can interact with each other forming higher molecular weight oligomers. Besides the α-crystallin domain, both the C-terminal and the N-terminal regions are involved in the stabilization of the oligomers (Kim et al. 1998; Lambert et al. 1999; Van Montfort et al. 2001a, b). However, besides these different aspects, ten members of the HSPB family share the same properties and so alteration of these properties represent a key mechanism that can lead to diseases onset, also at cardiovascular level.

Cardiac and skeletal muscle cells express the largest variety of HSPBs: HSPB1, HSPB2, HSPB3, HSPB5, HSPB6, HSPB7 and HSPB8. These family members are commonly considered the main focus of interest of CVDs researches and new scientific evidences emphasize the functions of these molecules in cardioprotection (Carra et al. 2017; Charmpilas et al. 2017; Edwards et al. 2011; Fan et al. 2005; Martin et al. 2014; Weintraub and Rubinstein 2013).

HSPB1 is a ubiquitously expressed, multifunctional protein chaperone. A decrease in HSPB1 secretion was shown in human atherosclerotic plaques respect to control vessels and also plasmatic HSPB1 levels are reduced in atherosclerotic patients compared to healthy subjects (Martin-Ventura et al. 2004). Moreover,

Seibert et al. (2013) showed that in human low serum HSPB1 levels are associated with coronary artery disease and prognostic of adverse clinical events; while in mouse models of atherosclerosis the increase in HSPB1 level in aorta induce a reduction of atherosclerotic lesion progression and promote plaque stability. It was also shown that HSPB1 overexpression in atherosclerotic-prone mouse models (apolipoproteinE knock out mice – $ApoE^{-/-}$) modulates plaque formation preventing inflammation, foam cells formation and reducing atherosclerotic plaque area (Rayner et al. 2008). In heart tissue HSPB1 expression was associated with sarcomeres and so it was found to be cardioprotective (Brundel et al. 2006a, b). Moreover, Ghayour-Mobarhan et al. (2012) reported that increased expression of myocardial HSPB1 is a prognostic marker of myocardial ischemia. HSPB1 has been demonstrated to play also an important role in regulating intracellular redox homeostasis and anti-apoptotic pathway (Arrigo et al. 2005; Zhang et al. 2010). Panneerselvam et al. (2017) observed that differential expression of HSP in the rat myocardium could serve as a balance between pro-survival and death signal during acute fluoride-induced heart failure. In detail, the Authors suggested that increased expression of myocardial HSPB1 serves as a prognostic marker for myocardial ischemia and this increase is correlated with myocardial necrosis, impaired contractile function and regulation of intracellular redox homeostasis.

HSPB2 has shown to be highly expressed in heart and it has protective effects against heart diseases such as cardiac hypertrophy and ischemia (Ishiwata et al. 2012; Nakagawa et al. 2001; Sugiyama et al. 2000). Moreover, HSPB2 was associated with the outer membrane of mitochondria and involved in mitochondria permeability transition and calcium uptake. Overexpression of this sHSP was found to conserve ATP synthesis during ischemic/reperfusion injury in mice (Nakagawa et al. 2001). Moreover, mice knockout for HSPB2 with ischemic stress show altered mitochondria respiration rates and reduced ATP production as well as modification in expression of several metabolic and mitochondrial regulators (Ishiwata et al. 2012). These findings suggest that HSPB2 has cardioprotective effects maintaining mitochondrial function and metabolic activity during cardiac stress.

Golenhofen et al. (2006) showed that the cytosolic calcium increase due to the lack of HSPB2 in knockout animals may modify the calcium sensitivity of myofibrils altering the cardiac contractility, suggesting an involvement of HSPB2 in maintaining muscular elasticity during ischemic insult. Interestingly, mice with HSPB2 overexpression in heart revealed lower levels of cardiac injury biomarker troponin I in the blood after ischemia/reperfusion stress, confirming the involvement of HSPB2 in preserving contractile function of the heart (Grose et al. 2015).

HSPB3 was considered not expressed in the heart (Vos et al. 2009), but recently Carra et al. (2017) reported in patients with myopathy two novel mutations in the HSPB3 gene that lead respectively to protein aggregation or truncation and destabilization. Remarkably, these mutations could abolish HSPB2-HSPB3 complex formation, leading to free HSPB2 proteins that tend to mislocalize inside the cells. Moreover, the same authors reported that HSPB2-HSPB3 interaction would lead to deregulation of HSPB2, with potential consequences on myoblast function and viability. These studies suggested that alterations in expression and solubility of specific HSPBs due to aggregation propensity and deregulated association with other HSPBs could be the basis of complex diseases, such as myopathies.

HSPB5 was found in cardiomyocytes on the I-band and M-line region of sarcomeres (van de Klundert et al. 1998). It is known to bind and stabilize intermediate filaments, actin microfilaments, and sarcomeric proteins, including actin, desmin and titin in physiological conditions (Bullard et al. 2004; Ghosh et al. 2007; Perng et al. 1999). HSPB5 has shown to be involved also in stabilization of the cytoskeleton (Vicart et al. 1998). Interestingly, mutations in HSPB5 are associated with cardiac and muscular disorders. In fact, HSPB5 mutations result in an irregular protein structure and defective chaperone-like function (Bova et al. 1999), which leading to desmin-related myopathy and also early onset of cardiomyopathy (Selcen and Engel 2003; Vicart et al. 1998).

HSPB6 is abundantly expressed in skeletal muscle and heart in two complex formations: 43 kDa dimers and 470 kDa multimers and that this protein is able to bind itself and other HSPBs, like HBPB1, HSPB5 and HSPB8 (Pipkin et al. 2003). Recently, in a more recent study, the overexpression of HSPB6 resulted in enhanced cardiac function by interaction with protein phosphatase 1 and in turn inducing calcium (Ca^{2+}) cycling and sarcoplasmic reticulum Ca^{2+} load (Qian et al. 2011). In addition, this stress protein induces the Ca^{2+} cycling in the sarcoplasmic reticulum and increases the contractile function of the cardio myocyte (Qian et al. 2011). Further, in an experimental study in mice, the phosphorylation of HSPB6 at serine 16 level was shown to be needed for modulating cell injury during ischemia/reperfusion (Qian et al. 2009). So, HSPB6 acts maintaining the heart integrity in mice with ischemia/reperfusion injury (Fan et al. 2005).

HSPB7 is expressed in heart and skeletal muscle. It was shown that HSPB7 is significantly expressed in aged muscle like to HSPB5 (Doran et al. 2007). HSPB7 was also shown to be upregulated in the muscular dystrophy-affected diaphragm, suggesting its induction under stress conditions. Additionally, HSPB7 is involved in protections of cells from protein aggregation, probably by facilitating cargo delivery to autophagosomes (Vos et al. 2010). To note, HSPB4, HSPB6 or HSPB7 could not promote the cellular capacity to chaperone heat-denatured luciferase as shown by HSPB1, indicating different functions and properties of HSPB members (Vos et al. 2010, 2011). Additionally, in cardiomyocytes was observed the co-localization of HSPB7 on myofibrils, suggesting a protective role maintaining the sarcomeric structure (Golenhofen et al. 2004).

HSPB8 is showed in striated and smooth muscles, brain, and keratinocytes level (Vos et al. 2008). This protein can be phosphorylated in vitro, but, respect to HPB1 and HSPB5, HSB8 phosphorylation influences only marginally its tertiary and quaternary structure. HSPB8 exists, both in its wild type and phosphorylated form, as low molecular mass oligomers. Respect to HSPB1 and HSPB5 that showed reduced oligomeric size and increased chaperone activity after phosphorylation, the phosphorylation of HSPB8 results in larger oligomeric structures and decreased chaperone activity (Basha et al. 2006). In in vitro experiments, it has been shown that HSPB8 interacts with several proteins and forms stoichiometric complexes (Carra et al. 2008a). This complexes was found to induce autophagy, which may be beneficial in response to irreparable protein damage (Carra et al. 2008b, 2009). In addition, HSPB8 has been shown to be cardioprotective in experimental models of myocardial ischemia. In detail, Depre et al. 2006 showed that HSPB8 overexpression promotes cardiomyocyte survival after ischemia in mice, while Chen et al. 2011 indicated that this sHSP attenuates the myocardial damage and contractile dysfunction in experimental animals. On the other hand, it was demonstrated that depletion of HSPB8 in mice with pressure overload supports cardiac dysfunction and promotes transition to heart failure (Qiu et al. 2011). Other studies showed that HSPB8 maintains mitochondrial function and energy production that results in attenuation of oxidative stress in infarcted hearts (Marunouchi et al. 2014). In contrast these beneficial HSPB8 effects on cardiomyocyte function, HSPB8 overexpression, both in in vitro and in vivo model systems, was found to induce cardiac hypertrophy (Depre et al. 2002; Hedhli et al. 2008). Therefore, the function of HSPB8 seems to have two faces in heart diseases: HSPB8 acts as a cardioprotective protein during myocardial ischemia by conserving the mitochondrial function and energy production but, at the same time, HSPB8 is a mediator of cardiac hypertrophy and thereby results in heart failure.

HSP60

HSP60 protein is structurally highly conserved and abundantly expressed by prokaryotic and eukaryotic cells under stress conditions. It is considered a constitutively cytosolic protein that translocated to the mitochondria (Gething and Sambrook 1992), but it is also stress inducible, indeed ischemia has been demonstrated to be a potent inducer (Marber et al. 1993). During heart failure this protein is translocated to the plasma membrane and released into the plasma; in fact was shown that HSP60 levels were doubled in end-stage heart failure (Knowlton et al. 1998; Lin et al. 2007). Some studies reported that high titers of anti-HSP60 were correlated with coronary atherosclerosis (Prohászka et al. 2001; Zhu et al. 2001), moreover high levels of anti-HSP60 in children increase the probability to develop CVDs (Cohen and Young 1991). Other studies suggested that plasmatic levels of HSP60 was increased early in heart failure (Brundel et al. 2006a, b). During heart failure, HSP60 localizes to the plasma membrane and the cell surface, in addition to its normal distribution to the mitochondria and cytosol (Wang et al. 2010) and acts as a "danger signal". This localization of HSP60 could be linked with an increase in apoptosis of the affected cell, because the cell-surface expression of HSP60 may be able to interact with other cells to trigger the innate immune response, resulting in the release of pro-inflammatory cytokines which could induce myocyte loss and contribute to heart disease. Previous studies (Malik et al. 2013; Gupta and Knowlton 2007) demonstrated that HSP60, via exosomal pathway, was released by cardiomyocytes and this stress protein was increased by exposure of the cardiomyocytes to ethanol. Li et al. (2011) reported that, HSP60 induced inflammation in the heart, via activation of the TLR4- MyD88-IRAK-1 pathway, and led to cardiomyocyte apoptosis, promoting progression of heart failure.

Some studies in human associated HSP60 expression with atherosclerosis and its severity and it was detected on endothelium, smooth muscle cells (SMCs) and mononuclear cells (Kleindienst et al. 1993; Lamb et al. 2002). Other studies reported a positive association between titers of antibody to HSP60 and the extent of atherosclerosis (Burian et al. 2001; Veres et al. 2002; Xu et al. 2000; Zhao et al. 2015; Zhu et al. 2001). Xiao et al. (2005) reported that soluble HSP60 (sHSP60) is probably involved in activating proinflammatory processes associated with early vessel pathology providing the first prospective data confirming an association between elevated levels of sHSP60 and early carotid atherosclerosis. Moreover, studies in cholesterol-fed rabbits showed that plasma titers of anti-HSP60 are consistent with atherosclerotic formation and increased expression of HSP60 on the endothelium during atherogenesis (Ghayour-Mobarhan et al. 2007; Khan et al. 1998; Lamb et al. 2002; Pfister et al. 2005; Zhao et al. 2015). Recently, in vitro experiments have shown that HSP60 can activate the proliferation of vascular SMCs (VSMCs) (de Graaf et al. 2006; Fukuoka et al. 2004; Sasu et al. 2001; Zhao et al. 2015). Moreover, Zhao et al. (2015), in an experimental study, suggested that HSP60 is implicated in the VSMC migration during atherosclerosis. The authors reported that activation of HSP60 could be considered one of the most powerful methods of sending a 'danger signal' to the immune system to generate chemokine, such as interleukin-8, which is involved in stimulation of VSMC migration.

HSP70

HSP70 is one major member of HSP family, which is able to bind its substrate through the substrate binding domain, dephosphorylate substrate and in turn maintain its "normal" functions (Connarn et al. 2014). This HSP have some function in the cells including binding to partially synthesized peptide sequences, keeping translocation competent structures of endoplasmic reticulum and mitochondrial precursors in the cytosol and simplify translocation from inside the target compartment (Craig et al. 1994; Mamipour et al. 2017). Previous studies have provided evidences that during myocardial ischemia/reperfusion HSP70 had protective effects and it was able to reduce the deleterious effects characteristics of this pathological condition (Feng et al. 2014; Garrido et al. 2006; Vicencio et al. 2015). In particular, during myocardial ischemia/reperfusion HSP70 accelerates the dephosphorylation of the stress kinase JNK inducing cardiac protection (Gabai et al. 2000). Moreover, in this pathological condition HSP70 is able to prevent the nucleus translocation of apoptosis inducing factor (AIF), acting as a molecule chaperone, so it stabilizes nucleus and antagonizes cardiomyocyte apoptosis (Choudhury et al. 2011). Other data confirming this action showed that the deletion of the domain of AIF involved in interaction with HSP70 enhanced AIF translocation to nucleus inducing apoptosis (Gurbuxani et al. 2003; Zhang et al. 2017). Recently, Vicencio et al. (2015) showed that plasma exosomes, which carry HSP70 on their surfaces, attenuated myocardial reperfusion injury in rats. In in vitro studies, exosomes derived from plasma, containing HSP70, bound to toll-like receptors 4 (TLR4s) on cardiomyocytes and activated various kinases leading to cardioprotection and suggesting the therapeutic use of exosomes for reperfusion injury. Mayr et al. (1999) showed that, in human atherosclerotic aorta, HSP70 is expressed with a heterogeneous pattern respect to control vessels. This HSP70 altered pattern could be due to its release from damaged cells into the plaque. Mazzaferro et al. (2003) hypothesized that insufficient HSP70 in SMC of the aorta leads to their death and favors plaque rupture and thrombotic complications. Moreover atherosclerotic plaque thickness was correlated with the HSP70 expression to the center of the lesion (Pockley et al. 2003). Other studies attributed to HSP70 a protective role in atherosclerosis due to its pro-survival effect on VSMCs. It was demonstrated that HSP70 localization changed during plaque development and progression and was positively related to atherosclerosis severity (Dybdahl et al. 2005; Johnson et al. 1993; Pockley et al. 2003). Furthermore Zhu et al. (2003) have reported that there is an inverse relation between HSP70 expression and severity of coronary disease.

HSP90

The 90 kDa heat shock protein, HSP90, is one of the most abundantly expressed protein in eukaryotic cells, comprising 1–2% of cellular proteins under physiological conditions. It contributes to various cellular processes as signal transduction, protein folding and degradation and morphological evolution (Csermely et al. 1998; Pearl and Prodromou 2001; Picard 2002; Pratt and Toft 2003; Richter and Buchner 2001; Sreedhar et al. 2004). The activity of HSP90 is in cooperation with other co-chaperones, playing a key role in the folding of newly synthesized proteins and stabilization and refolding of denatured proteins during stress conditions (Sreedhar et al. 2004). Some studies indicated that HSP90 plays an important protective role in ischemic pre-conditioning and post-conditioning, acting as a cardioprotective and antiapoptotic protein (Amour et al. 2009; Zhong et al. 2014). Moreover, targeted overexpression of HSP90 protects myocardial reperfusion injury reducing infarct size and myocardial dysfunction (Kupatt et al. 2004). On the other hand, Wang et al. (2009) showed that HSP90 inhibition increases cardiomyocyte apoptosis. It has also been shown that, in isolated hearts, HSP90 attenuates ischemia/reperfusion-induced cardiomyocyte apoptosis and necrosis (Budas et al. 2010). Moreover, recent reports demonstrated that HSP90 is anti-apoptotic in cardiomyocytes acting as antioxidant and preserving mitochondrial function under stress conditions (Wu et al. 2012).

Previously, a human study demonstrated augmented systemic HSP90-specific cellular and humoral responses in carotid atherosclerosis patients (Mu et al. 2017; Rigano et al. 2007). Furthermore, other authors showed that HSP90 is highly expressed in plaque and serum from patients with atherosclerosis, in which immune responses are stimulated, suggesting that HSP90 is a potential target in the pathogenesis of carotid atherosclerosis (Businaro et al. 2009; Mu et al. 2017). Recently, Mu et al. (2017) showed

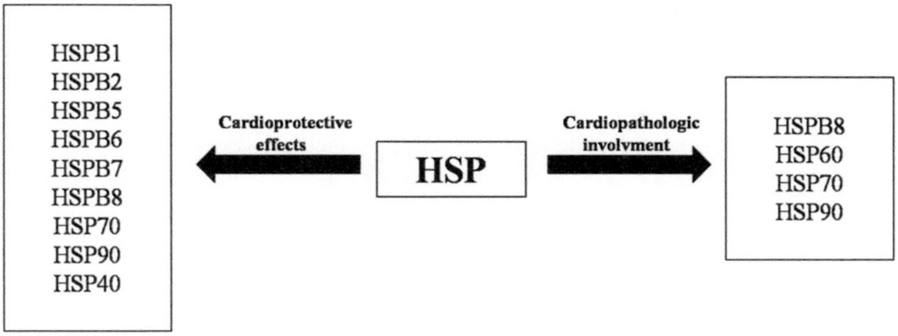

Fig. 2 Schematic representation of heat shock proteins involvement in cardioprotective and/or cardiopathological processes

that in a mouse model of atherosclerosis HSP90 overexpression in vulnerable carotid plaque induced a decrease in plaque stability, along with increased accumulation of lipids, macrophages, inflammation, as well as a decrease in VSMCs and collagen, while HSP90 inhibition exerted opposing effects. So, these authors suggested that regulation of the local HSP90 expression could be a new route for atheroprotection.

HSP40

HSP40 is the co-chaperone of HSP70. This heat shock protein has also been shown to be involved in the pathology of dilated cardiomyopathy that is an important cause of sudden cardiac death and heart failure (Hayashi et al. 2006; Henderson and Pockley 2012). In fact, inactivation of the gene encoding mitochondrial HSP40 in mice resulted in the development of severe cardiomyopathy which induced death probably due to the role played by HSP40 in mitochondrial biogenesis (Hayashi et al. 2006). Hayashi et al. (2006) showed that in cardiomyocytes lacking HSP40 there were progressive respiratory chain deficiency and reduced copy number of mitochondrial DNA, underlining the key role of this protein in CVDs. Figure 2 summarizes the HSP involvement in cardioprotective/cardiopathological processes.

Heat Shock Proteins Modulation as a Potential Therapeutic Target

The broad function on cardiovascular system discussed above strongly sustains the notion that strategies aimed at modulating HSP expression may be important for preserving CVDs onset and related symptoms. In fact, some experimental studies have tested the hypothesis that HSP modulation improves the outcome of CVDs, particularly in the context of ischemia/reperfusion injury. The modulations of HSP can be performed in different ways such as by the use of small molecules able to modulate HSP expression or through their activation by phosphorylation. The induction of chaperones in response to different stress conditions are mediated at the transcriptional level by the heat transcription factor 1 (Hsf1) that binds to its target sites (heat shock response elements) in the promoter regions of stress-induced genes and thus, following Hsf1 phosphorylation, induces HSP gene expression (Westerheide et al. 2012). In physiological conditions, without stress, Hsf1 is retained in an inactive state in the cytoplasm binded to HSP90. Stress conditions induce dissociation of HSP90 from Hsf1 that, in turn translocate to the nucleus where it up-regulates transcription of chaperone genes. Although the HSP were identified on the basis of their induction by stressful procedures, from a clinical standpoint there are several drugs and natural compounds that modulates HSP and could be beneficial in the treatment of CVDs.

Geranylgeranylacetone (GGA) is a cyclic polyisoprenoid used as a gastromucoprotective drug. In experimental conditions, it has been showed that this pharmacological compound can induce transcriptional activation of HSP70 (Chang et al. 2013; Hirakawa et al. 1996) and that ischemia/reperfusion-induced damage of myocardial cells was prevented in GGA-treated myocytes (Chang et al. 2013; Ooie et al. 2001). Moreover, GGA upregulates HSP expression in heart and attenuates ischemia/reperfusion injury, degradation of myofibrils, and atrial fibrillation genesis caused by rapid atrial pacing or atrial ischemia (Brundel et al. 2006a, b; Chang et al. 2013; Ooie et al. 2001). Recently, Chang et al. (2013) showed that HSP induced by GGA can regulate the atrial arrhythmogenesis in heart failure by modulation of sodium and potassium channels, as well as calcium homeostasis (Chang et al. 2013).

HSP co-inducer arimoclomol has been tested in a number of clinical trials (Cudkowicz et al. 2008; Lanka et al. 2009; Kirkegaard et al. 2016). Arimoclomol belongs to a group of HSP-modulating drugs that act as inducers of HSP70, whose mechanism of action involves stabilization of the interaction of HSF1 with heat shock elements (HSEs), the transcriptional elements controlling HSP production (Anckar and Sistonen 2011; Crul et al. 2013; Kieran et al. 2004; Kirkegaard et al. 2016; Neef et al. 2011; Parfitt et al. 2014; Vígh et al. 1997).

Celastrol (CEL), a quinone methide triterpene, is a derived from the Celastraceae family of plants, that is used in traditional Chinese medicine. This compound has been shown to induce HSF1 and HSP70 expression in skeletal muscle cells (Gwag et al. 2013; Trott et al. 2008; Westerheide et al. 2004). Furthermore, this compound has been demonstrated to exhibit a broad range of functions like antioxidant, anti-inflammatory (Gwag et al. 2013; Lee et al. 2006; Sassa et al. 1990; Trott et al. 2008) and neuroprotective activity (Franklin et al. 2005; Gwag et al. 2013). All of these activities could be related to its action s HSP inducer.

Moreover, Deane and Brown (2016) in a recent in vitro research on neuronal cells showed that co-application of celastrol and arimoclomol induced higher HSP levels compared to heat shock paired with arimoclomol. This co-application targets multiple alterations including protein misfolding, protein aggregation, inflammation and oxidative stress. Several studies demonstrated the beneficial effects of

Table 2 Pharmacological compounds that modulate heat shock proteins and relative beneficial effects

Compounds	Heat shock proteins modulation	Beneficial effects
GGA	↑HPS70	Prevents ischemia/reperfusion-induced damage at myocardial level
		Regulate the atrial arrhythmogenesis in heart failure
ARIMOCLOMOL	↑HSP70	Neuroprotective activity
CELASTROL	↑HSP70	Antioxidant and anti-inflammatory activities
		Neuroprotective activity
STATIN	↑HSP70	Contributes to plaque stability
	↑HSPB1	Reduces both mortality and the incidence of acute coronary syndrome
	↑HSP90	

statins in reducing both mortality and incidence of acute coronary syndrome (Noguchi et al. 2015; Smith et al. 2011 Weintraub et al. 2011). In addition, to reducing levels of serum low-density lipoprotein-cholesterol (LDL-C), statins also may contribute to plaque stability by reducing inflammation (Noguchi et al. 2015; Puato et al. 2010), improving endothelial function (Noguchi et al. 2015; Schönbeck and Libby 2004) and reinforcing the fibrous cap (Hattori et al. 2012, Komukai et al. 2014; Noguchi et al. 2015); these effects play crucial roles in the protection from CVDs (Fuster et al. 1992a, b; Noguchi et al. 2015). Remarkably, Frostegård et al. (2016) showed that statin, in particular atorvastatin, restores the proatherogenic HSP profile, characterized by an increase of pro-atherogenic HSP60 and HSP70 and a decrease of HSPB1, suggesting that statins could modulate immune reactions pivotal to atherosclerosis and CVDs pathogenesis. Simvastatin induces also vascular endothelial cells HSF1 translocate in nucleus and so cause the transcription of HSP70 and HSP90 (Uchiyama et al. 2007; Willis and Patterson 2010). All the above reported studies suggest that several drugs, have the potential to be cardioprotective because of their ability to modulate HSP in cardiovascular system. The effects of pharmacological compounds are summarized in Table 2.

In addition to the pharmacological induction of HSP, physical exercise was also discovered to induce HSPB levels and consequently has important cardioprotective effects. In fact, various studies demonstrated that gene and protein levels of HSPB1 and HSPB6 in heart are increased after physical exercise in animal models. In these studies, HSPB1 and HSPB6 were phosphorylated inducing myofilaments' stabilization, contractile damaged proteins restoration that resulted in improved contractile function of the heart (Boluyt et al. 2006; Burniston 2009; Campos et al. 2012; de Moraes et al. 2015; Rinaldi et al. 2006; Sakamoto et al. 2006; Hu et al. 2017). Moreover, it has been showed that physical exercise induced elevation of myocardial HSP expression (such as HSP70, HSP90 and HSPB5) and that regular endurance exercise protected the heart against ischemia/reperfusion and infarction (Harris and Starnes 2001; Powers et al. 2002). Therefore, physical exercise may represent a promising preventive therapeutic therapy against CVDs by also its activity in HSP modulation. Furthermore, gene therapy can also be considered as a possible option to modulate HSP levels and function. In particular, adeno-associated viral vectors, which have been successfully utilized in clinical trials for the treatment of cardiomyopathy with SERCA2 overexpression, are currently the most effective in vivo delivery system (Jessup et al. 2011).

Conclusions

During the development of CVDs there is a modulation of chaperones and co-chaperones. Molecular chaperones are members of a large family of proteins that provide physiological control of proteostasis. Lack of regulation of the physiological balance between protein synthesis, folding, and degradation results in accumulation of misfolded proteins, so inducing several CVDs. In particular, we reported in this chapter that different cellular stress proteins have opposed role in promotion or inhibition of CVDs or related symptoms. The data discussed in this manuscript suggest that modulation of molecular chaperone expression could have a protective function in CVDs. Indeed, molecules capable of regulating chaperone expression are currently used in clinical trials, principally for the treatment of neurodegenerative diseases. Therefore, the discovery and the development of novel strategies to modulate chaperone expression and functions could be considered a crucial step for expanding drugs against CVDs.

Acknowledgements This study was supported by the grant (ex-60%) of the University of Brescia, Italy. The Authors sincerely thanks also Fondazione Cariplo e Regione Lombardia "New opportunities and ways towards ERC" (Project 2014-2256).

References

Amour, J., Brzezinska, A. K., Weihrauch, D., et al. (2009). Role of heat shock protein 90 and endothelial nitric oxide synthase during early anaesthetic and ischemic preconditioning. *Anesthesiology, 110*, 317–325.

Anckar, J., & Sistonen, L. (2011). Regulation of HSF1 function in the heat stress response: Implications in aging and disease. *Annual Review of Biochemistry, 80*, 1089–1115.

Arrigo, A. P., Virot, S., Chaufour, S., et al. (2005). Hsp27 consolidates intracellular redox homeostasis by upholding glutathione in its reduced form and by decreasing iron intracellular levels. *Antioxidants & Redox Signaling, 7*, 414–422.

Baruah, K., Norouzitallab, P., Linayati, L., et al. (2014). Reactive oxygen species generated by a heat shock protein (Hsp) inducing product contributes to Hsp70 production and Hsp70-mediated protective immunity in *Artemia franciscana* against pathogenic vibrios. *Developmental and Comparative Immunology, 46*, 470–474.

Basha, E., Friedrich, K. L., & Vierling, E. (2006). The N-terminal arm of small heat shock proteins is important for both chaperone activity and substrate specificity. *The Journal of Biological Chemistry, 281*, 39943–39952.

Boluyt, M. O., Brevick, J. L., Rogers, D. S., et al. (2006). Changes in the rat heart proteome induced by exercise training: Increased abundance of heat shock protein hsp20. *Proteomics, 6*, 3154–3169.

Boncoraglio, A., Minoia, M., & Carra, S. (2012). The family of mammalian small heat shock proteins (HSPBs): Implications in protein deposit diseases and motor neuropathies. *The International Journal of Biochemistry & Cell Biology, 44*, 1657–1669.

Bond, U., & Schlesinger, M. J. (1985). Ubiquitin is a heat shock protein in chicken embryo fibroblasts. *Molecular and Cellular Biology, 5*, 949–956.

Bova, M. P., Yaron, O., Huang, Q., et al. (1999). Mutation R120G in alphaB-crystallin, which is linked to a desmin-related myopathy, results in an irregular structure and defective chaperone-like function. *Proceedings of the National Academy of Sciences of the United States of America, 96*, 6137–6142.

Brundel, B. J., Henning, R. H., Ke, L., et al. (2006a). Heat shock protein upregulation protects against pacing- induced myolysis in HL-1 atrial myocytes and in human atrial fibrillation. *Journal of Molecular and Cellular Cardiology, 41*, 555–562.

Brundel, B. J., Shiroshita-Takeshita, A., Qi, X., et al. (2006b). Induction of heat shock response protects the heart against atrial fibrillation. *Circulation Research, 99*, 1394–1402.

Budas, G. R., Churchill, E. N., Disatnik, M. H., et al. (2010). Mitochondrial import of PKCepsilon is mediated by HSP90: a role in cardioprotection from ischaemia and reperfusion injury. *Cardiovascular Research, 88*, 83–92.

Bullard, B., Ferguson, C., Minajeva, A., et al. (2004). Association of the chaperone alphaB-crystallin with titin in heart muscle. *The Journal of Biological Chemistry, 279*, 7917–7924.

Burian, K., Kis, Z., Virok, D., et al. (2001). Independent and joint effects of antibodies to human heat-shock protein 60 and Chlamydia pneumoniae infection in the development of coronary atherosclerosis. *Circulation, 103*, 1503–1508.

Burniston, J. G. (2009). Adaptation of the rat cardiac proteome in response to intensity-controlled endurance exercise. *Proteomics, 9*, 106–115.

Businaro, R., Profumo, E., Tagliani, A., et al. (2009). Heat-shock protein 90: A novel autoantigen in human carotid atherosclerosis. *Atherosclerosis, 207*, 74–83.

Campos, J. C., Queliconi, B. B., Dourado, P. M., et al. (2012). Exercise training restores cardiac protein quality control in heart failure. *PLoS One, 7*, e52764.

Carra, S., Seguin, S. J., Lambert, H., et al. (2008a). HspB8 chaperone activity toward poly (Q)-containing proteins depends on its association with Bag3, a stimulator of macroautophagy. *The Journal of Biological Chemistry, 283*, 1437–1444.

Carra, S., Seguin, S. J., & Landry, J. (2008b). HspB8 and Bag3: A new chaperone complex targeting misfolded proteins to macroautophagy. *Autophagy, 4*, 237–239.

Carra, S., Brunsting, J. F., Lambert, H., et al. (2009). HspB8 participates in protein quality control by a non-chaperone-like mechanism that requires eIF2 {alpha} phosphorylation. *The Journal of Biological Chemistry, 284*, 5523–5532.

Carra, S., Alberti, S., Arrigo, P. A., et al. (2017). The growing world of small heat shock proteins: From structure to functions. *Cell Stress & Chaperones, 22*, 601–611.

Chang, S. L., Chen, Y. C., Hsu, C. P., et al. (2013). Heat shock protein inducer modifies arrhythmogenic substrate and inhibits atrial fibrillation in the failing heart. *International Journal of Cardiology, 168*, 4019–4026.

Charmpilas, N., Kyriakakis, E., & Tavernarakis, N. (2017). Small heat shock proteins in ageing and age-related diseases. *Cell Stress & Chaperones, 22*, 481–492.

Chebotareva, N., Bobkova, I., & Shilov, E. (2017). Heat shock proteins and kidney disease: Perspectives of HSP therapy. *Cell Stress & Chaperones, 22*, 319–343.

Chen, Y. S., Chien, C. T., Ma, M. C., et al. (2005). Protection "outside the box" (skeletal remote preconditioning) in rat model is triggered by free radical pathway. *The Journal of Surgical Research, 126*, 92–101.

Chen, L., Lizano, P., Zhao, X., et al. (2011). Preemptive conditioning of the swine heart by H11 kinase/Hsp22 provides cardiac protection through inducible nitric oxide synthase. *American Journal of Physiology. Heart and Circulatory Physiology, 300*, H1303–H1310.
Chen, Y., Jiang, B., Zhuang, Y., et al. (2017). Differential effects of heat shock protein 90 and serine 1179 phosphorylation on endothelial nitric oxide synthase activity and on its cofactors. *PLoS One, 12*, e0179978.
Choudhury, S., Bae, S., Ke, Q., et al. (2011). Mitochondria to nucleus translocation of AIF in mice lacking Hsp70 during ischemia/reperfusion. *Basic Research in Cardiology, 106*, 397–407.
Clark, A. R., Lubsen, N. H., & Slingsby, C. (2012). sHSP in the eye lens: Crystallin mutations, cataract and proteostasis. *The International Journal of Biochemistry & Cell Biology, 44*, 1687–1697.
Cohen, I. R., & Young, D. B. (1991). Autoimmunity, microbial immunity and the immunological homunculus. *Immunology Today, 12*, 105–110.
Connarn, J. N., Assimon, V. A., Reed, R. A., et al. (2014). The molecular chaperone Hsp70 activates protein phosphatase 5 (PP5) by binding the tetratricopeptide repeat (TPR) domain. *The Journal of Biological Chemistry, 289*, 2908–2917.
Craig, E. A., Weissman, J. S., & Horwich, A. L. (1994). Heat shock proteins and molecular chaperones: Mediators of protein conformation and turnover in the cell. *Cell, 78*, 365–372.
Crul, T., Toth, N., Piotto, S., et al. (2013). Hydroximic acid derivatives: Pleiotropic HSP co-inducers restoring homeostasis and robustness. *Current Pharmaceutical Design, 19*, 309–346.
Csermely, P., Schnaider, T., Soti, C., et al. (1998). The 90-kDa molecular chaperone family: Structure, function, and clinical applications. A comprehensive review. *Pharmacology & Therapeutics, 79*, 129–168.
Cudkowicz, M. E., Shefner, J. M., Simpson, E., et al. (2008). Arimoclomol at dosages up to 300 mg/day is well tolerated and safe in amyotrophic lateral sclerosis. *Muscle & Nerve, 38*, 837–844.
de Graaf, R., Kloppenburg, G., Kitslaar, P. J., et al. (2006). Human heat shock protein 60 stimulates vascular smooth muscle cell proliferation through Toll-like receptors 2 and 4. *Microbes and Infection, 8*, 1859–1865.
de Moraes, W. M., Melara, T. P., de Souza, P. R., et al. (2015). Impact of leucine supplementation on exercise training induced anti-cardiac remodeling effect in heart failure mice. *Nutrients, 7*, 3751–3766.
Deane, C. A., & Brown, I. R. (2016). Induction of heat shock proteins in differentiated human neuronal cells following co-application of celastrol and arimoclomol. *Cell Stress & Chaperones, 21*, 837–848.
Depre, C., Hase, M., Gaussin, V., et al. (2002). H11 kinase is a novel mediator of myocardial hypertrophy in vivo. *Circulation Research, 91*, 1007–1014.
Depre, C., Wang, L., Sui, X., et al. (2006). H11 kinase prevents myocardial infarction by preemptive preconditioning of the heart. *Circulation Research, 98*, 280–288.
Doran, P., Gannon, J., O'Connell, K., et al. (2007). Aging skeletal muscle shows a drastic increase in the small heat shock proteins alphaB-crystallin/HspB5 and cvHsp/HspB7. *European Journal of Cell Biology, 86*, 629–640.
Dybdahl, B., Slørdahl, S. A., Waage, A., et al. (2005). Myocardial ischaemia and the inflammatory response: Release of heat shock protein 70 after myocardial infarction. *Heart, 91*, 299–304.
Edwards, H. V., Cameron, R. T., & Baillie, G. S. (2011). The emerging role of HSP20 as a multifunctional protective agent. *Cellular Signalling, 23*, 1447–1454.
Fan, G. C., Chu, G., Mitton, B., et al. (2004). Small heat-shock protein Hsp20 phosphorylation inhibits beta-agonist-induced cardiac apoptosis. *Circulation Research, 94*, 1474–1482.
Fan, G. C., Ren, X., Qian, J., et al. (2005). Novel cardioprotective role of a small heat-shock protein, Hsp20, against ischemia/reperfusion injury. *Circulation, 111*, 1792–1799.
Feng, Y., Huang, W., Meng, W., et al. (2014). Heat shock improves Sca-1 + stem cell survival and directs ischemic cardiomyocytes toward a pro survival phenotype via exosomal transfer: A critical role for HSF1/miR-34a/HSP70 pathway. *Stem Cells, 32*, 462–472.

Fontaine, J. M., Rest, J. S., Welsh, M. J., et al. (2003). The sperm outer dense fiber protein is the 10th member of the superfamily of mammalian small stress proteins. *Cell Stress & Chaperones, 8*, 62–69.

Franklin, T. B., Krueger-Naug, A. M., Clarke, D. B., et al. (2005). The role of heat shock proteins Hsp70 and Hsp27 in cellular protection of the central nervous system. *International Journal of Hyperthermia, 21*, 379–392.

Frostegård, J., Zhang, Y., Sun, J., et al. (2016). Oxidized low-density lipoprotein (OxLDL)-treated dendritic cells promote activation of T cells in human atherosclerotic plaque and blood, which is repressed by statins: MicroRNA let-7c is integral to the effect. *J Am Heart Assoc, 5*, e003976.

Fukuoka, K., Sawabe, A., & Sugimoto, T. (2004). Inhibitory actions of several natural products on proliferation of rat vascular smooth muscle cells induced by Hsp60 from Chlamydia pneumoniae J138. *Journal of Agricultural and Food Chemistry, 52*, 6326–6329.

Fuster, V., Badimon, L., Badimon, J. J., et al. (1992a). The pathogenesis of coronary artery disease and the acute coronary syndromes (1). *The New England Journal of Medicine, 326*, 242–250.

Fuster, V., Badimon, L., Badimon, J. J., et al. (1992b). The pathogenesis of coronary artery disease and the acute coronary syndromes (2). *The New England Journal of Medicine, 326*, 310–318.

Gabai, V. L., Meriin, A. B., Yaglom, J. A., et al. (2000). Suppression of stress kinase JNK is involved in HSP72-mediated protection of myogenic cells from transient energy deprivation. HSP72 alleviates the stress-induced inhibition of JNK dephosphorylation. *The Journal of Biological Chemistry, 275*, 38088–38094.

Garrido, C., Brunet, M., Didelot, C., et al. (2006). Heat shock proteins 27 and 70: Anti-apoptotic proteins with tumorigenic properties. *Cell Cycle, 5*, 2592–2601.

Gething, M. J., & Sambrook, J. (1992). Protein folding in the cell. *Nature, 355*, 33–45.

Ghayour-Mobarhan, M., Lamb, D. J., Tavallaie, S., et al. (2007). Relationship between plasma cholesterol, von Willebrand factor concentrations, extent of atherosclerosis and antibody titers to heat shock proteins-60, -65 and -70 in cholesterol-fed rabbits. *International Journal of Experimental Pathology, 88*, 249–255.

Ghayour-Mobarhan, M., Saber, H., & Ferns, G. A. (2012). The potential role of heat shock protein 27 in cardiovascular disease. *Clinica Chimica Acta, 413*, 15–24.

Ghosh, J. G., Houck, S. A., & Clark, J. I. (2007). Interactive domains in the molecular chaperone human alphaB crystallin modulate microtubule assembly and disassembly. *PLoS One, 2*, e498.

Golenhofen, N., Perng, M. D., Quinlan, R. A., et al. (2004). Comparison of the small heat shock proteins alphaB-crystallin, MKBP, HSP25, HSP20, and cvHSP in heart and skeletal muscle. *Histochemistry and Cell Biology, 122*, 415–425.

Golenhofen, N., Redel, A., Wawrousek, E. F., et al. (2006). Ischemia- induced increase of stiffness of alphaB-crystallin/HSPB2-deficient myocardium. *Pflügers Archiv, 451*, 518–525.

Grose, J. H., Langston, K., Wang, X., et al. (2015). Characterization of the cardiac overexpression of HSPB2 reveals mitochondrial and myogenic roles supported by a cardiac HspB2 interactome. *PLoS One, 10*, e0133994.

Gupta, S., & Knowlton, A. A. (2007). HSP60 trafficking in adult cardiac myocytes: role of the exosomal pathway. *American Journal of Physiology. Heart and Circulatory Physiology, 292*, H3052–H3056.

Gurbuxani, S., Schmitt, E., & Cande, C. (2003). Heat shock protein 70 binding inhibits the nuclear import of apoptosis-inducing factor. *Oncogene, 22*, 6669–6678.

Gwag, T., Park, K., Kim, E., et al. (2013). Inhibition of C2C12 myotube atrophy by a novel HSP70 inducer, celastrol, via activation of Akt1 and ERK1/2 pathways. *Archives of Biochemistry and Biophysics, 537*, 21–30.

Harris, M. B., & Starnes, J. W. (2001). Effects of body temperature during exercise training on myocardial adaptations. *American Journal of Physiology. Heart and Circulatory Physiology, 280*, H2271–H2280.

Hattori, K., Ozaki, Y., Ismail, T. F., et al. (2012). Impact of statin therapy on plaque characteristics as assessed by serial OCT, grayscale and integrated backscatter-IVUS. *JACC Cardiovascular Imaging, 5*, 169–177.

Hayashi, M., Imanaka-Yoshida, K., et al. (2006). A crucial role of mitochondrial Hsp40 in preventing dilated cardiomyopathy. *Nature Medicine, 12*, 128–132.
Hedhli, N., Lizano, P., Hong, C., et al. (2008). Proteasome inhibition decreases cardiac remodeling after initiation of pressure overload. *American Journal of Physiology. Heart and Circulatory Physiology, 295*, H1385–H1393.
Henderson, B., & Pockley, A. G. (2012). Proteotoxic stress and circulating cell stress proteins in the cardiovascular diseases. *Cell Stress & Chaperones, 17*, 303–311.
Hirakawa, T., Rokutan, K., Nikawa, T., et al. (1996). Geranylgeranylacetone induces heat shock proteins in cultured guinea pig gastric mucosal cells and rat gastric mucosa. *Gastroenterology, 111*, 345–357.
Hu, X., Van Marion, D. M. S., Wiersma, M., et al. (2017). The protective role of small heat shock proteins in cardiac diseases: Key role in atrial fibrillation. *Cell Stress & Chaperones, 22*, 665–674.
Ishiwata, T., Orosz, A., Wang, X., et al. (2012). HSPB2 is dispensable for the cardiac hypertrophic response but reduces mitochondrial energetics following pressure overload in mice. *PLoS One, 7*, e42118.
Jee, H. (2016). Size dependent classification of heat shock proteins: A mini-review. *Journal Exercise Rehabilitation, 12*, 255–259.
Jessup, M., Greenberg, B., Mancini, D., et al. (2011). Calcium upregulation by percutaneous administration of gene therapy in cardiac disease (CUPID) investigators. Calcium upregulation by percutaneous administration of gene therapy in cardiac disease (CUPID): A phase 2 trial of intracoronary gene therapy of sarcoplasmic reticulum Ca2+-ATPase in patients with advanced heart failure. *Circulation, 124*, 304–313.
Jimenez, S. K., Small, B. A., Hsu, A. K., et al. (2014). Heat shock proteins HSP90 and HSP70 mediate opioid- and GSK3β-induced cardioprotection. *Circulation Research, 115*, 340.
Johnson, A. D., Berberian, P. A., Tytell, M., et al. (1993). Atherosclerosis alters the localization of HSP70 in human and macaque aortas. *Experimental and Molecular Pathology, 58*, 155–168.
Kampinga, H. H., & Craig, E. A. (2010). The HSP70 chaperone machinery: J proteins as drivers of functional specificity. *Nature Reviews Molecular Cell Biology, 11*, 579–592.
Kappe, G., Franck, E., Verschuure, P., et al. (2003). The human genome encodes 10 alpha-crystallin-related small heat shock proteins: HspB1-10. *Cell Stress & Chaperones, 8*, 53–61.
Kessing, D., Denollet, J., Widdershoven, J., et al. (2016). Self-care and all-cause mortality in patients with chronic heart failure. *JACC Heart Failure, 4*, 176–183.
Khalil, A. A., Kabapy, N. F., Deraz, S. F., et al. (2011). Heat shock proteins in oncology: Diagnostic biomarkers or therapeutic targets? *Biochimica et Biophysica Acta, 1816*, 89–104.
Khan, I. U., Wallin, R., Gupta, R. S., et al. (1998). Protein kinase A-catalyzed phosphorylation of heat shock protein 60 chaperone regulates its attachment to histone 2B in the T lymphocyte plasma membrane. *Proceedings of the National Academy of Sciences of the United States of America, 95*, 10425–10430.
Kieran, D., Kalmar, B., Dick, J. R., et al. (2004). Treatment with arimoclomol, a coinducer of heat shock proteins, delays disease progression in ALS mice. *Nature Medicine, 10*, 402–405.
Kim, K. K., Kim, R., & Kim, S. H. (1998). Crystal structure of a small heat-shock protein. *Nature, 394*, 595–599.
Kirkegaard, T., Gray, J., Priestman, D. A., et al. (2016). Heat shock protein-based therapy as a potential candidate for treating the sphingolipidoses. *Science Translational Medicine, 8*, 355ra118.
Kleindienst, R., Xu, Q., Willeit, J., et al. (1993). Immunology of atherosclerosis. Demonstration of heat shock protein 60 expression and T lymphocytes bearing alpha/beta or gamma/delta receptor in human atherosclerotic lesions. *The American Journal of Pathology, 142*, 1927–1937.
Kliková, K., Pilchova, I., Stefanikova, A., et al. (2016). The role of heat shock proteins in leukemia. *Klinická Onkologie, 29*, 29–38.

Knowlton, A. A., Kapadia, S., & Torre-Amione, G. (1998). Differential expression of heat shock proteins in normal and failing human hearts. *Journal of Molecular and Cellular Cardiology, 30*, 811–818.

Komukai, K., Kubo, T., Kitabata, H., et al. (2014). Effect of atorvastatin therapy on fibrous cap thickness in coronary atherosclerotic plaque as assessed by optical coherence tomography: The EASY-FIT study. *Journal of the American College of Cardiology, 64*, 2207–2217.

Kupatt, C., Dessy, C., Hinkel, R., et al. (2004). Heat shock protein 90 transfection reduces ischemia-reperfusion-induced myocardial dysfunction via reciprocal endothelial NO synthase serine 1177 phosphorylation and threonine 495 dephosphorylation. *Arteriosclerosis, Thrombosis, and Vascular Biology, 24*, 1435–1441.

Lamb, D. J., El-Sankary, W., & Ferns, G. A. (2002). Molecular mimicry in atherosclerosis: A role for heat shock proteins in immunization. *Atherosclerosis, 167*, 177–185.

Lambert, H., Charette, S. J., Bernier, A. F., et al. (1999). HSP27 multimerization mediated by phosphorylation-sensitive intermolecular interactions at the amino terminus. *The Journal of Biological Chemistry, 274*, 9378–9385.

Lanka, V., Wieland, S., Barber, J., et al. (2009). Arimoclomol: A potential therapy under development for ALS. *Expert Opinion on Investigational Drugs, 18*, 1907–1918.

Lee, J. H., Koo, T. H., Yoon, H., et al. (2006). Inhibition of NF-kappa B activation through targeting I kappa B kinase by celastrol, a quinone methide triterpenoid. *Biochemical Pharmacology, 72*, 1311–1321.

Li, Y., Si, R., Feng, Y., et al. (2011). Myocardial ischemia activates an injurious innate immune signaling via cardiac heat shock protein 60 and Toll-like receptor 4. *The Journal of Biological Chemistry, 286*, 31308–31319.

Lin, L., Kim, S. C., Wang, Y., et al. (2007). HSP60 in heart failure: Abnormal distribution and role in cardiac myocyte apoptosis. *American Journal of Physiology. Heart and Circulatory Physiology, 293*, H2238–H2247.

Madamanchi, N. R., Patterson, C., Li, S., & Runge, S. M. (2001). Reactive oxygen species regulate heat-shock protein 70 via the JAK/STAT pathway. *Arteriosclerosis, Thrombosis, and Vascular Biology, 21*, 321–326.

Malik, Z. A., Kott, K. S., Poe, A. J., et al. (2013). Cardiac myocyte exosomes: Stability, HSP60, and proteomics. *American Journal of Physiology Heart and Circulatory Physiology, 304*, H954–H965.

Mamipour, M., Yousefi, M., & Hasanzadeh, M. (2017). An overview on molecular chaperones enhancing solubility of expressed recombinant proteins with correct folding. *International Journal of Biological Macromolecules, 102*, 367–375.

Marber, M. S., Latchman, D. S., Walker, J. M., et al. (1993). Cardiac stress protein elevation 24 hours after brief ischemia or heat stress is associated with resistance to myocardial infarction. *Circulation, 88*, 1264–1272.

Martin, T. P., Currie, S., & Baillie, G. S. (2014). The cardioprotective role of small heat-shock protein 20. *Biochemical Society Transactions, 42*, 270–273.

Martin-Ventura, J. L., Duran, M. C., Blanco-Colio, L. M., et al. (2004). Identification by a differential proteomic approach of heat shock protein 27 as a potential marker of atherosclerosis. *Circulation, 110*, 2216–2219.

Marunouchi, T., Inomata, S., Sanbe, A., et al. (2014). Protective effect of geranylgeranylacetone via enhanced induction of HSPB1 and HSPB8 in mitochondria of the failing heart following myocardial infarction in rats. *European Journal of Pharmacology, 730*, 140–147.

Mayr, M., Metzler, B., Kiechl, S., et al. (1999). Endothelial cytotoxicity mediated by serum antibodies to heat shock proteins of *Escherichia coli* and Chlamydia pneumoniae: Immune reactions to heat shock proteins as a possible link between infection and atherosclerosis. *Circulation, 99*, 1560–1566.

Mazzaferro, V., Coppa, J., Carrabba, M. G., et al. (2003). Vaccination with autologous tumor-derived heat-shock protein gp96after liver resection for metastatic colorectal cancer. *Clinical Cancer Research, 9*, 3235–3245.

Mu, H., Wang, L., & Zhao, L. (2017). HSP90 inhibition suppresses inflammatory response and reduces carotid atherosclerotic plaque formation in ApoE mice. *Cardiovascular Therapy, 35*(2), 1–9.

Multhoff, G., Pockley, A. G., Schmida, T. E., et al. (2015). The role of heat shock protein 70 (Hsp70) in radiation-induced immunomodulation. *Cancer Letters, 368*, 179–184.

Nakagawa, M., Tsujimoto, N., Nakagawa, H., et al. (2001). Association of HSPB2, a member of the small heat shock protein family, with mitochondria. *Experimental Cell Research, 271*, 161–168.

Nakai, A., & Ishikawa, T. (2001). Cell cycle transition under stress conditions controlled by vertebrate heat shock factors. *The EMBO Journal, 20*, 2885–2895.

Neef, D. W., Jaeger, A. M., & Thiele, D. J. (2011). Heat shock transcription factor 1 as a therapeutic target in neurodegenerative diseases. *Nature Reviews Drug Discovery, 10*, 930–944.

Noguchi, T., Tanaka, A., Kawasaki, T., et al. (2015). Effect of intensive statin therapy on coronary high-intensity plaques detected by noncontrast T1-weighted imaging: The AQUAMARINE pilot study. *Journal of the American College of Cardiology, 66*, 245–256.

Ooie, T., Takahashi, N., Saikawa, T., et al. (2001). Single oral dose of geranylgeranylacetone induces heat-shock protein 72 and renders protection against ischemia/reperfusion injury in rat heart. *Circulation, 104*, 1837–1843.

Panneerselvam, L., Raghunath, A., & Perumal, E. (2017). Differential expression of myocardial heat shock proteins in rats acutely exposed to fluoride. *Cell Stress Chaperones, 22*, 743–750. [Epub ahead of print].

Parfitt, D. A., Aguila, M., McCulley, C. H., et al. (2014). The heat-shock response co-inducer arimoclomol protects against retinal degeneration in rhodopsin retinitis pigmentosa. *Cell Death & Disease, 5*, e1236.

Pearl, L. H., & Prodromou, C. (2001). Structure, function, and mechanism of the Hsp90 molecular chaperone. *Advances in Protein Chemistry, 59*, 157–186.

Perng, M. D., Cairns, L., van den Ijssel, P., et al. (1999). Intermediate filament interactions can be altered by HSP27 and alphaB-crystallin. *Journal of Cell Science, 112*, 2099–2112.

Pfister, G., Stroh, C. M., & Perschinka, H. (2005). Detection of HSP60 on the membrane surface of stressed human endothelial cells by atomic force and confocal microscopy. *Journal of Cell Science, 118*, 1587–1594.

Picard, D. (2002). Heat-shock protein 90, a chaperone for folding and regulation. *Cellular and Molecular Life Sciences, 59*, 1640–1648.

Pipkin, W., Johnson, J. A., Creazzo, T. L., et al. (2003). Localization, macromolecular associations, and function of the small heat shock-related protein HSP20 in rat heart. *Circulation, 107*, 469–476.

Pockley, A. G. (2002). Heat shock proteins, inflammation, and cardiovascular disease. *Circulation, 105*, 1012–1017.

Pockley, A. G., Georgiades, A., Thulin, T., et al. (2003). Serum heat shock protein 70 levels predict the development of atherosclerosis in subjects with established hypertension. *Hypertension, 42*, 235–238.

Powers, S. K., Lennon, S. L., Quindry, J., et al. (2002). Exercise and cardioprotection. *Current Opinion in Cardiology, 17*, 495–502.

Pratt, W. B., & Toft, D. O. (2003). Regulation of signaling protein function and trafficking by the hsp90/hsp70-based chaperone machinery. *Experimental Biology and Medicine (Maywood, N.J.), 228*, 111–133.

Prohászka, Z., Duba, J., Horváth, L., et al. (2001). Comparative study on antibodies to human and bacterial 60 kDa heat shock proteins in a large cohort of patients with coronary heart disease and healthy subjects. *European Journal of Clinical Investigation, 31*, 285–292.

Puato, M., Faggin, E., Rattazzi, M., et al. (2010). Atorvastatin reduces macrophage accumulation in atherosclerotic plaques: A comparison of a nonstatin-based regimen in patients undergoing carotid endarterectomy. *Stroke, 41*, 1163–1168.

Qian, J., Ren, X., Wang, X., et al. (2009). Blockade of Hsp20 phosphorylation exacerbates cardiac ischemia/reperfusion injury by suppressed autophagy and increased cell death. *Circulation Research, 105*, 1223–1231.

Qian, J., Vafiadaki, E., Florea, S. M., et al. (2011). Small heat shock protein 20 interacts with protein phosphatase-1 and enhances sarcoplasmic reticulum calcium cycling. *Circulation Research, 108*, 1429–1438.

Qiu, H., Lizano, P., Laure, L., et al. (2011). H11 kinase/heat shock protein 22 deletion impairs both nuclear and mitochondrial functions of STAT3 and accelerates the transition into heart failure on cardiac overload. *Circulation, 124*, 406–415.

Rayner, K., Chen, Y. X., McNulty, M., et al. (2008). Extracellular release of the atheroprotective heat shock protein 27 is mediated by estrogen and competitively inhibits acLDL binding to scavenger receptor-A. *Circulation Research, 103*, 133–141.

Richter, K., & Buchner, J. (2001). Hsp90: Chaperoning signal transduction. *Journal of Cellular Physiology, 188*, 281–290.

Rigano, R., Profumo, E., Buttari, B., et al. (2007). Heat shock proteins and autoimmunity in patients with carotid atherosclerosis. *Annals of the New York Academy of Sciences, 1107*, 1–10.

Rinaldi, B., Corbi, G., Boccuti, S., et al. (2006). Exercise training affects age-induced changes in SOD and heat shock protein expression in rat heart. *Experimental Gerontology, 41*, 764–770.

Ritossa, F. (1962). A new puffing pattern induced by heat shock and DNP in drosophila. *Experientia, 18*, 571–573.

Sakamoto, M., Minamino, T., Toko, H., et al. (2006). Upregulation of heat shock transcription factor 1 plays a critical role in adaptive cardiac hypertrophy. *Circulation Research, 99*, 1411–1418.

Sassa, H., Takaishi, Y., & Terada, H. (1990). The triterpene celastrol as a very potent inhibitor of lipid peroxidation in mitochondria. *Biochemical Biophysical Research Communications, 172*, 890–897.

Sasu, S., LaVerda, D., Qureshi, N., et al. (2001). Chlamydia pneumoniae and chlamydial heat shock protein 60 stimulate proliferation of human vascular smooth muscle cells via toll-like receptor 4 and p44/p42 mitogen-activated protein kinase activation. *Circulation Research, 89*, 244–250.

Schönbeck, U., & Libby, P. (2004). Inflammation, immunity, and HMG-CoA reductase inhibitors: Statins as antiinflammatory agents? *Circulation, 109*, II18–II26.

Seibert, T. A., Hibbert, B., Chen, Y. X., et al. (2013). Serum heat shock protein 27 levels represent a potential therapeutic target for atherosclerosis: Observations from a human cohort and treatment of female mice. *Journal of the American College of Cardiology, 62*, 1446–1454.

Selcen, D., & Engel, A. G. (2003). Myofibrillar myopathy caused by novel dominant negative alpha B-crystallin mutations. *Annals of Neurology, 54*, 804–810.

Singh, L., Randhawa, P. K., Singh, N., et al. (2017). Redox signaling in remote ischemic preconditioning-induced cardioprotection: Evidences and mechanisms. *European Journal of Pharmacology, 809*, 151–155.

Smith, S. C., Jr., Benjamin, E. J., Bonow, R. O., et al. (2011). AHA/ACCF secondary prevention and risk reduction therapy for patients with coronary and other atherosclerotic vascular disease: 2011 update: a guideline from the American Heart Association and American College of Cardiology Foundation endorsed by the World Heart Federation and the Preventive Cardiovascular Nurses Association. *Journal of the American College of Cardiology, 58*, 2432–2446.

Sreedhar, A. S., Kalmár, E., Csermely, P., et al. (2004). Hsp90 isoforms: Functions, expression and clinical importance. *FEBS Letters, 562*, 11–15.

Sugiyama, Y., Suzuki, A., Kishikawa, M., et al. (2000). Muscle develops a specific form of small heat shock protein complex composed of MKBP/HSPB2 and HSPB3 during myogenic differentiation. *The Journal of Biological Chemistry, 275*, 1095–1104.

Suzuki, A., Sugiyama, Y., Hayashi, Y., et al. (1998). MKBP, a novel member of the small heat shock protein family, binds and activates the myotonic dystrophy protein kinase. *The Journal of Cell Biology, 140*, 1113–1124.

Trott, A., West, J. D., Klaić, L., et al. (2008). Activation of heat shock and antioxidant responses by the natural product celastrol: Transcriptional signatures of a thiol-targeted molecule. *Molecular Biology of the Cell, 19*, 1104–1112.

Uchiyama, T., Atsuta, H., Utsugi, T., et al. (2007). HSF1 and constitutively active HSF1 improve vascular endothelial function (heat shock proteins improve vascular endothelial function). *Atherosclerosis, 190*, 321–329.

van de Klundert, F. A., Gijsen, M. L., van den Ijssel, P. R., et al. (1998). alpha B-crystallin and hsp25 in neonatal cardiac cells-differences in cellular localization under stress conditions. *European Journal of Cell Biology, 75*, 38–45.

Van Montfort, R. L., Basha, E., Friedrich, K. L., et al. (2001a). Crystal structure and assembly of a eukaryotic small heat shock protein. *Nature Structural Biology, 8*, 1025–1030.

Van Montfort, R., Slingsby, C., & Vierling, E. (2001b). Structure and function of the small heat shock protein/alpha-crystallin family of molecular chaperones. *Advances in Protein Chemistry, 59*, 105–156.

Veres, A., Füst, G., Smieja, M., et al. (2002). Heart Outcomes Prevention Evaluation (HOPE) study investigators. Relationship of anti-60 kDa heat shock protein and anti-cholesterol antibodies to cardiovascular events. *Circulation, 106*, 2775–2780.

Verschuure, P., Tatard, C., Boelens, W. C., et al. (2003). Expression of small heat shock proteins HspB2, HspB8, Hsp20 and cvHsp in different tissues of the perinatal developing pig. *European Journal of Cell Biology, 82*, 523–530.

Vicart, P., Caron, A., Guicheney, P., et al. (1998). A missense mutation in the alphaB-crystallin chaperone gene causes a desmin-related myopathy. *Nature Genetics, 20*, 92–95.

Vicencio, J. M., Yellon, D. M., Sivaraman, V., et al. (2015). Plasma exosomes protect the myocardium from ischemia-reperfusion injury. *Journal of the American College of Cardiology, 65*, 1525–1536.

Vígh, L., Literáti, P. N., Horváth, I., et al. (1997). Bimoclomol: A nontoxic, hydroxylamine derivative with stress protein-inducing activity and cytoprotective effects. *Nature Medicine, 3*, 1150–1154.

Vos, M. J., Hageman, J., Carra, S., et al. (2008). Structural and functional diversities between members of the human HSPB, HSPH, HSPA, and DNAJ chaperone families. *Biochemistry, 47*, 7001–7011.

Vos, M. J., Kanon, B., & Kampinga, H. H. (2009). HSPB7 is a SC35 speckle resident small heat shock protein. *Biochimica et Biophysica Acta, 1793*, 1343–1353.

Vos, M. J., Zijlstra, M. P., & Kanon, B. (2010). HSPB7 is the most potent polyQ aggregation suppressor within the HSPB family of molecular chaperones. *Human Molecular Genetics, 19*, 4677–4693.

Vos, M. J., Zijlstra, M. P., Carra, S., et al. (2011). Small heat shock proteins, protein degradation and protein aggregation diseases. *Autophagy, 7*, 101–103.

Wang, W., Peng, Y., Wang, Y., et al. (2009). Anti-apoptotic effect of heat shock protein 90 on hypoxia-mediated cardiomyocyte damage is mediated via the phosphatidylinositol 3-kinase/AKT pathway. *Clinical and Experimental Pharmacology & Physiology, 36*, 899–903.

Wang, Y., Chen, L., Hagiwara, N., et al. (2010). Regulation of heat shock protein 60 and 72 expression in the failing heart. *Journal of Molecular and Cellular Cardiology, 48*, 360–366.

Wei, H., Campbell, W., & Vander Heide, R. S. (2006). Heat shock-induced cardioprotection activates cytoskeletal-based cell survival pathways. *American Journal of Physiology. Heart and Circulatory Physiology, 291*, H638–H647.

Weintraub, N. L., & Rubinstein, J. (2013). Cooling the fire of atherosclerosis with heat shock protein 27. *Journal of the American College of Cardiology, 62*, 1455–1456.

Weintraub, W. S., Daniels, S. R., Burke, L. E., et al. (2011). Value of primordial and primary prevention for cardiovascular disease: A policy statement from the American Heart Association. *Circulation, 124*, 967–990.

Westerheide, S. D., Bosman, J. D., Mbadugha, B. N., et al. (2004). Celastrols as inducers of the heat shock response and cytoprotection. *The Journal of Biological Chemistry, 279*, 56053–56060.

Westerheide, S. D., Raynes, R., Powell, C., et al. (2012). HSF transcription factor family, heat shock response, and protein intrinsic disorder. *Current Protein & Peptide Science, 13*, 86–103.
Willis, M. S., & Patterson, C. (2010). Hold me tight: Role of the heat shock protein family of chaperones in cardiac disease. *Circulation, 122*, 1740–1751.
Wu, K., Xu, W., You, Q., et al. (2012). Increased expression of heat shock protein 90 under chemical hypoxic conditions protects cardiomyocytes against injury induced by serum and glucose deprivation. *International Journal of Molecular Medicine, 30*, 1138–1144.
Xiao, Q., Mandal, K., Schett, G., et al. (2005). Association of serum-soluble heat shock protein 60 with carotid atherosclerosis: Clinical significance determined in a follow-up study. *Stroke, 36*, 2571–2576.
Xu, Q., Schett, G., Perschinka, H., et al. (2000). Serum soluble heat shock protein 60 is elevated in subjects with atherosclerosis in a general population. *Circulation, 102*, 14–20.
Zhang, X., Min, X., Li, C., et al. (2010). Involvement of reductive stress in the cardiomyopathy in transgenic mice with cardiac-specific overexpression of heat shock protein 27. *Hypertension, 55*, 1412–1417.
Zhang, C., Liu, X., Miao, J., et al. (2017). Heat shock protein 70 protects cardiomyocytes through suppressing SUMOylation and nucleus translocation of phosphorylated eukaryotic elongation factor 2 during myocardial ischemia and reperfusion. *Apoptosis, 22*, 608–625.
Zhao, Y., Zhang, C., Wei, X., et al. (2015). Heat shock protein 60 stimulates the migration of vascular smooth muscle cells via Toll-like receptor 4 and ERK MAPK activation. *Scientific Reports, 5*, 15352.
Zhong, G. Q., Tu, R. H., Zeng, Z. Y., et al. (2014). Novel functional role of heat shock protein 90 in protein kinase C-mediated ischemic post- conditioning. *The Journal of Surgical Research, 189*, 198–206.
Zhu, J., Quyyumi, A. A., Rott, D., et al. (2001). Antibodies to human heat-shock protein 60 are associated with the presence and severity of coronary artery disease: Evidence for an autoimmune component of atherogenesis. *Circulation, 103*, 1071–1075.
Zhu, J., Quyyumi, A. A., Wu, H., et al. (2003). Increased serum levels of heat shock protein 70 are associated with low risk of coronary artery disease. *Arteriosclerosis, Thrombosis, and Vascular Biology, 23*, 1055–1059.

Clinical Implication of Heat Shock Protein 70 in Kidney Disease

Kidney Diseases and HSP70

Alev Yilmaz and Zeynep Nagehan Yuruk Yildirim

Abstract Heat shock protein (HSP) 70 has been investigated from various aspects in experimental studies, since it is the most abundant form of HSP in cells. HSP 70 is involved in response to various acute and chronic insults of kidney and also other parts of the urinary tract. Although it is not a specific biomarker for any of the kidney diseases, HSP 70 level of body fluids may be beneficial as a biomarker in some specific circumstances, for example for the differential diagnosis of "a children with fever". The number of possible diseases in this situation is limited and an infection is most likely to be the reason of fever. Additionally, clarifying the exact role of HSP 70 in different kidney diseases enables to discover new therapeutic options in order to prevent renal fibrosis and chronic renal failure in the future. This chapter reviews experimental and clinical studies to find out the opportunities to utilize HSP 70 in daily clinical practice.

Keywords Acute kidney diseases · Biomarker · Chronic kidney diseases · Heat shock protein 70 · HSP70 · Kidney disease

Abbreviations

AGEs	advanced glycosylated-end products
AKI	acute kidney injury
AUC	area under the curve
CKD	chronic kidney disease
CSF	cerebrospinal fluid
EMT	epithelial-mesenchymal transition
ESRD	end stage renal disease
GGA	geranylgeranylacetone
HSP	heat shock protein

A. Yilmaz · Z. N. Y. Yildirim (✉)
Istanbul Faculty of Medicine, Pediatric Nephrology Department, Istanbul University, Fatih, Istanbul, Turkey

A. A. A. Asea, P. Kaur (eds.), *HSP70 in Human Diseases and Disorders*, Heat Shock Proteins 14, https://doi.org/10.1007/978-3-319-89551-2_12

IFNγ	interferon gamma
IL-18	interleukin 18
INS	idiopathic nephrotic syndrome
JAK/STAT	janus kinase/signal transducers and activators of transcription
KIM-1	kidney injury molecule type 1
MEK/ERK	Mitogen-activated protein kinase/ERK kinase/extracellular-signal-regulated kinase
NGAL	neutrophil gelatinase-associated lipocalin
NF-kB	nuclear factor kappa B
NOS	nitric oxide synthase
p-STAT3	phospho-signal transducer and activator of transcription 3
ROS	reactive oxygen species
SLK	Ste20-like kinase
STAT3	Signal transducer and activator of transcription 3
TGF-β	transforming growth factor beta
T1DM	type 1 diabetes mellitus
uHSP70	urine level of heat shock proetin 70
uHSP/Cr	urine heat shock protein/creatinine ratio
UTI	urinary tract infection
WT1	wilms tumor 1

Introduction

Heat shock proteins (HSP) are intracellular chaperones which are classified according to their molecular weight such as small HSP (16–40 kDa), HSP60 (60 kDa), HSP70 (70 kDa), HSP90 (90 kDa) (Beck et al. 2000; Musial and Zwolinska 2011). The main function of these chaperones is contributing to maintain cell survival by restoring proteins in cells (Beck et al. 2000). HSP are involved in correct folding of proteins, refolding of misfolded proteins and degradation of irrecoverable proteins to protect structure of cytoskeleton (Beck et al. 2000; Musial and Zwolinska 2011; Barutta et al. 2008; Buraczynska et al. 2009). They are expressed under normal conditions and can be induced by various stressors including thermal, osmotic, oxidative, toxic, ischemic injury (Beck et al. 2000; Mueller et al. 2003; O'Neill et al. 2014). Heat shock protein (HSP) 70 is a member of HSP protein families which is the most abundant form of HSP in cells (Beck et al. 2000). Hence, HSP 70 has been investigated from various aspects in experimental studies. HSP 70 usually acts as a protective part of the inflammatory and immune response in case of tissue damage (Manucha 2014). It has been demonstrated that HSP 70 contributes to development, infection, inflammation, apoptosis and fibrosis of kidney (Mazzei and Manucha 2017; Manucha 2014; Kennedy et al. 2014; Zhou et al. 2016). Based on these studies which are mostly experimental, our goal for the future should be to benefit from HSP 70 in favor of patients in our daily practice. One of the best ways to achieve this goal may be to use HSP 70 as a biomarker to diagnose kidney diseases. First of all, HSP 70 must be detectable in body fluids to be utilized as a biomarker, since blood, urine and the other body fluids have been employed to detect the presence of diagnostic biomarkers in practice.

Acute Kidney Diseases and HSP 70

Although HSP 70 is mostly considered to be an intracellular chaperone, several clinical studies have demonstrated that it might be found in extracellular space, serum and urine of human (Mambula et al. 2007). HSP 72, a member of HSP 70 protein family, has been shown to be elevated in body fluids secondary to bacterial infections (Tang et al. 2008; Varano Della Vergiliana et al. 2013). Tang et al (2008) evaluated level of HSP 72 in cerebrospinal fluid (CSF) of the patients who underwent lumbar puncture due to the suspicion of meningitis. They reported that HSP72 level in CSF of patients with meningitis was significantly higher than in the patients without meningitis according to assessment of CSF (Tang et al. 2008). Moreover, HSP 72 level was found to be higher in bacterial meningitis than in aseptic meningitis (Tang et al. 2008). Similarly, another study showed that level of HSP 72 was significantly increased in exudates than in transudates in the patients with pleural effusion (Varano Della Vergiliana et al. 2013). The authors claimed that the reason of increased HSP 72 was the stimulation of mesothelial cells by the pneumococcal products (Varano Della Vergiliana et al. 2013). Likewise, cell culture studies demonstrated that HSP 70 released from macrophages due to exposure of bacterial products of *Escherichia coli* which was the most common causative microorganism in urinary tract infection (UTI) (Davies et al. 2006). The prevailing opinion in the past was that HSP 70 was released to extracellular space due to the cell death. However, it has been shown that HSP 70 is secreted by kidney cells actively rather than because of cell death (Evdokimovskaya et al. 2010). Molinas et al (2010) reported that HSP 70 levels increased in urine and in renal cortex simultaneously in rats after receiving Acetaminophen which caused renal toxicity. This finding suggest not only that nephrotoxic agents stimulate HSP 70 expression in renal cortex but also that urine may be a good indicator of HSP 70 expression in kidney related to renal injury. All these evidences imply that urine level of HSP 70 may be an indicator for infection in the urinary tract.

Urinary tract infection is one of the most common infections in children and should be included in the differential diagnosis of children with the symptoms of systemic infections (Roberts 2011). Nowadays, UTI is diagnosed with significant bacterial growing in urine culture and positive of leukocyte esterase and/or nitrite tests in urinalysis. However, urinalysis is not specific for the UTI. There are some limitations of urine culture as well, although it is essential for the diagnosis. First, three days are required for identifying the bacterial growth in culture. Second, invasive methods are required to obtain sterile urine sample for culture because urine may be contaminated by microorganisms which are originated outside of the urinary

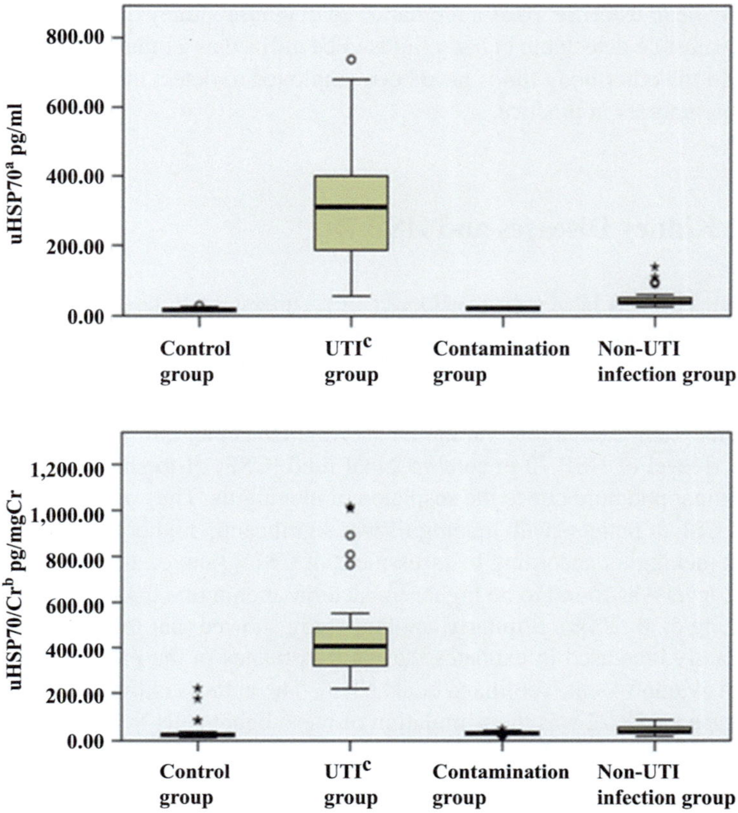

Fig. 1 Urine heat shock protein 70 level and urine heat shock protein 70/creatinine ratio in urinary tract infection and controls ([a]uHSP70: urine heat shock protein 70, [b]uHSP70/Cr: urine heat shock protein 70/Creatinine, [c]UTI: urinary tract infection) (published in Pediatr Nephrol 2016 Sep)

tract. A reliable biomarker is needed to distinguish UTI from contamination as well as other infections for the accurate diagnosis. Therefore, we evaluated urine level of HSP 70 (uHSP70) and urine HSP 70/creatinine ratio (uHSP70/Cr) to determine whether these parameters could indicate UTI in children. In this study, both uHSP70 and uHSP70/Cr were found to be elevated in patients with UTI at the onset of the infection and they decreased after antibiotic treatment proving that the elevation of HSP 70 in urine was secondary to UTI (Yilmaz et al. 2016a, b). Additionally, our results demonstrated that both these parameters were higher in the patients with UTI than in the contamination and non-UTI infection group including upper and lower respiratory tract infection, gastroenteritis etc. (Fig. 1) (Yilmaz et al. 2016a, b). All these findings demonstrated that increase of these parameters may help to distinguish UTI from not only the other infections but also bacterial contamination of the urine.

The possibility of having UTI is increased in children with unexplained fever when uHSP70/Cr is above a cut-off 158 pg/mgCr in the study. Urine HSP70/Cr has

100% sensitivity and specificity using this cut-off value to distinguish UTI from other infections. As an alternative, urinalysis has lower sensitivity and specificity (Yilmaz et al. 2016a, b). Previous reports showed that either leucocyte esterase or nitrite positivity has a sensitivity of 88% and a specificity of 79% (Williams et al. 2010). These promising results of our study suggest that urine level of HSP 70, especially uHSP70/Cr, may be used as a reliable biomarker in UTI in children.

HSP 70 in urine can be originated not only from kidney but also from bladder. Urine HSP 70 level was found to be increased in bladder cancer and the authors concluded that this increase might be related to HSP 70 secretion from the tumor cells to urine (Margel et al. 2011). However, uHSP70/Cr was significantly higher in pyelonephritis than in cystitis in our study implying that the reaction against the infection was stronger when the kidney was involved rather than the bladder only. Our study showed that bacterial infection of the urinary tract as an acute insult may give rise to increase of HSP 70 in urine. It has been notified that HSP 70 is a part of the response to tissue damage due to various acute conditions causing kidney injury. Studies that investigate the role of HSP70/72 in acute kidney injury (AKI) have intensified on ischemia-reperfusion models. Ischemia-reperfusion injury results from decreasing of oxygen supply of the kidney tissue due to reduction of blood flow, and subsequently sudden restoration of the blood flow to the kidney tissue. Ischemia and thereafter reperfusion cause to increase of reactive oxygen species (ROS), apoptotic molecules, necrosis, inflammatory cells (Friedewald and Rabb 2004). These cellular responses finally lead to tissue injury (Friedewald and Rabb 2004). Cellular ATP decreases after renal ischemia and HSP70 expression is induced when the ATP reduction reaches 35–50% (van Why et al. 1994, 1999).

Acute kidney injury results from an ischemic or toxic incident. Amongst kidney cells, renal tubular cells are the most vulnerable to AKI. Renal tubule cells loss their structural integrity and polarity in AKI and regain their functions through remodeling (Molitoris 1991). Primary mechanism of cellular repair is recycling of damaged proteins following ATP depletion (Molitoris et al. 1996). HSP 70, as an important chaperon, participates in restoring of damaged proteins (Mayer 2013). Detachment of Na-K-ATPase from the cytoskeleton causes loss of tubule polarity in ischemic renal injury (Riordan et al. 2005). HSP 70 binds to Na-K-ATPase and overexpression of HSP 70 reduces of detachment of Na-K-ATPase (Riordan et al. 2005). Binding of HSP70 to Na-K-ATPase increases in response to injury and decreases during recovery period (Riordan et al. 2005).

Besides, it has been reported that HSP 70 has various effects in AKI. Geranylgeranylacetone (GGA), an HSP 70 inducer, prevents morphine induced renal damage by inhibition of caspase-3 and caspase-9 (Luo et al. 2013). Also, HSP 70 reduces apoptosis in AKI inhibiting the c-Jun N-terminal kinase phosphorylation as well as caspase-3 inhibition (Kim et al. 2009). Overexpression of HSP 72 activates MEK/ERK signaling pathway via phosphorylation and provides cell survival form nephrotoxic injury (Wang et al. 2009). Expression or activation of Ste20-like kinase (SLK) activates HSP 70 induction (Cybulsky et al. 2016). Expression of HSP 70 mitigates SLK associated apoptosis. Conversely, inhibition of HSP 70 enhances the pro-apoptotic activity of SLK (Cybulsky et al. 2016). Selective COX2 inhibitors reduce papillary HSP 70 induction which is associated with caspase-3 activity and apoptotic cells (Neuhofer et al. 2004).

Mueller T et al. evaluated urine HSP 72 in rat with exposure heat and ischemia, and also first urine after the renal transplantation in six pediatric renal allograft recipients (Mueller et al. 2003). They reported that renal HSP 72 expression was stimulated both by renal ischemia or hyperthermia in rats, whereas only renal ischemia led to HSP 72 excretion into urine (Mueller et al. 2003). Urine HSP 72 is detectable in the first urine after renal transplantation and presence of HSP 72 in urine continues during the first 12 h following renal transplantation (Mueller et al. 2003). In another study, it has been demonstrated that both renal HSP 72 mRNA and urine HSP 72 level increase gradually in rats as renal ischemia is prolonged (Barrera-Chimal et al. 2011). Moreover, 9 patients with AKI were evaluated in the same study and increased urine HSP 72 was observed in clinical AKI before serum creatinine elevation (Barrera-Chimal et al. 2011). The authors concluded that urinary HSP 72 was a useful biomarker for detection of AKI before fulfilled of AKI criteria (Barrera-Chimal et al. 2011). Another study of the same group compared urine HSP 72 levels of 17 patients with AKI and 20 controls (Morales-Buenrostro et al. 2014). Urine HSP 72 levels remained stable at the level of 0.3 ng/ml in the patients without AKI, while they progressively increase before development of AKI beginning from -3. day and reach a level of up to 15fold higher in patients with AKI (Morales-Buenrostro et al. 2014). Sensitivity and specificity of urine HSP 72 levels were 100% and 83.3%, respectively, using the cut-off value as 1 ng/ml at -1. day (Morales-Buenrostro et al. 2014). Moreover, increase of HSP 72 was earlier than other markers including neutrophil gelatinase-associated lipocalin (NGAL), kidney injury molecule type 1 (KIM-1) and interleukin 18 (IL-18) (Morales-Buenrostro et al. 2014). Upregulation of HSP72 by renal ischemia is only seen in renal cortex, although it does not occur in renal medulla or in other tissues such as brain, liver, intestine, lung and heart (Ortega-Trejo et al. 2015). Urinary HSP 72 level is constant at room temperature for 48 h and remains stable at -80 C storage for 9 months, also is not affected by several freeze/thaw cycles (Ortega-Trejo et al. 2015). HSP 72 has become one of the most important candidates among biomarkers recommended for early detection of AKI according to all these studies. Conversely, Vaara et al did not find any differences between the critically ill patients with and without AKI in terms of urinary HSP70 level (Vaara et al. 2016). Ramirez-Sandoval et al. demonstrated that urinary HSP72 did not differentiate between immunologic rejection with AKI, pre-renal AKI and other causes of AKI in renal transplant recipients (Ramirez-Sandoval et al. 2014). Kierulf-Lassen et al. evaluated effects of remote ischemic conditioning on renal ischemia-reperfusion injury. They found that remote ischemic pre-conditioning or per-conditioning strategies did not alter HSP70 upregulation (Kierulf-Lassen et al. 2015).

It has been notified that toxic injury of kidney may affect HSP 70 levels. Ochratoxin A, a mycotoxin secreted by Aspergillus, prevents HSP 70 induction in rat kidney and cell culture (Barisic et al. 2002). The authors stated that kidney damage due to Ochratoxin A is enhanced in the absence of protective effect of HSP

70 (Barisic et al. 2002). Molinas et al. investigated HSP 70 levels in urine, kidney and liver samples of the rats receiving Acetaminophen (Molinas et al. 2010). The authors established that HSP 70 levels increased in urine and in renal cortex simultaneously, whereas HSP 70 levels in liver tissue did not increase despite the elevation of plasma ALT levels due to hepatic damage (Molinas et al. 2010). Injury of proximal tubular cells via gentamicin induces HSP 72 according to degree of proximal tubular injury (Wang et al. 2006).

Chronic Kidney Diseases and HSP 70

Oxidative stress, inflammation and apoptosis increase in chronic kidney disease (CKD) as well as AKI (Lebherz-Eichinger et al. 2013). All these factors have a potential inducing role for HSP 70 response in CKD (Lebherz-Eichinger et al. 2013). HSP70 can inhibit apoptosis in different ways (Gotoh et al. 2004; Stankiewicz et al. 2005; Gurbuxani et al. 2003; Ravagnan et al. 2001). Also, HSP72 prevents synthesis of pro-inflammatory cytokines by repressing NF-kB (Jo et al. 2006). Especially, inflammation is an important part of progression in CKD, in fact in all chronic diseases. It has been considered that intracellular HSP have a protective role, although extracelluar HSP have immunogenic and inflammatory effect (Sreedharan and Van Why 2016). It has been claimed that HSP70 may have long-term detrimental effects in chronic inflammation, although it has been shown to be beneficial in the acute phase (Sreedharan and Van Why 2016, Schmitt et al. 2007). On the contrary, cell culture and animal model studies revealed that HSP70 is a protective factor in the case of urea elevated environment (Maddock and Westenfelder 1996; Neuhofer et al. 2001). Kim et al. demonstrated that inhibition of HSP70 suppresses Treg cells as well as suppressed their renoprotective effects (Kim et al. 2014).

The main reason for the progression of chronic kidney diseases to end stage renal disease (ESRD) is development of renal fibrosis in glomeruli and tubulointerstitial area regardless of the underlying renal disease. The effect of HSP 70/72 on renal fibrosis was studied in obstructive nephropathy models as a good example for renal fibrosis. It has been observed that HSP 70 concentration of the kidney increases in a model of obstructive nephropathy and returns to baseline after the obstruction was removed (Lin et al. 1998). Similarly, HSP 70 induction was also demonstrated in patients with uretheropelvic junction obstruction (Valles et al. 2003). Induction of HSP 72 by GGA reduces sclerotic damage, tubulointerstitial fibrosis, interstitial fibroblast accumulation and collagen deposition (Mao et al. 2008). Additionally, HSP 72 induction decreases apoptosis and cell proliferation in tubular epithelium (Mao et al. 2008). Besides these effects, GGA as an HSP 72 inducer prevent tubular epithelial-mesenchmal transition (EMT), retaining E-cadherin and inhibiting α-smooth muscle actin (Mao et al. 2008). The effect of HSP 72 on EMT has been shown to be independent of apoptotic mechanisms (Mao et al. 2008). TGF-β signaling pathways, particularly phosphorylation of Smad3, have an important role on EMT and kidney fibrosis (Zhou et al. 2010). HSP 72 inhibits TGF-β1 induced

Smad3 activation and EMT. Peptide-binding domain of HSP 72 is required for these inhibitions (Zhou et al. 2010). Zhou et al confirmed that HSP 72 expression was increased in fibrotic human kidneys. This increase is accompanied by the expression of p-STAT3 and α-smooth muscle actin. HSP 72 deficiency significantly increases collagen and α-smooth muscle actin, aggravates STAT3 phosphorylation and kidney fibrosis in the fibrotic kidneys of HSP72 knockout mice (Zhou et al. 2016). Further, $CD4^{+}CD28^{null}$ and $CD4^{+}CD28^{+}$T-cells incubated with HSP 70 increases the expression of IFN- and cytotoxic molecule, perforin, granzyme B in CKD subjects unlike healthy controls (Yadav et al. 2013). Blockade of angiotensin II, a pro-fibrotic cytokine, with losartan suppresses oxidative stress via decreasing superoxide dismutase and this process is strongly associated with HSP70 increase (Manucha et al. 2005). Downregulation of endothelial nitric oxide synthase (NOS) and HSP 70 coexist in obstructed kidney and induce apoptosis via increased Bax/BcL2 and caspase-3 activity (Manucha et al. 2011). Conversely, induction of HSP 70 expression is linked to decrease apoptosis on tubular cells during obstruction via the augmentation of nitric oxide (Manucha et al. 2011). The interaction between NOS and HSP 70 modulates WT-1 expression in obtrusive kidney (Mazzei and Manucha 2017). It is known that the WT-1 gene has crucial functions in kidney formation and development (Menke and Schedl 2003). The relationship between WT-1 gene and HSP 70 suggests that HSP70 is a chaperone that also plays a role in kidney development (Mazzei and Manucha 2017). Taken all together, HSP 70/72 has an important role in the development kidney fibrosis by different ways. These features make HSP 70 an attractive drug target to prevent or mitigate progression of kidney fibrosis.

Clinical studies evaluating the role of HSP 70/72 in CKD are limited. Marzec et al. demonstrated that HSP 72 expression was significantly decreased in monocytes of adult pre-dialysis CKD patients and the level of HSP 72 was negatively correlated with serum creatinine levels (Marzec et al. 2009). In the same study, HSP 70 expression in monocytes found to be lower in hemodialysis patients than in the pre-dialysis patients (Marzec et al. 2009). On the contrary, Musial et al found that serum HSP 70 was not different whereas anti-HSP 70 was increased in children with CKD (Musial et al. 2010). In addition, anti-HSP 70 is a negative prognostic factor for cardiovascular complications in children with CKD (Musial et al. 2010). Urine HSP 70 in stage 4–5 CKD was found to be elevated approximately fourfold comparing the healthy controls (Lebherz-Eichinger et al. 2012).

Reuter et al. evaluated HSP 72 expression in monocyte of patients receiving hemodialysis due to end-stage renal disease (Reuter et al. 2009). They found no difference between monocytes of patients and controls although the response of HSP 72 to heat shock was decreased and apoptosis was enhanced in patient's monocytes but not affected by dialysis itself (Reuter et al. 2009). Crowe et al. observed that HSP 70 was higher in muscle specimen of hemodialysis patients than in the control group although the difference was not statistically significant (Crowe et al. 2007). Musial et al reported that pre-dialysis serum anti-HSP 70 levels in hemodialysis patients were not different from controls. After the hemodialysis session, the anti-HSP 70 levels decrease (Musial et al. 2009). It has been demonstrated that the expression of HSP 72/73 in renal tubules was higher in the hemodialysed patients than in the non-dialysed patients (Dinda et al. 1998). Shioshita et al. reported that expression of HSP 70 in peritoneal tissue of the patients on peritoneal dialysis was significantly higher than in the non-dialysed CKD patients (Shioshita et al. 2000). The expression of HSP 70 was much higher in patients with ultrafiltration loss, while there was no difference between in the patient with and without recurrent peritonitis (Shioshita et al. 2000).

Despite these studies, the role of HSP 70 in CKD is still unclear. Whether HSP 70 has positive or negative effects on the progression of CKD and whether therapeutic modulation of HSP 70 prevents the progression of CKD or not need to be clarified in order to the features of HSP 70 in practice.

One of the most important causes of CKD in adults is diabetic nephropathy (Caramori and Mauer 2009; Marcovecchio and Chiarelli 2009). However, it may emerge in childhood because the age of onset of the type 1 diabetes mellitus (T1DM) has decreased over the last decade (Marcovecchio and Chiarelli 2009). The first sign of diabetic nephropathy is accepted to be microalbuminuria which occurs 6–15 years after the T1DM diagnosis whereas diabetic injury begins before microalbuminuria emerges (Caramori and Mauer 2009; Marcovecchio and Chiarelli 2009). Exposure consistently to high plasma glucose concentration and other metabolic factors including advanced glycosylated-end products (AGEs), oxidative stress, growth factors and cytokines contributes to developing nephropathy in T1DM (Caramori and Mauer 2009; Marcovecchio and Chiarelli 2009). High glucose concentration directly induces HSP70 gene expression in renal proximal tubular cells (Qi et al. 2007). Moreover, AGEs induce cellular oxidative stress and proliferation of interstitial fibroblasts in rat kidney activating Janus kinase 2- signal transducers and activators of transcription (JAK2/STAT) pathway (Marcovecchio and Chiarelli 2009; Chen et al. 2007; Guh et al. 2001). HSP70 expression was increased by JAK/STAT pathway activation (Madamanchi et al. 2001). It has been reported that HSP27/25, HSP60 and HSP70/72 overexpressed in the outer medulla of rat kidney in a model of diabetic nephropathy (Barutta et al. 2008). Therefore, we evaluated urine HSP27, HSP40, HSP60, HSP70, HSP90 levels in 33 children with T1DM twice, a year apart (Yilmaz et al. 2016a, b). The results were expressed as urine HSP/creatinine ratios (uHSP/Cr). In the first year of the study, all of these parameters were significantly higher in diabetic children than in the controls indicating that diabetic kidney injury has already started before microalbuminuria and renal insufficiency begin (Yilmaz et al. 2016a, b). Comparing all HSP with each other, uHSP70/Cr appears to be the best indicator for diabetic kidney injury because of the highest AUC (0.957), sensitivity (85%) and specificity (96%) (Yilmaz et al. 2016a, b). The possibility of diabetic kidney injury increases 20fold, when uHSP70/Cr is higher than a cut-off 22.59 pg/mgCr (Yilmaz et al. 2016a, b). Comparing the results of first year, uHSP70/Cr significantly increased at the second year of the study whereas uHSP60/Cr decreased and the others remained stable (Yilmaz et al. 2016a, b). According to our results, uHSP70/Cr increases over time and may indicate progressive damage in kidney of diabetic children (Yilmaz et al. 2016a, b). Our study demonstrated that urine level of HSP 70 increases as a response to a chronic insult for kidney.

Another multicenter study including our patients showed that increased serum HSP 70 level may be associated with glomerular diseases. Patients with IgA nephropathy were compared with the patients with idiopathic nephrotic syndrome (INS) as a patient control group, and healthy controls in terms of serum levels of HSP70. Serum HSP70 level was significantly higher in the IgA nephropathy group than in the INS group and controls whereas it was almost comparable between the patients with INS and controls (unpublished data).

Conclusions

Obviously, HSP 70 is involved in response to various acute and chronic insults of kidney and also other parts of the urinary tract. In this point of view, it does not seem to be a specific biomarker for any of the kidney diseases. However, HSP 70 level of body fluids may be beneficial as a biomarker, if we use it in some specific circumstances, for example for the differential diagnosis of "a children with fever". The number of possible diseases in this situation is limited and an infection is most likely to be the reason of fever. Our study showed that UTI can be distinguished from other infections and contamination in "a children with fever" by measuring urine HSP 70 level. Thus, objective and accurate diagnosis of UTI will be possible. Another expectation from the studies on HSP 70 is to provide information to use this protein to ameliorate the kidney diseases. Clarifying the exact role of HSP 70 in different kidney diseases enables to discover new therapeutic options in order to prevent renal fibrosis and chronic renal failure in the future.

Acknowledgements None

References

Barisić, K., Petrik, J., Rumora, L., Cepelak, I., & Grubisić, T. Z. (2002). Expression of Hsp70 in kidney cells exposed to ochratoxin A. *Archives of Toxicology, 76*(4), 218–226.

Barrera-Chimal, J., Pérez-Villalva, R., Cortés-González, C., et al. (2011). Hsp72 is an early and sensitive biomarker to detect acute kidney injury. *EMBO Molecular Medicine, 3*(1), 5–20.

Barutta, F., Pinach, S., Giunti, S., et al. (2008). Heat shock protein expression in diabetic nephropathy. *American Journal of Physiology. Renal Physiology, 295*(6), 1817–1824.

Beck, F. X., Neuhofer, W., & Müller, E. (2000). Molecular chaperones in the kidney: Distribution, putative roles and regulation. *American Journal of Physiology. Renal Physiology, 279*(2), 203–215.

Buraczynska, M., Swatowski, A., Buraczynska, K., Dragan, M., & Ksiazek, A. (2009). Heat-shock protein gene polymorphisms and the risk of nephropathy in patients with Type 2 diabetes. *Clinical Science, 116*(1), 81–86.

Caramori, M. L., & Mauer, M. (2009). Pathogenesis and pathophysiology of diabetic nephropathy. In A. Greenberg (Ed.), *Primer on kidney diseases* (pp. 214–223). Philadelphia: Saunders.

Chen, S. C., Guh, J. Y., Chen, H. C., Yang, Y. L., Huang, J. S., & Chuang, L. Y. (2007). Advanced glycation end-product-induced mitogenesis is dependent on Janus kinase 2-induced heat shock protein 70 in normal rat kidney interstitial fibroblast cells. *Translational Research, 149*(5), 274–281.

Crowe, A. V., McArdle, A., McArdle, F., et al. (2007). Markers of oxidative stress in the skeletal muscle of patients on hemodialysis. *Nephrology, Dialysis, Transplantation, 22*, 1177–1183.

Cybulsky, A. V., Guillemette, J., & Papillon, J. (2016). Ste20-like kinase, SLK, activates the heat shock factor 1 – Hsp70 pathway. *Biochimica et Biophysica Acta, 1863*(9), 2147–2155.

Davies, E. L., Bacelar, M. M., Marshall, M. J., Johnson, E., Wardle, T. D., Andrew, S. M., & Williams, J. H. (2006). Heat shock proteins form part of danger signal cascade in response to lipopolysaccharide and GroEL. *Clinical and Experimental Immunology, 145*(1), 183–189.

Dinda, A. K., Mathur, M., Guleria, S., Saxena, S., Tiwari, S. C., & Dash, C. (1998). Heat shock protein (HSP) expression and proliferation of tubular cells in end stage renal disease with and without haemodialysis. *Nephrology, Dialysis, Transplantation, 13*(1), 99–105.

Evdokimovskaya, Y., Skarga, Y., Vrublevskaya, V., & Morenkov, O. (2010). Secretion of the heat shock proteins HSP70 and HSC70 by baby hamster kidney (BHK-21) cells. *Cell Biology International, 34*(10), 985–990.

Friedewald, J. J., & Rabb, H. (2004). Inflammatory cells in ischemic acute renal failure. *Kidney International, 66*(2), 486–491.

Gotoh, T., Terada, K., Oyadomari, S., & Mori, M. (2004). hsp70-DnaJ chaperone pair prevents nitric oxide- and CHOP-induced apoptosis by inhibitingtranslocation of Bax to mitochondria. *Cell Death and Differentiation, 11*(4), 390–402.

Guh, J. Y., Huang, J. S., Chen, H. C., Hung, W. C., Lai, Y. H., & Chuang, L. Y. (2001). Advanced glycation end product-induced proliferation in NRK-49F cells is dependent on the JAK2/STAT5 pathway and cyclin D1. *American Journal of Kidney Diseases, 38*(5), 1096–1104.

Gurbuxani, S., Schmitt, E., Cande, C., et al. (2003). Heat shock protein 70 binding inhibits the nuclear import of apoptosis-inducing factor. *Oncogene, 22*(43), 6669–6678.

Jo, S. K., Ko, G. J., Boo, C. S., Cho, W. Y., & Kim, H. K. (2006). Heat preconditioning attenuates renal injury in ischemic ARF in rats: Role of heat-shock protein70 on NF-kappaB-mediated inflammation and on tubular cell injury. *Journal of the American Society of Nephrology, 17*(11), 3082–3092.

Kennedy, D., Jäger, R., Mosser, D. D., & Samali, A. (2014). Regulation of apoptosis by heat shock proteins. *International Union of Biochemistry and Molecular Biology Life, 66*(5), 327–338.

Kierulf-Lassen, C., Kristensen, M. L., Birn, H., Jespersen, B., & Nørregaard, R. (2015). No effect of remote ischemic conditioning strategies on recovery from renal ischemia-reperfusion injury and protective molecular mediators. *PLoS One, 10*(12), e0146109.

Kim, M. G., Jung Cho, E., Won Lee, J., et al. (2014). The heat-shock protein-70-induced renoprotective effect is partially mediated by CD4+ CD25+ Foxp3 + regulatory T cells in ischemia/reperfusion-induced acute kidney injury. *Kidney International, 85*(1), 62–71.

Kim, Y. S., Jung, M. H., Choi, M. Y., et al. (2009). Glutamine attenuates tubular cell apoptosis in acute kidney injury via inhibition of the c-Jun N-terminal kinase phosphorylation of 14-3-3. *Critical Care Medicine, 37*(6), 2033–2044.

Lebherz-Eichinger, D., Ankersmit, H. J., Hacker, S., et al. (2012). HSP27 and HSP70 serum and urine levels in patients suffering from chronic kidney disease. *Clinica Chimica Acta, 413*(1–2), 282–286.

Lebherz-Eichinger, D., Krenn, C. G., & Roth, G. A. (2013). Keratin 18 and heat-shock protein in chronic kidney disease. *Advances in Clinical Chemistry, 62*, 123–149.

Lin, K. C., Krieg, R. J., Jr., Saborio, P., & Chan, J. C. (1998). Increased heat shock protein-70 in unilateral ureteral obstruction in rats. *Molecular Genetics and Metabolism, 65*(4), 303–310.

Luo, F. C., Zhao, L., Deng, J., et al. (2013). Geranylgeranylacetone protects against morphine-induced hepatic and renal damage in mice. *Molecular Medicine Reports, 7*(2), 694–700.

Madamanchi, N. R., Li, S., Patterson, C., & Runge, M. S. (2001). Reactive oxygen species regulate heat-shock protein 70 via the JAK/STAT pathway. *Arteriosclerosis, Thrombosis, and Vascular Biology, 21*(3), 321–326.

Maddock, A. L., & Westenfelder, C. (1996). Urea induces the heat shock response in human neuroblastoma cells. *Journal of the American Society of Nephrology, 7*, 275–282.

Mambula, S. S., Stevenson, M. A., Ogawa, K., & Calderwood, S. K. (2007). Mechanisms for Hsp70 secretion: Crossing membranes without a leader. *Methods, 43*(3), 168–175.

Manucha, W., Carrizo, L., Ruete, C., Molina, H., & Vallés, P. (2005). Angiotensin II type I antagonist on oxidative stress and heat shock protein 70 (HSP 70) expression in obstructive nephropathy. *Cellular and Molecular Biology (Noisy-le-Grand, France), 51*(6), 547–555.

Manucha, W., Kurbán, F., Mazzei, L., et al. (2011). eNOS/Hsp70 interaction on rosuvastatin cytoprotective effect in neonatal obstructive nephropathy. *European Journal of Pharmacology, 650*(2–3), 487–495.

Manucha, W. (2014). HSP70 family in the renal inflammatory response. *Inflammation & Allergy Drug Targets, 13*(4), 235–240.

Mao, H., Li, Z., Zhou, Y., et al. (2008). HSP72 attenuates renal tubular cell apoptosis and interstitial fibrosis in obstructive nephropathy. *Journal of the American Society of Nephrology, 295*, 202–214.

Marcovecchio, M. L., & Chiarelli, F. (2009). Diabetic nephropathy. In D. E. Avner, E. V. Harmon, P. Niaudet, & N. Yoshikawa (Eds.), *Pediatric nephrology* (pp. 1199–1217). Berlin/Heidelberg: Springer.

Margel, D., Pevsner-Fischer, M., Baniel, J., Yossepowitch, O., & Cohen, I. R. (2011). Stress proteins and cytokines are urinary biomarkers for diagnosis and staging of bladder cancer. *European Urology, 59*(1), 113–119.

Marzec, L., Zdrojewski, Z., Liberek, T., et al. (2009). Expression of Hsp72 protein in chronic kidney disease patients. *Scandinavian Journal of Urology and Nephrology, 43*, 400–408.

Mayer, M. P. (2013). Hsp70 chaperone dynamics and molecular mechanism. *Trends in Biochemical Sciences, 38*(10), 507–514.

Mazzei, L., & Manucha, W. (2017). Growing evidence suggests WT1 effects in the kidney development are modulated by Hsp70/NO interaction. *Journal of Nephrology, 30*, 11–18.

Menke, A. L., & Schedl, A. (2003). WT1 and glomerular function. *Seminars in Cell & Developmental Biology, 14*(4), 233–240.

Molinas, S. M., Rosso, M., & Wayllace, N. Z. (2010). Heat shock protein 70 induction and its urinary excretion in a model of acetaminophen nephrotoxicity. *Pediatric Nephrology, 25*(7), 1245–1253.

Molitoris, B. A. (1991). New insights into the cell biology of ischemic acute renal failure. *Journal of the American Society of Nephrology, 1*(12), 1263–1270.

Molitoris, B. A., Dahl, R., & Hosford, M. (1996). Cellular ATP depletion induces disruption of the spectrin cytoskeletal network. *The American Journal of Physiology, 271*, 790–798.

Morales-Buenrostro, L. E., Salas-Nolasco, O. I., Barrera-Chimal, J., et al. (2014). Hsp72 is a novel biomarker to predict acute kidney injury in critically ill patients. *PLoS One, 9*(10), e109407.

Mueller, T., Bidmon, B., Pichler, P., et al. (2003). Urinary heat shock protein-72 excretion in clinical and experimental renal ischemia. *Pediatric Nephrology, 18*(2), 97–99.

Musiał, K., Szprynger, K., Szczepańska, M., & Zwolińska, D. (2009). Heat shock proteins in children and young adults on chronic hemodialysis. *Pediatric Nephrology, 24*(10), 2029–2034.

Musial, K., Szprynger, K., Szczepanska, M., & Zwolinska, D. (2010). The heat shock protein profile in children with chronic kidney disease. *Peritoneal Dialysis International, 30*, 227–232.

Musial, K., & Zwolinska, D. (2011). Heat shock proteins in chronic kidney disease. *Pediatric Nephrology, 26*(7), 1031–1037.

Neuhofer, W., Lugmayr, K., Fraek, M. L., & Beck, F. X. (2001). Regulated overexpression of heat shock protein 72 protects Madin-Darby canine kidney cells from the detrimental effects of high urea concentration. *Journal of the American Society of Nephrology, 12*, 2565–2571.

Neuhofer, W., Holzapfel, K., Fraek, M. L., Ouyang, N., Lutz, J., & Beck, F. X. (2004). Chronic COX-2 inhibition reduces medullary HSP70 expression and induces papillary apoptosis in dehydrated rats. *Kidney International, 65*(2), 431–441.

O'Neill, S., Harrison, E. M., Ross, J. A., Wigmore, S. J., & Hughes, J. (2014). Heat-shock proteins and acute ischaemic kidney injury. *Nephron. Experimental Nephrology, 126*(4), 167–174.

Ortega-Trejo, J. A., Pérez-Villalva, R., Barrera-Chimal, J., et al. (2015). Heat shock protein 72 (Hsp72) specific induction and temporal stability in urine samples as a reliable biomarker of acute kidney injury (AKI). *Biomarkers, 20*(6–7), 453–459.

Qi, W., Chen, X., Gilbert, R. E., et al. (2007). High glucose-induced thioredoxin-interacting protein in renal proximal tubule cells is independent of transforming growth factor-beta1. *The American Journal of Pathology, 171*(3), 744–754.

Ramirez-Sandoval, J. C., Barrera-Chimal, J., Simancas, P. E., Correa-Rotter, R., Bobadilla, N. A., & Morales-Buenrostro, L. E. (2014). Tubular urinary biomarkers do not identify aetiology of acute kidney injury in kidney transplant recipients. *Nephrology (Carlton), 19*(6), 352–358.

Ravagnan, L., Gurbuxani, S., Susin, S. A., et al. (2001). Heat-shock protein 70 antagonizes apoptosis-inducing factor. *Nature Cell Biology, 3*(9), 839–843.

Reuter, S., Bangen, P., Edemir, B., et al. (2009). The HSP72 stress response of monocytes from patients on haemodialysis is impaired. *Nephrology, Dialysis, Transplantation, 24*, 2838–2846.

Riordan, M., Sreedharan, R., Wang, S., et al. (2005). HSP70 binding modulates detachment of Na-K-ATPase following energy deprivation in renal epithelial cells. *American Journal of Physiology. Renal Physiology, 288*, 1236–1242.

Roberts, K. B., & Subcommittee on Urinary Tract Infection, Steering Committee on Quality Improvement and Management. (2011). Urinary tract infection: Clinical practice guideline for the diagnosis and management of the initial UTI in febrile infants and children 2 to 24 months. *Pediatrics, 128*(3), 595–610.

Schmitt, E., Gehrmann, M., Brunet, M., Multhoff, G., & Garrido, C. (2007). Intracellular and extracellular functions of heat shock proteins: Repercussions in cancer therapy. *Journal of Leukocyte Biology, 81*(1), 15–27.

Shioshita, K., Miyazaki, M., Ozono, Y., et al. (2000). Expression of heat shock proteins 47 and 70 in the peritoneum of patients on continuousambulatory peritoneal dialysis. *Kidney International, 57*(2), 619–631.

Sreedharan, R., & Van Why, S. K. (2016). Heat shock proteins in the kidney. *Pediatric Nephrology, 31*(10), 1561–1570.

Stankiewicz, A. R., Lachapelle, G., Foo, C. P., Radicioni, S. M., & Mosser, D. D. (2005). Hsp70 inhibits heat-induced apoptosis upstream of mitochondria by preventing Bax translocation. *The Journal of Biological Chemistry, 280*(46), 38729–38739.

Tang, D., Kang, R., Cao, L., et al. (2008). A pilot study to detect high mobility group box 1 and heat shock protein 72 in cerebrospinal fluid of pediatric patients with meningitis. *Critical Care Medicine, 36*(1), 291–295.

Vaara, S. T., Lakkisto, P., Immonen, K., Tikkanen, I., Ala-Kokko, T., Pettilä, V., & FINNAKI Study Group. (2016). Urinary biomarkers indicative of apoptosis and acute kidney injury in the critically Ill. *PLoS One, 11*(2), e0149956.

Vallés, P., Jorro, F., Carrizo, L., Manucha, W., Oliva, J., Cuello-Carrión, F. D., & Ciocca, D. R. (2003). Heat shock proteins HSP27 and HSP70 in unilateral obstructed kidneys. *Pediatric Nephrology, 18*(6), 527–535.

Van Why, S. K., Mann, A. S., Thulin, G., Zhu, X. H., Kashgarian, M., & Siegel, N. J. (1994). Activation of heat-shock transcription factor by graded reductions in renal ATP, in vivo, in the rat. *The Journal of Clinical Investigation, 94*(4), 1518–1523.

Van Why, S. K., Kim, S., Geibel, J., Seebach, F. A., Kashgarian, M., & Siegel, N. J. (1999). Thresholds for cellular disruption and activation of the stress response in renal epithelia. *The American Journal of Physiology, 277*(2), 227–234.

Varano Della Vergiliana, J. F., Lansley, S. M., Porcel, J. M., et al. (2013). Bacterial infection elicits heat shock protein 72 release from pleural mesothelial cells. *PLoS One, 8*(5), e63873.

Wang, Z., Liu, L., Mei, Q., Liu, L., Ran, Y., & Zhang, R. (2006). Increased expression of heat shock protein 72 protects renal proximal tubular cells from gentamicin-induced injury. *Journal of Korean Medical Science, 21*(5), 904–910.

Wang, Z., Jin, H., Li, C., Hou, Y., Mei, Q., & Fan, D. (2009). Heat shock protein 72 protects kidney proximal tubule cells from injury induced by triptolide by means of activation of the MEK/ERK pathway. *International Journal of Toxicology, 28*(3), 177–189.

Williams, G. J., Macaskill, P., Chan, S. F., Turner, R. M., Hodson, E., & Craig, J. C. (2010). Absolute and relative accuracy of rapid urine tests for urinary tract infection in children: A meta-analysis. *The Lancet Infectious Diseases, 10*(4), 240–250.

Yadav, A. K., Kumar, V., & Jha, V. (2013). Heat shock proteins 60 and 70 specific proinflammatory and cytotoxic response of CD4+CD28null cells in chronic kidney disease. *Mediators of Inflammation, 2013*, 384807.

Yilmaz, A., Gedikbasi, A., Yuruk Yildirim, Z., et al. (2016a). Higher urine heat shock protein 70/creatinine ratio in type 1 diabetes mellitus. *Renal Failure, 38*(3), 404–410.

Yilmaz, A., Yildirim, Z. Y., Emre, S., et al. (2016b). Urine heat shock protein 70 levels as a marker of urinary tract infection in children. *Pediatric Nephrology, 31*(9), 1469–1476.

Zhou, Y., Mao, H., Li, S., et al. (2010). HSP72 inhibits Smad3 activation and nuclear translocation in renal epithelial-to-mesenchymal transition. *Journal of the American Society of Nephrology, 21*(4), 598–609.

Zhou, Y., Cao, S., Li, H., et al. (2016). Heat shock protein 72 antagonizes STAT3 signaling to inhibit fibroblast accumulation in renal fibrogenesis. *The American Journal of Pathology, 186*(4), 816–882.

Heat Shock Proteins in the Kidney: What Is Known About Their Role in Kidney Disease

Shobhana Nayak Rao

Abstract Heat Shock Proteins (HSP) belongs to the family of intracellular proteins that are constitutively expressed and are markedly upregulated by various stressors including heat, oxidative and chemical stress. Expression of HSP helps in numerous reparative processes including the refolding of damaged proteins and removal of irreparably damaged proteins that would accumulate and initiate cellular death or apoptosis. A growing body of evidence in recent times has expanded the role of HSP and defined their physiological and pathological role in various diseases such as neurodegenerative disorders, cancer, ischemic heart disease as well as in kidney diseases. The protective role of HSP in ischemic renal injury has been described recently and HSP impairment has been noted in other forms of kidney injury as well as in post- transplant situation. Further research and better understanding of the role of HSP in prevention of kidney injury will be crucial if translation from the laboratory to patient bedside has to occur. This article aims to be a review of heat shock protein and its relevance to kidney disease.

Keywords Apoptosis · Dialysis · Heat shock protein · Ischemia reperfusion injury · Kidney disease

Abbreviations

AKI	acute kidney injury
APC	antigen presenting cell
CKD	chronic kidney disease
HSP	heat shock protein
IL	interleukin
PKD	polycystic kidney disease

S. N. Rao (✉)
Department of Nephrology, K.S. Hegde Medical Academy Medical Sciences Complex, Mangalore, Karnataka, India

A. A. A. Asea, P. Kaur (eds.), *HSP70 in Human Diseases and Disorders*, Heat Shock Proteins 14, https://doi.org/10.1007/978-3-319-89551-2_13

PLS peritoneal dialysis fluid
SLE systemic lupus erythematosus
THP Tamm- Horsfall Protein
TLR toll like receptor
TNF tumor necrosis factor

Introduction

Heat Shock Proteins (HSP) belong to the family of intracellular proteins that are phylogenetically highly conserved. HSPS are constitutively expressed and are markedly upregulated by various stressors including heat, oxidative and chemical stress. Expression of HSP helps in numerous reparative processes including the refolding of damaged proteins and removal of irreparably damaged proteins that would accumulate and initiate cellular death or apoptosis (O'Neill et al. 2014; Kampinga and Craig 2010). A growing body of evidence has expanded the role of HSP which have been regarded as intracellular chaperones beyond their cytoprotective function. The main goal of the chaperones discussed extensively in recent reviews (Joly et al. 2010; Calderwood et al. 2007) is to preserve cell survival by controlling the three –dimensional structure of the synthesized proteins, preventing misfolding or degradation. Therefore HSP regulates the response to any detrimental factors such as temperature, radiation, hypoxia, toxins or infectious agents. Stress response may also evoke the release of HSP outside the cell as an effect of an active transport or cell disintegration due to damage. The extracellular HSP thus released plays a pivotal role in innate adaptive immune responses. They also take part in many pathological processes. There are several HSP families each with specific properties that have been well established (as shown in Table 1). HSP have been classified according to their molecular weight into five groups: small HSP (incl HSP 27), HSP 60, HSP 70, HSP 90 and HSP110. Among these, HSP 60, 70 and 90 have been studied extensively (Joly et al.). In each of these groups, two types of HSP can be found: stress –induced isoforms (HSP 60, 70 and 90 alpha) as well as those constitutively and independent of stress conditions (HSP70, HSP90ß).

Table 1 Summary of HSP families

Mammalian HSP family	Location	Main established function
HSP90	Cytosol	Chaperone for a regulator of protein complex formation
HSP70	Cytosol/nucleus/ mitochondria	Protein trafficking, anti-apoptotic properties and degradation of denatured proteins under stress
HSP60	Mitochondria	Mitochondrial protein folding and assembly
HSP27	Cytosol	Preventing unfolded protein aggregation

Role of extracellular HSP: Although the existence of extracellular HSP has been known for close to two decades (Pockley et al. 1998, 1999) there is still much uncertainty and controversy regarding the role of the extracellular form of these proteins. Extracellular HSP has been considered as a danger signal (Matzinger 1994), where the molecules are released under pathological or physiological stress giving a warning to the immune system. The cellular response comes from the antigen –presenting cell (APC) which recognizes HSP through the Toll-like receptors (TLR). That interaction subsequently triggers the APC to release inflammatory cytokines and to activate the nuclear facto r(NF-KB), thus initiating the adaptive immune response and presentation of antigens to the cytotoxic T-cells (Srivatsava 2002). HSP can also stimulate the production of cytokines by monocytes and macrophages as well as the expression of adhesion molecules on endothelial cells. HSP 60 is pro-inflammatory certainly while the ambiguous function of HSP 70, inducing both pro-inflammatory 9IL-6, TNF-ɑ and anti-inflammatory (IL-10) cytokines creates some confusion.

The role of HSP in kidney health and disease is varied and HSP induction may be detrimental or beneficial to the kidney, depending on the specific HSP, type of cell and the context. Initial reports suggested that HSP had protective effects in renal cells, however subsequent studies have shown that stress protein induction may in certain situations have adverse effects on cell injury and may worsen disease progression.

HSP in Kidney Development

It has long been recognized that stress proteins are critical to developing organisms and that HSP regulation and expression varies during development. Knock-out models of constitutively expressed chaperones in unicellular organisms made them highly susceptible to injury from even mild forms of stress and in some cases proved fatal. It was expected that knocking out individual stress proteins or ablating their induction might cause significant developmental abnormalities of the kidney. However this has not been reported till date. No developmental abnormalities have been reported in the few mouse models where inducible HSP were knocked out. HSP regulation appears to be however differing during renal development compared to later in life. HSF is constitutively more active in neonatal rat kidneys compared to mature animals. It has been seen that Hsp 70, and Hsp 90 are constitutively higher, while Hsp 60 is lower in immature rat kidneys compared to adult (Le Masson and Christians 2011). This may reflect the overlap and occasional redundancy of function of individual Hsps and the differential regulation of Hsps in the developing kidney appears to contribute to the long recognized resistance of the immature kidney to injury.

Role of HSP in Acute Kidney Injury (AKI)

The incidence of AKI is increasing and represents a significant health concern globally. In a hospital –based setting, AKI may result from multiple causes including hypoperfusion, surgery and sepsis. Apart from supportive therapy including providing renal replacement therapy when appropriate, no definite therapy exists for patients with AKI and present treatments include avoidance of further renal insults by avoidance of nephrotic medications and effective treatment of sepsis (Murugan and Kellum 2011).

HSP and Renal Ischemia Reperfusion Injury (IRI)**:** IRI is a complex event that involves numerous classes of cells and involves various biological processes such as apoptosis, microvascular dysfunction, immunological activation and altered transcription (Eltzchig and Eckle 2011). There is a marked increase in renal HSP expression with the HSP 70 gene showing a 43-fold increase and the HSP 27 a 12-fold increase. The concept that HSPs are cytoprotective after IRI is re-inforced by the evidence obtained from HSP-70–/– mice (Zhang et al. 2008; Wang et al. 2011)HSP –/– mice have worse kidney function, tubular injury and worse survival following renal IRI. The protective effect from renal IRI provided by the HSP-70 inducing agent geranyl geranylacetone is also abrogated in HSP-70 knockout mice. The main barrier to the translation of these treatments to clinical use is the lack of complete understanding of HSP 70 induction results in kidney protection. The putative modes of protection include cytoskeletal stabilization, anti-inflammatory effects, requirement in autophagy, anti-apoptotic properties, influence over macrophage phenotype and stimulation of regulatory T cells. In addition to potent ant-inflammatory effects, HSP 70 may potentially exert pro-inflammatory effects. It is thought that HSP 70 located in the cytosol reduces inflammatory signaling, however HSP 70 released into extracellular compartments displays immune-stimulating properties (O Neill, Ross et al. 2011). This however remains poorly understood and calls for the need for further studies aimed at more accurately characterizing HSP 70 and its overall effect on the inflammatory equilibrium. Future prospects for translational use of HSP 70 in prevention of kidney ischemia reperfusion injury will be helpful in patients predicted to be at high risk of developing post-operative AKI such as after cardiac surgery or after ionic contrast use. This type of preconditioning strategy has potential since it would allow prevention of IRI at the earliest stage. Better predictive models about which group of patients would likely benefit the most from such therapy are needed. This use of HSP will also be helpful prior to kidney transplantation since in this situation, delivery of the drug to organ donor prior to organ procurement could facilitate protection from IRI insult and may result in improved graft outcomes.

HSP in Polycystic Kidney Disease (PKD)

Primary cilia have a significant role in normal embryonic development and differentiation of kidney epithelia. Ciliary dysfunction has been linked to developmental disorders including PKD and is associated with abnormal chemical and mechanical signals. The actin cytoskeleton which mediates these signals and HSP 27 is integral to maintaining the normal ciliary structure and function. HSP 90 also enhances cell survival and proliferation and facilitates disease progression (Seeger-Nukpeza et al. 2013). Anti Hsp 90 therapy may help in treatment of PKD while enhancing HSP 27 expression and function could also potentially help in preventing development of cysts and disease progression.

HSP in Glomerular Disease

The initial studies of HSP were on the tubulointerstitial compartment of the kidney. But the role of HSP in glomerular diseases has been better defined of late and their role in contributing to the disease process identified. In contrast to the largely beneficial role of HSP in AKI, the actions of HSP are more varied. HSP 25/27 have been extensively studied in mouse models of glomerular disease especially the puromycin aminonucleoside model of nephrotic syndrome (Smoyer et al. 1996). Phosphorylation of Hsp 25 in glomeruli was increased after puromycin treatment and in the in-vitro model of podocyte injury Hsp 27 was protective against puromycin-induced microfilament disruption (Smoyer and Ransom 2002). Hsp 25/27 have known functions in regulating actin polymerization and in the maintenance of normal foot process structure thereby modulating the cytoskeletal changes that occur in glomerular epithelial cells subject to toxic injury leading to nephrotic syndrome. It has also been found that steroids especially dexamethasone used in treatment increased Hsp25/27 expression indicating a potential therapeutic pathway.

Heat shock proteins have a diverse effect in glomerulonephritis depending on the cause of the lesion and the protein involved. Hsp 90 is upregulated in mesangial cells in a rat model of mesangial proliferative glomerulonephritis where it is believed that Hsp 90 enhances the proliferative capacity of mesangial cells (Pieper et al. 2000). Hsp47 is a collagen synthesis/assembly regulating stress protein. Hsp47 co-expression with collagen is increased in kidneys of patients with diabetic nephropathy and Ig A nephropathy compared to control specimens (Razzaque et al. 1998). Hsp90 is also closely associated with auto-immune diseases such as systemic lupus erythematosus (SLE) and has been found to be deposited in glomeruli of lupus patients (Kenderov et al. 2002). Inhibition of Hsp90 in the mouse model MRL/pr mice decreased proteinuria and anit-dsDNA titre indicating that HSP 90 may have a direct detrimental effect in facilitating nephritis in lupus. The role of stress proteins in glomerular diseases is not straightforward but more complex. While some

Hsps may show some benefit, it appears that most stress proteins enhance inflammatory pathways to injury.

HSP in Malignancy

Malignancy is characterized by unregulated cell [proliferation and resistance to apoptotic cell death. The finding of increased expression of HSP in a variety of cancers suggests that their role as cytoprotectants may help in resistance of malignant cells. Suppression of the stress response resulted in apoptosis and decreased proliferation in renal cell carcinoma. This was achieved by an inhibitor of Hsp 90 17 –allylamino-17- demthoxygeldanamycin (17-AAG) (Sato et al. 2012). The increased immunogenicity and overexpression of HSP can also be used as a basis for immunotherapy.

Tubulointerstitial Diseases

Mutations in the gene for Tamm-Horsfall protein (THP) have been described in tubule-interstitial disorders encompassing medullary cystic kidney disease type 2, familial juvenile hyperuricemic nephropathy, glomerulocystic kidney disease that progress to end-stage kidney disease. Mutations decreased cytosolic transport of THP in cells of the thick ascending limb of Henle to the luminal surface. This effect appeared to be associated with decreased cytosolic Hsp 70 (Ma et al. 2012). Enhancing function of the chaperones could be an important therapeutic approach in THP –mediated genetic disorders which have limited options for therapy.

Role of HSP in Chronic Kidney Disease (CKD)

CKD is a complex condition characterized by a variety of underlying pathological causes and pathways culminating in the common end result of glomerulosclerosis and interstitial fibrosis. The potential components of this stress cocktail include the putative uremic toxins, mediators of inflammation, reactive oxygen species, apoptosis, infections etc. Therefore the discussion of HSP in chronic kidney disease can be conducted in two parallel directions. The first one is the possible impact of HSP, either protective or deleterious on progression of chronic kidney disease. After defining the role of HSP possible therapeutic interventions such as administration of anti – HSP proteins as well as modification in the biocompatibility of dialysis materials will open up future treatment perspectives in optimizing renal replacement therapy and improving patient outcomes. In vitro investigations performed in the late 1990s has shown increased HSP 72 expression in human neuroblastoma calls treated with urea at varying concentrations from 40 to 200 mg/dl. It was shown that HSP values rose after 30 min, obtained peak values after 10 h and reduced to zero after about 48 h (Maddock and Westenfelder 1996). A similar experiment conducted with creatinine as a medium however did not show any impact on HSP 72 expression. This probably reflects the selective influence of uremic toxins on the stress response and suggests that increase in HSP 72 expression may protect against apoptosis and lead to cell adaptation to noxious conditions. However these experimental conditions are of short duration and cannot exactly mimic conditions responsible for chronic kidney damage. Therefore interpretation of the adaptive response in the course of CKD should be viewed cautiously. Mao et al. (2008) studied the impact of HSP on chronic kidney damage in rats with obstructive uropathy. Selective activation of HSP 72 given orally inhibited the proliferation an apoptosis in tubular cells and diminished accumulation of fibroblasts and collagen- 1 generation in renal parenchymal cells, thus slowing the process of fibrosis. Research on HSP levels in humans with CKD is limited and very few studies have been done in this area. Mahgoub et al. (2012) looked at HSP 27 levels in Type-2 diabetic patients with and without diabetic nephropathy and found serum HSP 27 levels significantly higher in patients with diabetic nephropathy compared to diabetic control patients. Marzec et al (2009) described HSP 72 expression in the peripheral blood monocytes from adult patients with predialysis CKD when compared to controls (359 ± 83 AU vs 405 ± 51 AU, $p < 0.01$) and found negative correlations between HSP 72 protein and serum creatinine concentrations, thus suggesting exhaustion of adaptive mechanisms along with aggravated apoptosis and impaired immunity characteristic for CKD. However when compared to adults with CKD, children have better preser5ved HSP 70 levels, thus favoring better preservation of cellular adaptive processes. The same study revealed diminished serum concentration of AntiHSP-70, a negative factor in cardiovascular complications thereby increasing risk of atherosclerosis.

HSP in ESRD Patients on Dialysis

In contrast to the scarce data concerning HSP in CKD patients on conservative therapy, a lot more work has been done in the field of dialysis. In hemodialysis, data has focused mainly on the role of HSP as potential marker of biocompatibility of materials used. Studies of HSP in hemodialysis have concentrated mainly on HSP 72. It was shown that in regard to peripheral blood monocytes, mRNA amount in adults on HD was significantly lower than in controls (293 ± 62 AU vs 405 ± 51 AU, $P < 0.001$). Interestingly, the exposure of urea-treated macrophages harvested from healthy controls to heat stress (47 C for 40 mins) resulted in greater increase of HSP 72 expression than after incubation with urea only. This suggests that the stress response although altered in CKD is not entirely abolished. Levels of HSP 70 however in children have however differed from adult patients on dialysis. HSP 70 and antiHSP 70 serum concentrations between children on hemodialysis and controls did not significantly differ. Two studies assessing the impact of a single dialysis session with polysulphone membrane on HSP have revealed increase in HSP 72 expression, high HSP 60 as well as decreased antiHSP 60 and antiHSP 70 concentrations after HD (Reuter et al. 2009; Gowe et al. 2007). The HSP 72 activation is indicative of a stress induced reaction such as blood –dialysis contact activation. However decrease in antibody levels could result from adsorption to the dialyser membrane surface, creation of HSP-antiHSP complexes and elimination by dialysis itself. The clinical interpretation of these findings is yet unclear.

HSP in Peritoneal Dialysis

Investigation of HSP in peritoneal dialysis patients have mainly focused on the impact of the peritoneal fluid as a stress factor on the function of the peritoneum. Induction of HSP 72 expression has been found in both mesothelial cells incubated with peritoneal dialysis fluid and in macrophages from the dialysis effluent collected after one 4 hr dwell (Arbriter et al. 2003). HSP 72 can be used also as a possible marker of peritoneal fluid biocompatibility. In vitro exposure of human mesothelial cells to peritoneal dialysis solution (PDS) resulted in overexpression and shift of HSP 27 and HSP 72 from the noncytoskeletal to cytoskeletal fraction within the cell. Bender et al (Bender et al. 2010) successfully improved the status of mesothelial cells by the pharmacological manipulation of the PDS content in vitro by adding glutamine to the PDS. This improved the viability of human mesothelial cells by inducing HSP 27 and HSp 72 expression. In the rate model, this addition also reduced the detachment of mesothelial cells and also decreased the amount of protein lost in PDS. The question whether there is any difference between hemodialysis and peritoneal dialysis patients with regard to HSP is difficult to answer in adults. In children however it appears that HSP 60 and HSP 90α concentrations were similar in both groups (Musia et al. 2009). The above mentioned results have revealed the complexity of the HSP response to stressful conditions and although HSP disturbances were more evident in the case of hemodialysis, this does not imply an explicitly negative opinion of hemodialysis when compared to peritoneal dialysis. AntiHSP may become a useful marker of biocompatibility.

Role of HSP in Kidney Transplantation

The transplanted kidney whether from cadaveric organ donor or from a living kidney donor undergoes ischemia-reperfusion injury and HSP plays an essential role in that process. Studies in children after kidney transplantation has suggested that the urinary excretion of HSP 72 is characteristically seen only in the early post-transplant period while patients with stable grafts do not have detectable levels of HSP in urine (Mueller et al. 2003). Urinary HSP 72 may be a good marker of tubular cell integrity. In ischemic kidney rat models, overexpression of Hsp 70 and HSP 90 and relocation of Na-K ATP-ase from the apical to the basolateral membrane domain of the proximal tubule cells has been documented. The observed translocation of Hsp 27 from the medulla to the cortex is an adaptive response of the ischemic milieu. A major concern and application of HSP in organ transplantation is their potential role in preventing or delaying the process of rejection. Experimental data has shown that activation of heat shock protein makes cells preconditioned by subjecting them to sublethal stress to become resistant and survive subsequent otherwise lethal stimuli. A natural consequence of such results will imply that HSP stimulation in the transplanted kidney would preserve kidney function for a longer time. However this has not been replicated in experimental data and results from studies are ambiguous and have not shown any definitive protective effects (Trieb et al. 2001; Mueller et al. 2004) Some studies have even shown unpredictable effects such as enhanced immunogenicity (Pockley 2001) which can be dangerous in a transplant setting.

Conclusions

Knowledge of the role played by HSP in kidney disease is slowly increasing, although we are far from understanding their entire role and pathological implications. The intracellular forms of HSP, especially HSP 70 in cytoprotective action and delaying of apoptosis will be crucial. Increasing biocompatibility of dialysis membranes as well as peritoneal dialysis fluids will help in improving the quality of renal replacement therapy. Future investigation should concentrate on this aspect.

Acknowledgements None

References

Arbriter, K., Bidmon, B., Endenmann, M., et al. (2003). Induction of mesothelial HSP 72 upon in vivo exposure to peritoneal dialysis fluid. *Peritoneal Dialysis International, 23*, 499–501.

Bender, O., Bohm, M., Kratochwill, K., et al. (2010). HSP- mediated cytoprotection of mesothelial cells in experimental acute peritoneal dialysis. *Peritoneal Dialysis International, 30*, 294–299.

Calderwood, S. K., Gray, P. J., Jr., & Mambula, S. S. (2007). Extracellular heat shock proteins in cell signaling and immunity. *Annals of the New York Academy of Sciences, 1113*, 28–39.

Eltzchig, H. K., & Eckle, T. (2011). Ischemia and reperfusion-from mechanism to translation. *Nature Medicine, 17*, 1391–1401.

Gowe, A. V., Mc Ardle, A., Mc Ardle, F., et al. (2007). Markers of oxidative stress in the skeletal muscle of patients on hemodialysis. *Nephrology, Dialysis, Transplantation, 22*, 1177–1183.

Joly, A. L., Garrido, C., Mignot, G., Qhiringhelli, F., & Wettstein, G. (2010). Dual role of heat shock proteins as regulators of apoptosis and innate immunity. *Journal of Innate Immunity, 2*, 238–247.

Kampinga, H. H., & Craig, E. A. (2010). The HSP 70 chaperone machinery: J proteins as drivers of functional specificity. *Nature Reviews Molecular Cell Biology, 11*, 579–592.

Kenderov, A., Giltiay, N., Kyurkchiev, S., et al. (2002). Lupus –specific kidney deposits of HSP 90 are associated with altered Ig G idiotypic interactions of anti-HSP 90 autoantibodies. *Clinical and Experimental Immunology, 129*, 169–176.

Le Masson, F., & Christians, E. (2011). HSF and regulation of Hsp70 (19hspa1b) in oocyts and preimplantation embryos; new insights brought by transgenic and knockout mouse models. *Cell Stress and Chaperones, 16*, 275–285.

Ma, L., El-Ahar, T. M., Liu, Y., & Wu, X. R. (2012). Molecular and cellular effects of Tamm-Horsfall protein mutations and their rescue by chemical chaperones. *The Journal of Biological Chemistry, 287*, 1290–1305.

Maddock, A. L., & Westenfelder, C. (1996). Urea induces the heat shock response in human neuroblastoma cells. *Journals of the American Society of Nephrology, 7*, 275–282.

Mahgoub, S., Bassyounis, A., Hassan, Z., & Youns, M. (2012). Serum levels of heat shock protein 27 as a potential marker of diabetic nephropathy in Egyptians with Type-2 diabetes. *Journal of Applied Pharmaceutical Science, 2*(11), 14–20.

Mao, H., Li, Z., Zhou, Y., et al. (2008). HSP 72 attenuates renal tubular apoptosis and interstitial fibrosis in obstructive nephropathy. *American Journal of Physiology Renal Physiology, 295*, F202–F214.

Marzec, L., Liberek, T., Zdrojewski, Z., et al. (2009). Expression of Hsp 72 protein in chronic kidney disease patients. *Scandinavian Journal of Urology and Nephrology, 43*(5), 400–408.

Matzinger, P. (1994). Tolerance, danger and the extended family. *Annual Review of Immunology, 12*, 991–1045.

Mueller, T., Bidmon, B., Pichler, P., et al. (2003). Urinary heat shock protein –72 excretion in clinical and experimental renal ischemia. *Pediatric Nephrology, 18*, 97–99.

Mueller, T., Posch, M., Regele, H., et al. (2004). Hsp 72 expression in pre-transplant donor kidney biopsies and post-transplant outcome. *Transplantation, 78*, 292–295.

Murugan, R., & Kellum, J. A. (2011). Acute kidney injury. What is the prognosis? *Nature Reviews. Nephrology, 7*, 209–217.

Musia, 1. K., Szezepanska, M., Szprynger, K., & Zwolinska, D. (2009). The impact of dialysis modality on serum heat shock proteins in children and young adults with chronic kidney disease. *Kidney and Blood Pressure Research, 32*, 366–372.

O'Neill, S., Harrison, E. M., Ross, J. A., & Wigmore, S. J. (2011). The role of heat –shock protein 90 in modulating ischemia-reperfusion injury in the kidney. *Expert Opinion on Investigational Drugs, 21*, 1535–1548.

O'Neill, S., Harrison, E. M., Hughes, J., Ross, T. A., & Wignore, S. J. (2014). Heat –shock protein and acute ischemic kidney injury. *Nephron Experimental Nephrology, 126*, 167–174.

Pieper, M., Bruch, K. M., De Heer, E., Rupprecht, H. D., & Schocklmann, H. O. (2000). Requirement of heat shock protein 90 in mesangial cell mitogenesis. *Kidney International, 58*, 2377–2389.

Pockley, A. G. (2001). Heat shock proteins, anti-heat shock protein reactivity and allograft rejection. *Transplantation, 71*, 1503–1507.

Pockley, A. G., Corton, J. M., & Sheperd, J. (1998). Detection of heat shock protein 70(HSP70) and anti-HSP -70 antibodies in the serum of normal individuals. *Immunological Investigations, 27*, 367–377.

Pockley, A. G., Bulmer, J., Hanks, B. M., & Wright, B. H. (1999). Identification of human heat shock protein 60 (HSP60) and anti-HSP 60 antibodies in the peripheral circulation of normal individuals. *Cell Stress and Chaperones, 4*, 29–35.

Razzaque, M. S., Harada, T., Kumatori, A., & Taguchi, T. (1998). Co-expression of collegens and collgen-binding heat shock protein 47 in human diabetic nephropathy and Ig A nephropathy. *Nephron, 80*, 434–443.

Reuter, S., Banger, P., Edemic, et al. (2009). The HSP 72 stress response of monocytes from patients on hemodialysis is impaired. *Nephrology, Dialysis, Transplantation, 24*, 2838–2846.

Sato, A., Asano, T., & Ito, K. (2012). 17 –Allyamino-17 demethoxygeldanamycin and ritonavir inhibit renal cancer growth by inhibiting the expression of heat shock factor-1. *International Journal of Oncology, 41*, 46–52.

Seeger-Nukpeza, Cai, K. Q., Egleston, B. L., et al. (2013). Inhibiting the HSP 90 chaperone slows cyst growth in a mouse model of autosomal dominant polycystic kidney disease. *Proceedings of the National Academy of Sciences, 110*, 12786–12791.

Smoyer, W. E., & Ransom, R. F. (2002). Hsp 27 regulates podocyte cytoplasm changes in an in vitro model of podocyte process retraction. *The FASEB Journal, 16*, 315–326.

Smoyer, W. E., Ballew, J. D., Gupta, A., Mundel, P., & Welsh, M. J. (1996). Altered expression of glomerular heat shock protein 27 in experimental nephrotic syndrome. *The Journal of Clinical Investigation, 97*, 2697–2704.

Srivatsava, P. (2002). Interaction of heat shock protein with peptides and antigen presenting cells. Chaperoning of the innate and adaptive immune responses. *Annual Review of Immunology, 20*, 395–425.

Trieb, K., Dirnhofer, S., Krumbock, N., et al. (2001). Heat shock protein expression in the transplanted human kidneys. *Transplant International, 14*, 281–286.

Wang, Z., Benefrio, R. G., Gall, J. S., et al. (2011). Induction of heat shock protein 70 inhibits ischemic renal injury. *Kidney International, 79*, 861–870.

Zhang, P. L., Lun, M., Schworer, C. M., et al. (2008). Heat –shock protein expression is highly sensitive to ischemia –reperfusion injury in rat kidneys. *Annals of Clinical and Laboratory Science, 38*, 57–64.

Heat Shock Protein 70 and Other Heat Shock Proteins in Diseased Retina

Ting Zhang, Bobak Bahrami, and Ling Zhu

Abstract Heat shock proteins (HSP) belong to a family of stress-induced proteins essential to cell survival. HSP have multiple protective roles through assistance in protein folding, maintaining mitochondrial homeostasis, suppressing proinflammatory cytokines, resisting ischemic damage and protecting cells from apoptotic and necrotic death. This chapter discusses the important roles of HSP, particularly HSP70 in enhancing the survival of neurons in retinal disease through different pathways. Studies in various retinal cell lines, animal models and human tissue demonstrate altered HSP expression under different stresses and diseases. These findings implicate the critical role of. HSP in the diseased retina as well as providing support for translating the HSP' cellular defense strategy into therapy to protect and rescue injured retina from different retinal pathology.

Keywords Age-related macular degeneration · Cell death · Diabetic retinopathy · Glaucoma · Heat shock proteins · Inflammation · Mitochondria · Oxidative stress · Protein folding

Abbreviations

AMD	age-related macular degeneration
DR	diabetic retinopathy
GGA	geranylgeranylacetone
HSP	heat shock proteins
LC-MS/MS	liquid chromatography tandem-mass spectrometry
NFκB	nuclear factor-κB

T. Zhang
State Key Laboratory of Biotherapy and Cancer Center, Collaborative Innovation Center for Biotherapy, West China Hospital, Sichuan University, Chengdu, Sichuan, People's Republic of China

Save Sight Institute, The University of Sydney, Sydney, NSW, Australia

B. Bahrami · L. Zhu (✉)
Save Sight Institute, The University of Sydney, Sydney, NSW, Australia
e-mail: ling.zhu@sydney.edu.au

A. A. A. Asea, P. Kaur (eds.), *HSP70 in Human Diseases and Disorders*, Heat Shock Proteins 14, https://doi.org/10.1007/978-3-319-89551-2_14

RGCs	retinal ganglion cells
RPE	retinal pigment epithelium
STZ	streptozotocin
TNF	tumour necrosis factor
VEGF	vascular endothelial growth factor

Introduction

Heat shock proteins (HSP) were originally discovered as heat-inducible proteins (Ritossa 1962, 1996; Schlesinger 1990). Mammalian HSP can be broadly categorised into five families according to their molecular weight: the small HSP, HSP60, HSP70, HSP90 and HSP110 (Kregel 2002). In addition to physiological roles, these evolutionary conserved proteins are elevated in response to different types of cellular stress including thermal, hemodynamic, osmotic, ischaemic and oxidative (Heise and Fort 2011).

As molecular chaperones, HSP are essential components of normal cellular function, including metabolism, development, maturation and growth (Basu and Srivastava 2000; Christians et al. 2012, 2003; Kiang and Tsokos 1998; Vanmuylder et al. 1997). Under stress, HSP actively unfold misfolded or aggregated proteins to conformations from which they can refold spontaneously (Hubbard and Sander 1991). They also promote refolding of redox-modified proteins in response to mitochondrial stress (Deocaris et al. 2006). HSP regulate the secretion of both anti- or pro-inflammatory cytokines thereby controlling the immune response, increasing endotoxin tolerance and survival (Brenu et al. 2013; Gao and Tsan 2004; van Eden et al. 2005). The expression of HSP can also protect tissues from ischemic injury (Dillmann 1999; Przyklenk and Kloner 1998). HSP can suppress both apoptotic and necrotic pathways by regulating survival signalling, apoptosis signalling or tumour necrosis factor (TNF) family receptor signal transduction (Takayama et al. 2003).

In the retina, they may have a role in development. HSP70 and HSP90 have been reported constitutively expressed throughout development in rat retina, suggesting these proteins may be involved in the maturation of the primary visual pathway (Kojima et al. 1996). Chicken heat-shock cognate 70 protein was found to be precisely regulated during retinal neurogenesis, suggesting its important role in retinal development (Morales et al. 1998). HSP27 is regulated in development of rat retinal ganglion cells during a critical period of connectivity with the superior colliculus (Hawkes et al. 2004).

Importantly, altered expression of HSP has been demonstrated in common vision impairing retinal pathologies including diabetic retinopathy, age-related macular degeneration and glaucoma. Here, we review the role of HSP in these retinal pathologies and discuss the therapeutic potential of HSP targeting.

HSP in Diabetic Retinopathy

Diabetic retinopathy (DR) is a common complication of diabetes and a leading cause of blindness in working-age adults (Liew et al. 2014). The pathogenesis of DR is complex and incompletely understood. Chronic hyperglycaemia is thought to be a key component in the pathogenesis and may lead to alteration of biochemical pathways in the retina, resulting in inflammation and oxidative stress. This leads to the release of angiogenic and pro-inflammatory cytokines, such as vascular endothelial growth factor (VEGF-A), interleukin 6, tumour necrosis factor-α (TNF-α) and IL1-b (Rangasamy et al. 2012). The ultimate result is dysfunction of the neurovascular unit and consequent vision impairment.

HSP are important anti-inflammatory mediators. They may reduce the presence of inflammatory mediators such as TNF-α through interaction with nuclear factor-κB (NFκB), a nuclear transcription factor for multiple genes associated with inflammation (Yenari et al. 2005). HSP may also modify angiogenic factors contributing to neovascularisation (Augustin et al. 2011), the hallmark of advanced, proliferative DR. These angiogenic factors, especially VEGF-A, also contribute to diabetic macular edema, the most common cause of visual impairment in diabetic retinopathy.

There is conflicting evidence for up- and down-regulation of different HSP in DR. Some data suggest that altered expression of HSP is a primary factor in the pathogenesis of DR, whilst others present these changes as a secondary effect of the pathology.

HSP70

HSP70 has multiple pleiotropic functions including p53-mediated apoptosis (Wadhwa et al. 2002), iron-sulfur cluster biogenesis (Lill and Muhlenhoff 2006), mitochondrial calcium regulation (Szabadkai et al. 2006) and ATP-dependent import of nuclear-encoded proteins into the mitochondrial matrix (Stojanovski et al. 2006). The expression of HSP70 was discovered as 1 of 17 genes that were markedly lower in subjects with diabetes (Hooper and Hooper 2004). The expression level of HSP72, a stress induced member of the HSP70 family, has been found to be lower in muscle of type 2 diabetic patients compared with healthy control subjects (Kurucz et al. 2002).

Increased expression of HSP70 in the retina may help neurons to resist the oxidative stress and inflammation induced by hyperglycaemia. HSP70 subtypes 1A and 8 were two distinctively upregulated proteins in a proteomic study of the retina in streptozotocin (STZ) induced diabetic rats (Quin et al. 2007). Furthermore, Sayed et al. found that HSP70 was significantly higher in serum of diabetic patients with DR compared with those without (Sayed and Mahmoud 2016). Rather than a direct effect of hyperglycaemia stress, this upregulation of serum HSP70 may be a secondary effect of endothelial dysfunction in the retina leading to decreased release of nitric oxide into the circulation (Malyshev et al. 1995).

Other Heat Shock Proteins

Kim et al. conducted a proteomic study on tears collected from healthy volunteers and patients with diabetes, both with and without retinopathy. There was downregulation of HSP27 in diabetic patients with and without DR compared to healthy controls, with this finding most pronounced in patients with DR (Kim et al. 2012). HSP27 protein and mRNA expression was attenuated and contributed to apoptosis in oxidative stress and high-glucose environments in human retinal endothelial cells (Nahomi et al. 2014).

Animal models of diabetes have demonstrated conflicting results. Pinach et al. found retinal HSP25, the mouse homologue of the human HSP27, mRNA and protein expression was significantly increased in streptozotocin induced diabetic mouse compared to control. The observation was paralleled by overexpression of oxidative stress markers nitrotyrosine and superoxide dismutase SOD and enhanced apoptosis (Pinach et al. 2013). Brucklacher et al. also found similar upregulation of retinal HSP25 in streptozotocin induced diabetic rat compared with non-diabetic rat (Brucklacher et al. 2008). Although HSP25 upregulation was suggested to diminish neuronal apoptosis by preventing cytochrome C dependent procaspase-9 activation and sequester both procaspase-3 and cytochrome C (Paul et al. 2002), the upregulation of HSP25 in diabetic mice retina may only delay the ultimately apoptosis.

The immune response to HSP has been implicated in the pathogenesis of vascular disease (Mandal et al. 2004). However, the large, EURODIAB Prospective Complications Study found no correlation between anti-HSP27 antibodies and the presence of diabetic complications, including DR (Burt et al. 2009).

α-crystallins are small HSP which prevent aberrant protein interactions. Both αA- and αB-crystallins are increasingly expressed in animal models of diabetes as well as diabetic human eyes. In streptozotocin induced diabetic rats, retinal αA-crystallin expression is increased (Kumar et al. 2005). αB-crystallin is expressed in proliferative diabetic retinopathy membranes, and colocalizes with VEGF-positive neovessels in surgical samples (Dong et al. 2012). Despite increased expression of α-crystallins, the properties of these proteins may be dysfunctional. Losiewicz et al. demonstrated the solubility and interaction of α-crystallins with Bax, a proapoptotic protein, are impaired. These properties were shown to be critical for neuroprotection of retinal ganglion cells (RGC)s in cell culture.

HSP in Age Related Macular Degeneration

Age-related macular degeneration (AMD) is the leading cause of blindness among older adults in developed nations (Buch et al. 2004; Congdon et al. 2004). Early clinical features of AMD include drusen, an accumulation of extracellular material between the retinal pigment epithelium (RPE), and Bruch's membrane. Advanced clinical features of AMD include atrophy of the RPE, and/or development of new blood vessels (neovascularization) (Ferrington et al. 2016). Both of these sequelae result in photoreceptor death and central vision loss.

The RPE is a monolayer between the photoreceptors and choroid that supports retinal function and homeostasis. RPE dysfunction and degeneration are critical events in AMD pathogenesis (de Jong 2006). RPE cells endure high levels of oxidative stress from consumption of oxygen, accumulation of peroxidized lipid and exposure to light (Kaarniranta et al. 2013). Chronic exposure to these oxidative stresses can eventually lead to mitochondrial dysfunction, DNA damage as well as the aberrant folding or aggregation of proteins in the retina (Shaw et al. 2016). HSP are a major cellular defense against the accumulation of toxic aggregate proteins.

HSP70

Dysregulation of HSP70 may contribute to the increased susceptibility of the retina to age-related retinal disease. Levels of HSP70 are known to decline during normal aging in human eyes (Bernstein et al. 2000). Expression of heat shock cognate 70, a constitutively expressed member of the HSP70 family, progressively declines during normal aging of the primate retina (Bernstein et al. 2000). Mitochondrial HSP70 (mtHSP70) is critical in the aging of mitochondria. Knockdown of mHSP70 by RNA interference in *C. elegans* caused lower motility, defects in oogenesis, earlier accumulation of autofluorescent material, and a shorter life span. These are the major phenotypes observed during the aging of worms, suggesting that the reduction of HSP70 causes early aging or progeria-like phenotypes (Kimura et al. 2007).

Nordgaard et al. found a decrease of mtHSP70 in mitochondria isolated from AMD donors, worsening with the severity of AMD (Nordgaard et al. 2008). Based on the pleiotropic function of the HSP70, the decreased mtHSP70 level may cause the dysfunction of apoptosis, iron-sulfur cluster biogenesis, calcium and matrix-localization in RPE cells (Manicki et al. 2014; Nordgaard et al. 2008; Plafker et al. 2012).

Part of the pathogenesis of AMD is chronic oxidative stress induced by impaired degradation of proteins in RPE cells (Jarrett and Boulton 2012; Mullins et al. 2000). HSP70 plays important roles in maintenance of protein homeostasis, also referred to as “Proteostasis”. It regulates cellular proteolysis and is upregulated under stress conditions which cause protein damage and unfolding in RPE cells (Kivinen et al. 2014). The depletion of HSP70 significantly increased cell death in conjunction with proteasome inhibition in RPE cells (Ryhanen et al. 2009). Thus, downregulation of HSP70 can affect the protein turnover rate in human RPE cells, which may lead to degenerative processes in retinal cells.

Contrary to the above findings, Kang et al. identified an upregulation of HSP70 in isolated exosomes from aqueous humor samples from patients with neovascular AMD (Kang et al. 2014). This upregulation of HSP70 decreased in concentration following anti-VEGF treatment with ranibizumab. The finding was further confirmed using liquid chromatography tandem-mass spectrometry (LC-MS/MS) analysis. The upregulaiton of HSP70 could be the retinal response to the stress in neovascular AMD, secondary to the disease process. The molecular mechanism of the downregulation of HSP70 after treatment of ranibizumab is still not clear. It may be induced by reduced oxidative stress after treatment.

Other Heat Shock Proteins

HSP60 expression was found to be decreased in RPE cells from human donors with AMD in a proteomics study (Nordgaard et al. 2006). HSP60 mediates the folding of mitochondrial proteins encoded by nuclear genes after translation (Voos and Rottgers 2002). Decreased HSP60 expression, together with decreased ATP synthase levels (Nordgaard et al. 2006) may indicate that mitochondrial dysfunction is involved in the pathogenesis of AMD.

In another study, Decanini et al. found HSP27 and HSP90 exhibited a 300% and 250% increase in expression in human donor RPE respectively throughout four stages of AMD (Decanini et al. 2007). These findings are consistent with previous increased HSP27 expression in cultured RPE cells after exposure to oxidative stress (Arrigo et al. 2005; Strunnikova et al. 2001).

The small heat shock proteins, αA- and αB-crystallin, were found accumulating in Bruch's membrane and choroidal connective tissues in AMD (Nakata et al. 2005). The upregulation of the αB-crystallin expression was also reported in RPE and photoreceptors over areas of drusen in AMD (De et al. 2007; Johnson et al. 2005). The overexpression of α-crystallin is induced in the cytosol and mitochondria of RPE cells under oxidative or ER stress (Kannan et al. 2016). This suggests that the retinal accumulation of small HSP reflects a disease-related stress response during the progression of AMD. αB-crystallin regulates the protein level of VEGF-A and thus may play an important role in retinal angiogenesis (Kase et al. 2010). Intraocular angiogenesis was attenuated and the level of VEGF-A protein was remained low in αB-crystallin ($^{-/-}$) retina of laser-induced choroidal neovascularization versus wild-type mice (Kase et al. 2010). Immunoprecipitation demonstrated that αB-crystallin can bind to VEGF-A and modulate its ubiquitination. These results indicate α-crystallins may be a potential therapeutic target for neovascular AMD.

HSP in Glaucoma

Glaucoma is characterized by progressive loss of vision caused by degeneration of retinal ganglion cells (RGCs) and their axons in the optic nerve (Nickells 2012). Whilst not essential, elevated intraocular pressure (IOP) is a key risk factor for the development and progression of this optic neuropathy (Gardiner et al. 2012). Animal models generally induce the glaucoma phenotype through increasing IOP. Oxidative stress (Kumar and Agarwal 2007; Sacca et al. 2007) and TGF-β2 (Lutjen-Drecoll

2005; Tripathi et al. 1994) also play important roles in the pathogenesis. HSP have been reported to play multiple roles in the onset and progression of the glaucoma as well as in neuroprotection (Piri et al. 2016).

HSP70

HSP72, a stress induced HSP70 family member, was increasingly expressed in the nerve fibre layer of rat retinas with induced glaucoma compared with controls (Wang and Xing 2017; Windisch et al. 2009). This may suggest enhanced expression of endogenous HSP72 may play an important role in glaucomatous optic neuro-protection, but the details of the molecular mechanisms involved are not clear.

Qing et al. demonstrated that the induction of HSP72 in RGCs led to higher cell densities compared to controls in a glaucoma rat model (Qing et al. 2005). HSP72 was shown to protect RGCs by blocking activation of the stress-activated kinase/c-Jun N-terminal kinase apoptotic pathway in a molecular mechanism study (Li et al. 2014). In these studies, the induction of upregulation of endogenous HSP72 expression level by heat stress or zinc (Zn^{2+}) administration could be a novel therapeutic approach to glaucoma.

Other Heat Shock Proteins

The intensity of immunostaining for HSP27 was found greatly enhanced in the RGC and nerve fibre layers of experimentally induced glaucoma in primate retina. (Sakai et al. 2003). Yu et al. found oxidative stress and TGF-β2 can increase HSP27 expression in the glaucomatous optic nerve heads *in vitro* (Yu et al. 2008). Hydrogen peroxide or TGF-β2 induced HSP27 upregulation could be attenuated by pre-treatment with the p38MAP kinase specific inhibitor SB203580, suggesting this activation is through phosphorylated p38MAP kinase.

The phosphorylation state of HSP27 (pHSP27) is critical for its neuroprotective ability. Huang et al. showed a significant increase in total HSP27 and pHSP27 high IOP eyes in both experimental rat and mouse glaucoma models (Huang et al. 2007). Increased HSP27 and pHSP27 immunoreactivity were observed in glial cells of elevated IOP eyes suggesting that glial cells may contribute to the early retinal defence activation in experimental glaucoma.

Activation of HSP27 in animal glaucoma models have been confirmed through post-mortem study of human eyes with glaucoma. Tezel et al. found the intensity of the immunostaining and the number of labelled cells for HSP27 were greater in retina sections from glaucomatous eyes than in sections from normal eyes from age-matched donors. The increased immunostaining of HSP27 was prominent in the optic nerve heads, nerve fibre layer, ganglion cells as well as in the retinal vessels in glaucomatous eyes (Tezel et al. 2000).

The activation of HSP27 is known to increase cell survival in response to stress. On the other hand, it may also elevate serum antibody to HSP27. HSP27 autoantibody may enter neuronal cells in human retina through an endocytic mechanism and facilitate apoptotic cell death (Tezel and Wax 2000).

Other heat shock proteins, such as HSP90 and HSP60, were also found increased in experimentally induced primate glaucoma eyes compared with normal control. Prominent immunostaining for HSP90 and HSP60 were observed in the inner retinal layers, especially in the RGC and nerve fibre layers (Sakai et al. 2003). Similar reactivity immunostaining of HSP 60 was also found prominent in human retinal ganglion cells and photoreceptors (Tezel et al. 2000). HSP90 plays critical roles in cell survival (Wakakura and Foulds 1989) and HSP 60 is necessary for normal mitochondrial function (Koll et al. 1992; Martin et al. 1992). The activation of both HSP suggested that different protective response to IOP-related damage or injury in the retina.

Therapeutic Strategies for Retinal Diseases by Targeting Heat Shock Proteins

The enhancement of endogenous self-defense mechanisms represents a novel avenue for rescuing neurons in retinal diseases. Therapy directed at increasing expression of HSP may have many different applications, such as AAV mediated mtHSP70 gene therapy, which was reported to preserve vision function and prevent axonal degeneration in the experimental autoimmune encephalomyelitis mouse model (Talla et al. 2014).

HSP70

Kim et al. demonstrated that HSP72 is a novel angiogenic regulator, binding to the surface of endothelial cells and participating in extracellular signal-related kinase (ERK)-dependent angiogenesis (Kim et al. 2016). Thus, HSP72 may have a novel role in angiogenesis and could be a potential therapeutic target for angiogenesis in DR and neovascular AMD.

Subrizi et al. found recombinant human HSP70 (rhHSP70) protects RPE cells from oxidative stress. RPE cells take up exogenous rhHSP70 and localize it in endosomes and lysosomes (Subrizi et al. 2015). Ishii et al. induced the expression of HSP72 in RGCs through intraperitoneal injection of geranylgeranylacetone (GGA) in a rat model of glaucoma. This significantly reduced the loss of RGCs, lessened optic nerve damage and decreased the number of TUNEL-positive cells after administration of GGA. The coadministration of Quercetin, an inhibitor of HSP expression, abolished these protective effects (Ishii et al. 2003).

HSP70 could also be induced by valproic acid, a histone deacetylase inhibitor, to attenuate the photoreceptors death in a retinal degeneration mouse model. HSP70 induction in outer nuclear layer activates PI3K/Akt signalling, which protects photoreceptors from death (Koriyama et al. 2014).

Adeno-associated virus (AAV)-based gene therapy targeting upregulation of HSP70 was test in RGCs in an optic nerve crush (ONC) mouse model. The AAV2 mediated HSP70 expression was primarily observed in the RGCs and significantly increased survival of ONC-injured RGCs (Kwong et al. 2015). This study indicated the beneficial effect of upregulation of HSP70 in stressed retina, and also suggested a more specific way to deliver HSP in the retina compared with chemical compound induced expression.

Other Heat Shock Proteins

Modulation of HSP expression may lead to increased neuronal survival in retinal pathology. Statins are a class of drug commonly used to manage hypercholesterolemia. These function by inhibiting HMG-CoA reductase, the rate-limiting enzyme in cholesterol synthesis but also have multiple pleiotropic effects. Multiple drugs in this class were found to upregulate αB-crystallin after a transient period of induced retinal ischemia in rats. As a negative regulator of apoptosis, the induced αB-crystallin expression localized to end feet and processes of Muller cells. The modulation of αB-crystallin expression significantly enhanced RGC survival (Schmeer et al. 2008).

Furthermore, a 20 amino acid mini-peptide derived from residues 73–92 of αB-crystallin was re-engineered into a protein polymer nanoparticle and was used to protect RPE cells from oxidative stress. The anti-apoptotic effects were evaluated by reduced caspase 3 activation and TUNEL staining (Wang et al. 2014). This study demonstrates the therapeutic potential of using peptides derived from α-crystallin to prevent oxidant induced retinal cell death.

HSP may conversely have a negative effect on retinal neuronal survival. The inhibition of HSP90 was used to activate the heat shock response thus improve viability in the neurons. Aguila et al. showed a single low dose of the HSP90 inhibitor, HSP990, can enhance visual function and delay photoreceptor degeneration in a retinal degeneration rat model (Aguilà et al. 2014). Inhibition of Hsp90 can disrupt the chaperone complex with Heat Shock Factor 1 (HSF-1), thus inducing the activation of HSF-1 and HSP expression (Zou et al. 1998). Sustained HSP90 inhibition might adversely affect visual function since it will led to a posttranslational reduction in GRK1 and phosphodiesterase protein levels, which may cause night blindness and visual impairment (Sessa et al. 2013). Prolonged HSP90 suppression results in photoreceptor death in canine retina (Kanamaru et al. 2014). However, shorter-term suppression of HSP90 may improve outcomes in DR. Suppression of

HSP90 by two specific inhibitors led to attenuation of vascular leakage in STZ induced diabetic mice. Furthermore, hypoxia-mediated neovascularization was also suppressed with this therapy (Jo et al. 2014). Pre-treatment with a HSP90 inhibitor suppressed the expression and release of VEGF from RPE cells cultured in hypoxic conditions (Wu et al. 2007).

Conclusions

The induction of HSP in the retina often represents the early response to stress and may initially contribute to increased neuronal survival. Chronic stress in retinal pathology, such as DR, AMD and glaucoma may lead to persistent activation of this defense mechanism and ultimately to an imbalanced defense homeostasis. An increasing number of studies have provided a proof of concept evidence to the applicability of HSP therapy against oxidative stress, apoptosis, and angiogenesis in diseased retina. These therapies need further investigation in animal studies and subsequently human study for efficacy, as well as long-term evaluation for potential adverse effects.

Acknowledgements This work was supported in part by the University of Sydney Postgraduate Award (to B. B.) and Early Career Researcher Kick Start Grant from the Balnaves Foundation and Sydney Medical School (to L. Z.)

References

Aguilà, M., Bevilacqua, D., McCulley, C., et al. (2014). Hsp90 inhibition protects against inherited retinal degeneration. *Human Molecular Genetics, 23*, 2164–2175.

Arrigo, A. P., Virot, S., Chaufour, S., Firdaus, W., Kretz-Remy, C., & Diaz-Latoud, C. (2005). Hsp27 consolidates intracellular redox homeostasis by upholding glutathione in its reduced form and by decreasing iron intracellular levels. *Antioxidants & Redox Signaling, 7*, 414–422.

Augustin, M., Ali Asim Mahar, M., Lakkisto, P., et al. (2011). Heat shock attenuates VEGF expression in three-dimensional myoblast sheets deteriorating therapeutic efficacy in heart failure. *Medical Science Monitor, 17*, BR345–BR353.

Basu, S., & Srivastava, P. K. (2000). Heat shock proteins: The fountainhead of innate and adaptive immune responses. *Cell Stress & Chaperones, 5*, 443–451.

Bernstein, S. L., Liu, A. M., Hansen, B. C., & Somiari, R. I. (2000). Heat shock cognate-70 gene expression declines during normal aging of the primate retina. *Investigative Ophthalmology & Visual Science, 41*, 2857–2862.

Brenu, E. W., Staines, D. R., Tajouri, L., Huth, T., Ashton, K. J., & Marshall-Gradisnik, S. M. (2013). Heat shock proteins and regulatory T cells. *Autoimmune Diseases, 8*, 813256.

Brucklacher, R. M., Patel, K. M., VanGuilder, H. D., et al. (2008). Whole genome assessment of the retinal response to diabetes reveals a progressive neurovascular inflammatory response. *BMC Medical Genomics, 1*, 26.

Buch, H., Vinding, T., La Cour, M., Appleyard, M., Jensen, G. B., & Nielsen, N. V. (2004). Prevalence and causes of visual impairment and blindness among 9980 Scandinavian adults: The Copenhagen City Eye Study. *Ophthalmology, 111*, 53–61.

Burt, D., Bruno, G., Chaturvedi, N., et al. (2009). Anti-heat shock protein 27 antibody levels and diabetes complications in the EURODIAB study. *Diabetes Care, 32*, 1269–1271.

Christians, E. S., Zhou, Q., Renard, J., & Benjamin, I. J. (2003). Heat shock proteins in mammalian development. *Seminars in Cell & Developmental Biology, 14*, 283–290.

Christians, E. S., Ishiwata, T., & Benjamin, I. J. (2012). Small heat shock proteins in redox metabolism: Implications for cardiovascular diseases. *The International Journal of Biochemistry & Cell Biology, 44*, 1632–1645.

Congdon, N., O'Colmain, B., Klaver, C. C., et al. (2004). Causes and prevalence of visual impairment among adults in the United States. *Archives of Ophthalmology, 122*, 477–485.

de Jong, P. T. (2006). Age-related macular degeneration. *The New England Journal of Medicine, 355*, 1474–1485.

De, S., Rabin, D. M., Salero, E., Lederman, P. L., Temple, S., & Stern, J. H. (2007). Human retinal pigment epithelium cell changes and expression of alphaB-crystallin: A biomarker for retinal pigment epithelium cell change in age-related macular degeneration. *Archives of Ophthalmology, 125*, 641–645.

Decanini, A., Nordgaard, C. L., Feng, X., Ferrington, D. A., & Olsen, T. W. (2007). Changes in select redox proteins of the retinal pigment epithelium in age-related macular degeneration. *American Journal of Ophthalmology, 143*, 607–615.

Deocaris, C. C., Kaul, S. C., & Wadhwa, R. (2006). On the brotherhood of the mitochondrial chaperones mortalin and heat shock protein 60. *Cell Stress & Chaperones, 11*, 116–128.

Dillmann, W. H. (1999). Heat shock proteins and protection against ischemic injury. *Infectious Diseases in Obstetrics and Gynecology, 7*, 55–57.

Dong, Z., Kase, S., Ando, R., et al. (2012). Alphab-crystallin expression in epiretinal membrane of human proliferative diabetic retinopathy. *Retina, 32*, 1190–1196.

Ferrington, D. A., Sinha, D., & Kaarniranta, K. (2016). Defects in retinal pigment epithelial cell proteolysis and the pathology associated with age-related macular degeneration. *Progress in Retinal and Eye Research, 51*, 69–89.

Gao, B., & Tsan, M. F. (2004). Induction of cytokines by heat shock proteins and endotoxin in murine macrophages. *Biochemical and Biophysical Research Communications, 317*, 1149–1154.

Gardiner, S. K., Fortune, B., Wang, L., Downs, J. C., & Burgoyne, C. F. (2012). Intraocular pressure magnitude and variability as predictors of rates of structural change in non-human primate experimental glaucoma. *Experimental Eye Research, 103*, 1–8.

Hawkes, E. L., Krueger-Naug, A. M., Nickerson, P. E., Myers, T. L., Currie, R. W., & Clarke, D. B. (2004). Expression of Hsp27 in retinal ganglion cells of the rat during postnatal development. *The Journal of Comparative Neurology, 478*, 143–148.

Heise, E. A., & Fort, P. E. (2011). Impact of diabetes on alpha-crystallins and other heat shock proteins in the eye. *Journal of Ocular Biology Disease Information, 4*, 62–69.

Hooper, P. L., & Hooper, J. J. (2004). Is low-heat shock protein 70 a primary or a secondary event in the development of atherosclerosis? *Hypertension, 43*, e18–e19. author reply e18-9.

Huang, W., Fileta, J. B., Filippopoulos, T., Ray, A., Dobberfuhl, A., & Grosskreutz, C. L. (2007). Hsp27 phosphorylation in experimental glaucoma. *Investigative Ophthalmology & Visual Science, 48*, 4129–4135.

Hubbard, T. J., & Sander, C. (1991). The role of heat-shock and chaperone proteins in protein folding: Possible molecular mechanisms. *Protein Engineering, 4*, 711–717.

Ishii, Y., Kwong, J. M., & Caprioli, J. (2003). Retinal ganglion cell protection with geranylgeranylacetone, a heat shock protein inducer, in a rat glaucoma model. *Investigative Ophthalmology & Visual Science, 44*, 1982–1992.

Jarrett, S. G., & Boulton, M. E. (2012). Consequences of oxidative stress in age-related macular degeneration. *Molecular Aspects of Medicine, 33*, 399–417.

Jo, D. H., An, H., Chang, D. J., et al. (2014). Hypoxia-mediated retinal neovascularization and vascular leakage in diabetic retina is suppressed by HIF-1alpha destabilization by SH-1242 and SH-1280, novel hsp90 inhibitors. *Journal of Molecular Medicine (Berl), 92*, 1083–1092.

Johnson, P. T., Brown, M. N., Pulliam, B. C., Anderson, D. H., & Johnson, L. V. (2005). Synaptic pathology, altered gene expression, and degeneration in photoreceptors impacted by drusen. *Investigative Ophthalmology & Visual Science, 46*, 4788–4795.

Kaarniranta, K., Sinha, D., Blasiak, J., et al. (2013). Autophagy and heterophagy dysregulation leads to retinal pigment epithelium dysfunction and development of age-related macular degeneration. *Autophagy, 9*, 973–984.

Kanamaru, C., Yamada, Y., Hayashi, S., et al. (2014). Retinal toxicity induced by small-molecule Hsp90 inhibitors in beagle dogs. *The Journal of Toxicological Sciences, 39*, 59–69.

Kang, G. Y., Bang, J. Y., Choi, A. J., et al. (2014). Exosomal proteins in the aqueous humor as novel biomarkers in patients with neovascular age-related macular degeneration. *Journal of Proteome Research, 13*, 581–595.

Kannan, R., Sreekumar, P. G., & Hinton, D. R. (2016). Alpha crystallins in the retinal pigment epithelium and implications for the pathogenesis and treatment of age-related macular degeneration. *Biochimica et Biophysica Acta, 1860*, 258–268.

Kase, S., He, S., Sonoda, S., et al. (2010). alphaB-crystallin regulation of angiogenesis by modulation of VEGF. *Blood, 115*, 3398–3406.

Kiang, J. G., & Tsokos, G. C. (1998). Heat shock protein 70 kDa: Molecular biology, biochemistry, and physiology. *Pharmacology & Therapeutics, 80*, 183–201.

Kim, H. J., Kim, P. K., Yoo, H. S., & Kim, C. W. (2012). Comparison of tear proteins between healthy and early diabetic retinopathy patients. *Clinical Biochemistry, 45*, 60–67.

Kim, T. K., Na, H. J., Lee, W. R., Jeoung, M. H., & Lee, S. (2016). Heat shock protein 70-1A is a novel angiogenic regulator. *Biochemical and Biophysical Research Communications, 469*, 222–228.

Kimura, K., Tanaka, N., Nakamura, N., Takano, S., & Ohkuma, S. (2007). Knockdown of mitochondrial heat shock protein 70 promotes progeria-like phenotypes in caenorhabditis elegans. *The Journal of Biological Chemistry, 282*, 5910–5918.

Kivinen, N., Hyttinen, J., Viiri, J. et al. (2014). *Hsp70 binds reversibly to proteasome inhibitor–induced protein aggregates and evades autophagic clearance in ARPE-19 cells*, Vol. 2.

Kojima, M., Hoshimaru, M., Aoki, T., et al. (1996). Expression of heat shock proteins in the developing rat retina. *Neuroscience Letters, 205*, 215–217.

Koll, H., Guiard, B., Rassow, J., et al. (1992). Antifolding activity of hsp60 couples protein import into the mitochondrial matrix with export to the intermembrane space. *Cell, 68*, 1163–1175.

Koriyama, Y., Sugitani, K., Ogai, K., & Kato, S. (2014). Heat shock protein 70 induction by valproic acid delays photoreceptor cell death by N-methyl-N-nitrosourea in mice. *Journal of Neurochemistry, 130*, 707–719.

Kregel, K. C. (2002). Heat shock proteins: Modifying factors in physiological stress responses and acquired thermotolerance. *Journal of Applied Physiology (1985), 92*, 2177–2186.

Kumar, D. M., & Agarwal, N. (2007). Oxidative stress in glaucoma: A burden of evidence. *Journal of Glaucoma, 16*, 334–343.

Kumar, P. A., Haseeb, A., Suryanarayana, P., Ehtesham, N. Z., & Reddy, G. B. (2005). Elevated expression of alphaA- and alphaB-crystallins in streptozotocin-induced diabetic rat. *Archives of Biochemistry and Biophysics, 444*, 77–83.

Kurucz, I., Morva, A., Vaag, A., et al. (2002). Decreased expression of heat shock protein 72 in skeletal muscle of patients with type 2 diabetes correlates with insulin resistance. *Diabetes, 51*, 1102–1109.

Kwong, J. M., Gu, L., Nassiri, N., et al. (2015). AAV-mediated and pharmacological induction of Hsp70 expression stimulates survival of retinal ganglion cells following axonal injury. *Gene Therapy, 22*, 138–145.

Li, N., Li, Y., & Duan, X. (2014). Heat shock protein 72 confers protection in retinal ganglion cells and lateral geniculate nucleus neurons via blockade of the SAPK/JNK pathway in a chronic ocular-hypertensive rat model. *Neural Regeneration Research, 9*, 1395–1401.

Liew, G., Michaelides, M., & Bunce, C. (2014). A comparison of the causes of blindness certifications in England and Wales in working age adults (16–64 years), 1999–2000 with 2009–2010. *BMJ Open, 4*, e004015.

Lill, R., & Muhlenhoff, U. (2006). Iron-sulfur protein biogenesis in eukaryotes: Components and mechanisms. *Annual Review of Cell and Developmental Biology, 22*, 457–486.

Lutjen-Drecoll, E. (2005). Morphological changes in glaucomatous eyes and the role of TGFbeta2 for the pathogenesis of the disease. *Experimental Eye Research, 81*, 1–4.

Malyshev, I., Manukhina, E. B., Mikoyan, V. D., Kubrina, L. N., & Vanin, A. F. (1995). Nitric oxide is involved in heat-induced HSP70 accumulation. *FEBS Letters, 370*, 159–162.

Mandal, K., Jahangiri, M., & Xu, Q. (2004). Autoimmunity to heat shock proteins in atherosclerosis. *Autoimmunity Reviews, 3*, 31–37.

Manicki, M., Majewska, J., Ciesielski, S., et al. (2014). Overlapping binding sites of the frataxin homologue assembly factor and the heat shock protein 70 transfer factor on the Isu iron-sulfur cluster scaffold protein. *The Journal of Biological Chemistry, 289*, 30268–30278.

Martin, J., Horwich, A. L., & Hartl, F. U. (1992). Prevention of protein denaturation under heat stress by the chaperonin Hsp60. *Science, 258*, 995–998.

Morales, A. V., Hadjiargyrou, M., Díaz, B., Hernández-Sánchez, C., De Pablo, F., & De La Rosa, E. J. (1998). Heat shock proteins in retinal neurogenesis: Identification of the PM1 antigen as the chick Hsc70 and its expression in comparison to that of other chaperones. *European Journal of Neuroscience, 10*, 3237–3245.

Mullins, R. F., Russell, S. R., Anderson, D. H., & Hageman, G. S. (2000). Drusen associated with aging and age-related macular degeneration contain proteins common to extracellular deposits associated with atherosclerosis, elastosis, amyloidosis, and dense deposit disease. *The FASEB Journal, 14*, 835–846.

Nahomi, R. B., Palmer, A., Green, K. M., Fort, P. E., & Nagaraj, R. H. (2014). Pro-inflammatory cytokines downregulate Hsp27 and cause apoptosis of human retinal capillary endothelial cells. *Biochimica et Biophysica Acta (BBA) – Molecular Basis of Disease, 1842*, 164–174.

Nakata, K., Crabb, J. W., & Hollyfield, J. G. (2005). Crystallin distribution in Bruch's membrane-choroid complex from AMD and age-matched donor eyes. *Experimental Eye Research, 80*, 821–826.

Nickells, R. W. (2012). The cell and molecular biology of glaucoma: Mechanisms of retinal ganglion cell death. *Investigative Ophthalmology & Visual Science, 53*, 2476–2481.

Nordgaard, C. L., Berg, K. M., Kapphahn, R. J., et al. (2006). Proteomics of the retinal pigment epithelium reveals altered protein expression at progressive stages of age-related macular degeneration. *Investigative Ophthalmology & Visual Science, 47*, 815–822.

Nordgaard, C. L., Karunadharma, P. P., Feng, X., Olsen, T. W., & Ferrington, D. A. (2008). Mitochondrial proteomics of the retinal pigment epithelium at progressive stages of age-related macular degeneration. *Investigative Ophthalmology & Visual Science, 49*, 2848–2855.

Paul, C., Manero, F., Gonin, S., Kretz-Remy, C., Virot, S., & Arrigo, A. P. (2002). Hsp27 as a negative regulator of cytochrome C release. *Molecular and Cellular Biology, 22*, 816–834.

Pinach, S., Burt, D., Berrone, E., et al. (2013). Retinal heat shock protein 25 in early experimental diabetes. *Acta Diabetologica, 50*, 579–585.

Piri, N., Kwong, J. M., Gu, L., & Caprioli, J. (2016). Heat shock proteins in the retina: Focus on HSP70 and alpha crystallins in ganglion cell survival. *Progress in Retinal and Eye Research, 52*, 22–46.

Plafker, S. M., O'Mealey, G. B., & Szweda, L. I. (2012). Mechanisms for countering oxidative stress and damage in retinal pigment epithelium. *International Review of Cell and Molecular Biology, 298*, 135–177.

Przyklenk, K., & Kloner, R. A. (1998). Ischemic preconditioning: Exploring the paradox. *Progress in Cardiovascular Diseases, 40*, 517–547.

Qing, G., Duan, X., & Jiang, Y. (2005). Heat shock protein 72 protects retinal ganglion cells in rat model of acute glaucoma. *Yan Ke Xue Bao, 21*, 163–168.

Quin, G., Len, A. C., Billson, F. A., & Gillies, M. C. (2007). Proteome map of normal rat retina and comparison with the proteome of diabetic rat retina: New insight in the pathogenesis of diabetic retinopathy. *Proteomics, 7*, 2636–2650.

Rangasamy, S., McGuire, P. G., & Das, A. (2012). Diabetic retinopathy and inflammation: Novel therapeutic targets. *Middle East Africa Journal of Ophthalmology, 19*, 52–59.

Ritossa, F. (1962). A new puffing pattern induced by temperature shock and DNP in drosophila. *Experientia, 18*, 571–573.

Ritossa, F. (1996). Discovery of the heat shock response. *Cell Stress & Chaperones, 1*, 97–98.

Ryhanen, T., Hyttinen, J. M., Kopitz, J., et al. (2009). Crosstalk between Hsp70 molecular chaperone, lysosomes and proteasomes in autophagy-mediated proteolysis in human retinal pigment epithelial cells. *Journal of Cellular and Molecular Medicine, 13*, 3616–3631.

Sacca, S. C., Izzotti, A., Rossi, P., & Traverso, C. (2007). Glaucomatous outflow pathway and oxidative stress. *Experimental Eye Research, 84*, 389–399.

Sakai, M., Sakai, H., Nakamura, Y., Fukuchi, T., & Sawaguchi, S. (2003). Immunolocalization of heat shock proteins in the retina of normal monkey eyes and monkey eyes with laser-induced glaucoma. *Japanese Journal of Ophthalmology, 47*, 42–52.

Sayed, K. M., & Mahmoud, A. A. (2016). Heat shock protein-70 and hypoxia inducible factor-1alpha in type 2 diabetes mellitus patients complicated with retinopathy. *Acta Ophthalmologica, 94*, e361–e366.

Schlesinger, M. J. (1990). Heat shock proteins. *The Journal of Biological Chemistry, 265*, 12111–12114.

Schmeer, C., Gamez, A., Tausch, S., Witte, O. W., & Isenmann, S. (2008). Statins modulate heat shock protein expression and enhance retinal ganglion cell survival after transient retinal ischemia/reperfusion in vivo. *Investigative Ophthalmology & Visual Science, 49*, 4971–4981.

Sessa, C., Shapiro, G. I., Bhalla, K. N., et al. (2013). First-in-human phase I dose-escalation study of the HSP90 inhibitor AUY922 in patients with advanced solid tumors. *Clinical Cancer Research, 19*, 3671–3680.

Shaw, P. X., Stiles, T., Douglas, C., et al. (2016). Oxidative stress, innate immunity, and age-related macular degeneration. *AIMS Molecular Science, 3*, 196–221.

Stojanovski, D., Rissler, M., Pfanner, N., & Meisinger, C. (2006). Mitochondrial morphology and protein import – A tight connection? *Biochimica et Biophysica Acta, 1763*, 414–421.

Strunnikova, N., Baffi, J., Gonzalez, A., Silk, W., Cousins, S. W., & Csaky, K. G. (2001). Regulated heat shock protein 27 expression in human retinal pigment epithelium. *Investigative Ophthalmology & Visual Science, 42*, 2130–2138.

Subrizi, A., Toropainen, E., Ramsay, E., Airaksinen, A. J., Kaarniranta, K., & Urtti, A. (2015). Oxidative stress protection by exogenous delivery of rhHsp70 chaperone to the retinal pigment epithelium (RPE), a possible therapeutic strategy against RPE degeneration. *Pharmaceutical Research, 32*, 211–221.

Szabadkai, G., Bianchi, K., Varnai, P., et al. (2006). Chaperone-mediated coupling of endoplasmic reticulum and mitochondrial Ca2+ channels. *The Journal of Cell Biology, 175*, 901–911.

Takayama, S., Reed, J. C., & Homma, S. (2003). Heat-shock proteins as regulators of apoptosis. *Oncogene, 22*, 9041–9047.

Talla, V., Porciatti, V., Chiodo, V., Boye, S. L., Hauswirth, W. W., & Guy, J. (2014). Gene therapy with mitochondrial heat shock protein 70 suppresses visual loss and optic atrophy in experimental autoimmune encephalomyelitis. *Investigative Ophthalmology & Visual Science, 55*, 5214–5226.

Tezel, G., & Wax, M. B. (2000). The mechanisms of hsp27 antibody-mediated apoptosis in retinal neuronal cells. *The Journal of Neuroscience, 20*, 3552–3562.

Tezel, G., Hernandez, R., & Wax, M. B. (2000). Immunostaining of heat shock proteins in the retina and optic nerve head of normal and glaucomatous eyes. *Archives of Ophthalmology, 118*, 511–518.

Tripathi, R. C., Li, J., Chan, W. F., & Tripathi, B. J. (1994). Aqueous humor in glaucomatous eyes contains an increased level of TGF-beta 2. *Experimental Eye Research, 59*, 723–727.

van Eden, W., van der Zee, R., & Prakken, B. (2005). Heat-shock proteins induce T-cell regulation of chronic inflammation. *Nature Reviews. Immunology, 5*, 318–330.

Vanmuylder, N., Evrard, L., & Dourov, N. (1997). Strong expression of heat shock proteins in growth plate cartilage, an immunohistochemical study of HSP28, HSP70 and HSP110. *Anatomy and Embryology (Berlin), 195*, 359–362.

Voos, W., & Rottgers, K. (2002). Molecular chaperones as essential mediators of mitochondrial biogenesis. *Biochimica et Biophysica Acta, 1592*, 51–62.

Wadhwa, R., Yaguchi, T., Hasan, M. K., Mitsui, Y., Reddel, R. R., & Kaul, S. C. (2002). Hsp70 family member, mot-2/mthsp70/GRP75, binds to the cytoplasmic sequestration domain of the p53 protein. *Experimental Cell Research, 274*, 246–253.

Wakakura, M., & Foulds, W. S. (1989). Response of cultured Muller cells to heat shock – An immunocytochemical study of heat shock and intermediate filament proteins in response to temperature elevation. *Experimental Eye Research, 48*, 337–350.

Wang, G. H., & Xing, Y. Q. (2017). Evaluation of heat shock protein (HSP-72) expression in retinal ganglion cells of rats with glaucoma. *Experimental and Therapeutic Medicine, 14*, 1577–1581.

Wang, W., Sreekumar, P. G., Valluripalli, V., et al. (2014). Protein polymer nanoparticles engineered as chaperones protect against apoptosis in human retinal pigment epithelial cells. *Journal of Controlled Release, 191*, 4–14.

Windisch, B. K., LeVatte, T. L., Archibald, M. L., & Chauhan, B. C. (2009). Induction of heat shock proteins 27 and 72 in retinal ganglion cells after acute pressure-induced ischaemia. *Clinical & Experimental Ophthalmology, 37*, 299–307.

Wu, W. C., Kao, Y. H., Hu, P. S., & Chen, J. H. (2007). Geldanamycin, a HSP90 inhibitor, attenuates the hypoxia-induced vascular endothelial growth factor expression in retinal pigment epithelium cells in vitro. *Experimental Eye Research, 85*, 721–731.

Yenari, M. A., Liu, J., Zheng, Z., Vexler, Z. S., Lee, J. E., & Giffard, R. G. (2005). Antiapoptotic and anti-inflammatory mechanisms of heat-shock protein protection. *Annals of the New York Academy of Sciences, 1053*, 74–83.

Yu, A. L., Fuchshofer, R., Birke, M., Kampik, A., Bloemendal, H., & Welge-Lussen, U. (2008). Oxidative stress and TGF-beta2 increase heat shock protein 27 expression in human optic nerve head astrocytes. *Investigative Ophthalmology & Visual Science, 49*, 5403–5411.

Zou, J., Guo, Y., Guettouche, T., Smith, D. F., & Voellmy, R. (1998). Repression of heat shock transcription factor HSF1 activation by HSP90 (HSP90 complex) that forms a stress-sensitive complex with HSF1. *Cell, 94*, 471–480.

Part III
Hsp70 as a Therapeutic Target

Emerging Role of HSP70 in Human Diseases

Anjali Garg, Bandana Kumari, and Manish Kumar

Abstract HSP70 are prominent stress proteins, which also act like molecular chaperones. The synthesis of HSP70 increases when the cell is exposed to any form of stress physical, biological or chemical. Under stress conditions, HSP70 recognize and bind to the unstable protein substrates and protect them from denaturation and aggregation. Besides, HSP70 are also essential during normal growth where they assist in folding of nascent proteins, degradation of misfolded and truncated proteins and, in subcellular localizations of proteins and vesicles. Since HSP70 are involved in a plethora of cellular activities, their role been implicated with several pathological diseases primarily related to apoptosis, carcinogenesis, amyloidogenesis. Here, we summarize the current knowledge on the HSP70 and their relevance in diseases such as cancer, diabetes, seizures and many more. Further, the relevance of HSP70 to serve as biomarkers and/or therapeutics in human diseases is also discussed.

Keywords Chaperone · Heat shock proteins · Human disease · Protein aggregation · Protein refolding · Stress

Abbreviations

AFLD	Alcoholic fatty liver diseases
AIF	Apoptosis-inducing factor
ATP	Adenosine triphosphate
DISC	Death inducing signaling complex
HSP	Heat shock protein
iNOS	Inducible nitric oxide synthase
MMP	Matrix metalloproteinase

Author contributed equally Anjali Garg and Bandana Kumari

A. Garg · B. Kumari · M. Kumar (✉)
Department of Biophysics, University of Delhi South Campus, New Delhi, India
e-mail: manish@south.du.ac.in

A. A. A. Asea, P. Kaur (eds.), *HSP70 in Human Diseases and Disorders*, Heat Shock Proteins 14, https://doi.org/10.1007/978-3-319-89551-2_15

NAFLD	Nonalcoholic fatty liver diseases
NBD	Nucleotide binding domain
NEF	Nucleotide exchange factor
SBD	Substrate binding domain
TLR	Toll-like receptor

Introduction

Prokaryotes and eukaryotes are exposed to various environmental stresses. To counteract their effects, these have evolved a wide array of molecular and physiological processes. Stress proteins, also named as heat shock proteins, are one of the responses against the deleterious effects of many abiotic and biotic stresses including extreme temperatures, radiations, heavy metals, drought, hypoxia, ischemia and assaults of bacterial, viral and parasitic origin (Whitley *et al.* 1999). Heat shock proteins (HSP) are a class of evolutionarily conserved, functionally related cellular proteins which primarily act as chaperons (Verghese *et al.* 2012; Tóth *et al.* 2015). HSP are ubiquitous in nature and present in cytoplasm under normal conditions, but they are transferred to the nucleus and their expression is increased when cells are exposed to high temperature or shock (Xu *et al.* 2012). In 1962, Ferrucio Ritossa serendipitously discovered the response to heat shock in the form of heat-induced chromosomal puffing in salivary gland chromosomes of *Drosophila buscii* (Ritossa 1962). Later, Tissieres *et al.* (1974) observed that exposure to heat shock led to increased synthesis of a new kind of proteins in different tissues of *Drosophila melanogaster,* which were highly similar. However, it was observed that the concentrations of other proteins declined during heat shock (Tissieres *et al.* 1974). Based on initial studies in different organisms, HSP were considered to be upregulated in response to heat only, and were therefore named so. Later, it was found that in addition to temperature, several other stress factors are also responsible for higher expression of HSP (Kalmar and Greensmith 2009; Tóth *et al.* 2015). The most important function of HSP is to protect cells from stress by maintaining homeostasis and by assisting the folding of denatured proteins under stressed conditions (Hartl and Hayer-Hartl 2002). During heat-treatment, expression of cellular proteins is highly suppressed, while the expression of HSP mRNA is highly increased. HSP are essential to prevent the conformational changes in other proteins, prevent aggregation of misfolded proteins, refolding of misfolded proteins, support proteasomal removal of peptides that cannot be refolded and membrane protection. In addition, they are important for growth and development and have anti-apoptotic functions.

Types of Heat Shock Proteins

Initially, HSP were classified on the basis of their molecular weight into following groups:

(a) Small heat shock protein (sHSP) family
(b) HSP40 (J-proteins)
(c) Chaperonin (HSP60/GroEL) family
(d) 70-kDa heat shock protein (HSP70/DnaK) family
(e) HSP90 family
(f) HSP100/ClpB family.

It is pertinent to mention here that a few reports have also included ubiquitin (8.5 kDa) as one of the HSP class in eukaryotes (Vierling 1997). There is a distinct pattern of ATP usage in HSP. For example, high molecular weight HSP (hHSP; 27–110 kDa), i.e., HSP60, HSP70, HSP90 and HSP100 are ATP dependent while smaller HSP (sHSP; 15–42 kDa) are ATP independent. Expression of hHSP at euthermic or stress temperatures show distinct set of functions such as protein folding and translocation, cytoprotection, regulation of nuclear hormone receptors as well as regulation of apoptosis, whereas sHSP are mostly tissue specific, and play an important role as chaperone for protein folding as well as strong anti-apoptotic effectors. In 2009, Kampinga *et al.* proposed a new classification system for human HSP families and categorized them into (Kampinga *et al.* 2009):

(a) Small heat shock protein (HSPB)
(b) DNAJ (HSP40)
(c) HSPA (HSP70)
(d) HSPC (HSP90)
(e) HSPH (HSP110)
(f) Chaperonin family HSPD/E (HSP60/HSP10)
(g) CCT (TRiC)

In addition to the above HSP classes, human exclusively contain HSP33 (De Maio 1999). The detailed comparison between the different families of HSP is presented in the Table 1.

In the present chapter, we would be focusing primarily on the HSP of the HSP70 family, which are well characterized, ubiquitous and highly conserved ATP-dependent chaperones. In comparison to other HSP, HSP70 are the most prominent response to the heat stress, toxic chemicals and heavy metals. In stressed cells, HSP70 are mostly localized in the nucleus and nucleolus. Under stress, HSP70 are over-expressed, refolds denatured proteins and induces tolerance. In addition to helping the cell to survive in stress, HSP70 have several functions in unstressed conditions also, e.g., folding of nascent peptides, intracellular protein transport and apoptosis. It has also been shown that expression level profile of HSP70 vary for healthy and diseased conditions (Radons 2016).

Table 1 The heat shock protein families (Kaminga *et al.* 2009)

Characters	Family						
	Small heat shock protein (HSPB)	DNAJ (HSP40)	HSPA (HSP70)	HSPC (HSP90)	HSPH (HSP110)	HSPD/E (HSP60/HSP10)	CCT (TRiC)
Number of member in human	11 (mammals)	~50	13	5	4	–	–
Characteristics	HSPB have conserved α-crystalline C-terminal domain of 100 amino acids (de Jong *et al.* 1998)	HSP40 contains a conserved N-terminal J-domain that stimulates ATPase activity (Qiu *et al.* 2006)	HSPA contains N-terminal regulatory ATPase domain and C-terminal substrate binding domain (Gragerov *et al.* 1994; Zhu *et al.* 1996)	HSPC contain three regions: the N-terminal region, central region and C-terminal (Csermely *et al.* 1998)	The HSPH are homologous to HSPA. HSPH have longer linker region between N- and C-terminal domains (Kampinga *et al.* 2009)	HSPD has three domains i.e. apical domain, equatorial domain and intermediate domain these play important role in binding of substrate and co-chaperone, ATP binding and act as hinge prompting conformational change respectively (Fink 1999) and it is heat inducible protein	CCT is not upregulated during heat shock

Functions	**HSPB** proteins bind to unfolded, partially denatured and damaged proteins, and inhibiting their irreversible aggregation, retaining them in a refolding competent state, it provides protection aganist both apoptosis and oxidative stress	Folding, refolding, translocation, responsible for stimulation of HSPA ATPase activity	Folding of newly synthesized proteins, protein transport across intracellular membrane, DNA repair	Suppress aggregation of unfolded proteins, disaggregates loose protein aggregates, and enhances refolding of partially denatured proteins and cellular signaling	Folding, nucleotides exchange factor and removal of ADP after ATP hydrolysis	Folding of nascent and misfolded proteins in ATP-dependent manner	Folding of newly synthesized cytosolic proteins, preventing protein aggregation
Subcellular location	Cytoplasm, cytoskeleton and nucleus	Cytosol, nucleus, endoplasmic reticulum, mitochondria, endosomes and ribosomes	Mitochondria (HSPA9), endoplasmic reticulum (HSPA5)	Cytosol, endoplasmic reticulum and mitochondria	Cytosol and endoplasmic reticulum	Mitochondria and chloroplast	Cytosol
Comments	Present in both prokaryotes and eukaryotes that expressed under stress conditions	–	–	–	–	Represented by GroEL in prokaryotes and HSP60 in mitochondria	Present in eukaryotes

HSP70: Structure and Mechanism

There are three distinct regions of HSP70: (a) Conserved N-terminal ATPase domain or nucleotide binding domain (NBD) of ~40 kDa; composed of four subdomains (IA, IB, IIA, IIB) surrounding the ATP-binding pocket, (b) Substrate binding domain (SBD) of ~18-kDa and (c) Variable C-terminal of ~10-kDa. Each of the three domains has different functions. The SBD binds to the substrate proteins and their association is regulated by NBD. The variable C-terminal acts as a "lid" of HSP70 and helps to hold the substrates at SBD (Zhu *et al.* 1996). Human HSP70 (HSPA) is a dimer of N-terminal ATPase domain (45 kDa) (Flaherty *et al.* 1990) and a C-terminal peptide binding domain (25 kDa) (Zhu *et al.* 1996). A small linker domain separates the N-terminal domain and the C-terminal domain. In order to help in protein folding, HSPA repeatedly binds and release the unfolded protein. The binding occurs at hydrophobic regions, since they are exposed in unfolded proteins. This cyclic process is dependent on the ATPase activity of HSP70, which is assisted by co-chaperones J-proteins and nucleotide exchange factor (Miot *et al.* 2011). J-proteins induce the hydrolysis of ATP that is required for binding of solvent-exposed hydrophobic amino acids of substrate proteins whereas nucleotide exchange factor is associated with ATP–ADP exchange which release ADP from HSP70 and ultimately the substrate.

Functions of HSP70

The HSP70 family of proteins is housekeeping proteins and is highly conserved across all living domains. The major responsibilities of HSP70 are folding of nascent proteins in normal cells and refolding of denatured proteins under shock condition. Apart from this, HSP70 are also involved in multiple biological functions including import and translocation of proteins and vesicles into organelles across membranes, growth, apoptosis, proteolytic degradation of unstable proteins by targeting the proteins to lysosomes or proteasomes and the degradation of unwanted proteins. The functions of HSP70 family members highly depend on their cellular localization, and on the basis of their localization they are broadly classified in two types, intracellular and extracellular HSP70. The intracellular residing HSP70 protect cells against lethal damage induced by stress, and support folding and transport of newly synthesized polypeptides and aberrant proteins as well as assembly of multi-protein complexes. Further, extracellular HSP70 are considered as molecules with immunomodulatory functions, which act either as cross-presenters of immunogenic peptides via MHC antigen or in a peptide-free version as chaperokines or stimulators of innate immune responses. The major cellular functions of HSP and their molecular mechanism are as follows:

1. *Unfolding/refolding*: HSP70 family members under normal physiological conditions act as molecular chaperones. In response to the stress-induced damages,

intracellular HSP70 bind to the exposed hydrophobic amino acids of non-native conformation of proteins, thus protecting them against denaturation or aggregation until the cell attain the favorable condition (for reviews, see (Boston *et al.* 1996; Hartl 1996)). In conjunction with other chaperones i.e., dimeric HSP40 and co-chaperones i.e., nucleotide exchange factor (NEF), HSP70 recognizes stable misfolded polypeptides and convert them into native proteins by repeated cycle of binding, ATP-dependent unfolding, and spontaneous refolding. Improper unfolding/refolding phenomenon might lead to attachment of substrate to "holdases", including small HSP and HSP90, which maintain the substrate in a non-aggregated folding-component state and pass it to the HSP70 unfoldase machinery for refolding. Additionally, HSP70/HSP110 heterodimer converts protein aggregates into natively unfolded substrates and form NEFs by acting reciprocally on each other and, cooperatively, they efficiently disassemble stable protein aggregates.

2. *Anti-apoptotic Activity*: HSP70 are potent anti-apoptotic proteins and block apoptosis at many different levels. HSP70 block the mitochondrial translocation and activation of Bax, inhibiting mitochondrial membrane permeabilization and release of pro-apoptotic factors, and also inhibit assembly of death inducing signaling complex (DISC) (Gurbuxani *et al.* 2003; Lanneau *et al.* 2008).
3. *Repairing*: HSP70–1 in nucleus, assist the repairing machines of ssDNA by binding to poly (ADP-ribose) polymerase 1 (PARP-1), thus mediating their assembly and initiating their functions.
4. *Tumorigenic*: In cancer cells, a constitutive high-level expression of cytosolic HSP70 is observed frequently. Here, they provide resistance to stress-induced apoptosis, assist in suppressing default senescence, and are correlated with the development of metastasis and drug resistance. HSP70 stabilize the lysosomal membranes and affect autophagy, leading to the survival of cancerous cells. Cell senescence is initiated when HSP70-1 undergoes down-regulation via p53-dependent and p53-independent pathways (Yaglom *et al.* 2007). Another role of extracellular HSP70-1 in tumor invasion and metastasis comes from its ability to increase the MMP-9 expression by activating NF-κB and activating protein-1 (AP1) (Lee *et al.* 2006). However, cytosolic HSP70 have negative impact on cancer patients. Extracellular HSP70 are associated with cancer immunity and thus can be used as drug.
5. *Immunomodulation*: HSP70 act as stimulators of the adaptive immune response through their ability to bind antigenic peptide during intracellular antigen processing. The extracellular HSP70 may act as a danger signal to the innate immune system and is also relevant for the establishment of cancerous and autoimmune diseases. HSP70 exert anti-inflammatory properties, by modulation of cytokine production of dendritic cells that provide a link between innate and adaptive immune response.

HSP70 Superfamily in Mammals

In mammalian cells, there are four major isoforms of HSP70 localized in different organelles: the constitutively expressed heat shock cognate 70 (HSC70/HSPA8/HSP73) in the cytoplasm and nucleus, the stress-induced HSP70 (or HSP72/HSPA1A) in cytoplasm, the glucose-regulated BiP (or Grp78/HSPA5) in endoplasmic reticulum (ER) and mtHSP70 (Grp75/mortalin/HSPA9/mito-HSP70) in mitochondrion. Despite the difference in expression pattern of HSP70 and HSC70, their major functions are same i.e., to avoid protein aggregation; folding and assembly of nascent polypeptides, to refold misfolded or aggregated proteins, to enhance the ubiquitination and the degradation of misfolded protein. These proteins are also involved in translocation of protein through intracellular membrane and show interaction with signal transduction proteins. BiP is a major regulator of ER stress, which binds to the proteins transported to the ER and assist in the formation of quaternary structure. The mtHSP70 is mainly involved in protein transportation from mitochondria.

In humans, HSP70 has 13 members, which share several structural and functional features. For example, HSPA1 (HSP70) (reviewed in (Kampinga *et al.* 2009)) is induced by high temperature, has subfamilies called HSPA1A (HSP70-1) and HSPA1B (HSP70-2), which differ by only two amino acids. The gene sequence of another member, HSPA6 (HSP70B') is 77% similar to the HSPA1 gene. The expression of these proteins is transcriptionally controlled by Heat Shock Factors (HSF) which includes four members: HSF1, HSF2, HSF3 and HSF4. HSF have distinctive and overlapping functions and have tissue-specific patterns of expression. Among all HSF, HSF1 is the prime transcriptional regulator and is required for transactivation of HSP genes and maintenance of thermo-tolerance. During stress, HSF1 is induced and binds to the promoter of HSP70 to enhance its transcription.

Role of HSP70 in Human Diseases

As discussed earlier, the HSP70 chaperones are mainly involved in folding of translated proteins, intracellular localization and prevention of aggregation. Therefore, improper functioning of HSP lead to several diseases related to defects in protein folding or trafficking. The implications of malfunctioning of HSP70 in some of the major human diseases are discussed below:

HSP70 in Cancer

HSP70 act as an important factor in development of different types of cancers and can be used as potential tumor biomarker. Usually HSP70 is overexpressed on the cell surface of tumors. Because of their chaperonin activity as well as cell signaling regulation activity, HSP70 are involved in tumor cell proliferation, differentiation, invasion, metastasis and death. However, in some cancers (renal and cervix), survival is not correlated with the Hsp70 levels. The sequential increase in the level of HSP70 has a potential prognostic value in patients with chronic hepatitis, liver cirrhosis and liver carcinomas intrahepatic cholangiocarcinoma (IH-ChCa) and metastatic tumors (Yang *et al.* 2010). Hence, change in the expression level of HSP70 could be used as a biomarker and prognosis in cancers like colon cancer, breast cancer, melanoma, bladder cancer, cholangiocarcinoma and squamous cell carcinoma of the head and neck (SCCHN). The clinical outcome of radiotherapy can also be monitored by ascertaining the levels of HSP70 in SCCHN patients. HSP70 are considered as favorable target for treating several cancers. The association of HSP70 and Bag3 (nucleotide exchange factor of HSP70) changes the activity of certain transcription factors (NF-κB, FoxM1, Hif1α), the translation regulator (HuR) and the cell-cycle regulators (p21, survivin) (Colvin *et al.* 2014). One of the ways to check the proliferation of tumors is to induce senescence; in some cases it is done by down-regulating HSP72 via p53-dependentand p53-independent pathways. HSP70 are also associated to base pair excision system, therefore inhibition of HSP70-based DNA repair in cancer cell might be important in chemotherapeutic regimens. Furthermore, the combination of the two chaperones HSP70 and HSP90 along with conventional anti-cancer drugs is a favorable therapeutic selection for patients suffering with advanced bladder cancer.

HSP70 in Apoptosis

Besides their role as molecular chaperones, HSP70 are also anti-apoptotic proteins. HSP70 inhibit the apoptosis at multiple points in intrinsic as well as extrinsic pathways. HSP70 interact with stress-induced kinases and inhibit their functions in apoptosis. In the intrinsic pathway, HSP70 inhibit the disruption of the mitochondrial membrane potential and help to prevent the release of pro-apoptotic factors such as cytochrome c and apoptosis-inducing factor (AIF) (Gurbuxani *et al.* 2003). In extrinsic pathway, it responds to the apoptotic stimulus by inhibiting the assembly of DISC (Lanneau *et al.* 2008). HSP70 also provide protection against hypoxia/reoxygenation-induced apoptosis and maintaining intestinal epithelial cells, with the increase in expression level of BCL-2.

HSP70 in Diabetes Mellitus

The expression of HSP70 is reported to high in type I diabetes mellitus (TIDM) and type II diabetes mellitus (TIIDM) (Nakhjavani *et al.* 2010). Generally, the pancreas regulates the levels of HSP70 where they protect the susceptible beta cells from exocrine pancreatic damage and from the stress associated with insulin hypersecretion. Recently, it has been shown that if the expression of HSP decreases in TIIDM patients, their wound healing process is impaired (Singh *et al.* 2015). It has also been proposed that one of the most effective and feasible strategy to improve the glucose tolerance in hyperglycemic (i.e. high blood sugar) condition is to increase the HSP70 level, potentially by targeting hyperglycemia-related deficits in HSF1 (Kavanagh *et al.* 2011). HSP70 have a direct correlation with several molecules and can be used as an indicator for variety of diseases. For example, (i) increased serum HSP70 and hemoglobin A1c (HbA1c) levels in women indicates gestational diabetes mellitus, (ii) In patients with high C-reactive protein (CRP) and in case of hunger inhibiting hormone such as leptin, higher levels of HSP70 and asymmetric dimethyl arginine (ADMA) were reported. Furthermore, a relationship between chronic inflammation with diabetes mellitus and diabetes mellitus-associated albuminuria can be postulated from the higher levels of HSP70 observed in diabetic patients with albuminuria (i.e. presence of albumin in the urine).

HSP70 in Obesity, Non-alcoholic Fatty Liver Disease, Alcoholic Fatty Liver Disease and Hepatic Steatosis

The nonalcoholic fatty liver diseases (NAFLD) and hepatic steatosis (HS) induces the risk of type 2 diabetes (T2D) and cardio-cerebrovascular diseases (Qu *et al.* 2015a); moreover, obesity, NAFLD, alcoholic fatty liver disease (AFLD) and HS increase the inflammation. In case of NAFLD, there is a decrease in the expression level of HSF-1 of liver and adipose tissue, which affects the HSP70-dependent anti-inflammation. The HSP70 inhibition in NAFLD patients occurs in kupffer cells. Obese patients with NAFLD also have a lower HSP70 serum concentration (Di Naso *et al.* 2015). Additionally, in comparison to patients with mild alcoholic fatty liver disease (AFLD) or alcohol consuming individuals without AFLD, the lower level of HSP70 was found in AFLD patients (Qu *et al.* 2015b). In case of hepatocellular injury in AFLD patients, HSP70 shows increased positive immunoreactivity and could be used as a sensitive marker.

HSP70 in Chronic Glomerulonephritis

The expression of HSP70 in urine is also higher in case of higher chronic glomerunephritis (CGN) activity and transient creatinine as compared to inactive nephritis, active CGN and preserved renal function, and persistent proteinuria and chronic renal failure (Chebotareva *et al.* 2014).

HSP70 in Stroke and Seizure-Related Pathological Events

One of the vital functions of HSP70 is to prevent the occurrence of apoptosis in brain. The neuro-protective effect is achieved via anti-apoptotic mechanism in association with the overexpression of HSP70 (Zhao *et al.* 2014). Extracellular HSP70 facilitates the production of cytotoxic levels of tumor necrosis factor alpha via

TLR4/MyD88 signaling cascade, which results in increased neuronal death (Dvoriantchikova *et al.* 2014). In case of seizure related pathologic events also, HSP70 has a potential value as a sensitive and specific biomarker.

HSP70 in Helicobacter pylori Infection

Helicobacter pylori (*H. pylori*) are important causative agents of gastritis, peptic ulcer diseases, and mucosa associated lymphoid tissue (MALT) lymphoma and gastric cancer. It has been reported that HSP70 level changes significantly during *H. pylori* infection, viz. H. pylori-associated chronic gastritis, ulcerative colitis, and glutamine-treated patients (Leri *et al.* 1996). Studies also suggested that during *H. pylori* infection, HSP70 expression level decreases. This might involve the initiation of HSP70 expression for cytoprotection against *H. pylori* infection to prevent the expression of inducible nitric oxide synthase (iNOS) (Yeo *et al.* 2004). Pierzchalski *et al.* suggested that HSP70 protects cytoplasmic and nuclear proteins from the damaging effects of bacterial products by delaying the apoptosis of monocytes (Pierzchalski *et al.* 2014).

HSP70 in Atherosclerosis

Atherosclerosis is an inflammatory disease that affects a large human population. HSP70 concentration changes during the progression of atherosclerosis, and thus it is an effective biomarker to monitor atherosclerosis. However, there are contradictory examples over the correlation of HSP70 in atherosclerosis disease: Dulin *et al.* measured a significantly lower concentration of extracellular HSP70 in atherosclerosis patients (Dulin *et al.* 2010) while other group has reported that in patients suffering from carotid artery disease and chronic lower limb ischemia, the concentration of serum HSP70 changed depending on the severity of atherosclerosis (Krepuska *et al.* 2011).

Conclusions

HSP70 are expressed in normal cellular conditions where they regulate protein homeostasis facilitating protein folding and degradation. However, their expression is increased manifold when the cell is exposed to stress. HSP70 protect the cell from stress-induced protein unfolding and other adversities. Thus, their expression has important implications on progression of several human diseases. Therefore, HSP70 have promising role as bio-molecular marker in diagnosis of several diseases and as potential drug targets.

Acknowledgements The work was supported by grants from Indian Council of Medical Research, India to AG (ICMR JRF: 3/1/3/JRF-2016/LS/HRD-3 (32262) and BK (ICMR SRF: BIC/11(33)/2014). Authors also acknowledge efforts of Dr. Neelja Singhal for critical reading of manuscript.

References

Boston, R. S., Viitanen, P. V., & Vierling, E. (1996). Molecular chaperones and protein folding in plants. *Plant Molecular Biology, 32*, 191–222.

Chebotareva, N. V., Neprintseva, N. V., Bobkova, I. N., & Kozlovskaia, L. V. (2014). Investigation of 70-kDa heat shock protein in the serum and urine of patients with chronic glomerulonephritis. *Terapevticheskiĭ Arkhiv, 86*, 18–23.

Colvin, T. A., Gabai, V. L., Gong, J., Calderwood, S. K., Li, H., Gummuluru, S., Matchuk, O. N., Smirnova, S. G., Orlova, N. V., Zamulaeva, I. A., et al. (2014). Hsp70-Bag3 interactions regulate cancer-related signaling networks. *Cancer Research, 74*, 4731–4740.

Csermely, P., Schnaider, T., Soti, C., Prohaszka, Z., & Nardai, G. (1998). The 90-kDa molecular chaperone family: Structure, function, and clinical applications. A comprehensive review. *Pharmacology & Therapeutics, 79*, 129–168.

De Jong, W. W., Caspers, G. J., & Leunissen, J. A. (1998). Genealogy of the alpha-crystallin–small heat-shock protein superfamily. *International Journal of Biological Macromolecules, 22*, 151–162.

De Maio, A. (1999). Heat shock proteins: Facts, thoughts, and dreams. *Shock, 11*, 1–12.

Di Naso, F. C., Porto, R. R., Fillmann, H. S., Maggioni, L., Padoin, A. V., Ramos, R. J., Mottin, C. C., Bittencourt, A., Marroni, N. A., de Bittencourt, P. I., & Jr. (2015). Obesity depresses the anti-inflammatory HSP70 pathway, contributing to NAFLD progression. *Obesity (Silver Spring), 23*, 120–129.

Dulin, E., Garcia-Barreno, P., & Guisasola, M. C. (2010). Extracellular heat shock protein 70 (HSPA1A) and classical vascular risk factors in a general population. *Cell Stress & Chaperones, 15*, 929–937.

Dvoriantchikova, G., Santos, A. R., Saeed, A. M., Dvoriantchikova, X., & Ivanov, D. (2014). Putative role of protein kinase C in neurotoxic inflammation mediated by extracellular heat shock protein 70 after ischemia-reperfusion. *Journal of Neuroinflammation, 11*, 81.

Fink, A. L. (1999). Chaperone-mediated protein folding. *Physiological Reviews, 79*, 425–449.

Flaherty, K. M., DeLuca-Flaherty, C., & McKay, D. B. (1990). Three-dimensional structure of the ATPase fragment of a 70K heat-shock cognate protein. *Nature, 346*, 623–628.

Gragerov, A., Zeng, L., Zhao, X., Burkholder, W., & Gottesman, M. E. (1994). Specificity of DnaK-peptide binding. *Journal of Molecular Biology, 235*, 848–854.

Gurbuxani, S., Schmitt, E., Cande, C., Parcellier, A., Hammann, A., Daugas, E., Kouranti, I., Spahr, C., Pance, A., Kroemer, G., et al. (2003). Heat shock protein 70 binding inhibits the nuclear import of apoptosis-inducing factor. *Oncogene, 22*, 6669–6678.

Hartl, F. U. (1996). Molecular chaperones in cellular protein folding. *Nature, 381*, 571–579.

Hartl, F. U., & Hayer-Hartl, M. (2002). Molecular chaperones in the cytosol: From nascent chain to folded protein. *Science, 295*, 1852–1858.

Kalmar, B., & Greensmith, L. (2009). Induction of heat shock proteins for protection against oxidative stress. *Advanced Drug Delivery Reviews, 61*, 310–318.

Kampinga, H. H., Hageman, J., Vos, M. J., Kubota, H., Tanguay, R. M., Bruford, E. A., Cheetham, M. E., Chen, B., & Hightower, L. E. (2009). Guidelines for the nomenclature of the human heat shock proteins. *Cell Stress & Chaperones, 14*, 105–111.

Kavanagh, K., Flynn, D. M., Jenkins, K. A., Zhang, L., & Wagner, J. D. (2011). Restoring HSP70 deficiencies improves glucose tolerance in diabetic monkeys. *American Journal of Physiology. Endocrinology and Metabolism, 300*, E894–E901.

Krepuska, M., Szeberin, Z., Sótonyi, P., Sarkadi, H., Fehérvári, M., Apor, A., Rimely, E., Prohászka, Z., & Acsády, G. (2011). Serum level of soluble HSP70 is associated with vascular calcification. *Cell Stress & Chaperones, 16*, 257–265.

Lanneau, D., Brunet, M., Frisan, E., Solary, E., Fontenay, M., & Garrido, C. (2008). Heat shock proteins: Essential proteins for apoptosis regulation. *Journal of Cellular and Molecular Medicine, 12*, 743–761.

Lee, K. J., Kim, Y. M., Kim, D. Y., Jeoung, D., Han, K., Lee, S. T., Lee, Y. S., Park, K. H., Park, J. H., Kim, D. J., et al. (2006). Release of heat shock protein 70 (Hsp70) and the effects of extracellular Hsp70 on matric metalloproteinase-9 expression in human monocytic U937 cells. *Experimental & Molecular Medicine, 38*, 364–374.

Leri, O., Teichner, A., Sinopoli, M. T., Abbolito, M. R., Pustorino, R., Nicosia, R., & Paparo Barbaro, S. (1996). Heat-shock-proteins-antibodies in patients with Helicobacter pylori associated chronic gastritis. *Rivista Europea per le Scienze Mediche e Farmacologiche, 18*, 45–47.

Miot, M., Reidy, M., Doyle, S. M., Hoskins, J. R., Johnston, D. M., Genest, O., Vitery, M. C., Masison, D. C., & Wickner, S. (2011). Species-specific collaboration of heat shock proteins (Hsp) 70 and 100 in thermotolerance and protein disaggregation. *Proceedings of the National Academy of Sciences of the United States of America, 108*, 6915–6920.

Nakhjavani, M., Morteza, A., Khajeali, L., Esteghamati, A., Khalilzadeh, O., Asgarani, F., & Outeiro, T. F. (2010). Increased serum HSP70 levels are associated with the duration of diabetes. *Cell Stress & Chaperones, 15*, 959–964.

Pierzchalski, P., Jastrzebska, M., Link-Lenczowski, P., Leja-Szpak, A., Bonior, J., Jaworek, J., Okon, K., & Wojcik, P. (2014). The dynamics of heat shock system activation in Monomac-6 cells upon Helicobacter pylori infection. *Journal of Physiology and Pharmacology, 65*, 791–800.

Qiu, X. B., Shao, Y. M., Miao, S., & Wang, L. (2006). The diversity of the DnaJ/Hsp40 family, the crucial partners for Hsp70 chaperones. *Cellular and Molecular Life Sciences, 63*, 2560–2570.

Qu, B., Jia, Y., Liu, Y., Wang, H., Ren, G., & Wang, H. (2015a). The detection and role of heat shock protein 70 in various nondisease conditions and disease conditions: A literature review. *Cell Stress & Chaperones, 20*, 885–892.

Qu, B. G., Wang, H., Jia, Y. G., Su, J. L., Wang, Z. D., Wang, Y. F., Han, X. H., Liu, Y. X., Pan, J. D., & Ren, G. Y. (2015b). Changes in tumor necrosis factor-alpha, heat shock protein 70, malondialdehyde, and superoxide dismutase in patients with different severities of alcoholic fatty liver disease: A prospective observational study. *Medicine (Baltimore), 94*, e643.

Radons, J. (2016). The human HSP70 family of chaperones: Where do we stand? *Cell Stress & Chaperones, 21*, 379–404.

Ritossa, F. (1962). A new puffing pattern induced by temperature and DNP in Drosophila. *Experientia, 18*, 571–573.

Singh, K., Agrawal, N. K., Gupta, S. K., Mohan, G., Chaturvedi, S., & Singh, K. (2015). Decreased expression of heat shock proteins may lead to compromised wound healing in type 2 diabetes mellitus patients. *Journal of Diabetes and its Complications, 29*, 578–588.

Tissieres, A., Mitchell, H. K., & Tracy, U. M. (1974). Protein synthesis in salivary glands of Drosophila melanogaster: Relation to chromosome puffs. *Journal of Molecular Biology, 84*, 389–398.

Tóth, M. E., Gombos, I., & Sántha, M. (2015). Heat shock proteins and their role in human disease. *Acta Biologica Szegediensis, 59*, 21–141.

Verghese, J., Abrams, J., Wang, Y., & Morano, K. A. (2012). Biology of the heat shock response and protein chaperones: Budding yeast (Saccharomyces cerevisiae) as a model system. *Microbiology and Molecular Biology Reviews, 76*, 115–158.

Vierling, E. (1997). The small heat shock proteins in plants are members of an ancient family of heat induced proteins. *Acta Physiologiae Plantarum, 19*, 539–547.

Whitley, D., Goldberg, S. P., & Jordan, W. D. (1999). Heat shock proteins: A review of the molecular chaperones. *Journal of Vascular Surgery, 29*, 748–751.

Xu, Z. S., Li, Z. Y., Chen, Y., Chen, M., Li, L. C., & Ma, Y. Z. (2012). Heat shock protein 90 in plants: Molecular mechanisms and roles in stress responses. *International Journal of Molecular Sciences, 13*, 15706–15723.

Yaglom, J. A., Gabai, V. L., & Sherman, M. Y. (2007). High levels of heat shock protein Hsp72 in cancer cells suppress default senescence pathways. *Cancer Research, 67*, 2373–2381.

Yang, X., He, H., Yang, W., Song, T., Guo, C., Zheng, X., & Liu, Q. (2010). Effects of HSP70 antisense oligonucleotide on the proliferation and apoptosis of human hepatocellular carcinoma cells. *Journal of Huazhong University of Science and Technology. Medical Sciences, 30*, 337–343.

Yeo, M., Park, H. K., Kim, D. K., Cho, S. W., Kim, Y. S., Cho, S. Y., Paik, Y. K., & Hahm, K. B. (2004). Restoration of heat shock protein 70 suppresses gastric mucosal inducible nitric oxide synthase expression induced by Helicobacter pylori. *Proteomics, 4*, 3335–3342.

Zhao, J. H., Meng, X. L., Zhang, J., Li, Y. L., Li, Y. J., & Fan, Z. M. (2014). Oxygen glucose deprivation post-conditioning protects cortical neurons against oxygen-glucose deprivation injury: Role of HSP70 and inhibition of apoptosis. *Journal of Huazhong University of Science and Technology. Medical Sciences, 34*, 18–22.

Zhu, X., Zhao, X., Burkholder, W. F., Gragerov, A., Ogata, C. M., Gottesman, M. E., & Hendrickson, W. A. (1996). Structural analysis of substrate binding by the molecular chaperone DnaK. *Science, 272*, 1606–1614.

Intranasal Administration of Hsp70: Molecular and Therapeutic Consequences

Michael B. Evgen'ev, David G. Garbuz, Alexei V. Morozov, and Natalia V. Bobkova

Abstract Hsp70 and other molecular chaperones function as a complex neuroprotective system, which fails in the brains of aged people and Alzheimer's disease (AD)-type neuropathologies. It was demonstrated that intranasally injected exogenous Hsp70 (eHsp70) effectively bypassed the blood-brain barrier and penetrates brain regions of the model animals. It was shown that chronic administration of eHsp70 decreases beta-amyloid level and the number of Aβ-plaques in two mouse models of AD. In both cases eHsp70 restored learning and memory parameters as well as functional state of neurons. Characteristically, eHsp70 treatment increased synaptophysin level and protects neurons in brain areas most affected in AD patients such as hippocampus and neocortex. It was also demonstrated that eHsp70 can promote longevity and life quality in male mice. The eHsp70 treatment decreased accumulation of aging marker lipofuscin and modulates the activity of UPS by increasing expression of several proteasome subunits including immunoproteasome subunit β5i. Deep sequencing studies exploring brain regions of AD-model – 5XFAD mice treated with eHsp70 revealed candidate genes and signal pathways probably underlying beneficial effects of eHsp70 treatment. Taken together, our findings establish intranasal administration of exogenous human Hsp70 as a practical therapeutic approach for the treatment of various neurodegenerative diseases and aging.

Keywords Aging · Deep sequencing · Hsp70 · Intranasal administration · Mouse model of AD

M. B. Evgen'ev (✉)
Engelhardt Institute of Molecular Biology, Moscow, Russia

Institute of Cell Biophysics, RAS, Moscow, Russia

D. G. Garbuz · A. V. Morozov
Engelhardt Institute of Molecular Biology, Moscow, Russia

N. V. Bobkova
Institute of Cell Biophysics, RAS, Moscow, Russia

A. A. A. Asea, P. Kaur (eds.), *HSP70 in Human Diseases and Disorders*, Heat Shock Proteins 14, https://doi.org/10.1007/978-3-319-89551-2_16

Abbreviations

AD	Alzheimer's disease
APP	Amyloid precursor protein
BAG	Bcl-2-associated athanogene-1
BBB	Blood-brain barrier
BSA	Bovine serum albumin
CHIP	Carboxy-terminus of Hsc70-interacting protein
CNS	Central nervous system
Hsp	Heat shock protein
LPS	Lipopolysaccharide
LTA	Lipoteichoic acid
NMDA	N-methyl-D-aspartate receptor
PTZ	Pentylenetetrazole
ROS	Reactive oxygen species
TLR	Toll-like receptor
TNF	Tumor necrosis factor
UPS	Ubiquitin-proteasome system

Introduction

Heat shock protein 70, encoded by the HSPA1A gene in humans functioning in cooperation with co-chaperones, is a key component of the machinery protecting cells from various pathological and stress conditions, including inflammation and degeneration (Asea and Brown 2008; Duncan et al. 2015; Radons 2016). Briefly, Hsp70 binds partially unfolded or damaged proteins and either assists in their refolding or directs them to a safe disposal (Morimoto 2011; Duncan et al. 2015). Hsp70 also has multiple additional functions, including acting as regulatory or cytokine-like molecules (Calderwood et al. 2016; Fleshner and Johnson 2005; Multhoff and Hightower 2011; Radons 2016). The activity of endogenous Hsp70 appears insufficient in many pathological states, notably in various neurodegenerative disorders and in aging (Calderwood and Murshid 2017; Murshid et al. 2013; Radons 2016). In other words Hsp70 and other chaperones are apparently responsible for proteins homeostasis supporting physiological status of various cells including CNS (Morimoto 2011; Morimoto and Cuervo 2014). Normally, Hsp70 is a cytoplasmic protein. However, it may exit the cell via unusual not properly described mechanisms, different from regular secretory processes. Frequently, Hsp70 exits the cell under stress conditions such as heat shock or comes from transformed malignant or damaged cells. The usefulness of exogenously produced or artificially induced Hsp70 has been demonstrated by our and other groups exploring several animal model in vivo and in vitro systems associated with inflammation and formation of reactive oxygen species (ROS) (Bobkova et al. 2014, 2015; Ekimova et al. 2010; Hoshino et al. 2011; Margulis et al. 2006; Vinokurov et al. 2012).

Indeed in recent decades exogenous Hsp70 (eHsp70) had emerged as a critical regulator of path-ways associated with a wide spectrum of neurological and inflammatory diseases (Guzhova and Margulis 2006; Ekimova et al. 2010; Gifondorwa et al. 2008; Hoshino et al. 2011; Johnson and Fleshner, 2006; Kirkegaard et al. 2016). Thus, it was demonstrated that eHsp70 significantly decreased the mortality in a rat model of endotoxin shock (Kustanova et al. 2006). Furthermore, the exogenous Hsp70 reduces the generation of ROS by neutrophils under the influence of LPS and LTA (Vinokurov et al. 2012; Yurinskaya et al. 2016, 2017). There are various lines of evidence demonstrating that exogenous Hsp70 may interact with the cells using TLR4, TLR2 or scavenger receptors (Asea et al. 2002; Gong et al. 2009). It was also shown that Hsp70 may decrease the level of TLR4 mRNA induced by LPS presence (Yurinskaya et al. 2016). Based on all these data, eHsp70 is considered as a promising drug that can be used to protect cells and organisms as a whole from the action of bacterial pathogens of different origin. Because of the ability of the members of the heat shock protein 70 (Hsp70) family to protect pathologically challenged cells, HSP70-based therapies are rapidly developing as highly promising treatments for many pathologies including neurodegeneration, inflammation etc. (Kirkegaard et al. 2016). An important aspect of therapeutic application of Hsp is to do this in a safe and non-invasive way, particularly for ageing people and patients with neurodegenerative diseases such as AD. In view of the demonstrated vital role of various Hsp and in particular Hsp70 in protecting a cell and organism as a whole (Evgen'ev et al. 2014) against various challenges and pathological states multiple attempts have been performed to either induce endogenous Hsp70 or introduce the recombinant protein for therapy (Abkin et al. 2016; Cudkowicz et al. 2008; Gehrig et al. 2012; Jinwal et al. 2010; Duncan et al. 2015; Pratt et al. 2015; Hoshino et al. 2013; Shevtsov et al. 2015). It is of note that samples of recombinant Hsp70 used in these studies were produced in various expression systems including Cos and Spodoptera cells, *E. coli*, yeast, milk of transgenic animals etc. (Ekimova et al. 2010; Gurskiy et al. 2016; Zheng et al. 2010; Kustanova et al. 2006; Kirkegaard et al. 2016).

Different Routes of HSP70 Administration

In our original studies involving recombinant Hsp70 and Hsp70 isolated from muscles we used intravenous injection of the protein, which reduced LPS-induced mortality of rats and ameliorated other manifestations of sepsis (Kustanova et al. 2006). It was demonstrated that under normal conditions concentration of endogenous Hsp70 in the animal blood of rats is very low and while intravenous administration of eHsp70 significantly increased the original concentration of Hsp70, probably more than 90% of injected protein is eliminated from the blood within the first 20 min after injection. Later, Dr. Margulis has obtained similar data on Hsp70 kinetics in rats subjected to moderate hypoxia: Hsp70 appeared in blood flow within 30 min after the treatment and virtually disappeared during the next 15–20 min (personal communications). Rapid disappearance of introduced Hsp70 from the blood flow has been also reported by other authors (e.g. Fleshner and Johnson 2005).

Other modes of eHsp70 injections also resulted in positive therapeutic outcomes. Thus, intraperitoneal injections of recombinant eHsp70 in the transgenic model of amyotrophic lateral sclerosis decreased the level of motor neuron death and increased the life-span. Furthermore, intracerebroventicular microinjections of purified Hsp70/Hsc70 isolated from muscle tissue resulted in the reduced severity of generalized seizures induced by NMDA or PTZ in epilepsy model (Ekimova et al. 2010). Recently, intravenous and intraperitoneal injections of recombinant Hsp70 gave positive results applied in several lysosomal storage disease models. eHsp70 penetrated effectively murine tissues and the treatment improved motor neurons function and life span of the model animals (Kirkegaard et al. 2016). Furthermore, the ability of Hsp70 to penetrate inside a living cell after administration prompted to develop a non-invasive method for the treatment of surface tumors. Margulis' group designed hydrogel-containing gel with human recombinant Hsp70 and demonstrated that based on histochemical and biochemical data Hsp70 diffused through the skin layer inside the melanoma. The application of Hsp70-containing gel reduced the rate of tumor growth and significantly prolonged the life of animals. Taken together, the data confirm the anti-tumor effect of pure recombinant Hsp70 delivered intratumorally and demonstrate the relevance of this non-invasive technology of Hsp70-based therapy (Abkin et al. 2013).

Multiple successful attempts have been performed to introduce recombinant Hsp70 into the cells using artificially produced containers of different size from nanoparticles to microcapsules (Gendelman et al. 2015; Shevtsov et al. 2016; Yurinskaya et al. 2016). Specifically, these findings demonstrated that targeted nanoparticles containing Hsp70 or other proteins represent an interesting and very promising platform for noninvasive therapy of many neurodegenerative or inflammatory diseases (Agulla et al. 2014; Gendelman et al. 2015). Along these lines, we experimentally inserted human Hsp70 into polyelectrolyte microcapsules and demonstrated that encapsulated Hsp70 reduced LPS-induced tumor necrosis factor alpha and LPS-induced reactive oxygen species production by neutrophils in the course of its release from the microcapsules. Thus, such microcapsules can be used as containers for the effective delivery of recombinant Hsp70 up to neutrophils and monocytes to improve the efficacy of the innate immunity system (Yurinskaya et al. 2017). Subchronic intranasal administration represents another promising noninvasive approach to deliver proteins and small- and large-sized drugs into CNS of model animals bypassing the BBB (Falcone et al. 2014; Hanson and Frey II 2008; Ying 2008). Thus, it was demonstrated that labeled bovine serum albumin (125-I) was present throughout the brain within 5 min after intranasal injection. The authors conclude that injected [(125)I] BSA reaches all parts of the brain through a dose-dependent mechanism that may involve fluid-phase transcytosis (Falcone et al. 2014).

The Fate of Intranasally Injected HSP70

All our experiments exploring the effects of intranasally injected human Hsp70 stemmed from the original observation that the surgical damage of the olfactory bulbs (olfactory bulbectomy, OBX) in mice triggers the development of several pathologies, including amyloid-β accumulation and strong decrease of neuron density in the cortex and hippocampus as well as significant cognitive disturbances and, hence, OBX mice represent valid model of AD-like neurodegeneration (Bobkova et al. 2005, 2008, 2013). Furthermore, these harmful consequences of the olfactory bulbectomy have maximal manifestation in periods of 1–1.5 months and 8 months after the surgery and, hence, exhibit biphasic pattern with almost complete recovery period taking place at 5–6 months after the operation. Surprisingly, the quantitative determination of endogenous inducible form of Hsp70 in different brain areas of OBX mice demonstrated maximal induction of Hsp70 synthesis in the hippocampus with clear-cut coincidence with the recovery period in OBX animals (Bobkova et al. 2013). The observed correlation enables to suggest curing effect of endogenous Hsp70 synthesis at certain periods of this pathology development and establishes it as a possible therapeutic agent for consequences of bulbectomy, such as AD-like degeneration and cognitive disturbances observed in the OBX mice at certain time points after the operation.

However, at the first stage it was necessary to find out whether exogenous recombinant human Hsp70 can penetrate brain after intranasal injections. The performed experiments exploring Hsp70 conjugated with Alexa Fluor 647 demonstrated that recombinant human Hsp70 after intranasal administration can penetrate various brain regions of mice Fig. 1. Three hours after the injection fluorescently-labeled Hsp70 was detected in the olfactory bulbs, neocortex, hippocampus, n.raphe dorsalis, locus coeruleus, and cerebellum in intact (non operated) and OBX mice (Bobkova et al. 2014). In most cases, it was concentrated in the perinuclear zone (e.g. Fig. 1b, c). Furthermore, the investigation of the fate of iodinated eHsp70 in the hippocampus and cortex after intranasal administration demonstrated that while the labeled protein is found in these areas in its native form after injection it undergoes degradation with kinetics similar to that of labeled albumin (BSA). The degradation of the introduced recombinant eHSP70 after entering the cells is attenuated in the presence of proteasome inhibitor and varies significantly depending on the cells type and origin (Yurinskaya et al. 2015). Precise mechanisms providing HSP70 transport are not known and may involve simple diffusion, neuron terminals, bulk flow, or involvement lymphatic channels directly connected to brain areas (Ying 2008; Falcone et al. 2014). The mechanism of Hsp70 transport after intranasal injection likely requires an as yet unidentified cellular transport machinery specific to Hsp70. It is noteworthy that labeled Hsp70 injected directly into the brain of rats is localized in essentially the same brain regions (Ekimova et al. 2010).

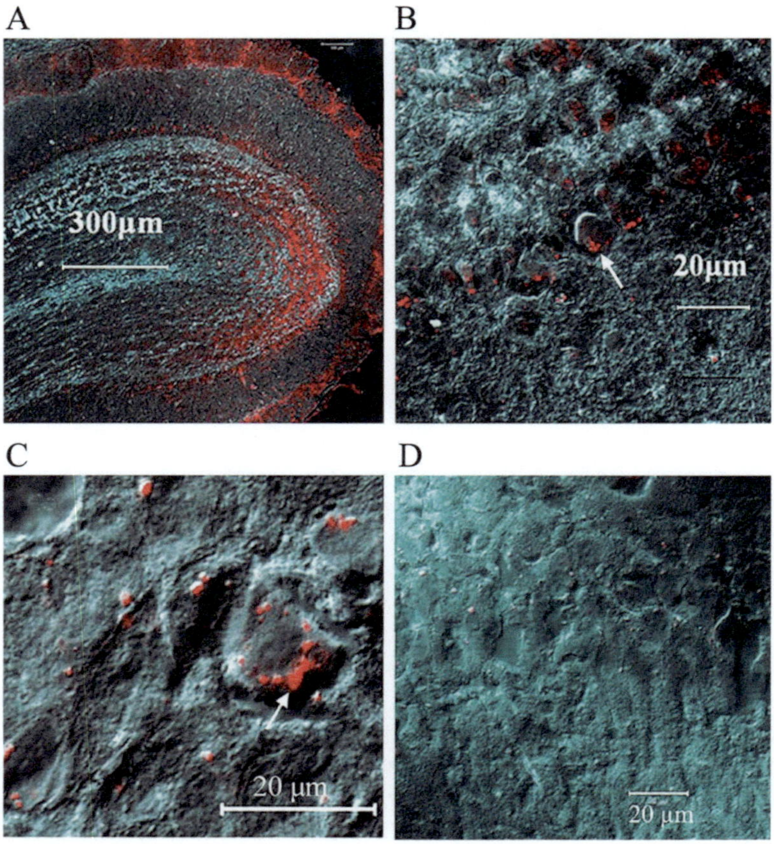

Fig. 1 Localization of exogenous Hsp70 in different brain regions of control mice after intranasal administration. The images show Hsp70 fluorescently labeled with the Alexafluor 647 dye (red) in the brain sections of NMRI mice 3 h after intranasal injection. Hsp70 is distributed non-randomly and concentrated in a few brain regions. Photos show its localization in the olfactory bulbs (**a** and **b**) and areas of the hippocampus (**c**). In most cases Hsp70 has an intracellular localization (e.g. **b**) and is restricted to the perinuclear zone. (**d**) Control mice treated with unlabeled Hsp70. Scale bars A – 300 µm; B, C, D – 20 µm. (Reprinted from "Therapeutic Effect of Exogenous Hsp70 in Mouse Models of Alzheimer's Disease" by Bobkova et al. 2014 Copyright (2014), with permission from IOS Press)

The Effect of Chronic HSP70 Intranasal Administration in Two Mouse Models of AD-Type Neurodegeneration and in the Course of Normal Aging

Although a large body of evidences indicates a neuroprotective role of chaperones and especially, Hsp70 (Asea and Brown 2008; Franklin et al. 2005; Hoshino et al. 2011; Pratt et al. 2015) there has been no direct demonstration that any of them could either delay the progression of AD or exert any therapeutic effect. The above-mentioned data accumulated by our group and other studies promoted us to check whether the intranasal administration of human recombinant Hsp70 results in some beneficial effects at molecular and cognitive levels in two complementary mouse models of AD-like neurodegeneration. Therefore, we investigated in detail the consequences of recombinant human Hsp70 sub-chronic administration exploring above mentioned bilateral olfactory bulbectomy model (OBX mice) and AD mouse model of transgenic 5XFAD mice.

Importantly, OBX mice exhibit cognitive, morphological, and biochemical characteristics very similar to those described for AD patients (Bobkova et al. 2005, 2008, 2013, 2014). Similar to AD patients, OBX mice display an increased level of brain amyloid precursor protein (APP) and Aβ (Bobkova et al. 2014). Most importantly, in contrast to many widely used transgenic mice models of AD (Morrissette et al. 2009), OBX animals are characterized by a substantial loss of neurons in the hippocampus and temporal cortex, i.e. the brain areas most affected in AD individuals. However, in OBX mice due to the differences in the Aβ amino acid sequence with the human orthologue, intracellular accumulation of Aβ in neurons of the cortex does not results in the formation of extracellular amyloid plaques (Bobkova et al. 2014). Therefore, for the sake of comparison we used transgenic 5XFAD mice that represent a widely used conventional model of AD-like neurodegeneration (Morrissette et al. 2009; Oakley et al. 2006). In contrast to OBX animals, 5XFAD mice are characterized by a high level of amyloid plaques that appear early in their lifespan in different brain regions. These transgenic mice co-express three mutations in human APP, and two mutations in the presenelin gene PS1, with expression of both transgenes driven by Thy1 promoter (Oakley et al. 2006).

We show that intranasally administrated Hsp70 mitigates multiple AD-like morphological and cognitive abnormalities observed in both model animals Fig. 2. In particular, in both cases it normalizes the density of neurons and decreases the manifestations of pathology in neurons such as pyknosis, karyolysis etc. in the hippocampus and neocortex which correlates with the diminished accumulation of amyloid β(Aβ) peptide and, in the case of 5XFAD mice, reduction of Aβ-plaque formation in all brain regions Fig. 3. In particular, significant protection was evident in the temporal cortex and two areas of the hippocampus (CA1 and CA3), the regions predominantly responsible for cognitive abilities strongly disturbed in both AD-models. Furthermore, Hsp70 sub-chronic treatment efficiently protects spatial memory in OBX and 5XFAD mice. Fig. 2a, b illustrates the effect of Hsp70 on spatial memory as judged by the time spent in the target sector of Morris water maze in comparison to other indifferent sectors. Our results show that the applied eHsp70 treatment is highly effective in ameliorating all major manifestations of AD-type neurodegeneration developed in OBX and 5XFAD animals at histological and cognitive levels.

It is well known that in animals and humans Hsp70 activity and levels in neuronal tissues significantly decrease with age (Morimoto 2011; Murshid et al. 2013). To further characterize the protective role of eHsp70 at the cellular and whole organism levels we performed detailed analysis of eHsp70 intranasal administration consequences exploring the model of normal aging in mice. The performed

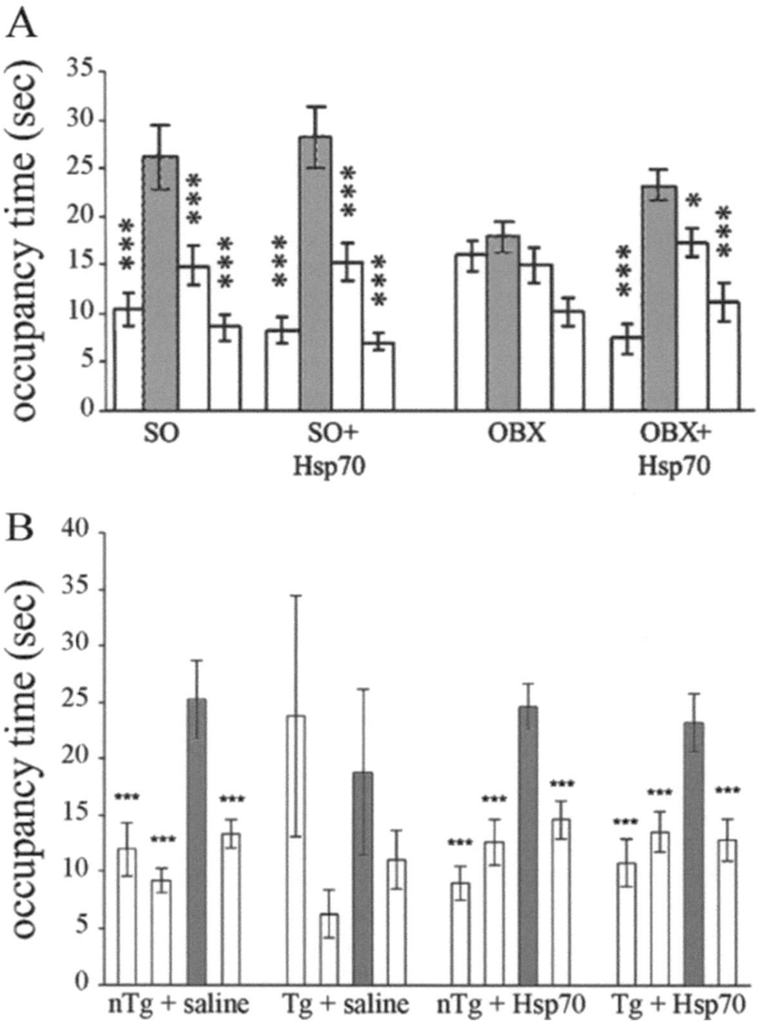

Fig. 2 Hsp70 improves memory in OBX and 5XFAD mice. The protective effects of sub-chronic intranasal administration of eHsp70 on spatial memory of OBX (**a**) and 5XFAD (**b**) mice. Results are presented as post-hoc data of ANOVA analysis. Hatched bars represent time (in seconds) spent by different groups of mice in the target sectors during probe trial (mean ± SEM); other bars represent time spent in indifferent sectors. (**a**) Probe trials with independent groups of OBX and sham-operated (SO) animals were performed five weeks after bulbectomy and after Hsp70 treatment for 3 weeks. (**b**) The effect of Hsp70 administration (3 weeks) on spatial memory of transgenic (Tg) 5XFAD mice (3 months age). Asterisks indicate significant differences (*$p < 0.05$; **$p < 0.01$; ***$p < 0.001$) in comparison with the target quadrant. (Reprinted from "Therapeutic Effect of Exogenous Hsp70 in Mouse Models of Alzheimer's Disease" by Bobkova et al. 2014 Copyright (2014), with permission from IOS Press)

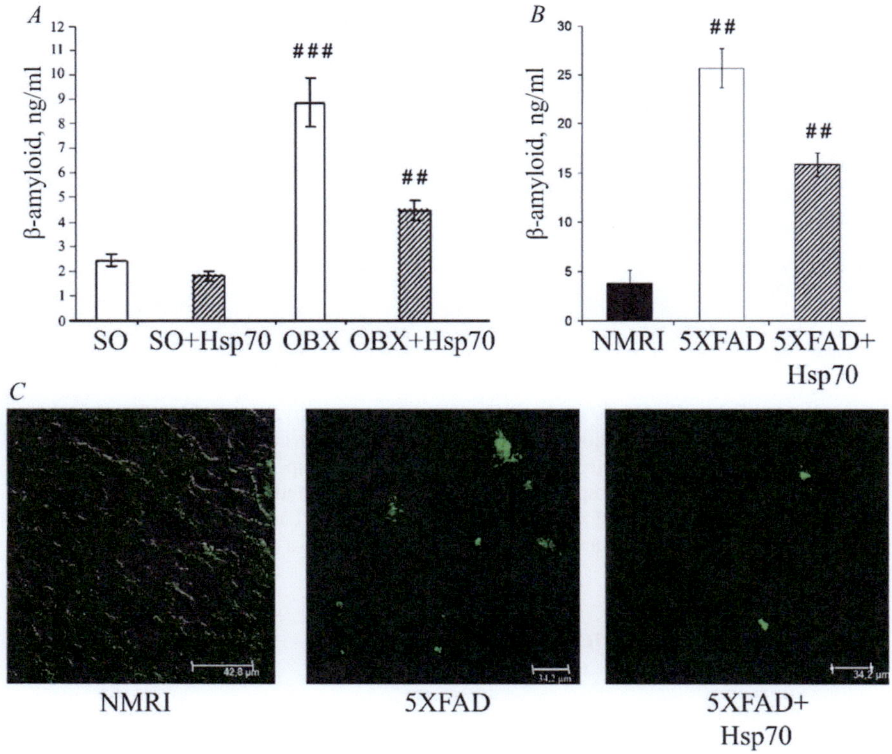

Fig. 3 Exogenous Hsp70 protects the brain of OBX and 5XFAD mice from Aβ accumulation and amyloid plaque formation. (**a** and **b**) OBX and 5XFAD mice treated with intranasal Hsp70 are characterized by decreased level of Aβ in the brain. Bars indicate the level of Aβ peptides (mean ± SEM) determined by ELISA in the brain tissue (cortex + hippocampus) in the groups of OBX and 5XFAD mice (n = 6–11 animals per group). Differences between the levels of Aβ in the groups were determined by using the two-tailed Student's t-test. ($^{\#}p < 0.05$; $^{\#\#}p < 0.01$; $^{\#\#\#}p < 0.001$) (**c**) The representative microphotographs of neocortical slices in control NMRI mice and 5XFAD mice exhibiting Aβ plaques treated and non-treated with Hsp70. (Reprinted from "Therapeutic Effect of Exogenous Hsp70 in Mouse Models of Alzheimer's Disease" by Bobkova et al. 2014 Copyright (2014), with permission from IOS Press)

experiments exploring NMRI wild type mice demonstrated that chronic administration of eHsp70 significantly enhanced the lifespan Fig. 4 and improves life quality of animals of different age groups after treatment for five and nine months Fig. 4. Interestingly, long-term (chronic) introduction of HSP70 apparently did not significantly affect the longevity of females while increased the life span of males by about 20% (Bobkova et al. 2015). Taking into account the observed sex-dependent differences in the response to HSP70 treatment in all future experiments on the effects of HSP70 on aged animals we used only males. The introduced eHsp70 improved learning and memory in middle-age and old mice. Furthermore, Hsp70 treatment demonstrated a significant therapeutic effect on locomotor and exploratory activities, significantly decreased in old mice.

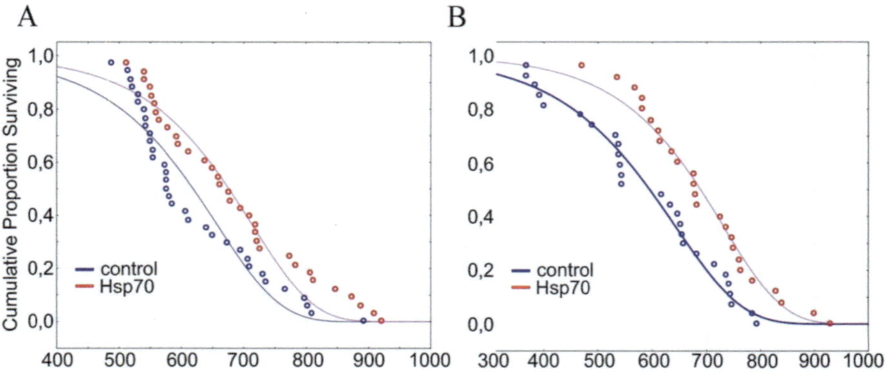

Fig. 4 Effect of chronic eHsp70 treatment on longevity and survival rates of NMRI mice belonging to different age groups. (**a**) Longevity of old mice treated with eHsp70 at the age of 17 months until animal's death. (**b**) Longevity of middle-aged mice when eHsp70 treatment starts at 12 months and lasted until animals death. Importantly, the treatment significantly increased not only longevity but survival rates in both tested age groups. (Reprinted from: "Exogenous Hsp70 delays senescence and improves cognitive function in aging mice" by Bobkova et al. 2015 with permission)

Effects of HSP70 at the Molecular Level

Synaptophysin is a predominant component of synaptic vesicles and plays a major role in synaptic vesicle trafficking in the brain (Bertoni-Freddari et al. 1996; Tonnies and Trushina 2017). The level of synaptophysin directly resembles the state of synaptic machinery in the brain. Measurements of synaptophysin demonstrated that eHsp70 treatment in old mice resulted in larger synaptophysin-immunopositive areas compared with control animals of the same age Fig. 5a. These results corroborated the data of another group which demonstrated that Hsp70 injected intracerebroventicularly co-localized with synaptophysin in brain tissue (Ekimova et al. 2010). Importantly, in our experiments long-term Hsp70 treatment decreased accumulation of lipofuscin, an aging-related marker, in the brain of aged animals Fig. 5b. Apparently, the demonstrated potential of eHsp70 intranasal treatment to protect synaptic machinery in old animals offers a unique pharmacological approach for various neurodegenerative disorders associated with human aging including AD-like pathologies.

Together with oxidative stress and chronic inflammation, aging is associated with decrease in protein synthesis, but an increase in accumulation of altered and abnormal proteins, indicating perturbations in protein degradation as one of the possible reasons (Hipkiss 2006, Rattan 1996). Thus, many reports reveal the decline in amount and/or proteolytic capacity of proteasomes – multisubunit cellular protease complexes, in different aged tissues (Ferrington et al. 2005; Keller et al. 2000; Li et al. 2008; Viteri et al. 2004). In contrast, fibroblast cultures from centenarians display proteasome activities comparable with those in cells of young donors (Chondrogianni et al. 2000). Proteasomes are central elements of the

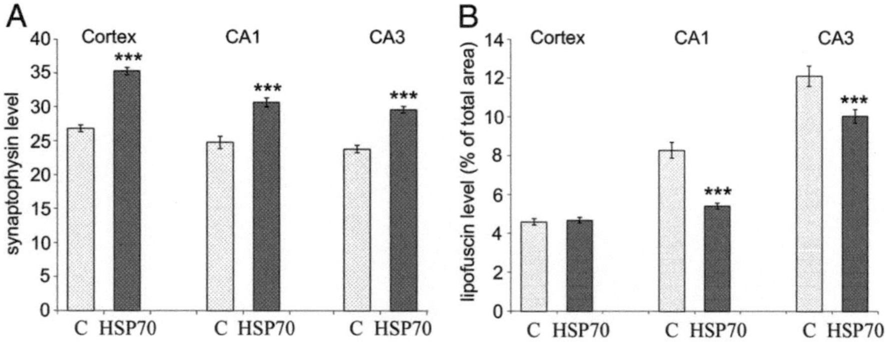

Fig. 5 Effect of eHsp70 intranasal administration on molecular markers of aging (treatment duration-5 months). (**a**) and (**b**) levels of synaptophysin and lipofuscin correspondently measured in the brains of eHsp70-treated and age-matched control untreated NMRI animals. Differences between the levels of synaptophysin or lipofuscin in the control and experimental groups were determined by using the two-tailed Student's t-test. It is evident that treatment increased synaptophysin immunostaining and decreased lipofuscin autofluorescence. (Reprinted from: "Exogenous Hsp70 delays senescence and improves cognitive function in aging mice" by Bobkova et al. 2015 with permission)

ubiquitin-proteasome system (UPS), which degrades most intracellular proteins, hence their functionality determines cellular homeostasis and to a certain extent the rate of aging (Chondrogianni and Gonos 2010; Glickman and Chiechanover 2002). Therefore, proteasome activation may represent a promising strategy to increase life and healthspan (Chondrogianni et al. 2015). Along these lines, increased chymotrypsin-like proteolytic activity of proteasomes was revealed in the frontal cortex of eHsp70-treated aged NMRI mice Fig. 6a (Bobkova et al. 2015). Moreover, in comparison to controls increased expression of several proteasome subunits was observed in brain tissue lysates of animals administrated with eHsp70 Fig. 6b. Interestingly, immunoproteasome subunit β5i, which possesses chymotrypsin-like activity, was elevated by 139% Fig. 6b (Bobkova et al. 2015).

In addition to their role in antigen presentation, immunoproteasomes (iP) (proteasomes with so called "immune" catalytic subunits) are involved in a variety of pathways maintaining cellular homeostasis regulating survival and differentiation (Ferrington and Gregerson 2012). Immunoproteasomes can limit inflammatory damage and were shown to be important for the degradation of oxidized proteins (Pickering et al. 2010; Seifert et al. 2010), accumulation of which is a common feature of aging and neurodegeneration (Hipkiss 2006; Choi et al. 2002). Interestingly, immunoproteasome subunit expression correlated with longevity in primates and rodents (Rodriguez et al. 2012, Pickering et al. 2015). Thus, in eHsp70-treated NMRI mice increased immunoproteasome content may be beneficial and facilitate withstand consequences of age-associated oxidative stress. However, it is not clear which cell type/s have increased proteasome activity and iP content. Immunoproteasome subunit expression may be activated in glia via NF-κB-dependent pathway, which was shown to be stimulated by interaction of

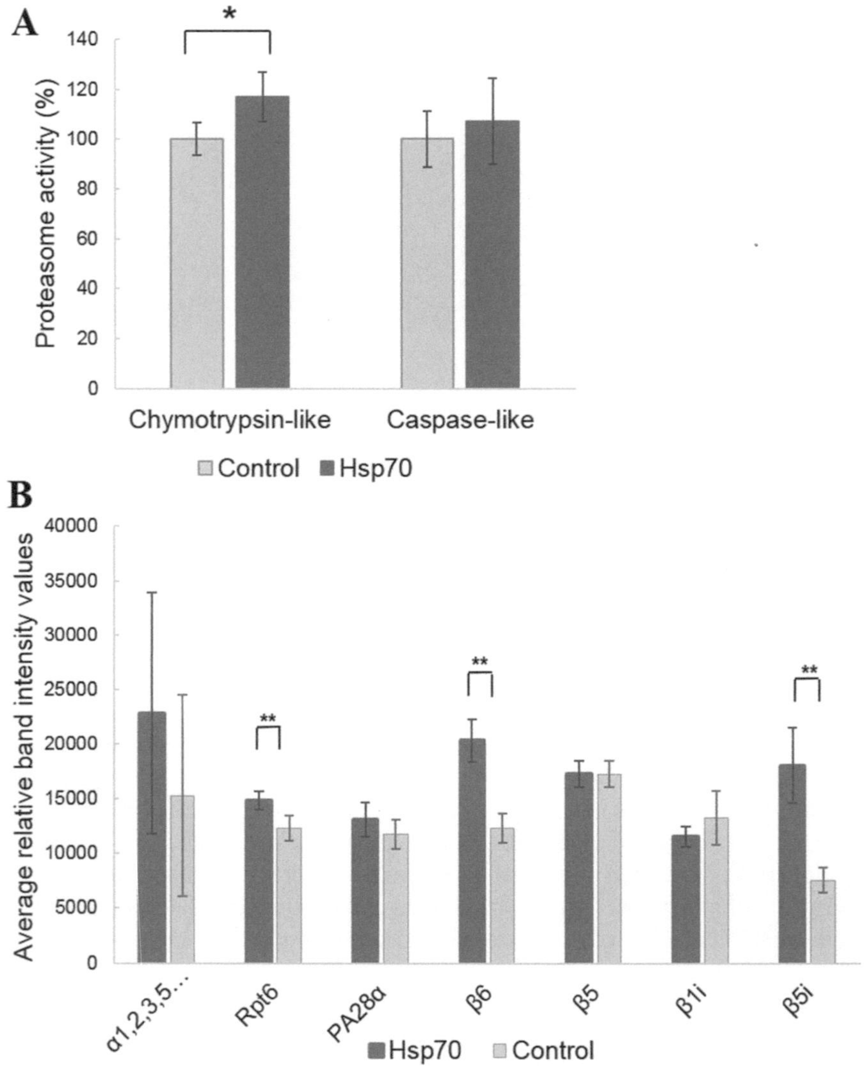

Fig. 6 Chymotrypsin-like proteasome activity and β5i subunit expression in the cerebral cortex of eHsp70-treated aged NMRI mice. (**a**) Proteolytic activities of proteasomes in the cerebral cortex lysates of experimental mice. Average relative activity levels in tissue lysate of eHsp70 (eight animals) and control (five animals) groups are shown. Chymotrypsin-like and caspase-like activities in control group were considered as 100%. Chymotrypsin-like activity is increased by 17% in tissue lysates of eHsp70-treated mice. (**b**) Expression of proteasome subunits in the cortex lysates. Columns represent average (in group) normalized specific protein band intensity values obtained after quantification of revealed Western blots using ImageJ software. Error bars represent standard deviation of the mean. Significance tested using Student's t-test (*$p < 0.05$; **$p < 0.01$). (Reprinted from: "Exogenous Hsp70 delays senescence and improves cognitive function in aging mice" by Bobkova et al. 2015 with permission)

eHsp70 with toll-like receptors (Asea et al. 2002; Ferrington and Gregerson 2012; Gong et al. 2009; Wright et al. 1995). However, other possible effects of eHsp70 on proteasome system should not be ruled out. It is known that with aging cells became less capable to upregulate Hsp70 synthesis in response to oxidative stress (Soti and Csermely 2002). eHsp70 was shown to efficiently penetrate cells (Guzhova et al. 2001; Bobkova et al. 2014; Yurinskaya et al. 2015). Therefore, constantly "imported" eHsp70 can perform functions of endogenous protein and modulate proteasome activity by refolding or preserving damaged proteins in soluble form and thus, reducing amounts of toxic protein aggregates, which can otherwise negatively influence proteasome functionality (Bence et al. 2001). In addition, Hsp70 can stimulate damaged protein turnover. Grune et al. demonstrated that immediately after mild oxidative stress Hsp70 participates in rearrangement of cellular proteasome pool in order to mediate degradation of damaged proteins (Grune et al. 2011; Pickering et al. 2010). Hsp70 can also support ubiquitin-dependent degradation of substrates via association with co-chaperone/E3 ubiquitin ligase CHIP and BAG-1 protein (Kastle and Grune 2012). Moreover, recent findings indicate that Hsp70 directly assists ubiquitin-independent degradation of oxidized proteins by the proteasomes (Reeg et al. 2016). More recently Hsp70 was shown to directly influence the activity of proteasomes in vitro (Morozov et al. 2017). As indicated in the previous section, in hipocampii of eHsp70 treated mice, decreased amounts of lipofuscin were observed Fig. 5b (Bobkova et al. 2015); lipofuscin can inhibit proteasomes (Sitte et al. 2000), and, hence, eHsp70 which induced reduction of lipofuscin can also positively influence proteasome functionality. Taken together, eHsp70 may potentially influence UPS functional state through several pathways. Long-term administration of Hsp70 can interfere with aging-associated accumulation of damaged proteins and facilitate reduction of oxidative stress, partially explaining beneficial effects of eHsp70 treatment. Considering presented data, the influence of eHsp70 administration on cellular proteasome pool deserves further investigation.

At the next step, we decided to perform deep sequencing experiments to monitor the effects of eHsp70 intranasal administration on transcriptome because the specific signaling pathways activated by such treatment are unknown. The analysis of transcriptome of transgenic and non-transgenic 5XFAD mice revealed significant changes in their hippocampus after long-term intranasal administration of eHsp70 Fig. 7. The changes represented up-regulation of many genes responsible for transmission of nerve impulse, including the dopamine biosynthetic process, neuropeptide signalling pathway, neurotransmitter transport, amine transport, G-protein coupled receptor signalling pathways and other pathways impaired in transgenic mice Fig. 7. These data corroborate our above-mentioned results indicting the beneficial effect of Hsp70 on the synaptophysin levels in the brains of aged mice. In addition, our transcriptomic studies revealed significant up-regulation of genes involved in the glutathione metabolic process. Glutathione is a well-known antioxidant and provides intracellular protection against free radicals such as

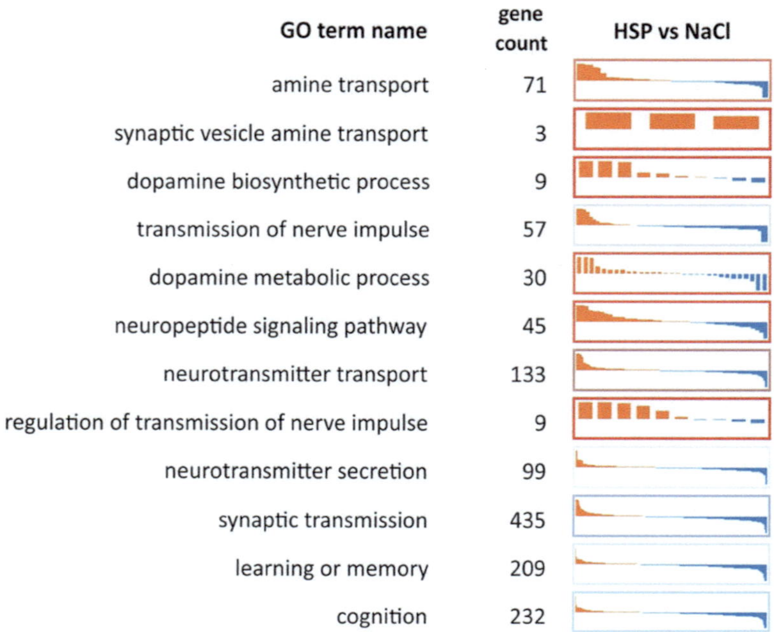

Fig. 7 Differentially expressed genes (DEGs) in the hippocampus of the compared groups of 5XFAD mice (eHsp70 vs. NaCl). Expression profiles of genes participating in various biological processes (Gene Ontology terms). For each term, genes are collected (irrespective of p-value) and sorted by decreasing LogFC. Overexpression is marked with red, while downregulation is marked in blue. Colour frames indicate GSEA *p*-values across the following statistically significantly enriched GO terms: top 40, 80, 250, 500 and 1000 up-regulated (red frames) or down-regulated (blue frames, see border colour legend). (Reprinted from: "Molecular mechanisms underlying neuroprotective effect of eHsp70 in transgenic 5XFAD mice" by Evgen'ev MB et al. Journal of Alzheimer's Desease 2017 (59) 1415–1426 Copyright (2014), with permission from IOS Press)

ROS. The level of glutathione is decreased in the brain of aged animals which results in increased oxidative stress representing one of the major factors leading to AD-type neuropathology (Zhang et al. 2012). Previously, using various in vitro systems and different cell lines we demonstrated that recombinant human eHsp70 efficiently decreases endotoxin-induced production of ROS and TNFα (Vinokurov et al. 2012; Yurinskaya et al. 2016; 2017). At the present time we are not able to state what are the primary upstream targets activated by eHsp70 and define the precise mechanism of its neuroprotective action. It is clear, however, that intranasal eHsp70 administration apparently influences multiple pathways involved in the regulation of transmission of nerve impulse, synaptic function, UPS and activates genes inhibiting ROS production and neuroinflammation common for AD-like pathologies and normal aging.

Conclusions

Taken together, the present study corroborates the data accumulated by many other groups on the neuroprotective action of eHsp70 and emphasized a new non-invasive way of Hsp70 delivery into the brain. Intranasally administrated recombinant human Hsp70 rapidly penetrated various brain regions of the model mice and exerts multiple beneficial effects including neuronal functional state and density. eHsp70 treatment also improved the major cognitive parameters disturbed in AD-mouse models and aged mice and increased longevity and quality of life in the aged males. Transcriptomic analysis suggests that eHsp70 treatment achieves its effects by inhibiting oxidative stress and suppressing neuroinflammation in Tg animals. Chronic eHsp70 administration, hence, leads to the amendment of various pathologies in the brain of Tg and aged mice associated with increased ROS production, synaptic dysfunction and impaired transmission of nerve impulses observed in various neurodegeneration models. The suggested nasal delivery pathway of recombinant eHsp70 has several obvious advantages for treating neurodegenerative disorders, including convenience of administration and non-invasiveness.

Acknowledgments The work was supported by Russian Science Foundation Grant №14-50-00060 and Program of fundamental research for state academies for 2013-2020 years (N-01201363817).

References

Abkin, S. V., Pankratova, K. M., Komarova, E. Y., Guzhova, I. V., & Margulis, B. A. (2013). Hsp70 chaperone-based gel composition as a novel immunotherapeutic anti-tumor tool. *Cell Stress & Chaperones, 18*(3), 391–396.

Abkin, S. V., Ostroumova, O. S., Komarova, E. Y., et al. (2016). Phloretin increases the anti-tumor efficacy of intratumorally delivered heat-shock protein 70 kDa (HSP70) in a murine model of melanoma. *Cancer Immunology, Immunotherapy, 65*(1), 83–92.

Agulla, J., Brea, D., Argibay, B., et al. (2014). Quick adjustment of imaging tracer payload, for in vivo applications of theranostic nanostructures in the brain. *Nanomedicine, 10*(4), 851–858.

Asea, A., & Brown, J. (Eds.). (2008). *Heat shock proteins and the brain: Implications for neurodegenerative diseases and neuroprotection*. Dordrecht: Springer.

Asea, A., Rehli, M., Kabingu, E., et al. (2002). Novel signal transduction pathway utilized by extracellular HSP70: Role of toll-like receptor (TLR) 2 and TLR4. *The Journal of Biological Chemistry, 277*(17), 15028–15034.

Bence, N. F., Sampat, R. M., & Kopito, R. R. (2001). Impairment of the ubiquitin-proteasome system by protein aggregation. *Science, 292*(5521), 1552–1555.

Bertoni-Freddari, C., Fattoretti, P., Casoli, T., Caselli, U., & Meier-Ruge, W. (1996). Deterioration threshold of synaptic morphology in aging and senile dementia of Alzheimer's type. *Analytical and Quantitative Cytology and Histology, 18*(3), 209–213.

Bobkova, N. V., Nesterova, I. V., Medvinskaya, N. I., et al. (2005). Possible role of the olfactory system in Alzheimer's disease genesis. In A. Fisher, I. Hanin, M. Memo, & F. Stocchi (Eds.), *New trends in Alzheimer and Parkinson disorders: ADPD 2005* (pp. 91–95). Sorento: Medimond.

Bobkova, N., Vorobyov, V., Medvinskaya, N., Aleksandrova, I., & Nesterova, I. (2008). Interhemispheric EEG differences in olfactory bulbectomized rats with different cognitive abilities and brain beta-amyloid levels. *Brain Research, 1232*, 185–194.

Bobkova, N., Guzhova, I., Margulis, B., et al. (2013). Dynamics of endogenous Hsp70 synthesis in the brain of olfactory bulbectomized mice. *Cell Stress & Chaperones, 18*(1), 109–118.

Bobkova, N. V., Garbuz, D. G., Nesterova, I., et al. (2014). Therapeutic effect of exogenous hsp70 in mouse models of Alzheimer's disease. *Journal of Alzheimer's Disease, 38*(2), 425–435.

Bobkova, N. V., Evgen'ev, M., Garbuz, D. G., et al. (2015). Exogenous Hsp70 delays senescence and improves cognitive function in aging mice. *Proceedings of the National Academy of Sciences of the United States of America, 112*(52), 16006–16011.

Calderwood, S. K., & Murshid, A. (2017). Molecular chaperone accumulation in cancer and decrease in Alzheimer's disease: The potential roles of HSF1. *Frontiers in Neuroscience, 11*, 192.

Calderwood, S. K., Gong, J., & Murshid, A. (2016). Extracellular HSPs: The complicated roles of extracellular HSPs in immunity. *Frontiers in Immunology, 7*, 159.

Choi, J., Malakowsky, C. A., Talent, J. M., Conrad, C. C., & Gracy, R. W. (2002). Identification of oxidized plasma proteins in Alzheimer's disease. *Biochemical and Biophysical Research Communications, 293*(5), 1566–1570.

Chondrogianni, N., & Gonos, E. S. (2010). Proteasome function determines cellular homeostasis and the rate of aging. *Advances in Experimental Medicine and Biology, 694*, 38–46.

Chondrogianni, N., Petropoulos, I., Franceschi, C., Friguet, B., & Gonos, E. S. (2000). Fibroblast cultures from healthy centenarians have an active proteasome. *Experimental Gerontology, 35*(6–7), 721–728.

Chondrogianni, N., Voutetakis, K., Kapetanou, M., et al. (2015). Proteasome activation: An innovative promising approach for delaying aging and retarding age-related diseases. *Ageing Research Reviews, 23*(Pt A), 37–55.

Cudkowicz, M. E., Shefner, J. M., Simpson, E., et al. (2008). Arimoclomol at dosages up to 300 mg/day is well tolerated and safe in amyotrophic lateral sclerosis. *Muscle & Nerve, 38*(1), 837–844.

Duncan, E. J., Cheetham, M. E., Chapple, J. P., & van der Spuy, J. (2015). The role of HSP70 and its co-chaperones in protein misfolding, aggregation and disease. *Sub-Cellular Biochemistry, 78*, 243–273.

Ekimova, I. V., Nitsinskaya, L. E., Romanova, I. V., Pastukhov, Y. F., Margulis, B. A., & Guzhova, I. V. (2010). Exogenous protein Hsp70/Hsc70 can penetrate into brain structures and attenuate the severity of chemically-induced seizures. *Journal of Neurochemistry, 115*(4), 1035–1044.

Evgen'Ev, M. B., Garbuz, D. G., & Zatsepina, O. G. (2014). *Heat shock proteins and whole body adaptation to extreme environments* (Vol. XVII, p. 218). Dordrecht: Springer.

Falcone, J. A., Salameh, T. S., Yi, X., et al. (2014). Intranasal administration as a route for drug delivery to the brain: Evidence for a unique pathway for albumin. *The Journal of Pharmacology and Experimental Therapeutics, 351*(1), 54–60.

Ferrington, D. A., & Gregerson, D. S. (2012). Immunoproteasomes: Structure, function, and antigen presentation. In T. Grune (Ed.), *The proteasomal system in aging and disease* (Progress in molecular biology and translational science, Vol. 109, pp. 75–112). Amsterdam: Academic.

Ferrington, D. A., Husom, A. D., & Thompson, L. V. (2005). Altered proteasome structure, function, and oxidation in aged muscle. *The FASEB Journal, 19*(6), 644–646.

Fleshner, M., & Johnson, J. D. (2005). Endogenous extra-cellular heat shock protein 72: Releasing signal(s) and function. *International Journal of Hyperthermia, 21*(5), 457–471.

Franklin, T. B., Krueger-Naug, A. M., Clarke, D. B., Arrigo, A. P., & Currie, R. W. (2005). The role of heat shock proteins Hsp70 and Hsp27 in cellular protection of the central nervous system. *International Journal of Hyperthermia, 21*(5), 379–392.

Gehrig, S. M., van der Poel, C., Sayer, T. A., et al. (2012). Hsp72 preserves muscle function and slows progression of severe muscular dystrophy. *Nature, 484*(7394), 394–398.

Gendelman, H. E., Anantharam, V., Bronich, T., et al. (2015). Nanoneuromedicines for degenerative, inflammatory, and infectious nervous system diseases. *Nanomedicine, 11*(3), 751–767.
Gifondorwa, D. J., Robinson, M. B., Hayes, C. D., et al. (2008). Exogenous delivery of heat shock protein 70 increases lifespan in a mouse model of amyotrophic lateral sclerosis. *The Journal of Neuroscience, 27*(48), 13173–13180.
Glickman, M. H., & Ciechanover, A. (2002). The ubiquitin-proteasome proteolytic pathway: Destruction for the sake of construction. *Physiological Reviews, 82*(2), 373–428.
Gong, J., Zhu, B., Murshid, A., et al. (2009). T cell activation by heat shock protein 70 vaccine requires TLR signaling and scavenger receptor expressed by endothelial cells-1. *Journal of Immunology, 183*(5), 3092–3098.
Grune, T., Catalgol, B., Licht, A., et al. (2011). HSP70 mediates dissociation and reassociation of the 26S proteasome during adaptation to oxidative stress. *Free Radical Biology & Medicine, 51*(7), 1355–1364.
Gurskiy, Y. G., Garbuz, D. G., Soshnikova, N. V., et al. (2016). The development of modified human Hsp70 (HSPA1A) and its production in the milk of transgenic mice. *Cell Stress & Chaperones, 21*(6), 1055–1064.
Guzhova, I., Kislyakova, K., Moskaliova, O., et al. (2001). In vitro studies show that Hsp70 can be released by glia and that exogenous Hsp70 can enhance neuronal stress tolerance. *Brain Research, 914*(1–2), 66–73.
Guzhova, I., & Margulis, B. (2006). Hsp70 chaperone as a survival factor in cell pathology. *International Review of Cytology, 254*, 101–149.
Hanson, L. R., & Frey, W. H., 2nd. (2008). Intranasal delivery bypasses the blood-brain barrier to target therapeutic agents to the central nervous system and treat neurodegenerative disease. *BMC Neuroscience, 9*(Suppl. 3), S5.
Hipkiss, A. R. (2006). Accumulation of altered proteins and ageing: Causes and effects. *Experimental Gerontology, 41*(5), 464–473.
Hoshino, T., Murao, N., Namba, T., et al. (2011). Suppression of Alzheimer's disease-related phenotypes by expression of heat shock protein 70 in mice. *The Journal of Neuroscience, 31*(14), 5225–5234.
Hoshino, T., Suzuki, K., Matsushima, T., Yamakawa, N., Suzuki, T., & Mizushima, T. (2013). Suppression of Alzheimer's disease-related phenotypes by geranylgeranylacetone in mice. *PLoS One, 8*(10), e76306.
Jinwal, U. K., Koren, J., O'Leary, J. C., Jones, J. R., Abisambra, J. F., & Dickey, C. A. (2010). Hsp70 ATPase modulators as therapeutics for Alzheimer's and other neurodegenerative diseases. *Molecular and Cellular Pharmacology, 2*(2), 43–46.
Johnson, J. D., & Fleshner, M. (2006). Releasing signals, secretory pathways, and immune function of endogenous extracellular heat shock protein 72. *Journal of Leukocyte Biology, 79*(3), 425–434.
Kästle, M., & Grune, T. (2012). Interactions of the proteasomal system with chaperones: Protein triage and protein quality control. In T. Grune (Ed.), *The proteasomal system in aging and disease* (Progress in molecular biology and translational science, Vol. 109, pp. 113–160). Amsterdam: Academic.
Keller, J. N., Huang, F. F., & Markesbery, W. R. (2000). Decreased levels of proteasome activity and proteasome expression in aging spinal cord. *Neuroscience, 98*(1), 149–156.
Kirkegaard, T., Gray, J., Priestman, D. A., et al. (2016). Heat shock protein-based therapy as a potential candidate for treating the sphingolipidoses. *Science Translational Medicine, 8*(355), 355ra118.
Kustanova, G. A., Murashev, A. N., Karpov, V. L., et al. (2006). Exogenous heat shock protein 70 mediates sepsis manifestations and decreases the mortality rate in rats. *Cell Stress & Chaperones, 11*(3), 276–286.
Li, Y., Wang, Y. S., Shen, X. F., et al. (2008). Alterations of activity and intracellular distribution of the 20S proteasome in ageing retinal pigment epithelial cells. *Experimental Gerontology, 43*(12), 1114–1122.

Margulis, B., Kinev, A., & Guzhova, I. (2006). Chaperones in neurodegenerative pathologies: Therapeutic prospects. In J. Radons & G. Multhoff (Eds.), *Heat shock proteins in biology and medicine* (pp. 305–329). Trivandrum: Research Singpost.

Morimoto, R. I. (2011). The heat shock response: Systems biology of proteotoxic stress in aging and disease. *Cold Spring Harbor Symposia on Quantitative Biology, 76*, 91–99.

Morimoto, R. I., & Cuervo, A. M. (2014). Proteostasis and the aging proteome in health and disease. *The Journals of Gerontology. Series A, Biological Sciences and Medical Sciences, 69*(Suppl 1), S33–S38.

Morozov, A. V., Astakhova, T. M., Garbuz, D. G., et al. (2017). Interplay between recombinant Hsp70 and proteasomes: Proteasome activity modulation and ubiquitin-independent cleavage of Hsp70. *Cell Stress & Chaperones*. https://doi.org/10.1007/s12192-017-0792-y.

Morrissette, D. A., Parachikova, A., Green, K. N., & LaFerla, F. M. (2009). Relevance of transgenic mouse models to human Alzheimer disease. *The Journal of Biological Chemistry, 284*(10), 6033–6037.

Multhoff, G., & Hightower, L. E. (2011). Distinguishing integral and receptor-bound heat shock protein 70 (Hsp70) on the cell surface by Hsp70-specific antibodies. *Cell Stress & Chaperones, 16*(3), 251–255.

Murshid, A., Eguchi, T., & Calderwood, S. K. (2013). Stress proteins in aging and life span. *International Journal of Hyperthermia, 29*(5), 442–447.

Oakley, H., Cole, S. L., Logan, S., et al. (2006). Intraneuronal beta-amyloid aggregates, neurodegeneration, and neuron loss in transgenic mice with five familial Alzheimer's disease mutations: Potential factors in amyloid plaque formation. *The Journal of Neuroscience, 26*(40), 10129–10140.

Pickering, A. M., Koop, A. L., Teoh, C. Y., Ermak, G., Grune, T., & Davies, K. J. (2010). The immunoproteasome, the 20S proteasome and the PA28αβ proteasome regulator are oxidative-stress-adaptive proteolytic complexes. *The Biochemical Journal, 432*(3), 585–594.

Pickering, A. M., Lehr, M., & Miller, R. A. (2015). Lifespan of mice and primates correlates with immunoproteasome expression. *The Journal of Clinical Investigation, 125*(5), 2059–2068.

Pratt, W. B., Gestwicki, J. E., Osawa, Y., & Lieberman, A. P. (2015). Targeting Hsp90/Hsp70-based protein quality control for treatment of adult onset neurodegenerative diseases. *Annual Review of Pharmacology and Toxicology, 55*, 353–371.

Radons, J. (2016). The human HSP70 family of chaperones: Where do we stand? *Cell Stress & Chaperones, 21*(3), 379–404.

Rattan, S. I. (1996). Synthesis, modifications, and turnover of proteins during aging. *Experimental Gerontology, 31*(1–2), 33–47.

Reeg, S., Jung, T., Castro, J. P., Davies, K. J. A., Henze, A., & Grune, T. (2016). The molecular chaperone Hsp70 promotes the proteolytic removal of oxidatively damaged proteins by the proteasome. *Free Radical Biology & Medicine, 99*, 153–166.

Rodriguez, K. A., Edrey, Y. H., Osmulski, P., Gaczynska, M., & Buffenstein, R. (2012). Altered composition of liver proteasome assemblies contributes to enhanced proteasome activity in the exceptionally long-lived naked mole-rat. *PLoS One, 7*(5), e35890.

Seifert, U., Bialy, L. P., Ebstein, F., et al. (2010). Immunoproteasomes preserve protein homeostasis upon interferon-induced oxidative stress. *Cell, 142*, 613–624.

Shevtsov, M. A., Nikolaev, B. P., Yakovleva, L. Y., Parr, M. A., Marchenko, Y. Y., Eliseev, I., Yudenko, A., Dobrodumov, A. V., Zlobina, O., Zhakhov, A., Ischenko, A. M., Pitkin, E., & Multhoff, G. (2015). 70-kDa heat shock protein coated magnetic nanocarriers as a nanovaccine for induction of anti-tumor immune response in experimental glioma. *Journal of Controlled Release, 220*(Pt A), 329–340.

Shevtsov, M. A., Nikolaev, B. P., Ryzhov, V. A., et al. (2016). Detection of experimental myocardium infarction in rats by MRI using heat shock protein 70 conjugated superparamagnetic iron oxide nanoparticle. *Nanomedicine, 12*(3), 611–621.

Sitte, N., Huber, M., Grune, T., et al. (2000). Proteasome inhibition by lipofuscin/ceroid during postmitotic aging of fibroblasts. *The FASEB Journal, 14*(11), 1490–1498.

Soti, C., & Csermely, P. (2002). Chaperones come of age. *Cell Stress & Chaperones, 7*(2), 186–190.

Tönnies, E., & Trushina, E. (2017). Oxidative stress, synaptic dysfunction, and Alzheimer's disease. *Journal of Alzheimer's Disease, 57*(4), 1105–1121.

Vinokurov, M., Ostrov, V., Yurinskaya, M., et al. (2012). Recombinant human Hsp70 protects against lipoteichoic acid-induced inflammation manifestations at the cellular and organismal levels. *Cell Stress & Chaperones, 17*(1), 89–101.

Viteri, G., Carrard, G., Birlouez-Aragón, I., Silva, E., & Friguet, B. (2004). Age-dependent protein modifications and declining proteasome activity in the human lens. *Archives of Biochemistry and Biophysics, 427*(2), 197–203.

Wright, K. L., White, L. C., Kelly, A., Beck, S., Trowsdale, J., & Ting, J. P. (1995). Coordinate regulation of the human TAP1 and LMP2 genes from a shared bidirectional promoter. *The Journal of Experimental Medicine, 181*(4), 1459–1471.

Ying, W. (2008). The nose may help the brain: Intranasal drug delivery for treating neurological diseases. *Future Neurology, 3*(1), 1–4.

Yurinskaya, M., Zatsepina, O. G., Vinokurov, M. G., et al. (2015). The fate of exogenous human HSP70 introduced into animal cells by different means. *Current Drug Delivery, 12*(5), 524–532.

Yurinskaya, M. M., Funikov, S. Y., Evgen'ev, M. B., & Vinokurov, M. G. (2016). Exogenous heat shock protein HSP70 reduces response of human neuroblastoma cells to lipopolysaccharide. *Doklady Biochemistry and Biophysics, 469*(1), 239–243.

Yurinskaya, M. M., Kochetkova, O. Y., Shabarchina, et al. (2017). Encapsulated Hsp70 decreases endotoxin-induced production of ROS and TNFα in human phagocytes. *Cell Stress & Chaperones, 22*(2), 163–171.

Zhang, C., Rodriguez, C., Spaulding, J., AW, T. Y., & Feng, J. (2012). Age-dependent and tissue-related glutathione redox status in a mouse model of Alzheimer's disease. *Journal of Alzheimer's Disease, 28*(3), 655–666.

Zheng, H., Nagaraja, G. M., Kaur, P., Asea, E. E., & Asea, A. (2010). Chaperokine function of recombinant Hsp72 produced in insect cells using a baculovirus expression system is retained. *The Journal of Biological Chemistry, 285*(1), 349–356.

The Effectiveness of Antitumor Vaccine Enriched with a Heat Shock Protein 70

Gennadiy Didenko, Olena Kruts, Larysa Skivka, and Yuriy Prylutskyy

Abstract Intracellular heat shock proteins (HSP) are overexpressed in majority of malignantly transformed cells providing stress-tolerance of tumor cells and playing important role in pathophysiology of tumor growth. The discovery of this fact has led to the development of anticancer drugs targeting molecular chaperone neutralization to sensitize tumor cells to such stressors as chemo- and radiotherapy. However, the results of applying these preparations proved to be insufficiently efficient in inhibiting tumor growth and preventing tumor progression. The finding about membrane and extracellular HSP localization has initiated a new trend in the development of methods of active immunotherapy of cancer. This has become possible due to the molecular chaperone's ability to transform even the most tolerogenic tumor-associated antigens into immunogenic in the reaction of cross-presentation, as well as the HSP ability to function as endogenous alarmins – the agonists of pattern recognition receptor structures of the immune system – which stimulate functional maturation of antigen presenting cells. Thus, this chapter summarizes the reported as well as our own data concerning the application of HSP in cancer immune therapy.

Keywords Antigen cross-presentation · Heat shock protein · Hsp70 · HSP-based cancer vaccines · Membrane and extracellular HSP

Abbreviations

APC	Antigen presenting cells
CTL	Cytotoxic T-lymphocytes
FDA	Food and Drug Administration, USA
HSP	Heat shock proteins

G. Didenko · O. Kruts
R.E. Kavetsky Institute of Experimental Pathology, Oncology and Radiobiology, NAS of Ukraine, Kyiv, Ukraine

L. Skivka · Y. Prylutskyy (✉)
Taras Shevchenko National University of Kyiv, ESC "Institute of Biology and Medicine", Kyiv, Ukraine

A. A. A. Asea, P. Kaur (eds.), *HSP70 in Human Diseases and Disorders*, Heat Shock Proteins 14, https://doi.org/10.1007/978-3-319-89551-2_17

LOX-1	Lectin-like oxidized low-density lipoprotein receptor-1
MHC	The major *histocompatibility* complex
siRNA	Small interfering RNA
SREC-1	Scavenger receptor expressed by endothelial cells-1
TLR	Toll-like receptors
Treg	Regulatory T-cells

Introduction

In spite of considerable advances in the treatment of oncological pathology, cancer remains one of the main causes of mortality all over the world (Rocque and Cleary 2013; Yedjou et al. 2017). The number of new cases of cancer is steadily growing every year in different countries. According to GLOBOCAN 2012, in the year of 2025, the cancer incidence is expected to reach 19.3 m (Bray et al. 2013; Ferlay et al. 2013). The traditional methods of cancer treatment – surgery, chemical and radiation therapy – will remain in use. However, these types of therapy cause serious side effects, tumor cells become drug and radiation resistant, which causes a considerable interest to biological methods of cancer treatment, especially after new biological properties of malignantly transformed cells and cancer tumors were found. The development of immune therapy of oncological pathology has dramatically changed the paradigm of treating many kinds of solid tumors. The results of numerous clinical researches in immune therapy suggest that in the near future these methods may change considerably the therapy algorithm of treating most kinds of cancer (Tsiatas et al. 2016; Lynch and Murphy 2016). The immune therapy approaches available today include the methods of passive and active immune therapy. Also, these methods can be conventionally subdivided into two groups: application of targeting means and application of nonspecific immune modulators.

The first group includes the methods aimed at recovering the elements of antitumor resistance lost due to tumor associated immune suppression and the activation of immune response to tumor associated and tumor specific antigens: the usage of monoclonal antibodies, immune checkpoints therapy, anticancer vaccines and adoptive cell therapy. The methods of the second group include the usage of cytokines and immune modulators – agonists of pattern recognition receptors, such as fungal polysacharides, synthetic agonists of toll-like receptors (TLR) for non-specific enhancement of organism's immune response that is suppressed due to the tumor growth (Kamta et al. 2017). In some cases it is efficient to combine several immune therapy approaches or traditional method of cancer treatment with the methods of immune therapy (Goldberg and Sondel 2015; Beatty et al. 2017; Ni and Dong 2017; Chajon et al. 2017). The aim of traditional methods in cancer treatment is to destroy malignant cells. The tasks of immune therapy are more complicated: (*i*) preventing cancer tumor development; (*ii*) recovering the body immune response (reduced until causing the development of cancer tumor, or wakened in the result of immune suppression due to the tumor growth), which is necessary, first of all, to patrol disseminating tumor cells; (*iii*) Increasing tumor immunogenicity, for which it is reasonable to combine immune therapy strategy with the methods aimed at causing the death of tumor cells; (*iv*) Stimulating innate and adaptive antitumor immune responses for targeting immune mediated destruction of tumor cells and formation of immune memory to prevent the tumor recurrence. The first and the last of the above listed objectives of immunotherapy are achieved with the help of anti-cancer vaccines. The development of vaccines and their usage in cancer treatment both as an independent strategy and as an element of combined therapy is one of the priorities in oncology today.

Antitumor Vaccines

Antitumor vaccines are divided into prophylactic and therapeutic. The prophylactic vaccines have been developed to prevent malignant tumor formation. The most well-known among those used in clinical practice, FDA-approved vaccine preparations for the prophylaxis of malignant tumor development, are the vaccines against carcinogenic viruses: papillomavirus and hepatitis B virus. They trigger the synthesis of neutralizing antibodies which block the viruses and prevent the development of cervical and liver cancer the respective viruses cause (Finn and Beatty 2016; Wojtowicz et al. 2016). Another category of prophylactic vaccines are those for the patients with the increased risk of cancer development, which contain antigens, expressed in pre-malignant lesions and cancer stem cells, as well as oncogenes (Lohmueller and Finn 2017). The therapeutic vaccines are used to treat the existing oncologic pathology, i.e. to prevent tumor dissemination and development of relapses after the primary tumor is removed due to induction of tumor specific immunity and immune memory (Sayour and Mitchell 2017). There are several types of therapeutic antitumor vaccines which are structurally different: autologous tumor cell vaccines, allogeneic whole tumor cell vaccines, dendritic cell vaccines, protein/peptide-based cancer vaccines, gene vaccines, viral-based vaccines and some others (De Gruijl et al. 2008; Guo et al. 2013; Mohammed et al. 2016; Thomas and Prendergast 2016; Hirayama and Nishimura 2016; Kumar et al. 2017; Rivera et al. 2017; Larocca and Schlom 2017). The category of protein/peptide-based cancer vaccines includes vaccines based on heat shock proteins (HSP).

HSP and Tumor Growth, HSP Antagonists in Cancer Treatment

HSP comprise a considerable part of «chaperoning system». The concept of «chaperoning system» was first suggested in 2008. According to the concept, «chaperoning system» is a physiological system that unites molecular chaperones and functionally associated molecules (co-chaperones and co-factors) in all the tissues, organs and biological fluids. The purpose of the system is to control protein homeostasis and maintain the whole network of proteins in all the cells, tissues and biological fluids in a properly functioning conformation. Besides the maintenance of protein homeostasis, «chaperoning system» is involved in some other processes: antigen presentations, formation and dissociation of protein complexes, etc. The pathological states caused by disorders of «chaperoning system» have been called chaperonopathies, and the usage of chaperones to treat chaperonopathies – chaperone therapy (Rappa et al. 2012; Bellipanni et al. 2016). "Chaperonopathies by collaborationism" also include oncological pathology involving overexpression of intracellular HSP without causing their malfunctioning. The numerous experimental and clinical researches prove that overexpression of intracellular HSP by tumor cells is associated with poor prognosis in cancer-bearing organism and shorter survival time thereby serving as a marker of advanced disease and metastases in regional lymph nodes. In some cases the overexpression of chaperones by tumor cells is a marker of undifferentiated form of cancer that correlates with the number of tumor stem cells in the tumor tissue, which results in increase in the metastatic potential of malignancy.

In what way are HSP involved in cancer pathogenesis? Tumor hallmarks include: uncontrolled proliferation, disorder in cell cycle regulation, avoiding activation of the programmed cell death, avoiding cell aging, *de novo* angiogenesis, as well as invasion and metastasizing. Each of these processes requires the participation of intracellular chaperones. The enhanced proliferation occurs due to overexpression of growth factors, their receptors and components of signaling pathways. All of these proteins are HSP "clients" (Vahid et al. 2017). Tumor cells avoid proapoptic signals due to mutation of antioncogene p53. HSP stabilize mutated antioncogene, thereby helping tumor cells avoid not only the programmed death but also cell aging.

A most important role in antiapoptotic programs of tumor proteins belongs to HSP70. Today four major members of this family are known: constitutively expressed HSP70, GRP78/Bip localized in the endoplasmic reticulum, mitochondrial mtHSP70 and stress-inducible HSP72, or simply HSP70. The last one stabilizes death receptors in the tumor cell, thereby blocking TNFα-mediated apoptosis. It blocks Bid-dependent apoptotic pathway and stabilizes Bax preventing its translocation to mitochondria and slowing mitochondrial pathway of programmed cell death (Wang et al. 2014; Santos et al. 2017). HSP stabilizes HIF1α, which facilitate tumor neoangiogenesis and kinases of focal adhesion as well as receptor tyrosinkinases, which are required for invasive tumor growth and metastasizing. Intracellular HSP are important for the development of drug resistance of tumor cells (Calderwood and Gong 2016; Stope et al. 2016; Wu et al. 2017). In particular, HSP70 mediates the resistance of cancer cells to cisplatin in patients with prostate cancer, to imatinib in patients with chronic myeloid leukaemia, to topotecan and gemcitabine in the case of fibrosarcoma. In general, intracellular HSP overexpression is necessary for the transformed cell to protect itself from apoptosis and stress associated with aneuploidy and accumulated mutant proteins, as well as proteotoxic stress caused by uncontrolled proliferation.

The findings about protumoral properties of intracellular HSP have motivated the development of cancer therapy methods based on usage of antagonists of these chaperones. Cancer therapy methods based on inhibiting protumoral action of HSP may be subdivided into four groups: small molecules, such as 2-phenyl ethyne sulfonamide, which directly bind to the chaperones and inhibit their functions; protein aptamers, which connect specifically with HSP and inhibit their functions; antisense oligonucleotide, which block the translation of HSP, and finally, anti-HSP antibodies (McConnell and McAlpine 2013; Hendriks and Dingemans 2017).

HSP Immunomodulation

It has been believed for a long time that HSP are localized exceptionally inside the cell and in some cases represent organelle-specific molecules. Today, however, it has been proved that there are membrane-associated HSP and HSP that can be localized in extracellular space, where they are exocytosed in microvesicules or in exosomes. In the extracellular space, HSP display pleiotropy and a large number of paracrine effects, which have been investigated only episodically. In addition, HSP can be released from the cell in case of its death and function as alarmins – endogenic danger signals (or Dander-Associated Molecular Patterns, DAMP), which are recognized by pattern-recognition molecules of the immune system and can activate both pro- and anti-inflammatory immune responses (De Maio and Vazquez 2013; De Maio 2014; Kanegasaki and Tsuchiya 2014; Mendonça et al. 2016).

The role of extracellular HSP in tumor growth is nowadays a subject of active investigations by many research teams. This interest is caused by the HSP dual character: chaperones of this location can both facilitate the tumor progression and inhibit the tumor growth and metastasizing. The general conclusion that can be drawn from the data reported up to now is that the influence which the intracellular HSP have on tumor growth depends on the mechanism they are released into intracellular space. As mentioned above, tumor cells can actively secrete HSP as exosome components which perform the function of paracrine transmission of stress-tolerance to tumor cells. Moreover, exosomal HSP activate the differentiation of tumor-associated mononuclear phagocytes into myeloid-derived suppressor cells creating favorable microenvironment for tumor cell invasive growth and dissemination (Chalmin et al. 2010; Borges et al. 2012). In addition, HSP released into extracellular space in the result of cell exocytosis are capable of stimulating native T-cells to differentiate regulator cells (Treg), which are one of the main factors causing tumor-associated immune suppression (Koliński et al. 2016).

However, the presence of HSP in the extracellular space, especially as a result of necrotic or apoptotic tumor cell deaths including the destruction induced by chemo- or radiotherapy can cause pro-inflammatory activation of immunocytes both in the tumor microenvironment and in the immune system as a whole, thereby inhibiting the tumor growth and metastasizing. Functioning as endogenic alarmins, HSP facilitate the functional maturation of antigen presenting cells (APC) – dendritic cells and macrophages – which enhance the expression of histocompatibility (MHC) and co-stimulatory (CD40, CD83, CD86) molecules, mobilize NFκB-dependent transcription, which results in initiating the synthesis of cytokines, such as TNFα, IL-1β, IL-6, IL-12, and GM-CSF, generating reactive forms of oxygen and nitrogen and, finally, activating the adaptive immune responses. The receptor structures responsible for recognition of HSP and their complexes with antigens are alpha-2 macroglobulin receptor CD91, scavenger receptor expressed by endothelial cells-1 (SREC-1) and some other scavenger receptors, TLR-2/4, CD14, lectin-like oxidized low-density lipoprotein receptor-1 (LOX-1), Clever-1 etc. (Tamura et al. 2016; Zuo et al. 2016; Shevtsov and Multhoff 2017). The result of HSP recognition by the innate immunity cells, apart from functional maturation of APC, is the enhancement of migration and activation of cytotoxic T-lymphocytes (CTL) of natural killer cells (Werthmöller et al. 2016), activation of pro-inflammatory cytokine synthesis by the effector cells of adaptive immunity (McNulty et al. 2013; Binder 2014).

The most important immune modulatory effect of HSP is their ability to facilitate cross-presentation and cross-priming. According to canonical ideas about the principles of antigen presentation, the endogenous antigens (for instance, the self antigens, viral antigens in the infected cell as well as tumor-specific and tumor-associated antigens) are localized in the cell cytozol, processed into peptide fragments by proteasome and presented on the cell surface in the complex with MHC I molecules. The complex MHC I:antigen peptide is recognized by antigen receptor CD8+T-lymphocytes and activates these cells to differentiate into mature CTL. The presentation of endogenous antigens in the complex with MHC I is characteristic for all nucleated cells. The exogenous antigens that enter the cell via endocytosis are processed to peptides in phagolysosome. The antigen peptides generated in the phagolysosome are associated with MHC II molecules, which are then transported to the cell surface where they interact with CD4+T-lymphocytes. MHC II are expressed only by a limited number of cells, first of all by antigen-presenting ones, such as dendritic cells and macrophages (Janeway et al. 2002; Abbas and Lichtmann 2009; Paul 2013). However, there is a phenomenon of cross-priming – activation of CD8+T-lymphocytes by exogenous antigens, which involves cross-presentation as a component – processing and presentation by certain APC of exogenous antigens in the complex with MHC I to CD8+T-lymphocytes (Bevan 2006). HSP can facilitate cross-presentation of antigens internalized by APC to the effectors of adaptive immunity including the cross-presentation of tumor-associated antigens in antitumor immune responses.

The ability of molecular chaperones to induce antitumor immune response was first found for paralog HSP90 in endoplasmic reticulum – gp96 (Srivastava and Das 1984, Srivastava et al. 1986). The complex of the chaperone with tumor-specific client peptides isolated from tumor cells in nondenaturing conditions, when injected to animals *in vivo,* stimulated an efficient antitumor immune response. Later HSP70 was found to have the same ability. The mechanism underlying this ability is that HSP facilitates internalizing its complex with client peptides by APC via binding with the above mentioned receptor structures. HSP-complexed client peptide is then processed followed by the display of their fragment on MHC I. In addition to facilitating endocytosis, HSP activate APC migration to regional organized lymphoid structures to present HSP-complexed peptides to native CD8+T-lymphocytes, which further differentiate into CTL – the main effectors of antitumor adaptive immunity (Murshid et al. 2011; Zachova et al. 2016). The HSP ability to form complexes with a wide range of antigen peptides and thereby to facilitate the complex interaction with APC makes the molecular chaperones attractive candidates as components of antitumor vaccines.

Hyperthermia in Cancer Treatment and HSP-Based Cancer Vaccines

The name of HSP family is derived from the ability to be synthesized in the cell in response to increased temperature. The investigations of the role of HSP in tumor growth were initiated after tumor-inhibiting effect of local hyperthermia was revealed. In oncology the term hyperthermia suggests a whole set of methods of cancer treatment by using increased temperatures. As a rule, hyperthermia is used as a method complementing the traditional treatment modalities, such as chemo- and radiotherapy (the so-called thermal chemo- and radiosensitization). Hyperthermia can be local or regional and includes such modern modifications as magnetic nanoparticle hyperthermia, whole-body hyperthermia, methods of hyperthermia perfusion, etc. (Sottile et al. 2015; Sohail et al. 2017; Hu et al. 2017). Hyperthermia causes the death of tumor cells via both apoptosis and necrosis, depending upon the temperature. Hyperthermia increases the cytoplasmic membrane fluidity in tumor cell and destroys the cell cytoskeleton (Hildebrandt et al. 2002; Rao et al. 2010; Burlaka et al. 2010). Moreover, tumor cell death caused by hyperthermia is immunogenic. Because of this fact, hyperthermia is included in the number of physical methods which induce immunogenic death of tumor cells (Adkins et al. 2015).

Immunogenic death is the form of cell death accompanied by releasing endogenic alarmins that can activate inflammatory immune response (Galluzzi et al. 2017; Ladoire et al. 2014; Skivka 2013). Among alarmins that are released during immunogenic tumor cell death due to hyperthermia, there are HSP and HSP complexes with antigenic tumor-specific and tumor-associated peptides. Cancer vaccines prepared from tumor tissue after hyperthermia stimulate antitumor immune responses, increase the life span of experimental animals and inhibit metastasizing more efficiently in comparison with similar vaccines produced from autologous tumor cells without prior hyperthermia (Dvorshchenko et al. 2008). The next step in this direction was the preparation of antitumor vaccines based on HSP from autologous tumor cells. HSP are used as antigens, chaperones or adjuvants to prepare DNA- or peptide-based cancer vaccines for applying in prophylactic and therapeutic protocols (Bolhassani and Rafati 2008). HSP-based antitumor vaccines are unique because most existing vaccines mainly cause the activation of humoral immune response and antibody-dependent immune reactions, while HSP-based vaccine preparations stimulate cell antitumor immunity, which is the basis of antitumor resistance of the organism.

Another property that belongs exclusively to HSP-based vaccines is that they decrease the expression of molecular chaperones in tumor cells thereby weakening their tumorigenic properties and malignant potency (Boliukh et al. 2014). At present several antitumor HSP-based vaccines are under clinical study. Vitespen – the vaccine based on gp96 – is at phase II and III of clinical research in the treatment of liver, kidneys, ovary, colon and skin (melanoma) malignancies (Randazzo et al. 2012). Vitespen is used as a single treatment and in combination with other biotherapy agents. HSP-based vaccines are also tested for treating cancer of pancreas, stomach and lungs (Tosti et al. 2014; Papaioannou et al. 2016; Weller et al. 2017). Also the efficiency of HSP-based vaccines is studied in the treatment of brain tumors (Bloch and Parsa 2014; Bloch et al. 2014; Ampie et al. 2015). In spite of the impressing results in HSP-based vaccines usage for treating experimental oncological pathologies, the clinical tests show considerably smaller efficacy. First of all, the reason is that tumor cells display high antigen heterogeneity in human cancers in comparison with antigen profile of tumors transplanted to animals. In addition, the biology of HSP in conditions of tumor growth as well as immunological and biological effects of HSP-based vaccine preparations requires more detailed study. It is known, for example, that different tumor cells including those used in modelling tumor growth in animals, express HSP at different levels. In particular, the level of HSP70 expression in the cells of Lewis lung carcinoma is higher than in the cells of Erlih's carcinoma (Boliukh et al. 2013). It is believed that in this field the combination of HSP-based vaccines with other methods of cancer adjuvant therapy is promising along with the improvement of constructions of this type of vaccine preparations.

HSP-Based Vaccines in Combination with Bacterial Adjuvants

One of the most common ways to make the efficiency of antitumor vaccines higher is to enhance vaccine immunogenicity. To achieve this goal adjuvants – immune response modifiers – are used, the most powerful among which are agonists of pattern-recognition receptors (Khong and Overwijk 2016; Li et al. 2017). The most commonly used are bacterial agonists of TLR, which have been found to inhibit tumor growth when used in combination with the traditional method of cancer adjuvant therapy (Chamoto et al. 2009; Dowling and Mansell 2016; Mikulandra et al. 2017). Bacterial adjuvants are used as components of prophylactic antitumor vaccines based on antigens induced by oncogenic viruses (Nguyen et al. 2013). The addition of bacterial adjuvants, in particular bacterial ghosts generated from *E. coli* Nissle 1917 or *B. subtilis* 70 kDa protein, to the vaccines based on syngeneic tumor cells lyzates makes them more therapeutically efficient in the animal models of tumor growth (Didenko et al. 2010; Kraśko et al. 2017). As mentioned above, HPS are capable of performing the role of adjuvants since they are also agonists of pattern-recognition receptors, and for this reason they are used in developing vaccine constructs that combine tumor proteins/peptides complexes with HSP, tumor antigen/HSP gene fusion, complexes or gene fusions of oncogenic virus peptides with HSP etc. (Ciocca et al. 2012). Also, there are reports about the attempts to combine adjuvant potency of HSP and bacterial agonists of TLR on the basis of the hypothesis about synergic interaction between endogenous (HSP) and exogenous (bacterial molecules) agonists of these pattern-recognizing receptors (Ludgate 2012; Yang et al. 2012). In particular, the combination of Erlih's carcinoma cells or Lewis lung carcinoma cells and 70 kDa protein-containing component extracted from the cultural fluid of *B. subtilis* B-7025 in vaccine preparation prolonged the life of immunized tumor-bearing animals with both kinds of model tumors (Boliukh et al. 2014). The vaccine based on molecular BCG chaperone HSP65 with human papiloma virus is being tested at the moment in treating oncological diseases of cervix uteri (Einstein et al. 2007).

Xenogeneic and Recombinant HSP in Cancer Vaccines

The technical problems concerning the preparation of sufficient amount of HSP complexes with tumor peptides obtained from the patient tumor tissue made the researchers to seek for alternative sources for preparing molecular chaperones. Therefore the next step in HSP-based cancer immunothreapy was the usage of xenogeneic and recombinant HSP preparations in combination with tumor antigens, especially those with low immunogenicity (Dong et al. 2013). Anticancer vaccines based on tumor proteins/peptides are intended to educate the immune system to recognize tumor cell-specific peptidome. However, in most cases tumor cell peptide does not contain neo-epithopes. Instead, tumor-associated antigens are antigens that do not differ from the analogous structures of untransformed cell and do not induce immune response, which causes tolerance. To overcome this essential obstacle in applying peptide-based cancer vaccines it was proposed to use orthologous xenogeneic (originating from the individuals of different species) proteins/peptides since they differ from homologous proteins/peptides in amino acid composition by one or several epithopes, which makes them immunogenic for the tumor-bearing organism (Cavallo et al. 2014; Strioga et al. 2014).

From the phylogenetic point of view, HSP are ancient intracellular proteins found in all nucleated cells. HSP function is the same in practically all the existing organisms – from bacteria to human – because their activity is a part of cell reparation system. The degree of homology between HSP from eucaryotes and procaryotes comprises over 50%, and some domains are totally identical (Lindquist and Craig 1988). This fact provides the grounds for preparing antitumor vaccines using HSP from different than tumor tissue sourses: either xenogeneic or recombination HSP. The usage of bacterial HSP in the composition of antitumor vaccines has shown that they are highly efficient in animal models of tumor growth. The combination of mycobacterial HSP65 with cell lyzate of Lewis lung carcinoma in the vaccine preparation caused statistically significant inhibition of primary tumor growth and metastatic dissemination (Dong et al. 2010). The vaccine construct that combines mycobacterial $HSP70_{407-426}$ (M2) peptide and tumor-derived autophagome (DRibble) has induces a powerful generation of tumor-specific CTL, which inhibit tumor growth and metastasizing in animals with syngeneic tumor model (Li et al. 2016a). The combination of mycobacterial HSP70 and endothelial cell antigen peptides with the same tandemic sequences resulted in inhibiting tumor angiogenesis in mice with hepatoma, slowing tumor growth and reliable prolonged life of the tumor-bearing animal (Xu et al. 2015).

One of the sources of xenogeneic antigens for using in antitumor vaccine compositions is embryo tissues including those of chicken embryos. For example, the usage of vaccines based on chicken homologous Tie-2 and matrix metalloproteinase-2, which are extremely important for tumor neoangiogenesis, has shown a great efficiency in animal models of tumor growth (Luo et al. 2006; Yi et al. 2007). The vaccine preparations based on xenogeneic neural embryonal tissue caused powerful activation of antitumor immune response in mice with melanoma B16, and inhibited primary tumor growth and metastatic nodules formed in lungs (Voeykova et al. 2014; Fedosova et al. 2015).

One of the sources of orthologous HSP that are also used in antitumor vaccines is chicken embryos. The yield of orthologous HSP from chicken embryos exceeds several times the yield from tumor tissue. The researches show that tumor-inhibitory effect was higher for vaccines based on tumor cell HSP in combination with HSP from chicken embryos of the 7th day of incubation in comparison with vaccine preparations enriched by tumor cell HSP only. It is interesting that HSP from tumor cells and chicken embryos of the 7th day of incubation, when used together in the vaccine composition, induced even more powerful retardation of experimental tumor growth than when these chaperones were used alone (Potebnya et al. 2013). The cytotoxic activity of lymphocytes in tumor-bearing animals treated with the combinational vaccines was higher than that in unvaccinated animals and 1.5 times higher in the animals treated with vaccine based on HSP from syngeneic tumor cells. Cytotoxic activity of macrophages from tumor-bearing mice immunized with combinational vaccine formulation was 2.5 times higher than in control untreated animal and mice treated with the vaccine based on tumor cell HSP. These facts prove the ability of xenogeneic chaperones to cause the shift of macrophage metabolic profile to M1 (proinflammatory) phanotype, which is extremely important in activation of antitumor adaptive immune response (Skivka et al. 2009; Chen and Bonaldo 2013). It should be noted that more vivid antitumor and immune stimulating effect of tumor cell HSP-based vaccine in combination with HSP from chicken embryos of the 7th day of incubation was found in the animals with Lewis lung carcinoma in comparison with animals with melanoma B-16, which indicates that the effect produced by combinational vaccine preparations depends on tumor origin and on tumor cell immunogenicity (Ashley and Kotlarski 1987; Overwijk and Restifo 2001).

Another alternative source of HSP for the development of immunogenic peptide-based antitumor vaccines is recombinant chaperones. The experience in using recombinant HSP-complexed tumor-associated antigens in combination with bacterial adjuvants – agonists of pattern-recognizing receptors – in the composition of antitumor vaccines for prophylaxis of tumor recurrences and metastasizing in animal models is positive. The complex of recombinant chaperone HSP65 with tumor antigen MUC1 (HSP65-MUC1) supplemented with TLR9 agonist C-type CpG oligodeoxynucleotide in the composition of antitumor vaccines considerably enhanced antibody-dependent cell immune responses in mice with melanoma B16, inhibited the growth of primary tumor and prolonged the life span of tumor-bearing animals (Yang et al. 2012).

HSP-Based Cancer Vaccine and Nanoparticles

Immunotherapy by using antitumor vaccines has three essential advantages over the other types of biotherapy of cancer: (*i*) targeted action without harmful effect on healthy cells; (*ii*) the activation of systemic immune reactivity, which prevents metastatic propagation of tumor cells because of the patrolling function of immunocytes in peripheral blood facilitating elimination of circulating tumor cells; (*iii*) Formation of immune memory which diminishes the risk of tumor recurrence. To realize these advantages of cancer immunotherapy it is necessary to provide a prolonged contact of native vaccine preparation with professional APC – dendritic cells that induce tumor-specific reactions of adaptive immunity thereby initiating systematic immune response and activating immune memory formation. The existing formulations of antitumor vaccines including those based on HSP, when systematically administered to the organism with tumor, rapidly become the target of fermentative degradation, which has negative effect on vaccine therapeutic and prophylactic efficacy. The usage of nanoparticles as vaccine preparation carriers has become an important step in overcoming this disadvantage. This method enhances bioavailability of vaccine preparation, prolonges vaccine components persistence in native state and improves the efficacy of vaccine treatment (Saleh and Shojaosadati 2016; Qiu et al. 2017; Mizrahy et al. 2017). In addition, vaccine nanoformulations allowed combining easily tumor-associated antigens in one construct together with the agents increasing their immunogenicity and serving as adjuvants for non-specific activation of APC, which results in activation of adaptive antitumor immune responses.

In general, vaccine preparations supplied with nanocarriers acquire additional advantages: (*i*) protection from fermentative degradation; (*ii*) selective accumulation of vaccine in tumor tissue; (*iii*) Possibility to control pharmacokinetics and tissue distribution of vaccine material; (*iv*) Possibility to activate immunogenic death of tumor cells simultaneously with vaccine material exposition. The sorption ability of nanomaterials allows increasing the density of targeted delivery of vaccine preparation. One of the antitumor vaccines that have demonstrated promising results in experiments with animal models of tumor growth is DNA-inorganic hybrid nanovaccine which includes TLR9 agonist – synthetic CpG as an active agent. The developed nanovaccine construct considerably prolongs the time of tissue persistence CpG and minimizes systemic pro-inflammatory side effects (Zhu et al. 2016).

Another example of antitumor vaccine that has been developed and tested in the animal models is the preparation that consists of nanoparticles and tumor-associated antigen. This unique preparation provides the delivery of tumor antigen to cytosol of antigen-presenting cells, which is principle for the antigen presentation in complex with MHC I to activate CTL. The targeted delivery is accompanied by activation of transcription of type I interferon-stimulated genes whose products are required to stimulate functional maturation of antigen-presenting cells (Luo et al. 2017).

HSP is one of the interesting objects of research in nanomedicine, first, because of the extremely important role HSP play in tumor growth and, second, due to the ability to increase immunogenicity of tumor antigens. It has been reported about the development of nanoparticles that contain small interfering RNA (siRNA) for HSP70 that are expected to selectively block inducible synthesis of HSP70 by tumor cells which thereby would make them vulnerable to other methods of adjuvant cancer therapy (Li et al. 2016b). The conjugate of magnetic nanoparticles with HSP70 are considered to be perspective agents in cancer theranostics. When administered systemically to animals with C6 glioma, the conjugates selectively accumulate in tumor tissue and increase considerably the magnetoresonance imaging contrast of tumor. The authors of the nanovaccine believe that the conjugates will make glioma surgery more successful and also can be used in therapy as a targeted vaccine preparation (Shevtsov et al. 2014, 2015).

There are also reports about successful tests of nanoliposomal vaccines based on HSP70 in animal models of tumor growth. These nanoformulations enhance considerably the immunogenicity of the chaperone complex with tumor antigens and make the immunogen more bioavailable, which is proved by efficient functional maturation of APC in organized lymphoid tissues followed by activation of tumor-specific adaptive immune response (Zhang et al. 2015). However, the combination of molecular chaperones with nanocarriers does not always show positive effect in tumor growth retardation. In particular, HSP was absorbed by aerosil, but the composite used for vaccination of animals with experimental tumor demonstrated lower efficiency in comparison with non-absorbed HSP-based vaccine. At the same time the aerosil-based composites combined with bacterial adjuvants proved efficient in inhibiting experimental tumor growth and stimulating antitumor resistance in experimental animals (Didenko et al. 2003).

Nanoparticles have many advantages in preparation of antitumor vaccines, but there are a number of requirements they should meet. For the vaccines to perform their antitumor and immune stimulating action it is important from which type of nanomaterial they are prepared, what size and shape of nanoparticles are used, what effect the nanoparticles produce on tumor cells and immune response when used separately, what surface charge the nanoparticles have, since cationic nanoparticles are preferred, in what way they are functionalized, etc. Many of these requirements are met, for instance, by water soluble biocompatible nanoparticles of C_{60} fullerene (Prilutski et al. 1998, 1999; Bulavin et al, 2000; Prylutskyy et al. 2001, 2003, 2013, 2014b; Scharff et al. 2004a, b; Ritter et al. 2015). Therefore, carbon nanoalotropes have been actively investigated in nanomedicine, in particular in oncology. When used individually, C_{60} fullerenes are capable of slowing down growth and metastasizing of experimental tumors, as well as enhancing immune response in tumor-carriers (Prylutska et al. 2011a, b, c, 2014a, b; Lynchak et al. 2017). When conjugated with antitumor chemical therapeutic preparations, C_{60} fullerenes decrease the side effects of these preparations, increase their bioavailability and enhance their therapeutic effect (Evstigneev et al. 2013; Prylutska et al. 2014a, b, 2015a, b, 2017a, b; Prylutskyy et al. 2014a, 2015a, b, 2016; Skamrova et al. 2014, Afanasieva et al. 2015; Panchuk et al. 2015). However, pilot research aimed at development of nanovaccines that combine HSP-enriched lyzates of tumor cells and C_{60} fullerenes has not only shown lower efficacy of the nanovaccines in inhibiting tumor growth and metastasizing but also weaker antitumor immune response they induced in tumor-carriers in comparison with vaccines containing only HSP-enriched lyzates or C_{60} fullerenes (Didenko et al. 2013). The probable cause of the revealed phenomenon is improper construction of nanoformulation, which requires further research *in vitro* and *in vivo*.

Conclusions

Thus, intracellular HSP play important role in tumor development and progression facilitating tumor angiogenesis, tumor cell proliferation and tumor stem cell viability by decreasing the sensitivity of transformed cells to apoptosis and increasing their stress-tolerance which can result in treatment resistance of tumors. In contrast, membrane HSP and especially HSP localized in extracellular space are extremely attractive targets for active antitumor immune therapy for two main reasons. Firstly, these proteins perform the functions of endogenous alarmins and are capable of binding to pattern-recognizing receptors of APC. The binding results in activation of functional maturation of APC accompanied by the enhanced expression of MHC and co-stimulatory molecules which are required to activate effectors of antitumor adaptive immunity. Secondly, both membrane and extracellular HSP form complexes with tumor-associated peptides and proteins turning them into immunogenic antigens and targets attacked by effectors of antitumor immunity. The especially important effect HSP: tumor peptide complexes have on immune system is their ability to activate cross-presentation of tumor antigens to cytotoxic T-cells, which are the most powerful effectors of antitumor immunity. The above properties of HSP make them practically perfect material for development of peptide-based cancer vaccines. However, an essential obstacle in realizing the prophylactic and therapeutic potential of these vaccines is tumor-associated immune suppression, to overcome which should become the primary challenge in this field. A promising way to do this could be the combination of HSP-based vaccines with T-cell checkpoint blockade therapies (Hellmann et al. 2016; D'Errico et al. 2017). The elimination of tumor-associated immune suppression will allow HSP-based vaccine preparations to realize their activating potential and target the immune attack even at those cancer tumor antigens which possess low immunogenicity, which will return the immune system the capacity to activate appropriate adaptive immune response.

Acknowledgements This work was supported by R.E. Kavetsky Institute of Experimental Pathology, Oncology and Radiobiology, NAS of Ukraine and Taras Shevchenko National University of Kyiv.

References

Abbas, A. K., & Lichtmann, A. H. (2009). In S. Pillai (Ed.), *Cellular and molecular immunology* (6th ed.). Philadelphia: Saunreds Elsevier.

Adkins, I., Fucikova, J., Garg, A. D., Agostinis, P., & Špíšek, R. (2015). Physical modalities inducing immunogenic tumor cell death for cancer immunotherapy. *OncoImmunology, 3*, 968434.

Afanasieva, K. S., Prylutska, S. V., Lozovik, A. V., Bogutska, K. I., Sivolob, A. V., Prylutskyy, Y. I., et al. (2015). C_{60} fullerene prevents genotoxic effect of doxorubicin on human lymphocytes *in vitro*. *Ukrainian Biochemical Journal, 87*, 91.

Ampie, L., Choy, W., Lamano, J. B., Fakurnejad, S., Bloch, O., & Parsa, A. T. (2015). Heat shock protein vaccines against glioblastoma: From bench to bedside. *Journal of Neuro-Oncology, 123*, 441.

Ashley, M. P., & Kotlarski, I. (1987). *In vivo* H-2K and H-2D antigen expression in two allogeneic mouse tumours of low immunogenicity. *Immunology and Cell Biology, 65*, 323.

Beatty, G. L., Li, Y., & Long, K. B. (2017). Cancer immunotherapy: Activating innate and adaptive immunity through CD40 agonists. *Expert Review of Anticancer Therapy, 17*, 175.

Bellipanni, G., Cappello, F., Scalia, F., Conway de Macario, E., Macario, A. J., & Giordano, A. (2016). Zebrafish as a model for the study of chaperonopathies. *Journal of Cellular Physiology, 231*, 2107.

Bevan, M. J. (2006). Cross-priming. *Nature Immunology, 7*, 363.

Binder, R. J. (2014). Functions of heat shock proteins in pathways of the innate and adaptive immune system. *Journal of Immunology, 193*, 5765.

Bloch, O., & Parsa, A. T. (2014). Heat shock protein peptide complex-96 (HSPPC-96) vaccination for recurrent glioblastoma: A phase II, single arm trial. *Neuro-Oncology, 16*, 758.

Bloch, O., Crane, C. A., Fuks, Y., Kaur, R., Aghi, M. K., Berger, M. S., et al. (2014). Heat-shock protein peptide complex-96 vaccination for recurrent glioblastoma: A phase II, single-arm trial. *Neuro-Oncology, 16*, 274.

Bolhassani, A., & Rafati, S. (2008). Heat-shock proteins as powerful weapons in vaccine development. *Expert Review of Vaccines, 7*, 1185.

Boliukh, I. A., Didenko, G. V., Shpak, E. G., Kuzmenko, O. P., Lisovenko, G. S., & Potebnya, G. P. (2013). Experimental study of the use of heat shocks in vaccine therapy of malignant neoplasms. *Clinical Oncology, 2*, 143.

Boliukh, I. A., Didenko, G. V., Shpak, E. G., Kuzmenko, O. P., Lisovenko, G. S., & Potebnya, G. P. (2014). The role of HSP-peptide complexes in the construction of antitumor vaccines. *Reports of the National Academy of Sciences of Ukraine, 2*, 146.

Borges, T. J., Wieten, L., van Herwijnen, M. J., Broere, F., van der Zee, R., Bonorino, C., & van Eden, W. (2012). The anti-inflammatory mechanisms of Hsp70. *Frontiers in Immunology, 3*, 95.

Bray, F., Ren, J. S., Masuyer, E., & Ferlay, J. (2013). Global estimates of cancer prevalence for 27 sites in the adult population in 2008. *International Journal of Cancer, 132*, 1133.

Bulavin, L., Adamenko, I., Prylutskyy, Y., Durov, S., Graja, A., Bogucki, A., & Scharff, P. (2000). Structure of fullerene C_{60} in aqueous solution. *Physical Chemistry Chemical Physics, 2*, 1627.

Burlaka, A., Lukin, S., Prylutska, S., Remeniak, O., Prylutskyy, Y., Shuba, M., et al. (2010). Hyperthermic effect of multi-walled carbon nanotubes stimulated with near infrared irradiation for anticancer therapy: *In vitro* studies. *Experimental Oncology, 32*, 48.

Calderwood, S. K., & Gong, J. (2016). Heat shock proteins promote cancer: It's a protection racket. *Trends in Biochemical Sciences, 41*, 311.

Cavallo, F., Aurisicchio, L., Mancini, R., & Ciliberto, G. (2014). Xenogene vaccination in the therapy of cancer. *Expert Opinion on Biological Therapy, 14*, 1427.

Chajon, E., Castelli, J., Marsiglia, H., & De Crevoisier, R. (2017). The synergistic effect of radiotherapy and immunotherapy: A promising but not simple partnership. *Critical Reviews in Oncology/Hematology, 111*, 124.

Chalmin, F., Ladoire, S., Mignot, G., Vincent, J., Bruchard, M., Remy-Martin, J. P., et al. (2010). Membrane-associated Hsp72 from tumor-derived exosomes mediates STAT3-dependent immunosuppressive function of mouse and human myeloid-derived suppressor cells. *The Journal of Clinical Investigation, 120*, 457.

Chamoto, K., Takeshima, T., Wakita, D., Ohkuri, T., Ashino, S., Omatsu, T., et al. (2009). Combination immunotherapy with radiation and CpG-based tumor vaccination for the eradication of radio- and immuno-resistant lung carcinoma cells. *Cancer Science, 100*, 934.

Chen, P., & Bonaldo, P. (2013). Role of macrophage polarization in tumor angiogenesis and vessel normalization: Implications for new anticancer therapies. *International Review of Cell and Molecular Biology, 301*, 1.

Ciocca, D. R., Cayado-Gutierrez, N., Maccioni, M., & Cuello-Carrion, F. D. (2012). Heat shock proteins (HSPs) based anti-cancer vaccines. *Current Molecular Medicine, 12*, 1183.

De Gruijl, T. D., van den Eertwegh, A. J., Pinedo, H. M., & Scheper, R. J. (2008). Whole-cell cancer vaccination: From autologous to allogeneic tumor- and dendritic cell-based vaccines. *Cancer Immunology, Immunotherapy, 57*, 1569.

De Maio, A. (2014). Extracellular Hsp70: Export and function. *Current Protein & Peptide Science, 15*, 225.

De Maio, A., & Vazquez, D. (2013). Extracellular heat shock proteins: A new location, a new function. *Shock, 40*, 239.

D'Errico, G., Machado, H. L., & Sainz, B., Jr. (2017). A current perspective on cancer immune therapy: Step-by-step approach to constructing the magic bullet. *Clinical and Translational Medicine, 6*, 3.

Didenko, G. V., Dvorshchenko, O. S., Lisovenko, G. S., Kovalenko, N. G., Potebnya, G. P., Kikot, V. V., et al. (2003). The modification of cancer vaccine prepared on the base of metabolic products of *B. subtilis* 7025 with the use of sorbents and automacrophages. *Experimental Oncology, 25*, 116.

Didenko, G. V., Yevtushenko, O. I., Kuzmenko, A. P., Lisovenko, G. S., & Potebnia, G. P. (2010). Patent for utility model No. 52252 (Ukraine). *Substance with cytotoxic action*. Publ. 25.08.2010, Bull. N16.

Didenko, G., Prylutska, S., Kichmarenko, Y., Potebnya, G., Prylutskyy, Y., Slobodyanik, N., et al. (2013). Evaluation of the antitumor immune response to C_{60} fullerene. *Materialwissenschaft und Werkstofftechnik, 44*, 124.

Dong, B., Sun, L., Wu, X., Zhang, P., Wang, L., Wei, H., et al. (2010). Vaccination with TCL plus MHSP65 induces anti-lung cancer immunity in mice. *Cancer Immunology, Immunotherapy, 59*, 899.

Dong, L., Zhang, X., Ren, J., Wu, S., Yu, T., Hou, L., et al. (2013). Human prostate stem cell antigen and HSP70 fusion protein vaccine inhibits prostate stem cell antigen-expressing tumor growth in mice. *Cancer Biotherapy & Radiopharmaceuticals, 28*, 391.

Dowling, J. K., & Mansell, A. (2016). Toll-like receptors: The Swiss army knife of immunity and vaccine development. *Clinical & Translational Immunology, 5*, 85.

Dvorshchenko, O., Didenko, G., Pavluchenko, N., Kuzmenko, O., Golub, O., Radchenko, E., & Potebnya, G. (2008). Modulation of antitumor activity of autovaccine prepared on the basis of exogenous glycoproteids of *B. subtilis* by nanocomposites of aerosol. Annales UMCS. *Pharmacia, 21*, 321.

Einstein, M. H., Kadish, A. S., Burk, R. D., Kim, M. Y., Wadler, S., Streicher, H., et al. (2007). Heat shock fusion protein-based immunotherapy for treatment of cervical intraepithelial neoplasia III. *Gynecologic Oncology, 106*, 453.

Evstigneev, M. P., Buchelnikov, A. S., Voronin, D. P., Rubin, Y. V., Belous, L. F., Prylutskyy, Y. I., & Ritter, U. (2013). Complexation of C_{60} fullerene with aromatic drugs. *Current Medicinal Chemistry, 14*, 568.

Fedosova, N. I., Voeykova, I. M., Karaman, O. M., Symchych, T. V., Didenko, G. V., Lisovenko, G. S., et al. (2015). Cytotoxic activity of immune cells following administration of xenogeneic cancer vaccine in mice with melanoma B-16. *Experimental Oncology, 37*, 130.

Ferlay, J., Soerjomataram, I., Ervik, M., Dikshit, R., Eser, S., Mathers, C., et al. (2013). *GLOBOCAN 2012 v1.0, Cancer Incidence and Mortality Worldwide: IARC Cancer Base* (Vol. 11). Lyon: International Agency for Research on Cancer.

Finn, O. J., & Beatty, P. L. (2016). Cancer immunoprevention. *Current Opinion in Immunology, 39*, 52.

Galluzzi, L., Buqué, A., Kepp, O., Zitvogel, L., & Kroemer, G. (2017). Immunogenic cell death in cancer and infectious disease. *Nature Reviews. Immunology, 17*, 97.

Goldberg, J. L., & Sondel, P. M. (2015). Enhancing cancer immunotherapy via activation of innate immunity. *Seminars in Oncology, 42*, 562.

Guo, C., Manjili, M. H., Subjeck, J. R., Sarkar, D., Fisher, P. B., & Wang, X. Y. (2013). Therapeutic cancer vaccines: Past, present, and future. *Advances in Cancer Research, 119*, 421.

Hellmann, M. D., Friedman, C. F., & Wolchok, J. D. (2016). Combinatorial cancer immunotherapies. *Advances in Immunology, 130*, 251.

Hendriks, L. E. L., & Dingemans, A. C. (2017). Heat shock protein antagonists in early stage clinical trials for NSCLC. *Expert Opinion on Investigational Drugs, 26*, 541.

Hildebrandt, B., Wust, P., Ahlers, O., Dieing, A., Sreenivasa, G., Kerner, T., et al. (2002). The cellular and molecular basis of hyperthermia. *Critical Reviews in Oncology/Hematology, 43*, 33.

Hirayama, M., & Nishimura, Y. (2016). The present status and future prospects of peptide-based cancer vaccines. *International Immunology, 28*, 319.

Hu, Y., Li, Z., Mi, D. H., Cao, N., Zu, S. W., Wen, Z. Z., et al. (2017). Chemoradiation combined with regional hyperthermia for advanced oesophageal cancer: A systematic review and meta-analysis. *Journal of Clinical Pharmacy and Therapeutics, 42*, 155.

Janeway, C. A., Travers, P., Walport, M., & Shlomchik, M. (2002). *Immunology: The immune system in health and disease* (5th ed.). New York: Garlandpress.

Kamta, J., Chaar, M., Ande, A., Altomare, D. A., & Ait-Oudhia, S. (2017). Advancing cancer therapy with present and emerging immuno-oncology approaches. *Frontiers in Oncology, 7*, 64.

Kanegasaki, S., & Tsuchiya, T. (2014). Alarmins released during local antitumor treatments play an essential role in enhancing tumor growth inhibition at treated and non-treated sites via a derivative of CCL3. *OncoImmunology, 3*, 958956.

Khong, H., & Overwijk, W. W. (2016). Adjuvants for peptide-based cancer vaccines. *Journal for ImmunoTherapy of Cancer, 4*, 56.

Koliński, T., Marek-Trzonkowska, N., Trzonkowski, P., & Siebert, J. (2016). Heat shock proteins (HSPs) in the homeostasis of regulatory T cells (Tregs). *Central European Journal of Immunology, 41*, 317.

Kraśko, J. A., Žilionytė, K., Darinskas, A., Strioga, M., Rjabceva, S., Zalutsky, I., et al. (2017). Bacterial ghosts as adjuvants in syngeneic tumour cell lysate-based anticancer vaccination in a murine lung carcinoma model. *Oncology Reports, 37*, 171.

Kumar, C., Kohli, S., Bapsy, P. P., Vaid, A. K., Jain, M., Attili, V. S., & Sharan, B. (2017). Immune modulation by dendritic-cell-based cancer vaccines. *Journal of Biosciences, 42*, 161.

Ladoire, S., Hannani, D., Vetizou, M., Locher, C., Aymeric, L., Apetoh, L., et al. (2014). Cell-death-associated molecular patterns as determinants of cancer immunogenicity. *Antioxidants & Redox Signaling, 20*, 1098.

Larocca, C., & Schlom, J. (2017). Viral vector-based therapeutic cancer vaccines. *Cancer Journal, 17*, 359.

Li, J., Xing, Y., Zhou, Z., Yao, W., Cao, R., Li, T., Xu, M., & Wu, J. (2016a). Microbial HSP70 peptide epitope 407-426 as adjuvant in tumor-derived autophagosome vaccine therapy of mouse lung cancer. *Tumour Biology, 37*, 15097.

Li, Y., Lin, Z., Zhao, M., Xu, T., Wang, C., Xia, H., Wang, H., & Zhu, B. (2016b). Multifunctional selenium nanoparticles as carriers of HSP70 siRNA to induce apoptosis of HepG2 cells. *International Journal of Nanomedicine, 11*, 3065.

Li, K., Qu, S., Chen, X., Wu, Q., & Shi, M. (2017). Promising targets for cancer immunotherapy: TLRs, RLRs, and STING-mediated innate immune pathways. *International Journal of Molecular Sciences, 18*, 404.

Lindquist, S., & Craig, E. (1988). The heat-shock proteins. *Annual Review of Genetics, 22*, 631.

Lohmueller, J., & Finn, O. J. (2017). Current modalities in cancer immunotherapy: Immunomodulatory antibodies, CARs and vaccines. *Pharmacology & Therapeutics*. https://doi.org/10.1016/j.pharmthera.2017.03.008.

Ludgate, C. M. (2012). Optimizing cancer treatments to induce an acute immune response: Radiation abscopal effects, PAMPs, and DAMPs. *Clinical Cancer Research, 18*, 4522.

Luo, Y., Wen, Y. J., Ding, Z. Y., Fu, C. H., Wu, Y., Liu, J. Y., et al. (2006). Immunotherapy of tumors with protein vaccine based on chicken homologous Tie-2. *Clinical Cancer Research, 12*, 1813.

Luo, M., Wang, H., Wang, Z., Cai, H., Lu, Z., Li, Y., et al. (2017). A STING-activating nanovaccine for cancer immunotherapy. *Nature Nanotechnology*. https://doi.org/10.1038/nnano.2017.52.

Lynch, D., & Murphy, A. (2016). The emerging role of immunotherapy in colorectal cancer. *Annals of Translational Medicine, 4*, 305.

Lynchak, O. V., Prylutskyy, Y. I., Rybalchenko, V. K., Kyzyma, O. A., Soloviov, D., Kostjukov, V. V., et al. (2017). Comparative analysis of the antineoplastic activity of C_{60} fullerene with 5-fluorouracil and pyrrole derivative *in vivo*. *Nanoscale Research Letters, 12*, 8.

McConnell, J. R., & McAlpine, S. R. (2013). Heat shock proteins 27, 40, and 70 as combinational and dual therapeutic cancer targets. *Bioorganic & Medicinal Chemistry Letters, 23*, 1923.

McNulty, S., Colaco, C. A., Blandford, L. E., Bailey, C. R., Baschieri, S., & Todryk, S. (2013). Heat-shock proteins as dendritic cell-targeting vaccines-getting warmer. *Immunology, 139*, 407.

Mendonça, R., Silveira, A. A., & Conran, N. (2016). Red cell DAMPs and inflammation. *Inflammation Research, 65*, 665.

Mikulandra, M., Pavelic, J., & Glavan, T. M. (2017). Recent findings on the application of toll-like receptors agonists in cancer therapy. *Current Medicinal Chemistry*. https://doi.org/10.2174/09 29867324666170320114359.

Mizrahy, S., Hazan-Halevy, I., Landesman-Milo, D., Ng, B. D., & Peer, D. (2017). Advanced strategies in immune modulation of cancer using lipid-based nanoparticles. *Frontiers in Immunology, 8*, 69.

Mohammed, S., Bakshi, N., Chaudri, N., Akhter, J., & Akhtar, M. (2016). Cancer vaccines: Past, resent, and future. *Advances in Anatomic Pathology, 23*, 180.

Murshid, A., Gong, J., Stevenson, M. A., & Calderwood, S. K. (2011). Heat shock proteins and cancer vaccines: Developments in the past decade and chaperoning in the decade to come. *Expert Review of Vaccines, 10*, 1553.

Nguyen, C. T., Hong, S. H., Sin, J. I., Vu, H. V., Jeong, K., Cho, K. O., et al. (2013). Flagellin enhances tumor-specific CD8+T cell immune responses through TLR5 stimulation in a therapeutic cancer vaccine model. *Vaccine, 31*, 3879.

Ni, L., & Dong, C. (2017). New checkpoints in cancer immunotherapy. *Immunological Reviews, 276*, 52.

Overwijk, W. W., & Restifo, N. P. (2001). B16 as a mouse model for human melanoma. *Current Protocols in Immunology*. https://doi.org/10.1002/0471142735.im2001s39.

Panchuk, R. R., Prylutska, S. V., Chumak, V. V., Skorokhyd, N. R., Lehka, L. V., Evstigneev, M. P., et al. (2015). Application of C_{60} fullerene-doxorubicin complex for tumor cell treatment *in vitro* and *in vivo*. *Journal of Biomedical Nanotechnology, 11*, 1139.

Papaioannou, T. G., Karamanou, M., Protogerou, A. D., & Tousoulis, D. (2016). Heat therapy: An ancient concept re-examined in the era of advanced biomedical technologies. *The Journal of Physiology, 594*, 7141.

Paul, W. E. (2013). *Fundamental immunology*. Philadelphia: Lippincott Williams & Wilkins.
Potebnya, G. P., Bolykh, I. A., Didenko, G. V., Kuzmenko, O. P., Shpak, E. G., Lisovenko, G. S., & Chekhun, V. F. (2013). Patent for utility model no. 83120 (Ukraine). *Method of constructing an antitumor vaccine*. Publ. 27.08.2013, Bull. N16.
Prilutski, Y., Durov, S., Bulavin, L., Pogorelov, V., Astashkin, Y., Yashchuk, V., et al. (1998). Study of structure of colloidal particles of fullerenes in water solution. *Molecular Crystals and Liquid Crystals, 324*, 65.
Prilutski, Y. I., Durov, S. S., Yashchuk, V. N., Ogul'chansky, T. Y., Pogorelov, V. E., Astashkin, Y. A., et al. (1999). Theoretical predictions and experimental studies of self-organization C_{60} nanoparticles in water solution and on the support. *The European Physical Journal D, 9*, 341.
Prylutska, S. V., Burlaka, A. P., Klymenko, P. P., Grynyuk, I. I., Prylutskyy, Y. I., Schuetze, C., & Ritter, U. (2011a). Using water-soluble C_{60} fullerenes in anticancer therapy. *Cancer Nanotechnology, 2*, 105.
Prylutska, S. V., Burlaka, A. P., Prylutskyy, Y. I., Ritter, U., & Scharff, P. (2011b). Pristine C_{60} fullerenes inhibit the rate of tumor growth and metastasis. *Experimental Oncology, 33*, 162.
Prylutska, S. V., Burlaka, A. P., Prylutskyy, Y. I., Ritter, U., & Scharff, P. (2011c). Comparative study of antitumor effect of pristine C_{60} fullerenes and doxorubicin. *Biotechnologia Acta, 4*, 82.
Prylutska, S. V., Didenko, G. V., Potebnya, G. P., Bogutska, K. I., Prylutskyy, Y. I., Ritter, U., & Scharff, P. (2014a). Toxic effect of C_{60} fullerene-doxorubicin complex towards normal and tumor cells *in vitro*. *Biopolymers and Cell, 30*, 372.
Prylutska, S., Grynyuk, I., Matyshevska, O., Prylutskyy, Y., Evstigneev, M., Scharff, P., & Ritter, U. (2014b). C_{60} fullerene as synergistic agent in tumor-inhibitory doxorubicin treatment. *Drugs in R&D, 14*, 333.
Prylutska, S. V., Korolovych, V. F., Prylutskyy, Y. I., Evstigneev, M. P., Ritter, U., & Scharff, P. (2015a). Tumor-inhibitory effect of C_{60} fullerene complex with doxorubicin. *Nanomedicine and Nanobiology, 2*, 49.
Prylutska, S., Skivka, L., Didenko, G., Prylutskyy, Y., Evstigneev, M., Potebnya, G., et al. (2015b). Complex of C_{60} fullerene with doxorubicin as a promising agent in antitumor therapy. *Nanoscale Research Letters, 10*, 499.
Prylutska, S. V., Politenkova, S. V., Afanasieva, K. S., Korolovych, V. F., Bogutska, K. I., Sivolob, A. V., et al. (2017a). Nanocomplex of C_{60} fullerene with cisplatin: Design, characterization and toxicity. *Beilstein Journal of Nanotechnology, 8*, 1494.
Prylutska, S., Panchuk, R., Gołuński, G., Skivka, L., Prylutskyy, Y., Hurmach, V., et al. (2017b). C_{60} fullerene enhances cisplatin anticancer activity and overcomes tumor cells drug resistance. *Nano Research, 10*, 652.
Prylutskyy, Y. I., Durov, S. S., Bulavin, L. A., Adamenko, I. I., Moroz, K. O., Geru, I. I., et al. (2001). Structure and thermophysical properties of fullerene C_{60} aqueous solutions. *International Journal of Thermophysics, 22*, 943.
Prylutskyy, Y. I., Yashchuk, V. M., Kushnir, K. M., Golub, A. A., Kudrenko, V. A., Prylutska, S. V., et al. (2003). Biophysical studies of fullerene-based composite for bio-nanotechnology. *Materials Science and Engineering: C, 23*, 109.
Prylutskyy, Y. I., Buchelnikov, A. S., Voronin, D. P., Kostjukov, V. V., Ritter, U., Parkinson, J. A., & Evstigneev, M. P. (2013). C_{60} fullerene aggregation in aqueous solution. *Physical Chemistry Chemical Physics, 15*, 9351.
Prylutskyy, Y. I., Evstigneev, M. P., Pashkova, I. S., Wyrzykowski, D., Woziwodzka, A., Gołuński, G., et al. (2014a). Characterization of C_{60} fullerene complexation with antibiotic doxorubicin. *Physical Chemistry Chemical Physics, 16*, 23164.
Prylutskyy, Y. I., Petrenko, V. I., Ivankov, O. I., Kyzyma, O. A., Bulavin, L. A., Litsis, O. O., et al. (2014b). On the origin of C_{60} fullerene solubility in aqueous solution. *Langmuir, 30*, 3967.
Prylutskyy, Y. I., Cherepanov, V. V., Evstigneev, M. P., Kyzyma, O. A., Petrenko, V. I., Styopkin, V. I., et al. (2015a). Structural self-organization of C_{60} and cisplatin in physiological solution. *Physical Chemistry Chemical Physics, 17*, 26084.

Prylutskyy, Y. I., Evstigneev, M. P., Cherepanov, V. V., Kyzyma, O. A., Bulavin, L. A., Davidenko, N. A., & Scharff, P. (2015b). Structural organization of C_{60} fullerene, doxorubicin and their complex in physiological solution as promising antitumor agents. *Journal of Nanoparticle Research, 17*, 45.

Prylutskyy, Y. I., Cherepanov, V. V., Kostjukov, V. V., Evstigneev, M. P., Kyzyma, O. A., Bulavin, L. A., et al. (2016). Study of the complexation between Landomycin a and C_{60} fullerene in aqueous solution. *RSC Advances, 6*, 81231.

Qiu, H., Min, Y., Rodgers, Z., Zhang, L., & Wang, A. Z. (2017). Nanomedicine approaches to improve cancer immunotherapy. *Wiley Interdisciplinary Reviews. Nanomedicine and Nanobiotechnology*. https://doi.org/10.1002/wnan.1456.

Randazzo, M., Terness, P., Opelz, G., & Kleist, C. (2012). Active-specific immunotherapy of human cancers with the heat shock protein Gp96-revisited. *International Journal of Cancer, 130*, 2219.

Rao, W., Deng, Z. S., & Liu, J. (2010). A review of hyperthermia combined with radiotherapy/chemotherapy on malignant tumors. *Critical Reviews in Biomedical Engineering, 38*, 101.

Rappa, F., Farina, F., Zummo, G., David, S., Campanella, C., Carini, F., et al. (2012). HSP-molecular chaperones in cancer biogenesis and tumor therapy: An overview. *Anticancer Research, 32*, 5139.

Ritter, U., Prylutskyy, Y. I., Evstigneev, M. P., Davidenko, N. A., Cherepanov, V. V., Senenko, A. I., et al. (2015). Structural features of highly stable reproducible C_{60} fullerene aqueous colloid solution probed by various techniques. *Fullerenes, Nanotubes, and Carbon Nanostructures, 23*, 530.

Rivera, V. T., Benoit-Lizon, I., & Apetoh, L. (2017). Rationale for stimulator of interferon genes-targeted cancer immunotherapy. *European Journal of Cancer, 75*, 86.

Rocque, G. B., & Cleary, J. F. (2013). Palliative care reduces morbidity and mortality in cancer. *Nature Reviews. Clinical Oncology, 10*, 80.

Saleh, T., & Shojaosadati, S. A. (2016). Multifunctional nanoparticles for cancer immunotherapy. *Human Vaccines & Immunotherapeutics, 12*, 1863.

Santos, T. G., Martins, V. R., & Hajj, G. N. M. (2017). Unconventional secretion of heat shock proteins in cancer. *International Journal of Molecular Sciences, 18*, 946.

Sayour, E. J., & Mitchell, D. A. (2017). Manipulation of innate and adaptive immunity through cancer vaccines. *Journal of Immunology Research*. https://doi.org/10.1155/2017/3145742.

Scharff, P., Carta-Abelmann, L., Siegmund, C., Matyshevska, O. P., Prylutska, S. V., Koval, T. V., et al. (2004a). Effect of X-ray and UV irradiation of the C_{60} fullerene aqueous solution on biological samples. *Carbon, 42*, 1199.

Scharff, P., Risch, K., Carta-Abelmann, L., Dmytruk, I. M., Bilyi, M. M., Golub, O. A., et al. (2004b). Structure of C_{60} fullerene in water: Spectroscopic data. *Carbon, 42*, 1203.

Shevtsov, M., & Multhoff, G. (2017). Heat shock protein-peptide and HSP-based immunotherapies for the treatment of cancer. *Frontiers in Immunology, 7*, 171.

Shevtsov, M. A., Yakovleva, L. Y., Nikolaev, B. P., Marchenko, Y. Y., Dobrodumov, A. V., Onokhin, K. V., et al. (2014). Tumor targeting using magnetic nanoparticle Hsp70 conjugate in a model of C6 glioma. *Neuro-Oncology, 16*, 38.

Shevtsov, M. A., Nikolaev, B. P., Yakovleva, L. Y., Parr, M. A., Marchenko, Y. Y., Eliseev, I., et al. (2015). 70-kDa heat shock protein coated magnetic nanocarriers as a nanovaccine for induction of anti-tumor immune response in experimental glioma. *Journal of Controlled Release, 220*, 329.

Skamrova, G. B., Laponogov, I. V., Buchelnikov, A. S., Shckorbatov, Y. G., Prylutska, S. V., Ritter, U., et al. (2014). Interceptor effect of C_{60} fullerene on the *in vitro* action of aromatic drug molecules. *European Biophysics Journal, 43*, 265.

Skivka, L. M. (2013). Immunogenic cancer cell death: How it can be exploited. *Experimental Oncology, 35*, 131.

Skivka, L. M., Horbyk, H. V., Fedorchuk, O. H., & Pozur, V. V. (2009). Tumor-associated macrophages in the prospect of development of targeted cancer therapy. *Tsitologiia i Genetika, 43*, 71.

Sohail, A., Ahmad, Z., Bég, O. A., Arshad, S., & Sherin, L. (2017). A review on hyperthermia via nanoparticle-mediated therapy. *Bulletin du Cancer, 104*, 452.

Sottile, M. L., Losinno, A. D., Fanelli, M. A., Cuello-Carrión, F. D., Montt-Guevara, M. M., Vargas-Roig, L. M., & Nadin, S. B. (2015). Hyperthermia effects on Hsp27 and Hsp72 associations with mismatch repair (MMR) proteins and cisplatin toxicity in MMR-deficient/proficient colon cancer cell lines. *International Journal of Hyperthermia, 31*, 464.

Srivastava, P. K., & Das, M. R. (1984). The serologically unique cell-surface antigen of zajdela ascitic hepatoma is also its tumor-associated transplantation antigen. *International Journal of Cancer, 33*, 417.

Srivastava, P. K., Deleo, A. B., & Old, L. J. (1986). Tumor rejection antigens of chemically-induced sarcomas of inbred mice. *Proceedings of the National Academy of Sciences of the United States of America, 83*, 3407.

Stope, M. B., Koensgen, D., Burchardt, M., Concin, N., Zygmunt, M., & Mustea, A. (2016). Jump in the fire-heat shock proteins and their impact on ovarian cancer therapy. *Critical Reviews in Oncology/Hematology, 97*, 152.

Strioga, M. M., Darinskas, A., Pasukoniene, V., Mlynska, A., Ostapenko, V., & Schijns, V. (2014). Xenogeneic therapeutic cancer vaccines as breakers of immune tolerance for clinical application: To use or not to use? *Vaccine, 32*, 4015.

Tamura, Y., Yoneda, A., Takei, N., & Sawada, K. (2016). Spatiotemporal regulation of Hsp90-ligand complex leads to immune activation. *Frontiers in Immunology, 24*, 201.

Thomas, S., & Prendergast, G. C. (2016). Cancer vaccines: A brief overview. *Methods in Molecular Biology, 1403*, 755.

Tosti, G., Cocorocchio, E., Pennacchioli, E., Ferrucci, P. F., Testori, A., & Martinoli, C. (2014). Heat-shock proteins-based immunotherapy for advanced melanoma in the era of target therapies and immunomodulating agents. *Expert Opinion on Biological Therapy, 14*, 955.

Tsiatas, M., Mountzios, G., & Curigliano, G. (2016). Future perspectives in cancer immunotherapy. *Annals of Translational Medicine, 4*, 273.

Vahid, S., Thaper, D., & Zoubeidi, A. (2017). Chaperoning the cancer: The proteostatic functions of the heat shock proteins in cancer. *Recent Patents on Anti-Cancer Drug Discovery, 12*, 35.

Voeykova, I. M., Fedosova, N. I., Karaman, O. M., Yudina, O. Y., Didenko, G. V., Lisovenko, G. S., et al. (2014). Use of xenogeneic vaccine modified with embryonal nervous tissue antigens in the treatment of B16-melanoma-bearing mice. *Experimental Oncology, 36*, 24.

Wang, X., Chen, M., Zhou, J., & Zhang, X. (2014). HSP27, 70 and 90, anti-apoptotic proteins, in clinical cancer therapy. *International Journal of Oncology, 45*, 18.

Weller, M., Roth, P., Preusser, M., Wick, W., Reardon, D. A., Platten, M., & Sampson, J. H. (2017). Vaccine-based immunotherapeutic approaches to gliomas and beyond. *Nature Reviews. Neurology, 13*, 363.

Werthmöller, N., Frey, B., Rückert, M., Lotter, M., Fietkau, R., & Gaipl, U. S. (2016). Combination of ionising radiation with hyperthermia increases the immunogenic potential of B16-F10 melanoma cells *in vitro* and *in vivo*. *International Journal of Hyperthermia, 32*, 23.

Wojtowicz, M. E., Dunn, B. K., & Umar, A. (2016). Immunologic approaches to cancer prevention-current status, challenges, and future perspectives. *Seminars in Oncology, 43*, 161.

Wu, J., Liu, T., Rios, Z., Mei, Q., Lin, X., & Cao, S. (2017). Heat shock proteins and cancer. *Trends in Pharmacological Sciences, 38*, 226.

Xu, M., Zhou, L., Zhang, Y., Xie, Z., Zhang, J., Guo, L., et al. (2015). A fixed human umbilical vein endothelial cell vaccine with 2 tandem repeats of microbial HSP70 peptide epitope 407-426 as adjuvant for therapy of hepatoma in mice. *Journal of Immunotherapy, 38*, 276.

Yang, M., Yan, Y., Fang, M., Wan, M., Wu, X., Zhang, X., et al. (2012). MF59 formulated with CpG ODN as a potent adjuvant of recombinant HSP65-MUC1 for inducing anti-MUC1+tumor immunity in mice. *International Immunopharmacology, 13*, 408.

Yedjou, C. G., Tchounwou, P. B., Payton, M., Miele, L., Fonseca, D. D., Lowe, L., & Alo, R. A. (2017). Assessing the racial and ethnic disparities in breast cancer mortality in the United States. *International Journal of Environmental Research and Public Health, 14*, 486.

Yi, T., Wei, Y. Q., Tian, L., Zhao, X., Li, J., Deng, H. X., et al. (2007). Humoral and cellular immunity induced by tumor cell vaccine based on the chicken xenogeneic homologous matrix metalloproteinase-2. *Cancer Gene Therapy, 14*, 158.

Zachova, K., Krupka, M., & Raska, M. (2016). Antigen cross-presentation and heat shock protein-based vaccines. *Archivum Immunologiae et Therapiae Experimentalis, 64*, 1.

Zhang, Y., Luo, W., Wang, Y., Chen, J., Liu, Y., & Zhang, Y. (2015). Enhanced antitumor immunity of nanoliposome-encapsulated heat shock protein 70 peptide complex derived from dendritic tumor fusion cells. *Oncology Reports, 33*, 2695.

Zhu, G., Liu, Y., Yang, X., Kim, Y. H., Zhang, H., Jia, R., et al. (2016). DNA-inorganic hybrid nanovaccine for cancer immunotherapy. *Nanoscale, 8*, 6684.

Zuo, D., Subjeck, J., & Wang, X. Y. (2016). Unfolding the role of large heat shock proteins: New insights and therapeutic implications. *Frontiers in Immunology, 7*, 75.

Mammalian Heat Shock Protein Hsp105: The Hsp70 Inducer and a Potent Target for Cancer Therapy

Youhei Saito and Yuji Nakayama

Abstract Major heat shock protein Hsp70 prevents protein aggregation and assists protein folding as molecular chaperone. Hsp70 also regulates apoptosis, senescence, and autophagy. Increased expression of Hsp70 causes the drug resistance of cancer cells. Hsp105, a mammalian heat shock protein, consists of Hsp105α and its splicing isoform Hsp105β. Hsp105α constitutively expresses in cytoplasm and functions as molecular chaperone and apoptotic regulator. Hsp105β is specifically expressed in nucleus under stressed condition and induces Hsp70 expression through the activation of Stat3. Recently, we identified that the novel regulators of Hsp105β-mediated Hsp70 induction including the transcriptional co-activator of Stat3. Additionally, we reveled that Hsp105α but not Hsp105β interacts with HIF-1α in nucleus and affects to the transcriptional activation of HIF-1. Since Hsp105 is overexpressed in several tumors including solid tumor, these evidences indicate that the nuclear expression of Hsp105α seems to function as a malignant factor of cancer through the induction of Hsp70 and the activation of the tumorigenic Stat3 and HIF-1 signaling pathways. In this chapter, we will introduce the recent our observations and discuss the possibility that the nuclear overexpression of Hsp105 is a useful marker for cancer prognosis and diagnosis.

Keywords HIF-1 · Hsp70 · Hsp70 inducer · Hsp105/110 · Stat3 · Tumor marker

Abbreviations

5-FU	5-fluorouracil
ER	Endoplasmic reticulum
ESCC	Esophageal squamous cell carcinoma
JAK	Janus kinase
HIF	Hypoxia-inducible factor

Y. Saito (✉) · Y. Nakayama
Department of Biochemistry and Molecular Biology, Kyoto Pharmaceutical University, Kyoto, Japan
e-mail: ysaito@mb.kyoto-phu.ac.jp

A. A. A. Asea, P. Kaur (eds.), *HSP70 in Human Diseases and Disorders*, Heat Shock Proteins 14, https://doi.org/10.1007/978-3-319-89551-2_18

Hsc	Heat shock cognate
HSF	Heat shock factor
Hsp	Heat shock protein
NES	Nuclear export signal
NLS	Nuclear localization signal
Nmi	N-myc interactor
PHD	Prolyl hydroxylase
snRNP	Small nuclear ribonucleoprotein
Stat	Signal transducer and activator of transcription
VHL	Von Hippel-Lindau

Introduction

Hsp70 family, consists of multiple members (Kampinga et al. 2009), has well conserved three functional domains: N-terminal ATPase domain, central β-sheet domain, and C-terminal α-helix domain (Fig. 1). Hsp70, is one of the most characterized heat shock protein, functions as molecular chaperone to maintain protein homeostasis (Balchin et al. 2016). In addition, Hsp70 suppresses apoptosis through several pathways, and also regulates important cellular functions including senescence and autophagy. Since Hsp70 is overexpressed in various tumor tissues and are involved in drug resistance, Hsp70 is a target for anticancer therapy (Murphy 2013).

Hsp105/110 family proteins are conserved in several species, such as yeast, worms, and mammals. Mammalian Hsp105/110 family proteins are classified into four members, including Hsp105, Hsp110 (a rat homologue of Hsp105; the same as a human Apg-2), Apg-1 (a testis specific Hsp105), and ER chaperone Grp170 (Kampinga et al. 2009). As member of the mammalian Hsp105/110 family, Hsp105α

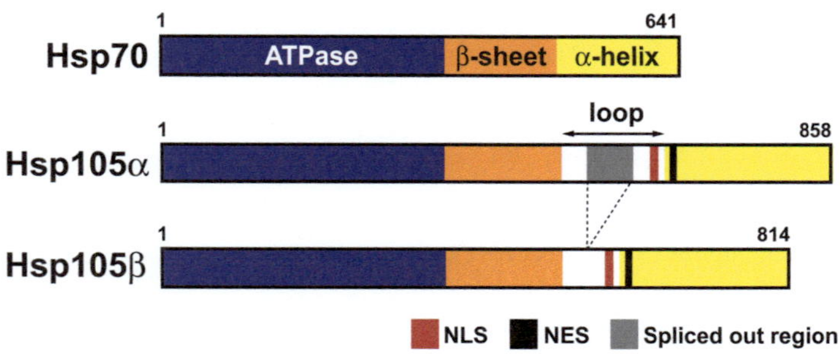

Fig. 1 The domain structure of Hsp70 and Hsp105. These proteins contain N-terminal ATPase, β-sheet, and C-terminal α-helix domains. Hsp105α and Hsp105β have an additional loop region between β-sheet and α-helix domains. NLS, nuclear localization signal; NES, nuclear export signal

and Hsp105β were identified as mouse and human *hsp105* genes (Yasuda et al. 1995; Ishihara et al. 1999). Hsp105 is relatively highly homologous to Hsp70 and has ATPase, β-sheet, and α-helix domains, similar to Hsp70 (Fig. 1). In this chapter, we mainly summarize the researches of mammalian Hsp105α and Hsp105β.

Hsp105α is constitutively expressed in most of tissues, especially in brain, and is induced by various stresses (Yasuda et al. 1995; Ishihara et al. 1999). Hsp105α interacts with Hsp70 and prevents Hsp70 chaperone activity (Yamagishi et al. 2004), whereas Hsp105α also acts as a nucleotide exchange factor for Hsp70 similar to yeast Hsp105/110 family protein and mammalian Hsp110 (Apg-2) (Bracher and Verghese 2015). Hsp105/110 family protein functions as a metazoan protein disaggregase cooperates with Hsp70 and Hsp40 (Nillegoda and Bukau 2015). Thus, Hsp105α (Hsp110) is an important regulator of Hsp70 chaperone machinery. In addition to the cooperative role of Hsp105, Hsp70-independent functions of Hsp105α have been suggested in the prevention of heat-denatured protein aggregation and the stress-induced apoptosis (Yamagishi et al. 2000, 2008).

Hsp105β, an alternative spliced form of Hsp105α, is lacked 44 amino acids (spliced out region) from Hsp105α. Hsp105β is specifically expressed during mild heat shock (42 °C), but not other stresses such as severe heat shock (45 °C), metal, azetidine, and arsenite (Ishihara et al. 1999). *Hsp105* gene consists 18 exons, which are separated by 17 introns, and the exon 12 encodes the spliced out sequence (Yasuda et al. 1999). Recently, the *cis*-regulatory element in exon 12 of *hsp105* gene and the regulators of the heat shock specific alternative splicing of *hsp105* gene were identified (Yamamoto et al. 2016), indicating that Hsp105β expression is mechanistically controlled under heat shock conditions. Since the differences of structure of Hsp105α and Hsp105β are only 44 amino acids, the function of Hsp105β has been considered similar to Hsp105α. In fact, an *in vitro* study showed that Hsp105β prevents the protein denaturation and the Hsc70 chaperone activity as same as Hsp105α (Yamagishi et al. 2000). However, previous study revealed that Hsp105β localizes to the nucleus in contrast to cytoplasmic Hsp105α, suggesting the intranuclear functions of Hsp105β (Saito et al. 2007). Strikingly, nuclear Hsp105β enhances Hsp70 expression at transcriptional levels (Saito et al. 2009), which involves the activation of Stat3 transcription factor (Yamagishi et al. 2009).

Recently, we screened and identified three proteins as an Hsp70 inducer (Saito et al. 2014, 2016). Furthermore, Hsp105α but not Hsp105β interacts with a hypoxia-inducible factor HIF-1α in nucleus and is required for the HIF-1 accumulation and transcriptional activation (Mikami et al. 2017), suggesting that Hsp105α may function in nucleus as well as Hsp105β. Hsp105 is overexpressed in many tissues of tumors, and the prognostic significance of Hsp105 expression in some tumor. In this chapter, we introduce that Hsp105 family protein functions as an Hsp70 inducer and modulates transcriptional activation of Stat3 and HIF-1, and discuss the possibility that the nuclear expression of Hsp105 is a useful marker and therapeutic target for cancer therapy.

HSP105β Localizes Nucleus and Enhances HSP70 Expression

Mechanisms of Hsp70 Gene Expression

The induction of Hsp70 expression is controlled by HSF1 in mammalian cells. HSF1 is constitutively expressed as inactive form under non-stressed conditions. Under stressed conditions, HSF1 is localized into the nucleus and activates the Hsp70 transcription (Morimoto et al. 1992). When Hsp70 accumulates in cells, Hsp70 interacts with the carboxyl-terminal activation domain of HSF1 and leads the negative feedback regulation of HSF1 (Shi et al. 1998). Several transcription factors and its regulators also regulate the *hsp70* gene expression in mammalian cells such as c-myc (Kingston et al. 1984; Kaddurah-Daouk et al. 1987), CCAAT box-binding protein NF-Y, (Taira et al. 1999), Stat1 (Stephanou et al. 1999), hematopoietic transcription factor GATA (Ray et al. 2004), and Histone deacetylase Sirt1 (Westerheide et al. 2009). Thus, Hsp70 expression is controlled multiple proteins. In addition, we identified that Hsp105β localizes in the nucleus and induces the Hsp70 expression (Saito et al. 2007, 2009) as described in detail below.

Different Localization of Hsp105α and Hsp105β

Previous studies showed that Hsp105α mainly localizes in cytoplasm (Hatayama et al. 1994a; Ishihara et al. 1999). However, the subcellular localization of Hsp105β had been not analyzed, because of the similarity of their structures and functions. When African monkey kidney COS-7 cells were transfected with the expression plasmid of mouse Hsp105α or Hsp105β, exogenously expressed Hsp105 could be detected by immunofluorescence analysis using anti-mouse Hsp105 antibody. Experiments using this method clearly showed that Hsp105α localizes mainly in cytoplasm, whereas Hsp105β localizes in the nucleus (Saito et al. 2007). Mutations of these proteins revealed that Hsp105α and Hsp105β have same nuclear localization signal (NLS) and nuclear export signal (NES) sequences (Fig. 1). However, the cytoplasmic localization of Hsp105α is dependent on its NES activity, and the spliced out sequence of Hsp105α suppress the NLS activity of Hsp105α (Saito et al. 2007). Furthermore, the NES activity of Hsp105β is suppressed by its N-terminal and C-terminal regions (Saito et al. 2009). Therefore, the subcellular localizations of Hsp105α and Hsp105β are determined by the balance of their NLS and NES activities.

Hsp105β Induces Hsp70 Expression

Hsp105β-mediated induction of Hsp70 has been reported in our laboratory (Saito et al. 2009, 2014, 2016; Yamagishi et al. 2009). In order to determine whether Hsp105β is required for heat shock-induced nuclear translocation of Hsp70, immunofluorescence analysis was performed using anti-Hsp70 antibody. Unexpectedly, higher expression of Hsp70 was observed in Hsp105β-transfected cells. (Saito Y., unpublished observations). Further analysis revealed that Hsp105β induces Hsp70 expression at transcription levels. The promoter activity of Hsp70 was increased in cells, which is overexpressing Hsp105β but not its NLS mutant, indicating that the nuclear localization of Hsp105β is required for Hsp70 induction. Consistent with this, cytoplasmic Hsp105α is less effective to the expression of Hsp70 compared with Hsp105β. Since Hsp105β specifically expresses under mild heat shock conditions (Hatayama et al. 1994a, b), Hsp105β may play an important cooperative role of in the protection of cells with Hsp70 under heat shock conditions. Further analysis of the Hsp105β-mediated Hsp70 expression revealed that Hsp105β induces the expression of Hsp70 through the activation of cytokine signal transcription factor Stat3 (Yamagishi et al. 2009). Stat family proteins exist as monomeric non-active form in cytoplasm, and are activated by JAK tyrosine kinase (Darnell 1997). The phosphorylated Stat is dimerized and localizes into the nucleus of cells to activate target gene expression. Interestingly, Hsp105β enhances the phosphorylation of Stat3, however, its mechanism is not understood (Yamagishi et al. 2009).

Regulators of HSP105β-Mediated HSP70 Expression

Subsequently analysis revealed that the region between amino acids 642 and 662 of Hsp105β and the sequence of the *hsp70* promoter between −206 and −187 are required for the Hsp105β-induced expression of *hsp70* gene (Yamagishi et al. 2009). In order to identify the regulators of the Hsp105β-mediated *hsp70* gene expression, we performed yeast genetic screenings and identified Nmi, AF9, and SNRPE as positive regulators of *hsp70* gene expression (Saito et al. 2014, 2016) (Fig. 2).

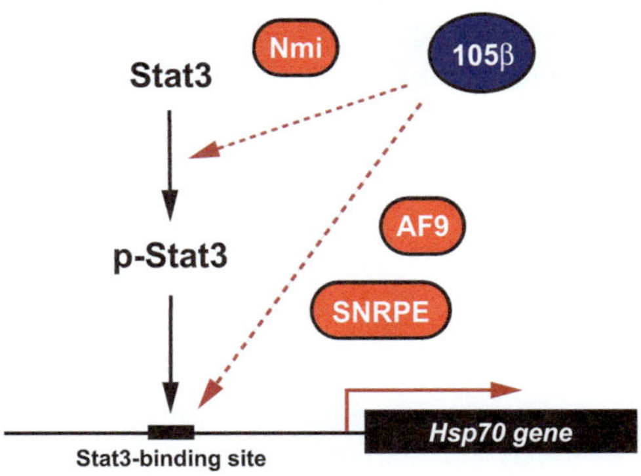

Fig. 2 **Regulators of Hsp105β-mediated *hsp70* gene expression**. Hsp105 enhances *hsp70* promoter activity through the Stat3 phosphorylation and activation. Nmi interacts with Hsp105β and enhances the Hsp105β-induced Stat3 phosphorylation. Hsp105β-induced activation of *hsp70* promoter is also enhanced by AF9 or SNRPE. These proteins seem to affect the *hsp70* promoter activation through Stat3-independent pathway

Nmi

Nmi was identified as an Hsp105β-binding protein by yeast two-hybrid screening using full-length Hsp105β as bait (Saito et al. 2014). Nmi interacted with Hsp105β and enhanced the Hsp105β-induced *hsp70* promoter activity. In addition, Nmi enhanced the Hsp105β-induced Stat3 phosphorylation. A previous study reported that Nmi interacts with Stat family proteins including Stat3 and promotes the transcriptional activation of Stat5 triggered by IL-2 and IFN-γ (Zhu et al. 1999). Nmi has been shown to promote the interaction between Stat family proteins including Stat3 and its transcriptional co-activator p300/CBP (Zhu et al. 1999). Nmi may promote the interaction between Stat3 and p300/CBP in order to activate the *hsp70* promoter. As mentioned above, Hsp105α and Hsp105β are mainly localized in the cytoplasm and nucleus, respectively. Hsp105α and Hsp105β have same NLS and NES sequences. The balance of activity of NLS and NES determines the localization of these proteins. The mutant of Hsp105α, which has a mutation in NES, was localized in the nucleus and enhanced the *hsp70* promoter activity (Saito Y., unpublished data). Interestingly, Nmi-overexpression induces the nuclear expression of Hsp105α. In this experimental conditions, *hsp70* promoter activity and the Hsp70 expression is increased (Saito et al. 2014), indicating that Nmi may affect the Hsp70 expression and the Stat3 activation through the interaction with not only Hsp105β but also Hsp105α.

AF9

AF9 was identified as an Hsp105β-binding protein by yeast two-hybrid screening using minimal region of Hsp105β, which requires for *hsp70* induction (Saito et al. 2016). AF9 itself activated *hsp70* promoter activity and further enhanced the Hsp105β-mediated *hsp70* promoter activity. AF9 is known as a component of the RNA polymerase II super elongation complex (SEC). AF9 interacts with RNA polymerase II elongation factors, ELL and p-TEFb (He et al. 2010; Lin et al. 2010), and histone methyltansferase DOT1L (Li et al. 2014). These findings suggest that AF9 may play a role in the induction of the *hsp70* gene during transcriptional elongation and histone modification.

SNRPE

SNRPE was identified as an *hsp70* promoter-binding protein by yeast one-hybrid screening using the region of *hsp70* promoter, which is responsible for the Hsp105β-mediated Hsp70 induction (Saito et al. 2016). Overexpression of SNRPE enhanced *hsp70* promoter activity in a manner that dependent on Hsp105β expression. Knockdown of SNRPE suppressed the expression of Hsp70 without Hsp105β expression, indicating that SNRPE essentially functions as a transcriptional activator of *hsp70* gene expression. However, the precious mechanism is unknown. SNRPE is a small nuclear ribonucleoprotein (snRNP), which is the core component of spliceosomes (Jurica and Moore 2003). Since the *hsp70* gene is an intronless gene (Silver and Noble 2012), SNRPE may affect the transcription initiation of Hsp70 mRNA, but not pre-mRNA processing. While the knockdown of SNRPE suppressed the steady state expression of the Hsp70 protein, the overexpression of SNRPE alone did not activate the *hsp70* promoter (Saito et al. 2016), suggesting that SNRPE may function cooperative with partner protein such as other snRNP (Leung et al. 2011).

HSP105α and HIF-1

Regulation of HIF-1 Expression and Transcriptional Activation

Tumor microenvironment is exposed to various stresses, such as hypoxia, low pH, and nutrient starvation. In these tumor tissues, hypoxia-inducible transcription factor HIF-1 induces the expression of malignant genes such as VEGF (Semenza 2009). HIF-1 functions as a heterodimer component of oxygen-sensitive HIF-1α, which is rapidly degraded under normoxic conditions, and of constitutively expressed HIF-1β (Wang and Semenza 1995; Wang et al. 1995). Under normoxic conditions, two proline residues of HIF-1α are hydroxylated by prolyl hydroxylases (PHDs), leading to ubiquitination and proteasomal degradation through binding with the von Hippel-Lindau (VHL) ubiquitin ligase complex (Safran and Kaelin 2003). The transcriptional activity of HIF-1α is regulated by hydroxylation of asparagine residue in its C-terminal transactivation domain. Factor-inhibiting HIF-1 (FIH-1) binds to this transactivation domain, which inhibits the binding of HIF-1α to the transcriptional co-activator p300/CBP (Lando et al. 2002). Hypoxia inhibits the hydroxylation of proline and asparagine residues and induces the transcription of HIF-1 target genes through the stabilization and transcriptional activation of HIF-1α. In addition to the hydroxylation-dependent regulation of HIF-1 activity, hydroxylation-independent regulation has been identified, and these are controlled by several proteins including heat shock proteins (Prabhakar and Semenza 2012). However, the role of Hsp105 in the regulation of transcriptional activation of HIF-1 has been not investigated until our recent report.

Hsp105α Interacts with HIF-1 and Regulates HIF-1 Transcriptional Activation

To elucidate the role of Hsp105 in HIF-1 function, we examined and revealed the effects of knockdown of Hsp105α on the HIF-1 transcriptional activation under chemical induced-hypoxic mimic conditions (Mikami et al. 2017). We used cobalt chloride ($CoCl_2$) as a HIF-1α inducer that inhibits the PHD-dependent hydroxylation of HIF-1α (Yuan et al. 2013). Luciferase reporter gene assay using the reporter gene containing the hypoxia responsive element revealed the requirement of Hsp105α for the $CoCl_2$-induced HIF-1 transcriptional activity. Further analysis

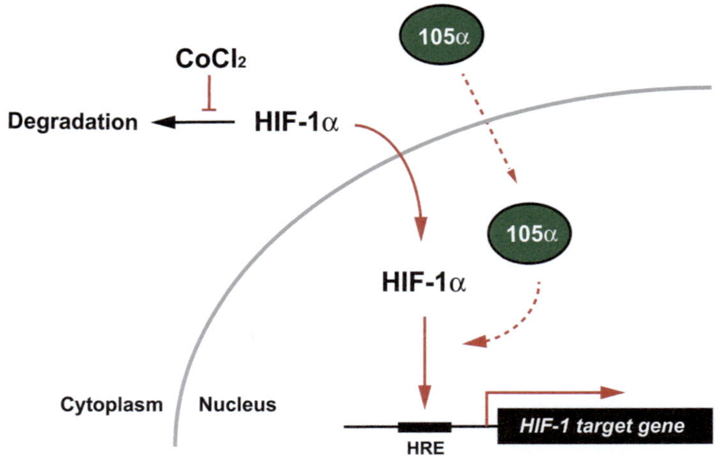

Fig. 3 $CoCl_2$ induces HIF-1α accumulation and nuclear localization of Hsp105α. In $CoCl_2$-treated cells, degradation of HIF-1α is inhibited and accumulates in nucleus. Under these conditions, Hsp105α translocates from cytoplasm to nucleus and interacts with HIF-1α. Hsp105α affects to the HIF-1 accumulation and transcriptional activation, although the exact mechanisms remain unknown

showed that Hsp105α is required for $CoCl_2$-induced accumulation of HIF-1α. $CoCl_2$-induced the nuclear localization of Hsp105α was correlated with HIF-1α expression levels. Importantly, overexpression of degradation resistant HIF-1α enhanced the nuclear localization of Hsp105α without the $CoCl_2$ treatment. In $CoCl_2$-treated cells, Hsp105α accumulated in the nucleus and interacted with HIF-1α, suggesting that this accumulation positively regulates the transcriptional activation of HIF-1 by association with the oxygen-sensitive HIF-1α subunit (Fig. 3). Under $CoCl_2$-induced hypoxic conditions, Hsp105β was not expressed (Mikami et al. 2017). As described above, Hsp105α interacts with Nmi and translocates into the nucleus, and enhances the activation of the *hsp70* promoter through the Stat3 signaling pathway. Nmi has been reported to enhance the recruitment of p300/CBP to nucleus. Since HIF-1α transcriptional activity is positively regulated by the C-terminal transactivation domain of HIF-1α through the binding of transcriptional co-activator p300/CBP (Lando et al. 2002), HIF-1-dependent nuclear translocation of Hsp105α may regulate the p300/CBP-mediated transcriptional activation of HIF-1α through its association with Nmi-associated p300/CBP complex.

HSP105 Overexpression in Tumor

Hsp105 is highly expressed in many types of tumors (Table 1). For example, Nakatsura and his colleagues have been identified Hsp105 as an immunogenic antigen of pancreatic cancer (Nakatsura et al. 2001), and further investigations revealed

Table 1 Overexpression of Hsp105 in tumor tissues

Pancreatic ductal adenocarcinoma	Nakatsura et al. (2001) and Kai et al. (2003)
Colon adenocarcinoma	Nakatsura et al. (2001) and Kai et al. (2003)
Colorectal carcinoma	Hwang et al. (2003)
Gastric carcinoma	Nakatsura et al. (2001) and Kimura et al. (2016)
Esophageal carcinoma	Nakatsura et al. (2001) and Kai et al. (2003)
Hepatocellular carcinoma	Nakatsura et al. (2001)
Breast	Kai et al. (2003)
Urinary bladder	Kai et al. (2003) and Kawai et al. (2014)
Skin squamous cell carcinoma	Muchemwa et al. (2006)
Melanoma	Muchemwa et al. (2008)
Lung adenocarcinoma	Oda et al. (2009)
Esophageal squamous cell carcinoma	Gao et al. (2014)
B-cell non-Hodgkin's lymphoma	Zappasodi et al. (2011)

that Hsp105 is overexpressed in many types of tumor tissues, including pancreas, colon, esophagus, breast, urinary bladder, and skin (Kai et al. 2003; Muchemwa et al. 2006, 2008). Interestingly, the nuclear localization of Hsp105 is observed in some kind of tumors, including esophageal squamous cell carcinoma (Gao et al. 2014), gastric cancer (Kimura et al. 2016), and urinary bladder (Kawai et al. 2014). Gao et al. (2014) reported that Hsp105 is a tumor-associated antigen identified by proteome analysis. Highly expression of Hsp105 was observed in the cytoplasm and nucleus of esophageal squamous cell carcinoma (ESCC) tissues. The authors said that the nuclear localization of Hsp105 was detected only tumor cells of ESCC but not normal cells, and suggested that Hsp105 is a potential tumor marker of early diagnosis.

A recent work of Kimura et al. (2016) showed that Hsp105 (authors called Hsp110 in this manuscript) overexpresses in the nucleus of gastric cancer tissues, and that nuclear expression of Hsp105 is associated with poor prognosis and chemotherapy resistance in gastric cancer. They performed immunohistochemical analysis using a tissue microarray, which were prepared form 210 paraffin blocks of gastric cancer patients. The nuclear expression Hsp105 is associated with invasion, indicating that Hsp105 affects to the cancer progression. Kaplan-Meier plot revealed that the higher nuclear expression of Hsp105 is correlated with poor prognosis. Furthermore, 5-FU or cisplatin sensitivity was increased by Hsp105 knockdown in human gastric adenocarcinoma cell lines. Thus, nuclear Hsp105 expression in gastric cancer may be a prognostic and drug sensitivity maker. In contrast to this report, highly expression of Hsp105 is found in nucleus of urinary bladder and associates with a favorable prognosis (Kawai et al. 2014), indicating that the nuclear overexpression of Hsp105 is not always associated with a poor prognostic.

Conclusions

We focused to examine the Hsp105β-mediated Hsp70 expression and identified three positive regulators including a transcriptional co-activator of Stat3 signaling pathway. We also revealed that Hsp105α interacts with HIF-1α and regulates HIF-1 expression and activation. Importantly, Hsp105α translocates into the nucleus in HIF-1α-expressing cells, suggesting that Hsp105α may function in nucleus under hypoxic conditions. Our preliminary observations suggest that Hsp105α may be localized in nucleus of some squamous cell carcinomas tissues. It would be interesting to know whether the nuclear expression of Hsp105α correlates with the Hsp70 expression and the activation of Stat3 in hypoxic tumor tissues and cells. A recent published paper shows that Hsp105α interacts with Stat3 and enhances Stat3 phosphorilation, and that the Stat3 phosphorylation is correlated with overexpression of Hsp105α in colon cancer patient samples (Berthenet et al. 2016). Given the findings that Hsp105α is involved in tumor progression and aggressiveness (Berthenet et al. 2016; Zappasodi et al. 2015; Yu et al. 2015), the insight into the Hsp105-mediated transcriptional activation of Stat3 and HIF-1 may provide novel strategy for cancer diagnosis and therapy.

Acknowledgements This work was supported, in part, by JSPS KAKENHI (Grant numbers: 26870701, YS and 16K08253, YN), the Kyoto Pharmaceutical University Fund for the Promotion of Scientific Research (YS), and the MEXT-Supported Program for the Strategic Research Foundation at Private Universities (Grant number: S1311035).

References

Balchin, D., Hayer-Hartl, M., & Hartl, F. U. (2016). In vivo aspects of protein folding and quality control. *Science, 353*, aac4354.

Berthenet, K., Bokhari, A., Lagrange, A., Marcion, G., Boudesco, C., Causse, S., De Thonel, A., Svrcek, M., & Goloudina. A. R. (2016). HSP110 promotes colorectal cancer growth through STAT3 activation. *Oncogene, 36*, 2328–2336.

Bracher, A., & Verghese, J. (2015). The nucleotide exchange factors of Hsp70 molecular chaperones. *Frontiers in Molecular Biosciences, 78*, 1–33.

Darnell, J. E. (1997). STATs and gene regulation. *Science, 277*, 1630–1635.

Gao, H., Zhaoxu, Z., Yousheng, M., Wei, W., Yuanyuan, Q., Lanping, Z., Fang, L., Hongzhi, H., & Xiaohang, Z. (2014). Identification of tumor antigens that elicit a humoral immune response in the sera of Chinese esophageal squamous cell carcinoma patients by modified serological proteome analysis. *Cancer Letters, 344*, 54–61.

Hatayama, T., Nishiyama, E., & Yasuda, K. (1994a). Cellular localization of high-molecular-mass heat shock proteins in murine cells. *Biochemical and Biophysical Research Communications, 200*, 1367–1373.

Hatayama, T., Yasuda, K., & Nishiyama, E. (1994b). Characterization of high-molecular-mass heat shock proteins and 42 degrees C-specific heat shock proteins of murine cells. *Biochemical and Biophysical Research Communications, 204*, 357–365.

He, N., Liu, M., Hsu, J., Xue, Y., Chou, S., Burlingame, A., Krogan, N. J., Alber, T., & Zhou, Q. (2010). HIV-1 Tat and host AFF4 recruit two transcription elongation factors into a bifunctional complex for coordinated activation of HIV-1 transcription. *Molecular Cell, 38*, 428–438.

Hwang, T. S., Han, H. S., Choi, H. K., Lee, Y. J., Kim, Y. J., Han, M. Y., & Park, Y. M. (2003). Differential, stage-dependent expression of Hsp70, Hsp110 and Bcl-2 in colorectal cancer. *Journal of Gastroenterology and Hepatology, 18*, 690–700.

Ishihara, K., Yasuda, K., & Hatayama, T. (1999). Molecular cloning, expression and localization of human 105 kDa heat shock protein, hsp105. *Biochimica et Biophysica Acta, 1444*, 138–142.

Jurica, M. S., & Moore, M. J. (2003). Pre-mRNA splicing: Awash in a sea of proteins. *Molecular Cell, 12*, 5–14.

Kaddurah-Daouk, R., Greene, J. M., Baldwin, A. S., Jr., & Kingston, R. E. (1987). Activation and repression of mammalian gene expression by the c-myc protein. *Genes & Development, 1*, 347–357.

Kai, M., Nakatsura, T., Egami, H., Senju, S., Nishimura, Y., & Ogawa, M. (2003). Heat shock protein 105 is overexpressed in a variety of human tumors. *Oncology Reports, 10*, 1777–1782.

Kampinga, H. H., Hageman, J., Vos, M. J., Kubota, H., Tanguay, R. M., Bruford, E. A., Cheethama, M. E., Chen, B., & Hightower, L. E. (2009). Guidelines for the nomenclature of the human heat shock proteins. *Cell Stress & Chaperones, 14*, 105–111.

Kawai, T., Enomoto, Y., Morikawa, T., Matsushita, H., Kume, H., Fukayama, M., Yamaguchi, H., Kakimi, K., & Homma, Y. (2014). High expression of heat shock protein 105 predicts a favorable prognosis for patients with urinary bladder cancer treated with radical cystectomy. *Molecular and Clinical Oncology, 2*, 38–42.

Kimura, A., Ogata, K., Altan, B., Yokobori, T., & Ide, M. (2016). Nuclear heat shock protein 110 expression is associated with poor prognosis and chemotherapy resistance in gastric cancer. *Oncotarget, 7*, 18415–18423.

Kingston, R. E., Baldwin, A. S., Jr., & Sharp, P. A. (1984). Regulation of heat shock protein 70 gene expression by c-myc. *Nature, 312*, 280–282.

Lando, D., Peet, D. J., Gorman, J. J., Whelan, D. A., Whitelaw, M. L., & Bruick, R. K. (2002). FIH-1 is an asparaginyl hydroxylase enzyme that regulates the transcriptional activity of hypoxia-inducible factor service FIH-1 is an asparaginyl hydroxylase enzyme that regulates the transcriptional activity of hypoxia-inducible factor. *Genes & Development, 214*, 1466–1471.

Leung, A. K. W., Nagai, K., & Li, J. (2011). Structure of the spliceosomal U4 snRNP core domain and its implication for snRNP biogenesis. *Nature, 473*, 536–539.

Li, Y., Wen, H., Xi, Y., Tanaka, K., Wang, H., Peng, D., Ren, Y., Jin, Q., Dent, S. Y. R., Li, W., Li, H., & Shi, X. (2014). AF9 YEATS domain links histone acetylation to DOT1L-mediated H3K79 methylation. *Cell, 159*, 558–571.

Lin, C., Smith, E. R., Takahashi, H., Lai, K. C., Martin-Brown, S., Florens, L., Washburn, M. P., Conaway, J. W., Conaway, R. C., & Shilatifard, A. (2010). AFF4, a component of the ELL/P--TEFb elongation complex and a shared subunit of MLL chimeras, can link transcription elongation to leukemia. *Molecular Cell, 37*, 429–437.

Mikami, H., Saito, Y., Okamoto, N., Kakihana, A., Kuga, T., & Nakayama, Y. (2017). Requirement of Hsp105 in $CoCl_2$-induced HIF-1α accumulation and transcriptional activation. *Experimental Cell Research, 352*, 225–233.

Morimoto, R. I., Sarge, K. D., & Abravaya, K. (1992). Transcriptional regulation of heat shock genes. *The Journal of Biological Chemistry, 267*, 21987–21990.

Muchemwa, F. C., Nakatsura, T., Ihn, H., & Kageshita, T. (2006). Heat shock protein 105 is overexpressed in squamous cell carcinoma and extramammary Paget disease but not in basal cell carcinoma. *The British Journal of Dermatology, 155*, 582–585.

Muchemwa, F. C., Nakatsura, T., Fukushima, S., Nishimura, Y., Kageshita, T., & Ihn, H. (2008). Differential expression of heat shock protein 105 in melanoma and melanocytic naevi. *Melanoma Research, 18*, 166–171.

Murphy, M. E. (2013). The HSP70 family and cancer. *Carcinogenesis, 34*, 1181–1188.

Nakatsura, T., Senju, S., Yamada, K., Jotsuka, T., Ogawa, M., & Nishimura, Y. (2001). Gene cloning of immunogenic antigens overexpressed in pancreatic cancer. *Biochemical and Biophysical Research Communications, 281*, 936–944.

Nillegoda, N. B., & Bukau, B. (2015). Metazoan Hsp70-based protein disaggregases: Emergence and mechanisms. *Frontiers in Molecular Biosciences, 2*, 57.

Oda, T., Morii, E., Inoue, M., Ikeda, J.-I., Aozasa, K., & Okumura, M. (2009). Prognostic significance of heat shock protein 105 in lung adenocarcinoma. *Molecular Medicine Reports, 2*, 603–607.

Prabhakar, N. R., & Semenza, G. L. (2012). Adaptive and maladaptive cardiorespiratory responses to continuous and intermittent hypoxia mediated by hypoxia-inducible factors 1 and 2. *Physiological Reviews, 92*, 967–1003.

Ray, S., Lu, Y., Kaufmann, S. H., Gustafson, W. C., Karp, J. E., Boldogh, I., Fields, A. P., & Brasier, A. R. (2004). Genomic mechanisms of p210BCR-ABL signaling: Induction of heat shock protein 70 through the GATA response element confers resistance to paclitaxel-induced apoptosis. *The Journal of Biological Chemistry, 279*, 35604–35615.

Safran, M., & Kaelin, W. G. (2003). HIF hydroxylation and the mammalian oxygen-sensing pathway. *The Journal of Clinical Investigation, 111*, 779–783.

Saito, Y., Yamagishi, N., & Hatayama, T. (2007). Different localization of Hsp105 family proteins in mammalian cells. *Experimental Cell Research, 313*, 3707–3717.

Saito, Y., Yamagishi, N., & Hatayama, T. (2009). Nuclear localization mechanism of Hsp105β and its possible function in mammalian cells. *Journal of Biochemistry, 145*, 185–191.

Saito, Y., Yukawa, A., Matozaki, M., Mikami, H., Yamagami, T., Yamagishi, N., Kuga, T., & Nakayama, Y. (2014). Nmi interacts with Hsp105β and enhances the Hsp105β-mediated Hsp70 expression. *Experimental Cell Research, 327*, 163–170.

Saito, Y., Nakagawa, T., Kakihana, A., Nakamura, Y., Nabika, T., Kasai, M., Takamori, M., Yamagishi, N., Kuga, T., Hatayama, T., & Nakayama, Y. (2016). Yeast two-hybrid and one-hybrid screenings identify regulators of *hsp70* gene expression. *Journal of Cellular Biochemistry, 117*, 2109–2117.

Semenza, G. L. (2009). Defining the role of hypoxia-inducible factor 1 in cancer biology and therapeutics. *Oncogene, 29*, 625–634.

Shi, Y., Mosser, D. D., & Morimoto, R. I. (1998). Molecular chaperones as HSF1-specific transcriptional repressors. *Genes & Development, 12*, 654–666.

Silver, J. T., & Noble, E. G. (2012). Regulation of survival gene hsp70. *Cell Stress & Chaperones, 17*, 1–9.

Stephanou, A., Isenberg, D. A., Nakajima, K., & Latchman, D. S. (1999). Signal transducer and activator of transcription-1 and heat shock factor-1 interact and activate the transcription of the Hsp-70 and Hsp-90β gene promoters. *The Journal of Biological Chemistry, 274*, 1723–1728.

Taira, T., Sawai, M., Ikeda, M., Tamai, K., Iguchi-Ariga, S. M. M., & Ariga, H. (1999). Cell cycle-dependent switch of up- and down-regulation of human hsp70 gene expression by interaction between c-Myc and CBF/NF-Y. *The Journal of Biological Chemistry, 274*, 24270–24279.

Wang, G. L., & Semenza, G. L. (1995). Purification and characterization of Hypoxia-inducible Factor 1. *The Journal of Biological Chemistry, 270*, 1230–1237.

Wang, G. L., Jiang, B. H., Rue, E. A., & Semenza, G. L. (1995). Hypoxia-inducible factor 1 is a basic-helix-loop-helix-PAS heterodimer regulated by cellular O_2 tension. *Genetics, 92*, 5510–5514.

Westerheide, S. D., Anckar, J., Stevens, S. M. J., Sistonen, L., & Morimoto, R. I. (2009). Stress-inducible regulation of Heat shock factor 1 by the deacetylase SIRT1. *Science, 323*, 1063–1067.

Yamagishi, N., Nishihori, H., Ishihara, K., Ohtsuka, K., & Hatayama, T. (2000). Modulation of the chaperone activities of Hsc70/Hsp40 by Hsp105α and Hsp105β. *Biochemical and Biophysical Research Communications, 272*, 850–855.

Yamagishi, N., Ishihara, K., & Hatayama, T. (2004). Hsp105α suppresses Hsc70 chaperone activity by inhibiting Hsc70 ATPase activity. *The Journal of Biological Chemistry, 279*, 41727–41733.

Yamagishi, N., Saito, Y., & Hatayama, T. (2008). Mammalian 105 kDa heat shock family proteins suppress hydrogen peroxide-induced apoptosis through a p38 MAPK-dependent mitochondrial pathway in HeLa cells. *The FEBS Journal, 275*, 4558–4570.

Yamagishi, N., Fujii, H., Saito, Y., & Hatayama, T. (2009). Hsp105β upregulates hsp70 gene expression through signal transducer and activator of transcription-3. *The FEBS Journal, 276*, 5870–5880.

Yamamoto, K., Furukawa, M. T., Fukumura, K., Kawamura, A., Yamada, T., Suzuki, H., Hirose, T., Sakamoto, H., & Inoue, K. (2016). Control of the heat stress-induced alternative splicing of a subset of genes by hnRNP K. *Genes to Cells, 21*, 1006–1014.

Yasuda, K., Nakai, A., Hatayama, T., & Nagata, K. (1995). Cloning and expression of murine high molecular mass heat shock proteins, HSP105. *The Journal of Biological Chemistry, 270*, 29718–29723.

Yasuda, K., Ishihara, K., Nakashima, K., & Hatayama, T. (1999). Genomic cloning and promoter analysis of the mouse 105-kDa heat shock protein (HSP105) gene. *Biochemical and Biophysical Research Communications, 256*, 75–80.

Yu, N., Kakunda, M., Pham, V., Lill, J. R., Du, P., Wongchenko, M., Matthew, Y., Yan, Y., Firestein, R., & Huang, X. (2015). HSP105 recruits protein phosphatase 2A to dephosphorylate β-catenin. *Molecular and Cellular Biology, 35*, 1390–1400.

Yuan, Y., Hilliard, G., Ferguson, T., & Millhorn, D. E. (2013). Cobalt inhibits the interaction between hypoxia-inducible factor-α and von Hippel-Lindau protein by direct binding to hypoxia-inducible factor-α. *The Journal of Biological Chemistry, 278*, 15911–15916.

Zappasodi, R., Bongarzone, I., Ghedini, G. C., Castagnoli, L., Cabras, A. D., Messina, A., Tortoreto, M., Tripodo, C., Magni, M., Carlo-Stella, C., Gianni, A. M., Pupa, S. M., & Di Nicola, M. (2011). Serological identification of HSP105 as a novel non-Hodgkin lymphoma therapeutic target. *Blood, 118*, 4421–4430.

Zappasodi, R., Ruggiero, G., Guarnotta, C., Tortoreto, M., Tringali, C., Cavanè, A., Cabras, A., Cabras, A. D., Castagnoli, L., Venerando, B., Zaffaroni, N., Gianni, A. M., De Braud, F., Tripodo, C., Pupa, S. M., & Di Nicola, M. (2015). HSPH1 inhibition downregulates Bcl-6 and c-Myc and hampers the growth of human aggressive B-cell non-hodgkin lymphoma. *Blood, 125*, 1768–1771.

Zhu, M., John, S., Berg, M., & Leonard, W. J. (1999). Functional association of Nmi with Stat5 and Stat1 in IL-2- and IFNgamma-mediated signaling. *Cell, 96*, 121–130.

Suppression of HSP70 Expression by Quercetin and Its Therapeutic Potential Against Cancer

Evren Önay Uçar, Aslıhan Şengelen, Elif Mertoğlu, Murat Pekmez, and Nazlı Arda

Abstract Heat shock response is one of several survival pathways that protects cells against harsh conditions. This response mechanism, which is evolutionarily protected in all organisms, enhances the expression of heat shock proteins (HSP) that show protective properties for cells under stress conditions. High expression of many HSP is observed in cancer, and their functions aides the advancement of disease. It is known that overexpression of HSP70, a member of HSP family, in cancerous cells has been closely associated with tumor cell proliferation, apoptosis inhibition, enhanced migration and metastasis and drug resistance promotion. Therefore, targeting HSP70 in cancer treatment is very important. One of the best-studied inhibitors known for HSP70 is quercetin that is widely distributed flavonoid in the plant kingdom. Several *in vivo* and *in vitro* studies have reported the efficacy of quercetin in reducing elevated HSP70 levels in cancer therapy. It has become a focal point as an anticancer agent because of the induction of apoptosis in many different cancer cells. In this chapter, we reviewed the role of HSP70 in different cancer types and the suppressive effect of quercetin on expression of HSP70 family members. Moreover, we emphasized molecular mechanisms targeted by quercetin in cancer and its relationship to Hsp70.

Keywords Apoptosis · Cancer · HSP70 · Quercetin · Stress proteins · Therapeutic target

---

E. Ö. Uçar (✉) · A. Şengelen · E. Mertoğlu · M. Pekmez N. Arda
Department of Molecular Biology & Genetics, Faculty of Science, Istanbul University, Istanbul, Turkey
e-mail: evrenonay@istanbul.edu.tr

A. A. A. Asea, P. Kaur (eds.), *HSP70 in Human Diseases and Disorders*, Heat Shock Proteins 14, https://doi.org/10.1007/978-3-319-89551-2_19

Abbreviations

Akts (or PKB)	protein kinase B
AMPK	AMP activated protein kinase
CaMKII	calcium/calmodulin-dependent protein kinase II
Cdk	cyclin-dependent kinases
Chk2	checkpoint kinase 2
CK2	casein kinase 2
ER	endoplasmic reticulum
ERK	extracellular signal-regulated kinase
Hsc	heat shock cognate
HSE	heat shock element
HSF	heat shock factor
HSP	heat shock protein
IL-6	interleukin-6
JAK	Janus kinase
JNK	C-Jun N-terminal kinase
MAPK	mitogen-activated protein kinase
MMP	matrix metalloproteinase
PI3K	phosphatidylinositol 3-kinase
pRb	retinoblastoma protein
ROS	reactive oxygen species
RSK2	ribosomal protein S6 kinase 2
S6K1	ribosomal protein S6 kinase beta-1
shRNA	short hairpin RNA
siRNA	small interfering RNA
STAT3	signal transducer and activator of transcription 3
VEGF	vascular endothelial growth factor

Introduction

Heat shock proteins (HSP, also called stress proteins) are a group of highly evolutionary conserved proteins whose expression are induced under many stress conditions such as heat shock, oxidative/ischemic stress, toxins, heavy metals, radiation, environmental pollutants and chemotherapy. They are also stimulated by the release of cytokines in the cell. Mammalian HSP are classified mainly into six families by their molecular weight: HSP100, HSP90, HSP70, HSP60, HSP40, and small HSP (Benjamin and McMillan 1998; Snoeckx et al. 2001). Many of them act as molecular chaperones and are responsible for maintaining protein homeostasis in normal cells under non-stressful conditions. However, high expression of many HSP is observed during the many diseases including cancer, and their functions aides the advancement of disease. In cancer which is one of the leading causes of death worldwide, induction of elevated HSP expression has crucial roles in tumor onset and progression processes such as rapid cell proliferation, evading apoptosis, and metastasis. Because of these properties, HSP have become one of the major therapeutic targets in cancer therapy (McConnell and McAlpine 2013; Lianos et al. 2015; Önay-Uçar 2015; Giri et al. 2017). In particular, increased expression of members of the HSP70 family in high grade malignant tumors has been reported. HSP70 has been shown to be highly expressed in many cancers including bladder, breast, colorectal, endometrial, gastric, lung, oral, uterine cervical and prostate cancer. Decreasing HSP70 levels in cancer cells will be beneficial, because increased HSP70 expression in cancerous cells is associated with cell proliferation, metastasis and poor prognosis (Ciocca and Calderwood 2005; Rohde et al. 2005; Evans et al. 2010; Murphy 2013; Alexiou et al. 2014).

Considering the possible therapeutic potential of suppressing of HSP70 expression in cancerous cells, a large number of studies in recent years have focused on quercetin, one of the most common bioflavonoids in the plant kingdom. Quercetin is well-known for its physiological functions and its role in the elimination of cancerous cells. It has important roles in the inhibition of increased HSP and in the activation of cell death pathways in cancer (Khan et al. 2016). In recent years, the number of studies on the activation of cell death pathways by quercetin-mediated suppression of HSP70s expression has increased. In this chapter, we reviewed the status of elevated HSP70 in various human cancers, and emphasized the effects of quercetin on the suppression of HSP70 expression and its therapeutic potential against cancer.

HSP70 and Cancer

The heat shock protein 70 (HSP70) family, one of the HSP families, consists of ATP dependent chaperones with molecular weight of approximately 70 kDa (in range 66–78 kDa). Their structure and function are evolutionary conserved and can be found in all organisms. Nowadays, there are 13 known homologous members of the human HSP70 family encoded by the HSPA gene family. They are transcriptionally regulated by four members of Heat Shock Factors (HSF): HSF1, HSF2, HSF3, and HSF4. The major transcriptional regulator HSF1 is induced during stress and is bound to the promoter of HSP70 to increase the transcription. The gene products differ by amino acid sequence, expression level, and cellular localization. Generally, these proteins, which exhibiting 52–99% amino acid sequence homology, function in protein homeostasis. It is known that HSP70 family members which play a role in correct protein folding and survival of the cells under stress conditions (Kampinga et al. 2009; Zuiderweg et al. 2013; Sherman and Gabai 2015; Giri et al. 2017).

The HSP70 family members are considerable important chaperones in cancer. There are eight well-studied members related to cancer. Six of them are mainly found in cytosol: Hsp70–1a and Hsp70-1b, collectively called Hsp70-1 (also known as Hsp70 or Hsp72), Hsp70-1t (also called Hsp70-hom), Hsp70-2, Hsp70-6 and heat shock cognate 70 (Hsc70 also called Hsp70-8 or Hsp73). One of them localizes to the endoplasmic reticulum (Hsp70-5, also known as BiP or Grp78) and one to the mitochondria (Hsp70-9, also called mtHsp70, Grp75 or mortalin) (Rohde et al. 2005; Daugaard et al. 2007a; Murphy 2013). They play a key role in the regulation of malignancy. Tumor cells that overexpress HSP70 are found to be associated with tumor cell proliferation, metastasis, resistance to therapies and poor prognosis in many human cancers (Rohde et al. 2005; Powers et al. 2008; Evans et al. 2010; Murphy 2013; Rodina et al. 2014). It is also known that the chaperones of the HSP70 family have a well-documented antiapoptotic function. Several studies in different tumor models indicated that overexpressed HSP70 blocks apoptosis by interacting with components of the apoptotic pathways, and by interfering in the apoptotic signaling such as apoptosome, caspases, and cathepsins (Mosser and Morimoto 2004; Garrido et al. 2006; Evans et al. 2010; Kumar et al. 2016).

The most published data in human cancers is on Hsp70. The two best studied members of the HSP70 family are ubiquitously expressed Hsc70 and Hsp70 (Hsp70-1). There is less information about the expression levels of other HSP70 family members. Hsc70 is the only cytosolic HSP70 protein expressed constitutively under normal physiologic conditions and found abundantly in all major intracellular compartments. It is required for normal cell growth (Liu et al. 2012). In addition, Hsc70 overexpressed in colon and esophageal cancers is known (Kubota et al. 2010; Moghanibashi et al. 2013). Hsc70 has been also known to regulate functions of tumor-associated genes and proteins (Wu et al. 2017). Whereas Hsp70 is generally expressed at basal levels in normal cells under non-stressful conditions, it is the major stress inducible protein and distributed predominantly in the cytoplasm, nucleus, and plasma membrane of various malignant tumor cells. Its essential role is to maintain cell survival under stressful conditions. It is known that high Hsp70 levels are correlated with an aggressive phenotype and poor prognosis in therapeutic responses in bladder, brain, breast, colorectal, endometrial, gastric, oral, uterine cervical and prostate cancers (Ciocca and Calderwood 2005; Evans et al. 2010; Giri et al. 2017).

Downregulation of Hsp70 expression is associated with reduced tumorigenicity, and cytotoxic to transformed cells but undetectable in non-transformed cells. For this reason, Hsp70 knockdown sensitizes or kills cancer cells rather than normal cells (Schmitt et al. 2006; Kumar et al. 2016). Nylandsted and co-workers (2000) have shown that elevated expression of Hsp70 is necessary for the survival of tumorigenic breast cancer cells, and that the reduction of Hsp70 levels activates the tumor-specific death program. In a study of MCF-10A cells exposed to NZ28 which is an inhibitor of heat shock response, the researchers developed this inhibitor to enhance main effect of HSF1 on tumor development via upregulation of Hsp70. Knockout of Hsp70 is sufficient to similarly prevent cancer development in the Her2-positive breast cancer (Meng et al. 2010). It has been also found that inducible Hsp70 is effective in resistance of cancer cells to gamma radiation (Lee et al. 2001). Increased radiosensitivity was observed in a study on Hsp70 knockout mice (Hunt et al. 2004). Hsp70 overexpression in breast cancer cells has been found to be associated with lymph node metastasis (Kluger et al. 2005). Similarly, it is shown that there is a correlation between vascular invasion and Hsp70 overexpression in gastric cancers (Canöz et al. 2002). Studies on patients with brain tumors have showed the presence of high expression of Hsp70 in gliomas, such as meningiomas and medulloblastomas (Alexiou et al. 2013, 2014). In a study performed on patients with prostate cancer, the Hsp70 expression of the patients was found to be higher than the control group, and Hsp70 was reported to be a potential biomarker for prostate cancer (Abe et al. 2004). Silencing of Hsp70 by short hairpin (sh)RNA in cervical and bladder cancer cells has been shown to suppress invasion and migration (Teng et al. 2012). Also, in the same study it has been demonstrated that Hsp70 interacted with Wiskott-Aldrich syndrome protein family 3 (WASF3), which is involved in prostate cancer invasion and metastasis. In bladder cancer cells, it has been found that the p63α protein -is a member of the p53 family- promotes the invasion and Hsp70 transcription, which has led to a new perspective in the understanding of Hsp70 function in high-invasive bladder cancer (Jin et al. 2017). In addition to studies on tissue Hsp70, studies on serum Hsp70 have also been carried out. Serum Hsp70s, which can be released by exocytotic traffic or cell disruption, have begun to gain importance in many studies. It has been found significantly high level of serum Hsp70 of patients with small cell lung cancer compared to healthy controls (Balázs et al. 2017). Similar findings observed in studies with colorectal cancer (Kocsis et al. 2010), chronic myeloid leukemia (Yeh et al. 2009), and head and neck squamous cell carcinoma (Ghermann et al. 2014).

There are few studies in the literature on other members of the HSP70 family associated with cancer except Hsc70 and Hsp70 proteins. Rohde et al. (2005) have reported that depletion of Hsp70 leaded to cancer cell detachment in many of human cancer cells. This data has shown that Hsp70 plays a role in the regulation of cancer cell adhesion. They have also reported for the first time that expression level of Hsp70-2, which is a member of HSP70 family, increases in breast, cervix, prostate and colon and liver cancers. In their study, it is shown that not only Hsp70-1 (Hsp70) but also Hsp70-2 are required for the survival of cancer cells. In another study on Hsp70-2, this protein depletion was found to induce lysosomal membrane permeabilization and cathepsin-dependent cell death in a variety of human cancers (Daugaard et al. 2007b). The level of Hsp70-5 (also known as BiP or Grp78), a member of the HSP70 family localized to the ER lumen, has been shown to be highly elevated associated with poor differentiation in many cancer types, such as breast cancer (Gazit et al. 1999; Fernandez et al. 2000), gastric cancer (Song et al. 2001), hepatocellular cancer (Shuda et al. 2003; Chen et al. 2014), lung cancer (Koomägi et al. 1999) and prostate cancer (Arap et al. 2004; Misra et al. 2005). For cancer cells subject to ER stress, Hsp70-5 acts as a survival factor, and plays an important role in drug resistance in addition to preventing apoptosis and autophagy (Lee 2007; Wu et al. 2017). Another member of the HSP70 family localized in the mitochondria, Hsp70-9 (also known as mtHsp70, Grp75 or mortalin) functions in human carcinogenesis, promotes proliferation and survival of cancer cells (Deocaris et al. 2013; Starenki et al. 2015). Elevated levels of this protein expression have been reported in human brain tumors (Takano et al. 1997), ovarian cancer (Hu et al. 2016), colorectal adenocarcinomas (Dundas et al. 2005), hepatocellular carcinoma (Yi et al. 2008) and medullary

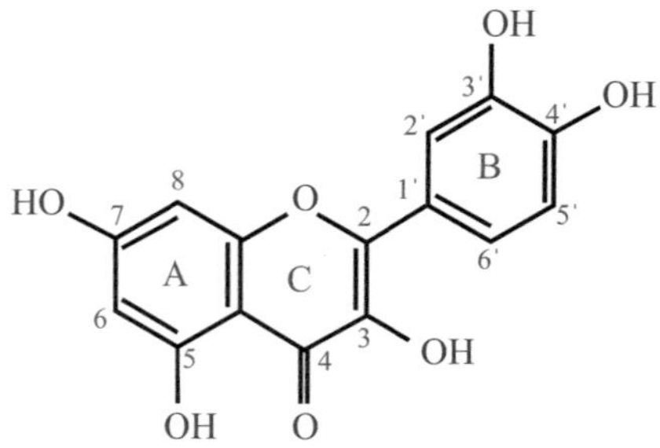

Fig. 1 Structure of quercetin

thyroid carcinomas (Starenki et al. 2015). Hsp70-9 is known to play a role in cancer formation by activating the MAPK/MEK/ERK pathway and binding to p53 in the cytoplasm to prevent translocation to the nucleus (Hu et al. 2016). Hsp70-6, another member of the HSP70 family, is stimulated at high rates by stress while the expression of it is very low under normal physiological conditions, and it is thought to have an anti-apoptotic function (Regeling et al. 2016; Wu et al. 2017).

In summary, all these studies support members of the HSP70 family as an attractive target in cancer therapy. The studies have shown that suppression of HSP70 expressions in tumor cells would become beneficial for treatment of cancers and the development of new approaches against cancer. Acting HSP70 as an apoptosis inhibitor in tumor cells supports HSP70 to be a potential target for apoptosis-based cancer therapy. However, other molecules that interact with HSP70, and mechanisms of the HSP70 family have not been fully elucidated. New studies are needed to understand how all members work.

Quercetin and Cancer

Quercetin (3,3′,4′,5,7-pentahydroxyflavone), an important member of the flavonoid family, is a highly abundant polyphenolic compound found in various vegetables and fruits, such as apples, berries, broccoli, cabbage, dill, grapes, lemons, onions, tomatoes, and in beverages such as tea, coffee and red wine. The daily intake of quercetin is estimated to range between 3 and 31 mg (Duthie et al. 2000; Khan et al. 2016). Quercetin has been shown to possess a wide range of biological and pharmacological activities, including antioxidant, anticarcinogenic, antiproliferative, antiinflammatory, antiviral and antiallergic properties (Harwood et al. 2007; Gupta et al. 2010; Vargas and Burd 2010; Gibellini et al. 2011; Khan et al. 2016). Compared to other flavonoids, quercetin has a very high antiradical property. Quercetin reveals this property through its three active functional groups in its structure as presented in Fig. 1. These are the o-dihydroxy (catechol) structure at the 3′- and 4′-position of the B ring, the 2,3-double bond in the conjugation with a 4-oxo group, and the hydroxyl groups in the 3- and 5-position (Bors et al. 1990; Silva et al. 2002; Wang et al. 2006).

Additionally, studies on quercetin have shown that it has antioxidant or prooxidant effects depending on its concentration. While quercetin has an antioxidant effect at low doses, it has the opposite effect (pro-oxidant effect) at high doses. The antioxidant and chemopreventive effects of quercetin are seen at 1–40 μM cellular concentrations of it. However, it has been reported that quercetin could act like a ROS (Reactive Oxygen Species) at concentrations higher than 40 μM after tumor formation, and thereby it could still be useful as an antitumoral agent by increasing the oxidative stress and cytotoxicity in tumor cells (Metodiewa et al. 1999; Awad et al. 2000; Vargas and Burd 2010).

Several literature studies report on cancer preventive and therapeutic effects of quercetin in different cell lines. It demonstrates anti-cancer properties by regulating various cell signaling mechanisms, by binding to cellular receptors and proteins, and by inhibiting various proteins that are effective in carcinogenesis. It is well known for its proapoptotic effect in various tumor cells (Murakami et al. 2008; Khan et al. 2016). Quercetin is also effective in the inhibition of Hsp, and quercetin-mediated Hsp inhibition has an important role in stimulation of cell death. It is known that quercetin inhibits HSF1 activation that induces Hsp70 expression (Vargas and Burd 2010; Kumar et al. 2016). Several *in vivo* and *in vitro* studies have demonstrated that quercetin has protective and preventive effects against cancers such as brain, breast, cervix, colorectum, lung and prostate (Table 1). Jakubowicz-Gil et al. (2002) have shown that quercetin reduces Hsp27 and Hsp72 expression and increases the number of apoptotic cells in human cervix and glioma cell lines. 15 μg/mL of quercetin was found to increase apoptosis by ~16% in human cervical carcinoma cell line (HeLa cells). In a study with MOG-G-CCM cells, human brain astrocytoma cells, co-administration of temozolomide with quercetin has been shown to be a useful, potent and promising combination for glioma treatment. In the study, it was shown that quercetin significantly reduced Hsp27 expression at low concentrations and that the combination of 100 μM temozolomide and 30 μM quercetin was highly effective in inducing apoptosis (Jakubowicz-Gil et al. 2010). In a similar study, co-administration of 50 μM quercetin with 100 μM temozolomide has been shown to significantly reduce the expression of Hsp27 and Hsp72 (Jakubowicz-Gil et al. 2013b). In a study performed on U-251 MG and U87 MG cells, it was determined that administration of quercetin at a concentration of 30 μM to cells reduced Hsp27 expression at a significant level (Li et al. 2016). *In vivo* studies in various animal models also show that quercetin reduces tumor growth and tumor volume.

In a study on colon carcinogenesis by Dihal et al. (2008), it was observed that the potential oncogenic MAPK signal decreased in F344 rats fed with 10 g quercetin/kg for 11 weeks. In another study by Jones et al. (2004), it has been shown that quercetin suppresses Hsp70 expression in PC-3 prostate cancer cells. Furthermore, apoptosis has been shown to have a negative correlation with the HSP70 expression. Hsp70 expression was increased in multiple pancreatic cancer cells compared with normal pancreatic ductal cells. Downregulation of HSP70 with siRNA in pancreatic cancer cells causes caspase dependent apoptotic cell death. Depletion of Hsp70 with quercetin decreased cell viability and induced apoptosis in cancer cells but not in normal pancreatic ductal cells. Quercetin treatment decreased tumor size and Hsp70

Table 1 *In vivo* and *in vitro* studies on the anti-cancer effects of quercetin

Cancer type	Animal model or cell line	Findings	References
Brain	U138MG cells	Inhibited proliferation, induced apoptosis (caspase-3/7 activation)	Braganhol et al. (2006)
	A172 cells	Inhibited proliferation, induced apoptosis (caspase-3 activation)	Kim et al. (2008)
	U87-MG and U251 cells	Reduced expression of Survivin and XIAP, induced apoptosis (TRAIL-Induced Apoptosis)	Siegelin et al. (2009)
		Reduced expression of Hsp27	Li et al. (2016)
	MOGGCCM cells	Reduced expression of Hsp27, induced apoptosis and autophagy	Jakubowicz-Gil et al. (2010)
	T98G cells	Reduced expression of Hsp27 and Hsp72, induced apoptosis (caspases 3/8/9 activation)	Jakubowicz-Gil et al. (2013a, b)
		Inhibited proliferation, induced apoptosis	Bądziul et al. (2014a)
Breast	MDA-MB-231 cells	Inhibited proliferation	Conklin et al. (2007)
	MCF-7 cells	Inhibited proliferation, induced apoptosis (Bcl-2 and Bax regulation)	Duo et al. (2012)
		Inhibited cell growth (G2/M arrest), induced apoptosis	Choi et al. (2001)
	4 T1 cells	Inhibited proliferation (Regulation of Wnt signaling activity)	Kim et al. (2013a)
Cervix	HeLa cells	Reduced expression of Hsp27 and Hsp72, induced apoptosis	Jakubowicz-Gil et al. (2002, 2005)
		Reduced expression of Hsp70, induced apoptosis (caspases 3 activation)	Jung et al. (2010)
		Reduced expression of Hsp72, induced apoptosis (caspases 3 activation)	Bądziul et al. (2014b)
Colorectum	F344 rats	Decreased oncogenic MAPK signal	Dihal et al. (2008)
	HCT116 cells	Inhibited proliferation	Shan et al. (2009)
	Mutant Apc mice	Loss of Hsp70, reduced tumor size, increased tumor cell death, increased apoptosis	Tao et al. (2016)
Leukemia	U937 cells	Reduced expression of Hsp70, induced apoptosis	Storniolo et al. (2015)
Lung	A549 cells	Reduced expression of Hsp70, induced apoptosis	Niu et al. (2006)
		Inhibited cell growth (G2/M arrest)	Yeh et al. (2011)
	A549 and H460 cells	Reduced cell viability, suppressed HSP70 expression, increased apoptosis (caspase-3 and caspase-9 activation)	Lee et al. (2015)

(continued)

Table 1 (continued)

Cancer type	Animal model or cell line	Findings	References
Pancreas	MiaPaCa-2 and Panc-1 cells	Reduced expression of Hsp70, induced apoptosis	Aghdassi et al. (2007)
	Nude mice	Decreased tumor size and Hsp70 levels	
Prostate	PC-3 cells	Reduced expression of Hsp70, induced apoptosis	Jones et al. (2004)
		Inhibited proliferation, reduced expression of matrix metalloproteinases 2 and 9 proteins	Vijayababu et al. (2006)
	PC-3 and LNCaP cells	Inhibited proliferation, reduced expression of Hsp90	Aalinkeel et al. (2008)
		Inhibited proliferation	Bhat et al. (2014)

levels in mice (Aghdassi et al. 2007). It has been found that reduced expression of HSP70 by both siRNA or by quercetin causes an enhancement in cytosolic calcium levels. Reduced expression of HSP70 in pancreatic cancer cells leads to release of lysosomal enzymes into the cytosol due to lysosomal membrane permeabilization. Lysosomal enzymes activate cell death pathways in the cytosol (Dudeja et al. 2009). Niu et al. (2006) have reported a significant increase in Hsp70 level after heat shock at 41 °C for 1 h in the study conducted in A549 cells, but the increase in Hsp70 was reported to be inhibited by quercetin treatment for 6 h before heat shock treatment. In addition to a significant decrease in Hsp70 levels, apoptosis has also been found to be induced. As a result, many studies have indicated that quercetin is an effective agent against cancer therapy by causing inhibition of cell growth, induction of apoptosis and suppression of many HSP expression in different cancer cells.

Molecular Mecanishms Targeted by Quercetin in Cancer and Its Relation to HSP70

As quercetin is a lipophilic molecule, it can trigger many intracellular pathways by crossing cellular membranes. Numerous studies have indicated that quercetin inhibits cell proliferation by inducing various apoptotic pathways and/or arresting cell cycle at different checkpoints. It has been shown that quercetin treatment regulates the expression of the cyclin-dependent kinases (Cdks), thus causing the cell cycle arrest at the G0/G1, G1/S and G2/M checkpoints in different cell types. p21, p27, p53, and Chk2 (Checkpoint kinase 2) upregulation, Cdk1 and cyclin B1 downregulation, and pRb (retinoblastoma protein) phosphorylation have been observed in the arrest of the cell cycle by quercetin (Yoshida et al. 1990; Csokay et al. 1997; Ong et al. 2004; Mu et al. 2007; Jeong et al. 2009; Vidya-Priyadarsin et al. 2010; Yeh et al. 2011; Atashpour et al. 2015; Srivastava et al. 2016; Nguyen et al. 2017). In a variety of studies quercetin has been shown to work as an activator in apoptotic pathways in

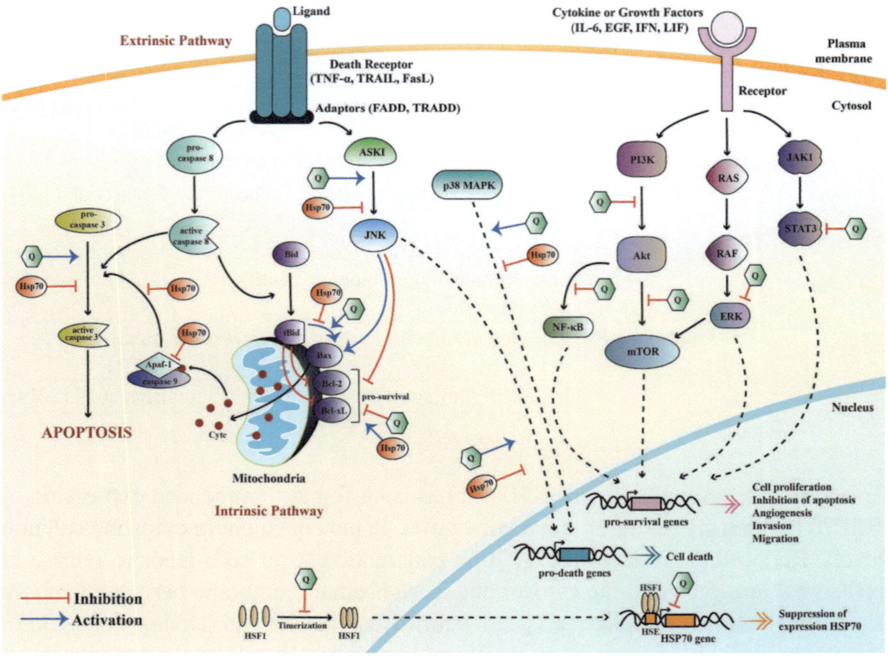

Fig. 2 Molecular mechanisms targeted by quercetin in cancer and its relation to Hsp70 (Q: Quercetin)

which overexpressed HSP70 acts as an inhibitor (Kashyap et al. 2016). It is known that quercetin is able to trigger both intrinsic (mitochondrial) and extrinsic (TNF-α, TRAIL, and Fas/FasL) apoptotic pathway in cancer cells (Elmore 2007; Kashyap et al. 2016; Khan et al. 2016). The apoptotic pathways targeted by quercetin in cancer and the relationship with Hsp70 are illustrated in Fig. 2. Increased expression of Hsp70 can be triggered by many factors such as HSF1, loss of p53 function and high expression of protooncogenes like HER2 and c-Myc (Balázs et al. 2017). It has been shown that overexpressed Hsp70 blocks the TNF-induced apoptosis (Jäättelä and Wissing 1993), inhibits caspase-3 induced apoptosis (Lee et al. 2005), blocks recruitment of procaspase-9 to the apoptosome and formation of the active apoptosome by binding to apoptotic protease-activating factor 1 (Apaf-1) (Beere et al. 2000; Saleh et al. 2000). Evidences have emphasized that quercetin induce apoptosis by direct activation of the caspase cascade in a variety of human cancer cell lines. In addition to apoptosis induction, it has been also stated that the expression of HSP70 is suppressed (Jung et al. 2010; Jakubowicz-Gil et al. 2013a, b; Bądziul et al. 2014b). Studies have also revealed that exposure of cancer cells to quercetin leads to an increment in the expression of Bax and Cyt-c (Cytochrome c), which are pro-apoptotic proteins (Chien et al. 2009; Zhang et al. 2013). Many studies have shown that quercetin suppress Bcl-xL and Bcl-2 anti-apoptotic proteins, and inhibit cellular-signaling proteins such as NF-κB and Cox-2 (Cyclooxygenase-2) (Banerjee et al. 2002; Cheong et al. 2004; Vijayababu et al. 2005; Kim et al. 2013b; Han et al. 2015).

Moreover, in the apoptotic pathway, Hsp70 also interacts with p38 mitogen-activated protein kinase (MAPK), c-Jun N-terminal kinase (JNK), and extracellular signal-regulated kinase (ERK) (Gabai et al. 1997; Lee et al. 2005). It has been shown that activation of JNK pathway and overexpression of Hsp70 in cancer cells show negative correlation, and JNK pathway in the presence of high Hsp70 levels is inactive (Li et al. 2007). Quercetin has been shown to activate JNK pathway, and increase levels of the JNK and p53 (Lee and Yoo 2013). It is known that cellular signal transduction pathways such as ERKs, Akts (also known as PKB, Protein Kinase B) and MAPKs, which affect cell survival modulate by quercetin (Kashyap et al. 2016). IL-6/JAK/STAT3 (Interleukin-6/Janus kinase/Signal Transducer and Activator of Transcription 3) and Akt/mTOR signaling pathways have been also shown to be downregulated by quercetin during cancer treatment (Mukherjee and Khuda-Bukhsh 2015; Chen et al. 2016). In addition, quercetin-mediated downregulation of Mcl-1 (myeloid cell leukemia-1), MMP-2 (matrix metalloproteinase-2), MMP-9 (matrix metalloproteinase-9), and VEGF (vascular endothelial growth factor) which are genes targeted by the STAT3 signal pathway have also been demonstrated (Cao et al. 2014). Quercetin has been shown to inhibit large survival signal pathways ERK and phosphatidylinositol 3-kinase (PI3K)/Akt in human hepatoma cells (Granado-Serrano et al. 2006). It has been also found that quercetin suppresses ERK signaling pathway resulting in the inhibition of angiogenesis in cancerous cells (Li et al. 2015). In a study on myeloma cells, the expression of HSF1 and inducible Hsp70 have been found to decrease by small interfering (si)RNA mediated inhibition of PI3K/Akt pathway (Chatterjee et al. 2013).

Transcription of HSP70 is generally regulated by HSF1. Activated HSF1 induces HSP70 expression. In this case it may be possible to block Hsp70 expression by inhibition of HSF1. Quercetin is well known inhibitor of HSP induction. The induction and regulation of HSP are highly complex processes and, in summary, involves the following steps: Initial release of heat shock transcription factor 1 (HSF1) from a chaperone complex including Hsp90 and Hsp70, trimerization, translocation to the nucleus, binding to heat shock element (HSE). The transcriptional activation of HSF1 localized as inactive monomers in the cytoplasm occurs by multiple phosphorylation by kinases (Önay-Uçar 2015). It is known that quercetin reduces HSP70 expression by suppressing HSF1 phosphorylation and transcriptional activity. It is one of the first inhibitors to be shown to be effective on expression of Hsp70 in this way. Quercetin has been also shown that inhibits a number of kinases, including AMPK (AMP-activated protein kinase), CK2 (Casein Kinase 2), CaMKII (Calcium/Calmodulin-Dependent Protein Kinase II), RSK2 (Ribosomal protein S6 kinase 2,) and S6 K1 (Ribosomal protein S6 kinase beta-1). Various studies have emphasized that quercetin suppress Hsp70 induction by blocking phosphorylation of HSF1 by CK2 and/or CaMKII kinases. There are also studies that demonstrate the effect of quercetin on HSF1 binding to HSE (Davies et al. 2000; Zorzi and Bonvini 2011). In a study investigating the effect of quercetin on HSP expression,150 μM quercetin has been shown to reduce Hsp70 expression by inhibiting both CK2 and CaMKII activity in Jurkat cells. Quercetin derivatives that poorly inhibit these kinases have been also shown to be weak inhibitors of Hsp70 induction, indicating that the activity of quercetin is due to its ability to inhibit both kinases. In addition to study, very low level of HSF1/HSE complex formed has been detected in quercetin treated cells 1 h before heat shock (Wang et al. 2009).

As a consequence, many studies have indicated that quercetin shows a negative correlation with elevated HSP70, which acts as an antiapoptotic. Evidence suggests that quercetin acts as an inhibitor of cell proliferation by inducing apoptosis and/or arresting cell cycle in cancerous cells, indicating that the activity of quercetin may be due to its ability to suppress the expression of HSP, especially Hsp70.

Conclusions

Today, the researchers working on cancer therapy have focused on downregulation of elevated HSP. It is known that HSP70 family plays a key role in the regulation of malignancy and have significant roles in various cancer types. High expression of HSP70 family members is related to tumor progression, aggressive phenotype, and a poor prognosis in therapeutic responses (Evans et al. 2010; Murphy 2013, Sherman and Gabai 2015; Giri et al. 2017). Especially, many studies have shown the prognostic value of the Hsp70 protein, a member of HSP70, is noteworthy in several human cancer (Nylandsted et al. 2000; Canöz et al. 2002; Abe et al. 2004; Hunt et al. 2004; Kluger et al. 2005; Yeh et al. 2009; Kocsis et al. 2010; Meng et al. 2010; Teng et al. 2012; Alexiou et al. 2013, 2014; Ghermann et al. 2014; Balázs et al. 2017; Jin et al. 2017). Therefore, the inhibition of expression of HSP70 family members, particularly Hsp70, and induction of apoptosis have become a novel strategy against cancer (Evans et al. 2010; Murphy 2013; Giri et al. 2017). Quercetin, a plant flavonoid, has long been known to suppress the expression of stress-induced Hsp70 (Jakubowicz-Gil et al. 2002, 2013b; Jones et al. 2004; Niu et al. 2006; Aghdassi et al. 2007, Bądziul et al. 2014b; Lee et al. 2015; Storniolo et al. 2015; Tao et al. 2016). Studies concerned with the role of quercetin in suppressing effect on expression of HSP70 family members have shown that this agent acts as an effective inhibitor of HSP70 expression in many cancer cell line. The number of studies investigating the effect of quercetin on the other members according to Hsp70 is relatively small, but nowadays studies are to ongoing (Vargas and Burd 2010; Khan et al. 2016; Kumar et al. 2016). As described in this chapter, all these findings indicate that quercetin alone or in combination with chemotherapeutic drugs causes suppression of HSP70 expression in different cancer cells, which makes this agent may be an effective adjuvant for cancer therapy.

Acknowledgements This study was supported by the Research Fund of Istanbul University (Project no. 57959 and 24987).

References

Aalinkeel, R., Bindukumar, B., Reynolds, J. L., et al. (2008). The dietary bioflavonoid, quercetin, selectively induces apoptosis of prostate cancer cells by down-regulating the expression of heat shock protein 90. *The Prostate, 68*, 1773–1789.

Abe, M., Manola, J. B., Oh, W. K., et al. (2004). Plasma levels of heat shock protein 70 in patients with prostate cancer: A potential biomarker for prostate cancer. *Clinical Prostate Cancer, 3*, 49–53.

Aghdassi, A., Phillips, P., & Dudeja, V. (2007). Heat shock protein 70 increases tumorigenicity and inhibits apoptosis in pancreatic adenocarcinoma. *Cancer Research, 67*, 616–625.

Alexiou, G. A., Vartholomatos, G., Stefanaki, K., et al. (2013). Expression of heat shock proteins in medulloblastoma. *Journal of Neurosurgery. Pediatrics, 12*, 452–457.

Alexiou, G. A., Karamoutsios, A., Lallas, G., et al. (2014). Expression of heat shock proteins in brain tumors. *Turkish Neurosurgery, 24*, 745–749.

Arap, M. A., Lahdenranta, J., Mintz, P. J., et al. (2004). Cell surface expression of the stress response chaperone GRP78 enables tumor targeting by circulating ligands. *Cancer Cell, 6*, 275–284.

Atashpour, S., Fouladdel, S., Movahhed, T. K., et al. (2015). Quercetin induces cell cycle arrest and apoptosis in CD133(+) cancer stem cells of human colorectal HT29 cancer cell line and enhances anticancer effects of doxorubicin. *Iranian Journal of Basic Medical Sciences, 18*, 635–643.

Awad, H. M., Boersma, M. G., Vervoort, J., & Rietjens, I. M. (2000). Peroxidase-catalyzed formation of quercetin quinone methide-glutathione adducts. *Archives of Biochemistry and Biophysics, 378*, 224–233.

Bądziul, D., Jakubowicz-Gil, J., Langner, E., Rzeski, W., Głowniak, K., & Gawron, A. (2014a). The effect of quercetin and imperatorin on programmed cell death induction in T98G cells in vitro. *Pharmacological Reports, 66*, 292–300.

Bądziul, D., Jakubowicz-Gil, J., Paduch, R., Głowniak, K., & Gawron, A. (2014b). Combined treatment with quercetin and imperatorin as a potent strategy for killing HeLa and Hep-2 cells. *Molecular and Cellular Biochemistry, 392*, 213–227.

Balázs, M., Zsolt, H., László, G., et al. (2017). Serum heat shock protein 70, as a potential biomarker for small cell lung cancer. *Pathology Oncology Research, 23*, 377–383.

Banerjee, T., Van der Vliet, A., & Ziboh, V. A. (2002). Downregulation of COX-2 and iNOS by amentoflavone and quercetin in A549 human lung adenocarcinoma cell line. *Prostaglandins, Leukotrienes, and Essential Fatty Acids, 66*, 485–492.

Beere, H. M., Wolf, B. B., Cain, K., et al. (2000). Heat-shock protein 70 inhibits apoptosis by preventing recruitment of procaspase-9 to the Apaf-1 apoptosome. *Nature Cell Biology, 2*, 469–475.

Benjamin, I. J., & McMillan, D. R. (1998). Stress (heat shock) proteins: molecular chaperones in cardiovascular biology and disease. *Circulation Research, 83*, 117–132.

Bhat, F. A., Sharmila, G., Balakrishnan, S., et al. (2014). Quercetin reverses EGF-induced epithelial to mesenchymal transition and invasiveness in prostate cancer (PC-3) cell line via EGFR/PI3K/Akt pathway. *The Journal of Nutritional Biochemistry, 25*, 1132–1139.

Bors, W., Heller, W., Michel, C., & Saran, M. (1990). Flavonoids as antioxidants: determination of radical-scavenging efficiencies. *Methods in Enzymology, 186*, 343–355.

Braganhol, E., Zamin, L. L., Canedo, A. D., et al. (2006). Antiproliferative effect of quercetin in the human U138MG glioma cell line. *Anti-Cancer Drugs, 17*, 663–671.

Canöz, O., Belenli, O., & Patiroglu, T. E. (2002). General features of gastric carcinomas and comparison of HSP70 and NK cell immunoreactivity with prognostic factors. *Pathology Oncology Research, 8*, 262–269.

Cao, H. H., Tse, A. K., Kwan, H. Y., et al. (2014). Quercetin exerts anti-melanoma activities and inhibits STAT3 signaling. *Biochemical Pharmacology, 87*, 424–434.

Chatterjee, M., Andrulis, M., Stühmer, T., et al. (2013). The PI3K/Akt signaling pathway regulates the expression of Hsp70, which critically contributes to Hsp90-chaperone function and tumor cell survival in multiple myeloma. *Haematologica, 98*, 1132–1141.

Chen, W. T., Zhu, G., Pfaffenbach, K., Kanel, G., Stiles, B., & Lee, A. S. (2014). GRP78 as a regulator of liver steatosis and cancer progression mediated by loss of the tumor suppressor PTEN. *Oncogene, 33*, 4997–5005.

Chen, X., Dong, X. S., Gao, H. Y., et al. (2016). Suppression of HSP27 increases the anti-tumor effects of quercetin in human leukemia U937 cells. *Molecular Medicine Reports, 13*, 689–696.

Cheong, E., Ivory, K., Doleman, J., Parker, M. L., Rhodes, M., & Johnson, I. T. (2004). Synthetic and naturally occurring COX-2 inhibitors suppress proliferation in a human oesophageal adenocarcinoma cell line (OE33) by inducing apoptosis and cell cycle arrest. *Carcinogenesis, 25*, 1945–1952.

Chien, S. Y., Wu, Y. C., Chung, J. G., et al. (2009). Quercetin-induced apoptosis acts through mitochondrial- and caspase-3-dependent pathways in human breast cancer MDA-MB-231 cells. *Human & Experimental Toxicology, 28*, 493–503.

Choi, J. A., Kim, J. Y., Lee, J. Y., et al. (2001). Induction of cell cycle arrest and apoptosis in human breast cancer cells by quercetin. *International Journal of Oncology, 19*, 837–844.

Ciocca, D. R., & Calderwood, S. K. (2005). Heat shock proteins in cancer: Diagnostic, prognostic, predictive, and treatment implications. *Cell Stress & Chaperones, 10*, 86–103.

Conklin, C. M., Bechberger, J. F., MacFabe, D., Guthrie, N., Kurowska, E. M., & Naus, C. C. (2007). Genistein and quercetin increase connexin43 and suppress growth of breast cancer cells. *Carcinogenesis, 28*, 93–100.

Csokay, B., Prajda, N., Weber, G., & Olah, E. (1997). Molecular mechanisms in the antiproliferative action of quercetin. *Life Sciences, 60*, 2157–2163.

Daugaard, M., Rohde, M., & Jäättelä, M. (2007a). The heat shock protein 70 family: Highly homologous proteins with overlapping and distinct functions. *FEBS Letters, 581*, 3702–3710.

Daugaard, M., Kirkegaard-Sørensen, T., Ostenfeld, M. S., et al. (2007b). Lens epithelium-derived growth factor is an Hsp70-2 regulated guardian of lysosomal stability in human cancer. *Cancer Research, 67*, 2559–2567.

Davies, S. P., Reddy, H., Caivano, M., & Cohen, P. (2000). Specificity and mechanism of action of some commonly used protein kinase inhibitors. *The Biochemical Journal, 351*, 95–105.

Deocaris, C. C., Lu, W. J., Kaul, S. C., & Wadhwa, R. (2013). Druggability of mortalin for cancer and neuro-degenerative disorders. *Current Pharmaceutical Design, 19*, 418–429.

Dihal, A. A., van der Woude, H., Hendriksen, P. J., et al. (2008). Transcriptome and proteome profiling of colon mucosa from quercetin fed F344 rats point to tumor preventive mechanisms, increased mitochondrial fatty acid degradation and decreased glycolysis. *Proteomics, 8*, 45–61.

Dudeja, V., Mujumdar, N., Phillips, P., et al. (2009). Heat shock protein 70 inhibits apoptosis in cancer cells through simultaneous and independent mechanisms. *Gastroenterology, 136*, 1772–1782.

Dundas, S. R., Lawrie, L. C., Rooney, P. H., & Murray, G. I. (2005). Mortalin is over-expressed by colorectal adenocarcinomas and correlates with poor survival. *The Journal of Pathology, 205*, 74–81.

Duo, J., Ying, G. G., Wang, G. W., & Zhang, L. (2012). Quercetin inhibits human breast cancer cell proliferation and induces apoptosis via Bcl-2 and Bax regulation. *Molecular Medicine Reports, 5*, 1453–1456.

Duthie, G. G., Duthie, S. J., & Kyle, J. A. (2000). Plant polyphenols in cancer and heart disease: Implications as nutritional antioxidants. *Nutrition Research Reviews, 13*, 79–106.

Elmore, S. (2007). Apoptosis: A review of programmed cell death. *Toxicologic Pathology, 35*, 495–516.

Evans, C. G., Chang, L., & Gestwicki, J. E. (2010). Heat shock protein 70 (hsp70) as an emerging drug target. *Journal of Medicinal Chemistry, 53*, 4585–4602.

Fernandez, P. M., Tabbara, S. O., Jacobs, L. K., et al. (2000). Overexpression of the glucose-regulated stress gene GRP78 in malignantbut not benign human breast lesions. *Breast Cancer Research and Treatment, 59*, 15–26.

Gabai, V. L., Meriin, A. B., & Mosser, D. D. (1997). Hsp70 prevents activation of stress kinases. A novel pathway of cellular thermotolerance. *Journal of Biological Chemistry, 272*, 18033–18037.

Garrido, C., Brunet, M., Didelot, C., Zermati, Y., Schmitt, E., & Kroemer, G. (2006). Heat shock proteins 27 and 70: Anti-apoptotic proteins with tumourigenic properties. *Cell Cycle, 5*, 2592–2601.

Gazit, G., Lu, J., & Lee, A. S. (1999). De-regulation of GRP stress protein expression in human breast cancer cell lines. *Breast Cancer Research and Treatment, 54*, 135–146.

Gehrmann, M., Specht, H. M., Bayer, C., et al. (2014). Hsp70-a biomarker for tumor detection and monitoring of outcome of radiation therapy in patients with squamous cell carcinoma of the head and neck. *Radiation Oncology, 9*, 131.

Gibellini, L., Pinti, M., Nasi, M., et al. (2011). Quercetin and cancer chemoprevention. *Evidence-based Complementary and Alternative Medicine, 2011*, 1–15.

Giri, B., Sethi, V., Modi, S., et al. (2017). Heat shock protein 70 in pancreatic diseases: Friend or foe. *Journal of Surgical Oncology, 116*, 114–122.

Granado-Serrano, A. B., Martín, M. A., Bravo, L., Goya, L., & Ramos, S. (2006). Quercetin induces apoptosis via caspase activation, regulation of Bcl-2, and inhibition of PI-3-kinase/Akt and ERK pathways in a human hepatoma cell line (HepG2). *The Journal of Nutrition, 136*, 2715–2721.

Gupta, C., Vikram, A., Tripathi, D. N., Ramarao, P., & Jena, G. B. (2010). Antioxidant and antimutagenic effect of quercetin against DEN induced hepatotoxicity in rat. *Phytotherapy Research, 24*, 119–128.

Han, Y., Yu, H., Wang, J., Ren, Y., Su, X., & Shi, Y. (2015). Quercetin alleviates myocyte toxic and sensitizes anti-leukemic effect of adriamycin. *Hematology, 20*, 276–283.

Harwood, M., Danielewska-Nikiel, B., Borzelleca, J. F., Flamm, G. W., Williams, G. M., & Lines, T. C. (2007). A critical review of the data related to the safety of quercetin and lack of evidence of in vivo toxicity, including lack of genotoxic/carcinogenic properties. *Food and Chemical Toxicology, 4*, 2179–2205.

Hu, Y., Yang, L., Yang, Y., et al. (2016). Oncogenic role of mortalin contributes to ovarian tumorigenesis by activating the MAPK-ERK pathway. *Journal of Cellular and Molecular Medicine, 20*, 2111–2121.

Hunt, C. R., Dix, D. J., Sharma, G. G., et al. (2004). Genomic instability and enhanced radiosensitivity in Hsp70.1- and Hsp70.3-deficient mice. *Molecular and Cellular Biology, 2*, 899–911.

Jäättelä, M., & Wissing, D. (1993). Heat-shock proteins protect cells from monocyte cytotoxicity: possible mechanism of self-protection. *The Journal of Experimental Medicine, 177*, 231–216.

Jakubowicz-Gil, J., Rzymowska, J., & Gawron, A. (2002). Quercetin, apoptosis, heat shock. *Biochemical Pharmacology, 64*, 1591–1595.

Jakubowicz-Gil, J., Paduch, R., Piersiak, T., Głowniak, K., Gawron, A., & Kandefer-Szerszeń, M. (2005). The effect of quercetin on pro-apoptotic activity of cisplatin in HeLa cells. *Biochemical Pharmacology, 69*, 1343–1350.

Jakubowicz-Gil, J., Langner, E., Wertel, I., Piersiak, T., & Rzeski, W. (2010). Temozolomide, quercetin and cell death in the MOGGCCM astrocytoma cell line. *Chemico-Biological Interactions, 188*, 190–203.

Jakubowicz-Gil, J., Langner, E., Bądziul, D., Wertel, I., & Rzeski, W. (2013a). Apoptosis induction in human glioblastoma multiforme T98G cells upon temozolomide and quercetin treatment. *Tumour Biology, 34*, 2367–2378.

Jakubowicz-Gil, J., Langner, E., Bądziul, D., Wertel, I., & Rzeski, W. (2013b). Silencing of Hsp27 and Hsp72 in glioma cells as a tool for programmed cell death induction upon temozolomide and quercetin treatment. *Toxicology and Applied Pharmacology, 273*, 580–589.

Jeong, J. H., An, J. Y., Kwon, Y. T., Rhee, J. G., & Lee, Y. J. (2009). Effects of low dose quercetin: Cancer cell-specific inhibition of cell cycle progression. *Journal of Cellular Biochemistry, 106*, 73–82.

Jin, H., Xie, Q., Guo, X., et al. (2017). p63α protein upregulates heat shock protein 70 expression via E2F1 transcription factor 1, promoting Wasf3/Wave3/MMP9 signaling and bladder cancer invasion. *Journal of Biological Chemistry, 292*(38), 15952–15963. https://doi.org/10.1074/jbc.M117.792010.

Jones, E. L., Zhau, M. J., Stevenson, M. A., & Calderwood, S. K. (2004). The 70 kilodalton heat shock protein is an inhibitor of apoptosis in prostate cancer. *International Journal of Hyperthermia, 20*, 835–849.

Jung, J. H., Lee, J. O., Kim, J. H., et al. (2010). Quercetin suppresses HeLa cell viability via AMPK-induced HSP70 and EGFR down-regulation. *Journal of Cellular Physiology, 223*, 408–414.

Kampinga, H. H., Hageman, J., Vos, M. J., et al. (2009). Guidelines for the nomenclature of the human heat shock proteins. *Cell Stress & Chaperones, 14*, 105–111.

Kashyap, D., Mittal, S., Sak, K., Singhal, P., & Tuli, H. S. (2016). Molecular mechanisms of action of quercetin in cancer: Recent advances. *Tumour Biology, 37*, 12927–12939.

Khan, F., Niaz, K., Maqbool, F., et al. (2016). Molecular targets underlying the anticancer effects of quercetin: An update. *Nutrients, 8*, 1–19.

Kim, E. J., Choi, C. H., Park, J. Y., Kang, S. K., & Kim, Y. K. (2008). Underlying mechanism of quercetin-induced cell death in human glioma cells. *Neurochemical Research, 33*, 971–979.

Kim, H., Seo, E. M., Sharma, A. R., et al. (2013a). Regulation of Wnt signaling activity for growth suppression induced by quercetin in 4T1 murine mammary cancer cells. *International Journal of Oncology, 43*, 1319–1325.

Kim, H., Moon, J. Y., Ahn, K. S., & Cho, S. K. (2013b). Quercetin induces mitochondrial mediated apoptosis and protective autophagy in human glioblastoma U373MG cells. *Oxidative Medicine and Cellular Longevity, 2013*, 596496.

Kluger, H. M., Chelouche Lev, D., & Kluger, Y. (2005). Using a xenograft model of human breast cancer metastasis to find genes associated with clinically aggressive disease. *Cancer Research, 65*, 5578–5587.

Kocsis, J., Madaras, B., Toth, E. K., Fust, G., & Prohaszka, Z. (2010). Serum level of soluble 70-kD heat shock protein is associated with high mortality in patients with colorectal cancer without distant metastasis. *Cell Stress & Chaperones, 15*, 143–151.

Koomägi, R., Mattern, J., & Volm, M. (1999). Glucose-related protein (GRP78) and its relationship to the drug-resistance proteins P170, GST-pi, LRP56 and angiogenesis in non-small cell lung carcinomas. *Anticancer Research, 19*, 4333–4336.

Kubota, H., Yamamoto, S., Itoh, E., et al. (2010). Increased expression of co-chaperone HOP with HSP90 and HSC70 and complex formation in human colonic carcinoma. *Cell Stress & Chaperones, 15*, 1003–1011.

Kumar, S., Stokes, J., 3rd., Singh, U. P., et al. (2016). Targeting Hsp70: A possible therapy for cancer. *Cancer Letters, 374*, 156–166.

Lee, A. S. (2007). GRP78 induction in cancer: therapeutic and prognostic implications. *Cancer Research, 67*, 3496–3499.

Lee, K. H., & Yoo, C. G. (2013). Simultaneous inactivation of GSK-3β suppresses quercetin-induced apoptosis by inhibiting the JNK pathway. *American Journal of Physiology. Lung Cellular and Molecular Physiology, 304*, 782–789.

Lee, S. J., Choi, S. A., Lee, K. H., et al. (2001). Role of inducible heat shock protein in radiation-induced cell death. *Cell Stress & Chaperones, 6*, 273–281.

Lee, J. S., Lee, J. J., & Seo, J. S. (2005). HSP70 deficiency results in activation of c-Jun N-terminal Kinase, extracellular signal-regulated kinase, and caspase-3 in hyperosmolarity-induced apoptosis. *The Journal of Biological Chemistry, 280*, 6634–6641.

Lee, S. H., Lee, E. J., & Min, K. H. (2015). Quercetin enhances chemosensitivity to gemcitabine in lung cancer cells by inhibiting heat shock protein 70 expression. *Clinical Lung Cancer, 16*, e235–e243.

Li, H., Sui, C., Kong, F., Zhang, H., Liu, J., & Dong, M. (2007). Expression of HSP70 and JNK-related proteins in human liver cancer: Potential effects on clinical outcome. *Digestive and Liver Disease, 39*, 663–670.

Li, F., Bai, Y., Zhao, M., et al. (2015). Quercetin inhibits vascular endothelial growth factor-induced choroidal and retinal angiogenesis in vitro. *Ophthalmic Research, 53*, 109–116.

Li, J., Tang, C., Li, L., Li, R., & Fan, Y. (2016). Quercetin blocks t-AUCB-induced autophagy by Hsp27 and Atg7 inhibition in glioblastoma cells in vitro. *Journal of Neuro-Oncology, 129*, 39–45.

Lianos, G. D., Alexiou, G. A., Mangano, A., et al. (2015). The role of heat shock proteins in cancer. *Cancer Letters, 360*, 114–118.

Liu, T., Daniels, C. K., & Cao, S. (2012). Comprehensive review on the HSC70 functions, interactions with related molecules and involvement in clinical diseases and therapeutic potential. *Pharmacology & Therapeutics, 136*, 354–374.

McConnell, J. R., & McAlpine, S. R. (2013). Heat shock proteins 27, 40, and 70 as combinational and dual therapeutic cancer targets. *Bioorganic & Medicinal Chemistry Letters, 23*, 1923–1928.

Meng, L., Gabai, V. L., & Sherman, M. Y. (2010). Heat-shock transcription factor HSF1 has a critical role in human epidermal growth factor receptor-2-induced cellular transformation and tumorigenesis. *Oncogene, 29*, 5204–5213.

Metodiewa, D., Jaiswal, A. K., Cenas, N., Dickancaité, E., & Segura-Aguilar, J. (1999). Quercetin may act as a cytotoxic prooxidant after its metabolic activation to semiquinone and quinoidal product. *Free Radical Biology & Medicine, 26*, 107–116.

Misra, U. K., Deedwania, R., & Pizzo, S. V. (2005). Binding of activated alpha2-macroglobulin to its cell surface receptor GRP78 in 1-LN prostate cancer cells regulates PAK-2-dependent activation of LIMK. *The Journal of Biological Chemistry, 280*, 26278–26286.

Moghanibashi, M., Rastgar-Jazii, F., Soheili, Z. S., et al. (2013). Esophageal cancer alters the expression of nuclear pore complex binding protein Hsc70 and eIF5A-1. *Functional & Integrative Genomics, 13*, 253–260.

Mosser, D. D., & Morimoto, R. I. (2004). Molecular chaperones and the stress of oncogenesis. *Oncogene, 23*, 2907–2918.

Mu, C., Jia, P., Yan, Z., Liu, X., Li, X., & Liu, H. (2007). Quercetin induces cell cycle G1 arrest through elevating Cdk inhibitors p21 and p27 in human hepatoma cell line (HepG2). *Methods and Findings in Experimental and Clinical Pharmacology, 29*, 179–183.

Mukherjee, A., & Khuda-Bukhsh, A. R. (2015). Quercetin down-regulates IL-6/STAT-3 signals to induce mitochondrial-mediated apoptosis in a nonsmall- cell lung-cancer cell line, A549. *Journal of Pharmacopuncture, 18*, 19–26.

Murakami, A., Ashida, H., & Terao, J. (2008). Multitargeted cancer prevention by quercetin. *Cancer Letters, 269*, 315–325.

Murphy, M. E. (2013). The HSP70 family and cancer. *Carcinogenesis, 34*, 1181–1188.

Nguyen, L. T., Lee, Y. H., Sharma, A. R., et al. (2017). Quercetin induces apoptosis and cell cycle arrest in triple-negative breast cancer cells through modulation of Foxo3a activity. *The Korean Journal of Physiology & Pharmacology, 21*, 205–213.

Niu, P., Liu, L., Gong, Z., et al. (2006). Overexpressed heat shock protein 70 protects cells against DNA damage caused by ultraviolet C in a dose-dependent manner. *Cell Stress & Chaperones, 11*, 162–169.

Nylandsted, J., Rohde, M., Brand, K., Bastholm, L., Elling, F., & Jäättelä, M. (2000). Selective depletion of heat shock protein 70 (Hsp70) activates a tumor-specific death program that is independent of caspases and bypasses Bcl-2. *Proceedings of the National Academy of Sciences of the United States of America, 97*, 7871–7876.

Önay-Uçar, E. (2015). Heat shock proteins and cancer: Plant based therapy. In A. A. A. Asea, N. N. Almasoud, S. Krishnan, & P. Kaur (Eds.), *Heat shock protein-based therapies* (pp. 27–48). Cham: Springer.

Ong, C. S., Tran, E., Nguyen, T. T., et al. (2004). Quercetin-induced growth inhibition and cell death in nasopharyngeal carcinoma cells are associated with increase in Bad and hypophosphorylated retinoblastoma expressions. *Oncology Reports, 11*, 727–733.

Powers, M. V., Clarke, P. A., & Workman, P. (2008). Dual targeting of HSC70 and HSP72 inhibits HSP90 function and induces tumor-specific apoptosis. *Cancer Cell, 14*, 250–262.

Regeling, A., Imhann, F., Volders, H. H., et al. (2016). HSPA6 is an ulcerative colitis susceptibility factor that is induced by cigarette smoke and protects intestinal epithelial cells by stabilizing anti-apoptotic Bcl-XL. *Biochimica et Biophysica Acta, 1862*, 788–796.

Rodina, A., Taldone, T., Kang, Y., et al. (2014). Affinity purification probes of potential use to investigate the endogenous Hsp70 interactome in cancer. *ACS Chemical Biology, 9*, 1698–1705.

Rohde, M., Daugaard, M., Jensen, M. H., Helin, K., Nylandsted, J., & Jäättelä, M. (2005). Members of the heat-shock protein 70 family promote cancer cell growth by distinct mechanisms. *Genes & Development, 19*, 570–582.

Saleh, A., Srinivasula, S. M., Balkir, L., Robbins, P. D., & Alnemri, E. S. (2000). Negative regulation of the Apaf-1 apoptosome by Hsp70. *Nature Cell Biology, 2*, 476–483.

Schmitt, E., Maingret, L., Puig, P. E., et al. (2006). Heat shock protein 70 neutralization exerts potent antitumor effects in animal models of colon cancer and melanoma. *Cancer Research, 15*, 4191–4197.

Shan, B. E., Wang, M. X., & Li, R. Q. (2009). Quercetin inhibit human SW480 colon cancer growth in association with inhibition of cyclin D1 and survivin expression through Wnt/beta-catenin signaling pathway. *Cancer Investigation, 27*, 604–612.

Sherman, M. Y., & Gabai, V. L. (2015). Hsp70 in cancer: Back to the future. *Oncogene, 34*, 4153–4161.

Shuda, M., Kondoh, N., Imazeki, N., et al. (2003). Activation of the ATF6, XBP1 and grp78 genes in human hepatocellular carcinoma: A possible involvement of the ER stress pathway in hepatocarcinogenesis. *Journal of Hepatology, 38*, 605–614.

Siegelin, M. D., Reuss, D. E., Habel, A., Rami, A., & von Deimling, A. (2009). Quercetin promotes degradation of survivin and thereby enhances death-receptor-mediated apoptosis in glioma cells. *Neuro-Oncology, 11*, 122–131.

Silva, M. M., Santos, M. R., Caroço, G., Rocha, R., Justino, G., & Mira, L. (2002). Structure-antioxidant activity relationships of flavonoids: A re-examination. *Free Radical Research, 36*, 1219–1227.

Snoeckx, L. H., Cornelussen, R. N., Van Nieuwenhoven, F. A., Reneman, R. S., & Van Der Vusse, G. J. (2001). Heat shock proteins and cardiovascular pathophysiology. *Physiological Reviews, 81*, 1461–1497.

Song, M. S., Park, Y. K., Lee, J. H., & Park, K. (2001). Induction of glucose-regulated protein 78 by chronic hypoxia in human gastric tumor cells through a protein kinase C-ε/ERK/AP-1 signaling cascade. *Cancer Research, 61*, 8322–8330.

Srivastava, S., Somasagara, R. R., Hegde, M., et al. (2016). Quercetin, a natural flavonoid interacts with DNA, arrests cell cycle and causes tumor regression by activating mitochondrial pathway of apoptosis. *Scientific Reports, 6*, 24049.

Starenki, D., Hong, S. K., Lloyd, R. V., & Park, J. I. (2015). Mortalin (GRP75/HSPA9) upregulation promotes survival and proliferation of medullary thyroid carcinoma cells. *Oncogene, 34*, 4624–4634.

Storniolo, A., Raciti, M., Cucina, A., Bizzarri, M., & Renzo, L. D. (2015). Quercetin affects Hsp70/IRE1α mediated protection from death Induced by endoplasmic reticulum stress. *Oxidative Medicine and Cellular Longevity, 2015*, 1–11.

Takano, S., Wadhwa, R., Yoshii, Y., Nose, T., Kaul, S. C., & Mitsui, Y. (1997). Elevated levels of mortalin expression in human brain tumors. *Experimental Cell Research, 237*, 38–45.

Tao, Y., Messer, J. S., Goss, K. H., Hart, J., Bissonnette, M., & Chang, E. B. (2016). Hsp70 exerts oncogenic activity in the Apc mutant Min mouse model. *Carcinogenesis, 37*, 731–739.

Teng, Y., Ngoka, L., Mei, Y., Lesoon, L., & Cowell, J. K. (2012). HSP90 and HSP70 proteins are essential for stabilization and activation of WASF3 metastasis-promoting protein. *The Journal of Biological Chemistry, 287*, 10051–10059.

Vargas, A., & Burd, R. (2010). Hormesis and synergy: pathways and mechanisms of quercetin in cancer prevention and management. *Nutrition Reviews, 68*, 418–428.

Vidya-Priyadarsini, R., Senthil-Murugan, R., Maitreyi, S., Ramalingam, K., Karunagaran, D., & Nagini, S. (2010). The flavonoid quercetin induces cell cycle arrest and mitochondria-mediated apoptosis in human cervical cancer (HeLa) cells through p53 induction and NF-κB inhibition. *European Journal of Pharmacology, 649*, 84–91.

Vijayababu, M. R., Kanagaraj, P., Arunkumar, A., Ilangovan, R., Aruldhas, M. M., & Arunakaran, J. (2005). Quercetin-induced growth inhibition and cell death in prostatic carcinoma cells (PC-3) are associated with increase in p21 and hypophosphorylated retinoblastoma proteins expression. *Journal of Cancer Research and Clinical Oncology, 131*, 765–771.

Vijayababu, M. R., Arunkumar, A., Kanagaraj, P., Venkataraman, P., Krishnamoorthy, G., & Arunakaran, J. (2006). Quercetin downregulates matrix metalloproteinases 2 and 9 proteins expression in prostate cancer cells (PC-3). *Molecular and Cellular Biochemistry, 287*, 109–116.

Wang, L., Tu, Y. C., Lian, T. W., Hung, J. T., Yen, J. H., & Wu, M. J. (2006). Distinctive antioxidant and antiinflammatory effects of flavonols. *Journal of Agricultural and Food Chemistry, 54*, 9798–9804.

Wang, R. E., Kao, J. L., Hilliard, C. A., et al. (2009). Inhibition of heat shock induction of heat shock protein 70 and enhancement of heat shock protein 27 phosphorylation by quercetin derivatives. *Journal of Medicinal Chemistry, 52*, 1912–1921.

Wu, J., Liu, T., Rios, Z., Mei, Q., Lin, X., & Cao, S. (2017). Heat shock proteins and cancer. *Trends in Pharmacological Sciences, 38*, 226–256.

Yeh, C. H., Tseng, R., Zhang, Z., et al. (2009). Circulating heat shock protein 70 and progression in patients with chronic myeloid leukemia. *Leukemia Research, 33*, 212–217.

Yeh, S. L., Yeh, C. L., Chan, S. T., & Chuang, C. H. (2011). Plasma rich in quercetin metabolites induces G2/M arrest by upregulating PPAR-γ expression in human A549 lung cancer cells. *Planta Medica, 77*, 992–998.

Yi, X., Luk, J. M., Lee, N. P., et al. (2008). Association of mortalin (HSPA9) with liver cancer metastasis and prediction for early tumor recurrence. *Molecular & Cellular Proteomics, 7*, 315–325.

Yoshida, M., Sakai, T., Hosokawa, N., et al. (1990). The effect of quercetin on cell cycle progression and growth of human gastric cancer cells. *FEBS Letters, 260*, 10–13.

Zhang, J. Y., Yi, T., Liu, J., Zhao, Z. Z., & Chen, H. B. (2013). Quercetin induces apoptosis via the mitochondrial pathway in KB and KBv200 cells. *Journal of Agricultural and Food Chemistry, 61*, 2188–2195.

Zorzi, E., & Bonvini, P. (2011). Inducible hsp70 in the regulation of cancer cell survival: analysis of chaperone induction, expression and activity. *Cancers (Basel), 3*, 3921–3956.

Zuiderweg, E. R., Bertelsen, E. B., Rousaki, A., Mayer, M. P., Gestwicki, J. E., & Ahmad, A. (2013). Allostery in the HSP70 chaperone proteins. *Topics in Current Chemistry, 328*, 99–153.

Hsp70 in Fungi: Evolution, Function and Vaccine Candidate

Shraddha Tiwari and Jata Shankar

Abstract In fungal system, Hsp70 protein being highly conserved in nature has played a major role in various stress conditions. Genes encoding for Hsp70 proteins in fungi are highly conserved. Hsp70 protein performs chaperone dependent or independent function, essential for growth and morphogenesis of fungi. Functional distinction of Hsp70 protein conjointly depends on the prevalence of Hsp70 in numerous cellular compartments. Fungal Hsp70 protein is involved in protein aggregation, folding as well as in degradation of nascent polypeptide. Additionally, Hsp70 protein has a vital role in the formation of prions in case of yeasts. Fungi showed expression of hsp70 mRNA during interaction with plant. Also, fungal hsp70 showed expression in human during various infections, and may provide lead as a potential bio-marker for disease conditions. This chapter summarizes our present knowledge on fungal Hsp70 proteins and their role in morphogenesis, stress responses and a potential candidate for vaccine.

Keywords Fungi · HSP70 · Morphogenesis · Stress responses · Vaccine candidate

Abbreviations

AmB	Amphotericin B
CFTR	cystic fibrosis transmembrane conductance regulator
Hsp	heat shock protein
kDa	kilo Dalton
NEF	nucleotide exchange factor

S. Tiwari · J. Shankar (✉)
Genomic Laboratory, Department of Biotechnology and Bioinformatics, Jaypee University of Information Technology, Solan, Himachal Pradesh, India
e-mail: jata.shankar@juit.ac.in

A. A. A. Asea, P. Kaur (eds.), *HSP70 in Human Diseases and Disorders*, Heat Shock Proteins 14, https://doi.org/10.1007/978-3-319-89551-2_20

Introduction

Kingdom Fungi, distributed worldwide, having a unique entity within living organisms belong to eukaryotes. These are a diverse taxonomy, classified based on their diversity, germination stages, reproduction, evolution, ability of inflicting contamination and toxins production (Guarro et al. 1999; Shankar 2013). Throughout the phylogenies, fungi have showed a great deal of variety in reproduction mode and adaptability towards their surrounding (Galagan et al. 2005). Optimum growth conditions required for fungal growth is warm (30–37 °C) (Shankar et al. 2004) and humid conditions, which if absent leads to spore formation (dormant stage), whereas extreme increase in temperature will cause fungal death (Dix 2012). Under certain conditions or stress, life cycle and cellular processes of fungi is affected. In stress conditions such as modulation of temperature, protein denaturation has been reported that leads to misfolding and aggregation of proteins inflicting loss of biological functions (Sharma et al. 2009a, b). A unique set of proteins are involved in stress related changes, hence involved in the fungal survival, termed as Heat Shock Proteins. These proteins could be present in mitochondria, cytosol, nucleus, endoplasmic reticulum and cell membrane (Kregel 2002). Hsp are extremely preserved biomolecules involve in activation of various intermediates of signal transduction pathway in fungi (Kregel 2002; Verghese et al. 2012). Fungal kingdom encompasses three predominant heat shock proteins, Hsp90, Hsp70, and Hsp20–40, that shows crucial role in stress adaptation and morphogenesis. The class of the 70-kDa Hsp family is considered as one of the potent immunogenic protein families involved in nascent and damaged intracellular protein refolding (Lindquist and Craig 1988; Daugaard et al. 2007; Cleare et al. 2017). Hsp70 plays a major role in folding of protein and newly synthesized oligomeric gathering, the transport of protein structures across membranes, misfolded proteins refolding to prevent aggregation and also the activation and regulation of signal of transduction proteins (Bukau and Horwich 1998; Tiwari et al. 2015). In addition, Hsp70 has a great impact in the field of medicine and has been proposed as a vaccine candidate against fungal pathogen (Wormley Jr. 2011; Blatzer et al. 2015). Thus, we reviewed Hsp70 in various stress responses and alternative fungal biological conditions. In this chapter, we focused on the role of 70-kDa heat shock protein in fungi. In addition application of Hsp70 as a vaccine candidate has also been discussed.

Genes Encoding HSP70 in Fungi

Hsp70 is evolutionary conserved protein within the eukaryotes that shares approximately 50% amino acid sequence similarity with prokaryotic Hsp70 protein (DnaK) (Gupta and Singh 1994; Daugaard et al. 2007). Fungal Hsp70 molecule contains an ATP domain at N-terminal region and, a domain specific for substrate-binding which has affinity for client proteins, present in C-terminal region and a C-terminal domain of variable length (Mayer and Bukau 2005). The deduced amino acid sequence of the Hsp70 corresponds to 636 amino acid sequences with a molecular mass of 70.56 kDa and a pI of 6.01 which exhibited a strong sequence homology with many other eukaryotic Hsp70 family proteins. Similarly, Hsp70 of B. emersonii has been cloned and showed it consists of 650 amino acids with molecular mass of 70.8 kDa. These observations suggested that yeast and fungi have conserved hsp70 genes (Stefani and Gomes 1995). Hsp70 subfamilies have shown to be conserved in terms of function and evolution. For example, *S. cerevisiae* Ssq1, involved in iron sulfur clustering which involved sub functioning as to mimic Ssc1 (Craig and Marszalek 2011). Ssa and Ssc are Hsp70 proteins, evolved due to higher change of copy number in gene evolution, while some Hsp70 family proteins such as Ssz, Lhs, Kar have evolved from change in few copy. Seven subfamilies of *S. cerevisiae* hsp70 blast analysis revealed similarity with 53 ascomycota genome associated 491 orthologs. Also, four different strains of Basidiomycota genomes analysis for hsp70 showed thirty hsp70 genes, hence studied for fungal hsp70 evolution (Kominek et al. 2013).

S. cerevisiae is the leading organism for hsp70 genomic studies, thus served as a model organism to understand hsp70 in fungi (Morano et al. 2012). *S. cerevisiae* exhibits multigene family of hsp70, which involves eight members, including constructive as well as stress inducible ones (Craig et al. 1993). There are 14 well characterized Hsp70 proteins of *S. cerevisiae* available that categorized into seven different subfamilies: four canonical-type Hsp70 chaperones (Kar, Ssa, Ssb, and Ssc), three atypical regulatory Hsp70s (Ssz, Sse, and Lhs), actively modulating Hsp70 partners activity (Kominek et al. 2013). Fungal Hsp70 have been ubiquitously distributed in all cellular compartments majorly cytosol, mitochondria, endoplasmic reticulum, and ribosome (Boorstein et al. 1994; Easton et al. 2000).From the seven Hsp70 homologues in *S. cerevisiae*, six have been observed to be located in cytosol, namely, Ssa1, Ssa2, Ssa3, Ssa4, Ssb1, and Ssb2 (Werner-Washburne et al. 1987; Černila et al. 2003). It is observed that cytosolic proteins have diversified and overlapping functions. For example, Ssb proteins did not showed substitution for Ssa protein survival in yeast (Boorstein et al. 1994). Ssd1/Kar2 is known to present in endoplasmic reticulum (Werner-Washburne and Craig 1989). Fungi have a specialized ribosomal Hsp70, have specific binding site on ribosome 36 position. Fungi also have a specialized ribosome-associated Hsp70, Ssb, which independently associates with the 60S subunit. Binding of Ssb in ribosome depends on conserved residue, which is specific in the peptide-binding domain at valine 442 region, allows flexibility (Pfund et al. 2001). In mitochondria, three different hsp70 encoded by the genes ecm10, ssc1 and ssq1 were observed. Deletion of ssc1 was lethal under all conditions. ssc1 have similar functions as of jac1 (Kampinga and Craig 2010). ssa1 and ssa2 are involved in transport of aminopeptidase-I from the cytoplasm into the vacuole (Satyanarayana et al. 2000). Recently, a variety of eukaryotes have showed a special class of protein known as Hsp110 protein which is found in the endoplamic reticulum and cytoplasm. At molecular level, they are large and vast protein but belonging to hsp70 gene superfamily as they have same functional and structural similarities (Easton et al. 2000; Nikolaidis and Nei 2004).

Role of HSP70 in Morphogenesis

Asexual reproduction in fungi accounts for fungal morphogenesis which comprises different stages, namely, conidia in dormant stage, vegetative hyphae and mycelia (Wang and Lin 2012). Hsp70 has a vital role in morphogenesis of fungi in stress related conditions (Tiwari et al. 2015). Transcripts of hsp70 were observed during lag and log phase followed by reduction at aerial hyphae in early stage and the hsp70 transcripts were found maximum at later stages of aerial hyphae. This is because of activation of hsp70 transcription or decrease mRNA degradation rate (Häfker et al. 1998). Xavier et al, also showed that expression of the heat shock inducible hsp70 gene depend on the various development stages of fungi and found maximum inducibility in late aerial hyphae due to lower mRNA degradation rate (Xavier 1998). Studies on *N. crassa* exhibited highest hsp70 transcripts at aerial and dormant conidia stage which fluctuate on later stages of germination (Fracella et al. 1997). Proteome profile of germinating conidia of *A. flavus* showed the expression of Hsp70, suggesting their role during transition from conidia to hyphae (Tiwari et al. 2016). With the increase in temperature (from room temperature to higher), Paracoccidioides converts from mycelia to yeast form, a pathogenic form. *P. brasiliensis* and *P. lutzii* showed high hsp70 transcripts in yeast form (Monteiro et al. 2009; Shankar et al. 2011a, b). Hyphal stage in *C. albicans* showed upregulated Ssa (Hsp70), suggesting the key role of Hsp70 protein in morphogenesis of fungi (Brand 2011). *C. albicans* showed higher accumulation of Hsp70 protein in its cell wall which was found to be expressed in all conditions (Lopez-Ribot et al. 1996). *A. flavus* (toxigenic and atoxigenic strains) showed expression of hsp70 transcripts at 30 °C, however, downregulated expression was observed in toxigenic strain in comparison to atoxigenic strain, thus suggesting lower expression of hsp70 transcripts favor aflatoxin biosynthesis (Thakur et al. 2016). Hsp70 induction and overexpression has been observed in the compensation of cwh41 gene, which codes for glucosidase-I in *A. fumigatus*, important for cell wall synthesis and morphology of fungi (Jin 2011).

During stress response, fungus such as *C. albicans, P. lutzii, H. capsulatum* etc existing in mycelia form converts to yeast form and vice versa, this phenomenon is termed as dimorphism (Brand 2011). *Penicillium marneffei* showed temperature dependent upregulation of hsp70 transcripts during the mycelium to yeast phase transition (increase at 39 °C followed by decrease at 42 °C) (Kummasook et al. 2007). Similar studies have also shown in *H. capsulatum* with higher expression of hsp70 from mycelia to yeast transition (Caruso et al. 1987; Shearer Jr. et al. 1987). This finding showed that hsp70 plays a very essential role in heat stress dependent fungal dimorphism.Hsp90 plays a major role in the fungal morphogenesis which requires the co-operation of Hsp70. Hsp70 acts as co-chaperon for Hsp90 and requires Sti1/Hop1 as an adapter protein for Hsp90 mediated polypeptide release (Rohl et al. 2015). Hsp70 also works in combination with Hsp30 and Hsp80 to interact with unfolded polypeptide, hence mimicking Hsp90 activity in fungal morphogenesis. Hsp70 interaction with Hsp30-Hsp80 complex has been observed in various yeasts such as *S. cerevisiae, N. crassa* and *C. albicans* in their dormant conidial stage (Ouimet and Kapoor 1999; Girvitz et al. 2000).

Role of HSP70 Co-factors

Hsp70 of fungi are the ubiquitous chaperonin family protein found in cellular compartments, and involved in various cellular functions such as protein synthesis, translocation and degradation of protein and also in protein folding (Frydman 2001). To perform the vital functions Hsp70 chaperone requires a specific binding to substrate protein which is controlled by ATPase cycle. The rate of substrate binding and release from ATP bound domains of Hsp70 is high and the affinity for substrates is low. This process is reversed by the hydrolysis of ATP bound Hsp70 which is converted to ADP-bound confirmation form, hence trapping the substrate with higher affinity. This whole cycle relies on interaction of Hsp70 with some specific co-factors which are regulatory molecules (Andréasson et al. 2010). Co- factors are involved in ATP hydrolysis or exchange of ATP for ADP. Some of the major co-factors of fungal Hsp70 chaperone protein have been described in this study.

Nucleotide Exchange Factors

Substrate release from Hsp70 is mediated by Nucleotide Exchange Factors (NEFs) which performs dissociation of nucleotides and rebinding of ATP. Hsp70 requires NEFs as they have low affinity for nucleotide binding. Fungi can possess multiple NEFs which may have specific interactions with different HSp70s. For example, multiple NEFs have been reported in *S. cerevisiae* such as Sls1, Snl1, Fes1, Sil1, Lhs1, and Sse1 (Kabani et al. 2000; Tyson and Stirling 2000; Sondermann et al. 2002). However, Sse1 stimulates yeast cytosolic Hsp70s whereas endoplasmic reticulum specific Hsp70 function is enhanced by Lhs1 (Steel et al. 2004; Shaner et al. 2006; Andréasson et al. 2010). Some NEFs may have functional overlapping. Lhs1 is more potent than Sil1 whereas, Sse1 is more effective with Ssa Hsp70s than Fes1 (Steel et al. 2004; Sharma and Masison 2009). Ssa1p and Hsp110 is the cytosolic Hsp70 chaperone of *S. cerevisiae* which when present in ADP bound form associates with Fes1p (nucleotide exchange factor) and favors nucleotide release. Translation related defects have been seen due to deletion of fes1p (Kabani et al. 2002). Grp170 (Lhs1) is the NEF of endoplasmic reticulum Hsp70 in yeast.

Hsp40 Co-chaperones

Hsp70 interacts with a client proteins known as J-proteins (also called Hsp40s) (Kampinga and Craig 2010). Hsp40s are critical companion of all Hsp70s in the regulation of Hsp70 activity (Steel et al. 2004). Hsp40s are present in multiple isoforms and regulate ATP dependent polypeptide binding of Hsp70 protein (Shen et al. 2002; Craig et al. 2006). Hsp40 is divided in three classes viz. class I, class II and class III which differs in their structural patterns (Cheetham and Caplan 1998; Hennessy et al. 2000). Yeast Hsp70 Sis1, a class II Hsp40 has structure similarity with Ydj1 which is a class I Hsp70 of yeast (Li et al. 2003a, b). *S. cerevisiae* Hsp70, that is, Ssa1 and Ssa2 interacts with Djp1, Ydj1, Sis, Hlj1 and Swa2, the binding/interaction of which is based on cellular location (Hettema et al. 1998; Sharma and Masison 2009). For example, Djp1 interacts with Ssa1 and 2 in peroxisomes biogenesis, Ydj1 in endoplasmic reticulum membrane, Sis1 in ribosome (Brodsky et al. 1998; Hettema et al. 1998; Horton et al. 2001). Another class of yeast Hsp40 co-chaperone Zou1 has shown interactions with Ssz1 (yeast Hsp70) (Craig et al. 2003).

Hsp70 Functions in Fungi

Hsp70 has been involved in various cellular processes. The following sections discuss the examples Hsp70 functions.

Protein Folding and Prevention of Aggregation

Hsp70 plays an important role in maintaining native conformations of folded proteins from their synthesis until their degradation. *S. cerevisiae* cytosolic Hsp70, Ssa and Ssb are highly identical (Craig et al. 1995). Ssb interacts at early stage of synthesis of polypeptide thus facilitates nascent polypeptide chains elongation whereas Ssa interacts with the secondary structure developed from nascent chain and helps in translocation of protein (Sharma and Masison 2009). Translation in fungi is sometimes hurdled by aggregation of peptides due to non-native (exposed) form of hydrophobic residues, so requires native contacts of all amino acids emerged from ribosome, which is mediated by Hsp70 (Sharma and Masison 2009).

Role of Hsp70 in Protein Degradation

Hsp70 that assists protein folding, also involved in misfolded protein degradation, in association with Hsp40. Genetic studies have shown that Ssa1, which is a yeast Hsp70 chaperone and yeast Hsp40 chaperone Sis1, together play an important role in degradation of misfolded substrate which hampers the quality of proteins (Shiber et al. 2013). Function of Sis1 is ubiquitinization of substrate that is carry forwarded by Ssa1 which functions by forwarding the ubiquitinilated substrate to proteosome (Shiber et al. 2013; Shiber and Ravid 2014). Hsp70 subfamilies are highly dependent on substrates hence, showing specificity towards protein degradation. For example, normal yeast strains containing Ssa1 showed cystic fibrosis transmembrane conductance regulator (CFTR) degradation whereas, 60% CFTR degradation was observed by ssa1 mutant yeast cells. Also, yeast Kar2 (BiP), which is known to have a regulatory domain which responded to unfolded proteins in endoplasmic reticulum, when mutated showed no effect on CFTR degradation. This suggests that CFTR degradation in yeast is dependent on Hsp70 chaperone, Ssa, and not on Kar2 (Molinari et al. 2002). Ssb and Ssa have also been involved in cytosolic protein degradation in yeast. Proteosomal protein degradation in yeast is mediated by Ssb1 (Luders et al. 2000).

Role of Hsp70s in Formation and Propagation of Yeast Prions

Prions are isoform of proteins which self-engendering and transmissible. Prions have been reported in several yeasts which exists in soluble form (normal state) or amyloid form (transmitting state) (Liebman and Chernoff 2012). To maintain themselves in fungal/yeast population, prions relies on Hsp70 function, which may depend on the abundance of Hsp70 for prion propagation (Jones et al. 2004). Extensively studied prions in *S. cerevisiae* are [PSI+] and [URE3], which are derived from Sup35 and Ure2 proteins in their amyloid form responsible to cause infection (Jung et al. 2000; Tibor Roberts et al. 2004). Hsp70 chaperons such as Hsp40s and Hsp104 have impact on propagation of prions, as studies showed that conjugated role of Ssa1 and Ydj1 is involved in inhibition of polymerization of Sup35 (Song and Masison 2005).In yeast, mutant strain of Ssa1 (Hsp70 chaperone) has shown impair mitotic stability leading to partial inhibition of [PSI+] mediated allosuppresion (Jung et al. 2000; Song et al. 2005). This inhibition of prion is also dependent on the variety of Hsp70 chaperons present, as if mutated Ssa1 is only present in cytosol, then it have more pronounced effect on prions weakening. Mutation in Ssa1 also weakens [URE3] activity whereas mutation in Ssa2 weakens both [PSI+] and [URE3] activity (Sharma et al. 2009a, b). In absence of Ssa1 polymers have shown to be self associated and self aggregated, so it can be said that Ssa1 may prevent higher polymer aggregation hence preventing aggregation of prions (Song et al. 2005). Apart from similar homology of Ssa and Ssb, both vary in yeast prion propagation, as Ssb increases [PSI+] formation and Ssb inhibits propagation of [PSI+] (Allen et al. 2005). Prions propagation is also influenced by Hsp70 specific co-factors. For example, Hsp40 over expression, TPR, or NEFs (Fes1 or Sse1) deletion, is involved in the impairment of propagation of [PSI+] (Jones et al. 2004; Qiu et al. 2006). Long term interaction of Hsp70 with Sup35 present in improper folded form may cause inhibition of conversion to prion state. So it is required for Hsp70 to perform proper activity alone or in co-operation with co-factors to prevent aggregation of polymers or prion regeneration (Sharma and Masison 2009).

Role of Hsp70 in Stress

Hsp are induced by two mechanisms in fungi, specific mechanism and general mechanism. Specific mechanism generally encompasses temperature related stress whereas, generalized mechanism by other stresses such as pH, oxidative stress, osmotic stress, starvation, or antifungal stress (Tereshina 2005). Thus, the roles of Hsp70 involved in stress responses have been categorized below.

Hsp70 Response Against Heat Stress/Thermotolerance

Hsp70 is a high molecular weight chaperone, which exists in fungi without any heat stress, but as the heat stress is given to fungi they starts increasing in number (Plesofsky-Vig and Brambl 1985a). The enhancement of fungal hsp70 mRNA due to different temperature gradients revealed a strong stress induced response. In response to heat stress, Hsp70 in association with Hsp40 and Hsp104 has been involved in unfolding of denatured protein, mediating thermotolerance in fungi (Sanchez et al. 1993). A temperature dependent increase of hsp70 expression was found in *A. fumigatus*, *A. terreus*, *C. cladosporioides* and *T. mentagrophytes*, but *P. chrysogenum* and *S. apiospermum* showed no stress induced response. At different incubation temperatures, *A. fumigatus* showed an induction of the hsp70 expression level. *A. fumigatus* showed upregulation of gene expression from 25 to 35 °C, while further increase in temperature resulted in lowering the expression of hsp70. Similarly, *A. terreus* and *C. cladosporioides* showed higher hsp70 expression at 40 °C. Hence it suggests that expression of hsp70 is most at temperature range between 35 and 40 °C. Other fungi such as *T. mentagrophytes* showed hsp70 expression at lower temperature of 25 °C. However, expression of hsp70 in these fungi (*P. chrysogenum* and *S. apiospermum*) is temperature independent (Salzer 2008). *Achlya ambisemalis*, an oomycete, in response to heat shock of 30–35 °C for 10–30 min, showed higher expression of hsp70 (Gwynne and Brandhorst 1982). Similar results were also seen during heat shock response in *A. nidulans*, at 37–43 °C for 1 h (Newbury and Peberdy 1996). *N. crassa* is a filamentous fungus widely studied for stress responses. Heat stress mediated response of *N. crassa* showed accumulation of nuclear associated 78 kDa heat shock protein (Kapoor 1983). Later, *N. crassa* conidia in response to heat stress at 45 °C, showed assembly of 70 kDa Hsp (Plesofsky-Vig and Brambl 1985b). Different morphological stages (mycelia or yeast) have shown no effect on the expression of hsp70 in few fungi, in response to stress (Kummasook et al. 2007). For example, *P. marneffei*, dimorphic fungi have shown same expression level of hsp70 when exposed to 39 °C (Kummasook et al. 2007). Also, *N. crassa* converts from mycelia to hyphae in response to heat stress (25–37 °C), with the higher expression of hsp70 in response to temperature but not during the transition state (Perlman and Feldman 1982). In in-vivo condition, expression of hsp70 plays a very important role, as it mediates resistance of fungi in host in response to high body temperature. Further, suggesting Hsp70 as a virulent factor in fungal mediated infection. To uphold this statement, studies on *N. crassa* treatment at 48 °C for 14 h, showed upregulation of hsp70 expression (Kapoor et al. 1995). Although, studies on *P. brasiliensis* showed that temperature above 42 °C have decreased the hsp70 mRNA expression (Da Silva et al. 1999). So, hsp70 mediates thermo-tolerance in fungi is thought to be species dependent.

Role of Hsp70 in Acid/Alkaline Stress

pH signaling is important for various cellular processes in fungi such as, gene expression regulation, virulence of fungi, secretion of nutritional enzymes, metabolic processes, etc. (Caddick et al. 1986; Ferreira-Nozawa et al. 2006). PacA/PacC pathway is involved in the pH mediated response in fungi (Penalva and Arst 2002). *A. nidulans* is known fungal model used for the study of pH response (Silva et al. 2008). Nucleotide sequence studies of hsp70 of *A. nidulans* revealed that it contains a binding site for PacC in their upstream region. It suggests that in normal conditions (acidic, optimum temperature) PacC with Hsp70 is responsible to maintain normal conditions in fungi. The transcription of hsp70 is mediated by palA (transcriptional factor) in normal growth condition (acidic environment), which is encoded by pacC (Tilburn et al. 1995). But, if the heat stress is given to fungi for 1–2 h in the absence of palA, the hsp70 response becomes independent of both acid and alkaline pH, so PalA in that condition does not play any role in hsp70 expression. *A. nidulans* showed high expression of hsp70 at pH 8 and the activation of palA (Freitas et al. 2011). Studies on pH mediated hsp70 response in *N. crassa* showed that hsp70 was highly expressed at acidic pH and optimum temperature (30 °C) but did not show any upregulation at alkaline pH, depending on PacC mediated signaling pathway. Also, Squina et al, showed that pH mediated expression of hsp70 vary depending upon culture conditions and heat stress. Some hsp70 genes respond to these conditions in addition to pH response (For example; hsp70-1 and hsp70-2 of *N. crassa*) while some hsp70 only depends on culture conditions or heat stress but not on pH (For example; hsp70-3 of *N. crassa*) (Squina et al. 2010). Another fungus, *Candida glabrata*, also showed same expression of hsp70 towards alkaline pH. Hsp70 (cytosolic Ssa1) protein was found to be downregulated at pH 7.4 and 8, which suggests that in fungi hsp70 expression is mediated by acidic pH (Schmidt et al. 2008).

Oxidative Stress and Osmotic Pressure Response

Oxidative stress and response towards osmotic pressure is a functional challenge for fungal cells that generally all fungus experience. Fungus overcomes the effect of reactive oxygen species by mediating activation of transcriptional processes and by producing stress response proteins. Very little information is available for the role of Hsp70 in oxidative stress response in fungi. Skn7 is a oxidative stress response mediated transcription factor which in co-ordination with heat shock transcriptional factor hsf1 induces heat shock genes (Raitt et al. 2000). Loss in Skn7 in *S. cerevisiae* showed inhibition of Hsp70 (Ssa1) (Morano et al. 2012). This finding might also related Hsp70 role in oxidative stress response. In response to H_2O_2 stress in Paracoccidioides yeast cells, hsp70 was found to be expressed which showed the important part of oxidative stress response fungal machinery (de Arruda Grossklaus et al. 2013). Kunal et al, showed that in response to oxidative stress in fungi, *Ascochyta rabiei*, hsp70 was found to be upregulated (Singh et al. 2012). Hypoxia is a state of deprived oxygen, causing oxidative stress. In fungi, the hypoxia conditions has led to heat shock responses, for example, C. *neoformans, Drosophila melanogaster, C. albicans* and *B. emersonii* in hypoxia condition, showed higher expression of hsp70 and the survivability of fungi was found to be increased significantly (Setiadi et al. 2006; Chun et al. 2007; de Castro Georg and Gomes 2007; Azad et al. 2009). *Metarhizium anisopliae*, an entomopathogenic fungus, is responsible for infection in insects. Osmotic stress response in this fungus is mediated by mos1 gene. When mos1 was knock down, showed the downregulated expression of hsp70 transcripts, possibly leading to delay in osmotic stress response (Wang et al. 2008).

Drug Response

To prevent the harmful effect of fungal species, various antifungal medications have been employed which have specific fungal targets for their action. Major antifungal drugs used are amphotericin, itraconazole, fluconazole, ketoconazole, clotrimazole, miconazole, terbinafine, etc. (Odds et al. 2003). Some of these antifungals have been studied for their effect on hsp70 mediated responses in fungi. In *A. terreus* hsp70 play a vital role in response to amphoteracin B (AmB). Transcriptome analysis of AmB resistance strain of *A. terreus* revealed relatively high ssa and ssb mRNA expression in comparison to AmB susceptible strains whereas mitochondrial and endoplasmic reticulum specific hsp70 members, kar and ssc, were downregulated. In addition, sse and ssz (nucleotide exchange factors) were also uplifted in response to AmB resistant *A. terreus*. Susceptibility of AmB towards A. terreus were found to increased in absence of Hsp70. AmB when combined with pifithrin (Hsp70 inhibitor) inhibited AmB resistant *A. terreus* (Blatzer et al. 2015), whereas *A. fumigatus* treated with AmB showed downregulation of mitochondrial Hsp70 protein (Gautam et al. 2008). 17β-estradiol is a female hormone that inhibits dimorphic fungus (Paracoccidioides) in their transition from mycelia to yeast, thus women are observed to be resistant to paracoccidioidomycosis (Shankar et al. 2011a, b). In Paracoccidioides, the transcriptional profile during temporal transition from mycelia to yeasts form showed differential regulation of hsp70 in response to 17β-estradiol. Expression of hsp70 was observed during 4–12 h of germination, which dropped further at 144 h, which suggested the role of hsp70 in delaying the morphological transitions from mycelium to yeast under the influence of 17β-estradiol (Shankar et al. 2011a, b). Genome wide studies of *C. albicans* treated with 17β-estradiol showed expression of hsp90, whereas the regulation of hsp70 was not onserved

(Burt et al. 2003; Banerjee et al. 2007; Clemons et al. 2010). Paromomycin is an antibiotic which is used as antifungal compound. *S. cerevisiae* when exposed to paromomycin showed induction of hsp70 mRNA translation (Grant and Tuite 1994). In other studies on *S. cerevisiae* treated with thiolutin (antifungal), induce the transcription of ssa4 (hsp70) (Adams and Gross 1991). These findings may signify that hsp70 is a novel target against pathogenic fungi.

Other Stress Responses

Apart from heat stress, Hsp70 in fungi have been found to play important role in salt, metal or ion stress etc. Weitzel et al, showed the increase in expression of hsp70 in response to dinitrophenol in *S. cerevisiae* (Weitzel et al. 1985). Also, *S. cerevisiae* has shown the increase in expression of hsp70 in response to NaCl ranging from 0.3 to 0.7 M (Antelo et al. 1992; Lewis et al. 1995). Thiolutin (RNA polymerase inhibitor) and parornomycin (antibiotic) was also showed high expression of ssa4 in *S. cerevisiae* (Grant et al. 1989; Adams and Gross 1991). In continuation, *S. cerevisiae* showed same expression level of hsp70 in response to ethanol treatment, which was not observed in case of *N. crassa* (Plesset et al. 1982; Roychowdhury and Kapoor 1988). *N. crassa* cells when treated with hydrogen peroxide showed the synthesis of Hsp70 protein (Kapoor and Lewis 1987). In several studies high concentration of sodium arsenite showed the increased transcripts of hsp70 in *N. crassa*. In addition, cadmium when inoculated with sodium arsenite, showed no effect on expression of hsp70 (Chakraborty et al. 1995). However, cadmium showed upregulation of hsp70 in *A. fumigatus*. Further, *A. fumigatus* also showed upregulation hsp70 on treatment with cadmium chloride (Damelin et al. 2000; Georg and Gomes 2007).

Role of HSP70 in Fungus-Plant Interaction

Fungus mediated plant pathogenesis and disease is responsible to cause loss of food products worldwide. So, studies are needed further to deduce the fungal targets which are essential for disease progression. In few studies on fungal-plant interaction, hsp70 has been observed to be upregulated, contributing to pathogenesis. For example, *Colletotrichum gloeosporioides*, showed higher expression of hsp70 in coffee plant in its infectious state (coffee berry disease) (Chen et al. 2003). *Magnaporthe oryzae* Hsp70 proteins (Lhs1p and Kar2p), were also found to accumulate in the endoplasmic reticulum of fungi and responsible for disease development in rice plant (Yi et al. 2009). *Rhizoctonia solani* is a plant pathogenic fungus causing various diseases in plants such as damping, root rot, collar rot, etc. (Zhu et al. 2016). Proteomic studies revealed that heat shock protein of 70 kDa was observed during cell wall degradation of soybean seedlings, thereby, assisting virulence (Lakshman et al. 2016).

HSP70 in Pathogenesis and Potential Diagnostic or a Vaccine Candidate

Fungal diseases, in recent years, have been taken into account due to their dramatic increase and aggressiveness that demands better medical diagnosis and therapies. Fungal diseases/contamination has made its impact on immune-compromised patients (Thakur et al. 2015) as well as on several food crops (Chauhan et al. 2016). Fungal resistance over major antifungal drugs has raised a need to develop new and efficient drug targets for pathogenic fungi and vaccine candidate. Fungal pathogen after entering into the host body suffers temperature shifts. Hence, to overcome temperature shift stress, fungus produces heat shock proteins that may play role in fungal survival/growth in the host. In various transcriptomic and proteomic studies, it has been shown that Hsp of fungi are up-regulated during infection. Hsp70 is a major chaperone that has been widely studied due to their role in infection/contamination in animals/plants. For example, Hsp70 in *C. albicans* has been reported in systemic murine candidiasis, which has been identified as an antigen by host immune system (Bromuro et al. 1998). Proteomic studies of *A. flavus* cultured on corn flour showed expression Hsp70 proteins (Tiwari and Shankar 2018). Thus, it suggests that fungal Hsp70 is important for both human and plant. Allergens from *C. cladosporioides* also showed expression of 72 kDa protein (Li et al. 2008). Systemic mycosis is prevalently caused by *P. brasiliensis* in yeast phase showed higher expression of hsp70 mRNA during infection (McEwen et al. 2000; Goldman et al. 2003; Cleare et al. 2017). *Cryptococcus neoformans* being a pathogenic fungus, is involved in impairing immunity of patients (HIV, defective T cells) causing severe infections. Studies on murine pulmonary cryptococcosis caused by *C. neoformans* have shown hsp70 as a target molecule for immune response (Kakeya et al. 1999). Hsp70 mediated immune response e.g., IL-12 and IFN-γ has been observed in immunological studies of *H. capsulatum* (Allendoerfer et al. 1996; Deepe Jr. and Gibbons 2002; Cutler et al. 2007). Immune response against Hsp70 also accounts production of tumor necrotic factors and interleukin 6, which have role in the prevention yeast infection (Bromuro et al. 1998). AntEroles et al., showed that heat stress induces the formation of *C. albicans* spores. These spores when reacted with polyclonal antibody of rabbit, exhibited hsp70 expression having homology with *S. cerevisiae* Ssa1 gene (Eroles et al. 1995). Studies from Li et al, showed that histatin 5 (a salivary protein) has antifungal properties and possess affinity towards Ssa1 and Ssa2 (Li et al. 2003a, b). Previously hsp70 of *Penicillium citrinum* has also been screened from asthmatic patients (Shen et al. 1997). Moreover, in other bacteria such as *P. carinii*, *P. rattus* and *P. jirovecii* also showed the expression of hsp70 during their pathogenic phase (Stedman et al. 1998; Latouche et al. 2001; Burnie et al. 2006). These literatures suggest that fungal Hsp70 plays a major role in pathogenesis and act as a potent antigen for mediating host immune response. Thus, Hsp70 has potential as a diagnostic candidate and may be a promising protein to study as a vaccine candidate (Cleare et al. 2017).

Conclusions

Fungi are the diverse genera produce Hsp70 in response to stress conditions for accomplishing various biological functions. Hsp70 has been observed to be highly conserved in the evolution of fungal kingdom. *S. cerevisiae* had been a best studied model for nucleotides and amino acid sequences of hsp70, which correlates with other fungal species. Majorly, Hsp70 protein has been reported in protein folding, protein aggregation and disaggregation, regulating transcription and post-transcription, morphogenesis and thermotolerance. Also, it has also showed role in other stress response such as osmotic, oxidative or response towards pH and stress response towards meal ion, antibiotics, alcohols etc. Hsp70 has emerged as a major heat shock protein expressed in fungi during the stress after Hsp90. Expression of fungal Hsp70 in the host system added a molecule to study for the development of vaccine candidate using Hsp70 protein/epitopes against various fungal infections.

Acknowledgements We are thankful to the Department of Biotechnology and Bioinformatics, Jaypee University of Information Technology, for providing facilities.

References

Adams, C. C., & Gross, D. S. (1991). The yeast heat shock response is induced by conversion of cells to spheroplasts and by potent transcriptional inhibitors. *Journal of Bacteriology, 173*, 7429–7435.

Allen, K. D., Wegrzyn, R. D., Chernova, T. A., Müller, S., Newnam, G. P., Winslett, P. A., Wittich, K. B., Wilkinson, K. D., & Chernoff, Y. O. (2005). Hsp70 chaperones as modulators of prion life cycle. *Genetics, 169*, 1227–1242.

Allendoerfer, R., Maresca, B., & Deepe, G. (1996). Cellular immune responses to recombinant heat shock protein 70 from *Histoplasma capsulatum*. *Infection and Immunity, 64*, 4123–4128.

Andréasson, C., Rampelt, H., Fiaux, J., Druffel-Augustin, S., & Bukau, B. (2010). The endoplasmic reticulum Grp170 acts as a nucleotide exchange factor of Hsp70 via a mechanism similar to that of the cytosolic Hsp110. *The Journal of Biological Chemistry, 285*, 12445–12453.

Antelo, J. M., Arce, F., Parajo, M., Rodriguez, P., & Varela, A. (1992). Disproportionation kinetics of NaCl isopropylamine and NaCl isopropylamine. *International Journal of Chemical Kinetics, 24*, 991–997.

Azad, P., Zhou, D., Russo, E., & Haddad, G. G. (2009). Distinct mechanisms underlying tolerance to intermittent and constant hypoxia in *Drosophila melanogaster*. *PLoS One, 4*, e5371.

Banerjee, D., Martin, N., Nandi, S., Shukla, S., Dominguez, A., Mukhopadhyay, G., & Prasad, R. (2007). A genome-wide steroid response study of the major human fungal pathogen *Candida albicans*. *Mycopathologia, 164*, 1–17.

Blatzer, M., Blum, G., Jukic, E., Posch, W., Gruber, P., Nagl, M., Binder, U., Maurer, E., Sarg, B., & Lindner, H. (2015). Blocking Hsp70 enhances the efficiency of amphotericin B treatment against resistant *Aspergillus terreus* strains. *Antimicrobial Agents and Chemotherapy, 59*, 3778–3788.

Boorstein, W. R., Ziegelhoffer, T., & Craig, E. A. (1994). Molecular evolution of the HSP70 multigene family. *Journal of Molecular Evolution, 38*, 1–17.

Brand, A. (2011). Hyphal growth in human fungal pathogens and its role in virulence. *International Journal of Microbiology, 2012*, 1–11.

Brodsky, J. L., Lawrence, J. G., & Caplan, A. J. (1998). Mutations in the cytosolic DnaJ homologue, YDJ1, delay and compromise the efficient translation of heterologous proteins in yeast. *Biochemistry, 37*, 18045–18055.

Bromuro, C., La Valle, R., Sandini, S., Urbani, F., Ausiello, C. M., Morelli, L., Féd'ostiani, C., Romani, L., & Cassone, A. (1998). A 70-kilodalton recombinant heat shock protein of *Candida albicans* is highly immunogenic and enhances systemic murine candidiasis. *Infection and Immunity, 66*, 2154–2162.

Bukau, B., & Horwich, A. L. (1998). The Hsp70 and Hsp60 chaperone machines. *Cell, 92*, 351–366.

Burnie, J. P., Carter, T. L., Hodgetts, S. J., & Matthews, R. C. (2006). Fungal heat shock proteins in human disease. *FEMS Microbiology Reviews, 30*, 53–88.

Burt, E. T., Daly, R., Hoganson, D., Tsirulnikov, Y., Essmann, M., & Larsen, B. (2003). Isolation and partial characterization of Hsp90 from *Candida albicans*. *Annals of Clinical and Laboratory Science, 33*, 86–93.

Caddick, M. X., Brownlee, A. G., & Arst, H. N. (1986). Regulation of gene expression by pH of the growth medium in *Aspergillus nidulans*. *Molecular & General Genetics, 203*, 346–353.

Caruso, M., Sacco, M., Medoff, G., & Maresca, B. (1987). Heat shock 70 gene is differentially expressed in *Histoplasma capsulatum* strains with different levels of thermotolerance and pathogenicity. *Molecular Microbiology, 1*, 151–158.

Černila, B., Črešnar, B., & Breskvar, K. (2003). Molecular characterization of genes encoding cytosolic Hsp70s in the zygomycete fungus *Rhizopus nigricans*. *Cell Stress & Chaperones, 8*, 317–328.

Chakraborty, B., Ouimet, P., Sreenivasan, G., Curle, C., & Kapoor, M. (1995). Sequence repeat-induced disruption of the major heat-inducible HSP70 gene of *Neurospora crassa*. *Current Genetics, 29*, 18–26.

Chauhan, N. M., Washe, A. P., & Minota, T. (2016). Fungal infection and aflatoxin contamination in maize collected from Gedeo zone, Ethiopia. *SpringerPlus, 5*, 753.

Cheetham, M. E., & Caplan, A. J. (1998). Structure, function and evolution of DnaJ: Conservation and adaptation of chaperone function. *Cell Stress & Chaperones, 3*, 28.

Chen, Z., Ribeiro, A., Silva, M., Santos, P., Guerra-Guimarães, L., Gouveia, M., Fernandez, D., & Rodrigues, C. (2003). Heat shock-induced susceptibility of green coffee leaves and berries to *Colletotrichum gloeosporioides* and its association to PR and hsp70 gene expression. *Physiological and Molecular Plant Pathology, 63*, 181–190.

Chun, C. D., Liu, O. W., & Madhani, H. D. (2007). A link between virulence and homeostatic responses to hypoxia during infection by the human fungal pathogen *Cryptococcus neoformans*. *PLoS Pathogens, 3*, e22.

Cleare, L. G., Miranda, D. Z., & Nosanchuk, J. D. (2017). Heat shock proteins in *Histoplasma* and *Paracoccidioides*. *Clinical and Vaccine Immunology, 24*, 00221–00217.

Clemons, K. V., Shankar, J., & Stevens, D. A. (2010). Mycologic endocrinology. In M. Lyte & P. P. E. Freestone (Eds.), *The microbial endocrinology* (pp. 269–290). New York: Springer.

Craig, E. A., & Marszalek, J. (2011). Hsp70 chaperones. In *The encyclopedia of life sciences (ELS)* (pp. 1–8). Chichester: Wiley.

Craig, E. A., Gambill, B. D., & Nelson, R. J. (1993). Heat shock proteins: molecular chaperones of protein biogenesis. *Microbiological Reviews, 57*, 402–414.

Craig, E., Ziegelhoffer, T., Nelson, J., Laloraya, S., & Halladay, J. (1995). Complex multigene family of functionally distinct Hsp70s of yeast. *Cold Spring Harbor Symposia on Quantitative Biology, 60*, 441–449.

Craig, E. A., Eisenman, H. C., & Hundley, H. A. (2003). Ribosome-tethered molecular chaperones: The first line of defense against protein misfolding? *Current Opinion in Microbiology, 6*, 157–162.

Craig, E., Huang, P., Aron, R., & Andrew, A. (2006). The diverse roles of J-proteins, the obligate Hsp70 co-chaperone. In N. Brend & G. Thomas (Eds.), *The reviews of physiology, biochemistry and pharmacology* (Vol. 156, pp. 1–21). Berlin/Heidelberg: Springer.

Cutler, J. E., Deepe, G. S., Jr., & Klein, B. S. (2007). Advances in combating fungal diseases: Vaccines on the threshold. *Nature Reviews. Microbiology, 5*, 13.

Da Silva, S. P., Borges-Walmsley, M. I., Pereira, I. S., Soares, C. M. A., Walmsley, A. R., & Felipe, M. S. S. (1999). Differential expression of an hsp70 gene during transition from the mycelial to the infective yeast form of the human pathogenic fungus *Paracoccidioides brasiliensis*. *Molecular Microbiology, 31*, 1039–1050.

Damelin, L., Vokes, S., Whitcutt, J., Damelin, S., & Alexander, J. (2000). Hormesis: A stress response in cells exposed to low levels of heavy metals. *Human & Experimental Toxicology, 19*, 420–430.

Daugaard, M., Rohde, M., & Jäättelä, M. (2007). The heat shock protein 70 family: Highly homologous proteins with overlapping and distinct functions. *FEBS Letters, 581*, 3702–3710.

De Arruda Grossklaus, D., Bailão, A. M., Rezende, T. C. V., Borges, C. L., de Oliveira, M. A. P., Parente, J. A., & de Almeida, S. C. M. (2013). Response to oxidative stress in Paracoccidioides yeast cells as determined by proteomic analysis. *Microbes and Infection, 15*, 347–364.

De Castro Georg, R., & Gomes, S. L. (2007). Comparative expression analysis of members of the Hsp70 family in the chytridiomycete *Blastocladiella emersonii*. *Gene, 386*, 24–34.

Deepe, G. S., Jr., & Gibbons, R. S. (2002). Cellular and molecular regulation of vaccination with heat shock protein 60 from *Histoplasma capsulatum*. *Infection and Immunity, 70*, 3759–3767.

Dix, N. J. (2012). The mycelium and substrates for growth. In Springer Science & Business Media & J. Dix Neville (Eds.), *The fungal ecology* (pp. 12–38). Dordrecht: Springer.

Easton, D. P., Kaneko, Y., & Subjeck, J. R. (2000). The Hsp110 and Grp170 stress proteins: newly recognized relatives of the Hsp70s. *Cell Stress & Chaperones, 5*, 276–290.

Eroles, P., Sentandreu, M., Elorza, M. V., & Sentandreu, R. (1995). Cloning of a DNA fragment encoding part of a 70-kDa heat shock protein of *Candida albicans*. *FEMS Microbiology Letters, 128*, 95–100.

Ferreira-Nozawa, M. S., Silveira, H. C., Ono, C. J., Fachin, A. L., Rossi, A., & Martinez-Rossi, N. M. (2006). The pH signaling transcription factor PacC mediates the growth of *Trichophyton rubrum* on human nail in vitro. *Medical Mycology, 44*, 641–645.

Fracella, F., Scholle, C., Kallies, A., Häfker, T., Schröder, T., & Rensing, L. (1997). Differential HSC70 expression during asexual development of *Neurospora crassa*. *Microbiology, 143*, 3615–3624.

Freitas, J. S., Silva, E. M., Leal, J., Gras, D. E., Martinez-Rossi, N. M., Dos Santos, L. D., Palma, M. S., & Rossi, A. (2011). Transcription of the Hsp30, Hsp70, and Hsp90 heat shock protein genes is modulated by the PalA protein in response to acid pH-sensing in the fungus *Aspergillus nidulans*. *Cell Stress & Chaperones, 16*, 565–572.

Frydman, J. (2001). Folding of newly translated proteins in vivo: the role of molecular chaperones. *Annual Review of Biochemistry, 70*, 603–647.

Galagan, J. E., Henn, M. R., Ma, L.-J., Cuomo, C. A., & Birren, B. (2005). Genomics of the fungal kingdom: insights into eukaryotic biology. *Genome Research, 15*, 1620–1631.

Gautam, P., Shankar, J., Madan, T., Sirdeshmukh, R., Sundaram, C. S., Gade, W. N., Basir, S. F., & Sarma, P. U. (2008). Proteomic and transcriptomic analysis of *Aspergillus fumigatus* on exposure to amphotericin B. *Antimicrobial Agents and Chemotherapy, 52*, 4220–4227.

Georg, R. C., & Gomes, S. L. (2007). Transcriptome analysis in response to heat shock and cadmium in the aquatic fungus *Blastocladiella emersonii*. *Eukaryotic Cell, 6*, 1053–1062.

Girvitz, T., Ouimet, P., & Kapoor, M. (2000). Heat shock protein 80 of Neurospora crassa: Sequence analysis of the gene and expression during the asexual phase. *Canadian Journal of Microbiology, 46*, 981–991.

Goldman, G. H., dos Reis Marques, E., Ribeiro, D. C. D., de Souza, B. L. A. N., Quiapin, A. C., Vitorelli, P. M., Savoldi, M., Semighini, C. P., de Oliveira, R. C., & Nunes, L. R. (2003). Expressed sequence tag analysis of the human pathogen *Paracoccidioides brasiliensis* yeast phase: identification of putative homologues of *Candida albicans* virulence and pathogenicity genes. *Eukaryotic Cell, 2*, 34–48.

Grant, C. M., & Tuite, M. F. (1994). Mistranslation of human phosphoglycerate kinase in yeast in the presence of paromomycin. *Current Genetics, 26*, 95–99.

Grant, C., Firoozan, M., & Tuite, M. (1989). Mistranslation induces the heat-shock response in the yeast *Saccharomyces cerevisiae*. *Molecular Microbiology, 3*, 215–220.

Guarro, J., Gené, J., & Stchigel, A. M. (1999). Developments in fungal taxonomy. *Clinical Microbiology Reviews, 12*, 454–500.

Gupta, R. S., & Singh, B. (1994). Phylogenetic analysis of 70 kD heat shock protein sequences suggests a chimeric origin for the eukaryotic cell nucleus. *Current Biology, 4*, 1104–1114.

Gwynne, D. I., & Brandhorst, B. (1982). Alterations in gene expression during heat shock of *Achlya ambisexualis*. *Journal of Bacteriology, 149*, 488–493.

Häfker, T., Techel, D., Steier, G., & Rensing, L. (1998). Differential expression of glucose-regulated (grp78) and heat-shock-inducible (hsp70) genes during asexual development of *Neurospora crassa*. *Microbiology, 144*, 37–43.

Hennessy, F., Cheetham, M. E., Dirr, H. W., & Blatch, G. L. (2000). Analysis of the levels of conservation of the J domain among the various types of DnaJ-like proteins. *Cell Stress & Chaperones, 5*, 347–358.

Hettema, E. H., Ruigrok, C. C., Koerkamp, M. G., Van, D., Berg, M., Tabak, H. F., Distel, B., & Braakman, I. (1998). The cytosolic DnaJ-like protein djp1p is involved specifically in peroxisomal protein import. *The Journal of Cell Biology, 142*, 421–434.

Horton, L. E., James, P., Craig, E. A., & Hensold, J. O. (2001). The yeast hsp70 homologue Ssa is required for translation and interacts with Sis1 and Pab1 on translating ribosomes. *The Journal of Biological Chemistry, 276*, 14426–14433.

Jin, C. (2011). Protein glycosylation in *Aspergillus fumigatus* is essential for cell wall synthesis and serves as a promising model of multicellular eukaryotic development. *International Journal of Microbiology, 2012*, 654251.

Jones, G., Song, Y., Chung, S., & Masison, D. C. (2004). Propagation of *Saccharomyces cerevisiae* [PSI+] prion is impaired by factors that regulate Hsp70 substrate binding. *Molecular and Cellular Biology, 24*, 3928–3937.

Jung, G., Jones, G., Wegrzyn, R. D., & Masison, D. C. (2000). A role for cytosolic Hsp70 in yeast [PSI+] prion propagation and [PSI+] as a cellular stress. *Genetics, 156*, 559–570.

Kabani, M., Beckerich, J.-M., & Gaillardin, C. (2000). Sls1p stimulates Sec63p-mediated activation of Kar2p in a conformation-dependent manner in the yeast endoplasmic reticulum. *Molecular and Cellular Biology, 20*, 6923–6934.

Kabani, M., Beckerich, J.-M., & Brodsky, J. L. (2002). Nucleotide exchange factor for the yeast Hsp70 molecular chaperone Ssa1p. *Molecular and Cellular Biology, 22*, 4677–4689.

Kakeya, H., Udono, H., Maesaki, S., Sasaki, E., Kawamura, S., Hossain, M., Yamamoto, Y., Sawai, T., Fukuda, M., & Mitsutake, K. (1999). Heat shock protein 70 (hsp70) as a major target of the antibody response in patients with pulmonary cryptococcosis. *Clinical and Experimental Immunology, 115*, 485.

Kampinga, H. H., & Craig, E. A. (2010). The Hsp70 chaperone machinery: J-proteins as drivers of functional specificity. *Nature Reviews. Molecular Cell Biology, 11*, 579.

Kapoor, M. (1983). A study of the heat-shock response in *Neurospora crassa*. *The International Journal of Biochemistry, 15*, 639–649.

Kapoor, M., & Lewis, J. (1987). Heat shock induces peroxidase activity in Neurospora crassa and confers tolerance toward oxidative stress. *Biochemical and Biophysical Research Communications, 147*, 904–910.

Kapoor, M., Curle, C., & Runham, C. (1995). The hsp70 gene family of *Neurospora crassa*: Cloning, sequence analysis, expression, and genetic mapping of the major stress-inducible member. *Journal of Bacteriology, 177*, 212–221.

Kominek, J., Marszalek, J., Neuvéglise, C., Craig, E. A., & Williams, B. L. (2013). The complex evolutionary dynamics of Hsp70s: A genomic and functional perspective. *Genome Biology and Evolution, 5*, 2460–2477.

Kregel, K. C. (2002). Invited review: Heat shock proteins: Modifying factors in physiological stress responses and acquired thermotolerance. *Journal of Applied Physiology, 92*, 2177–2186.

Kummasook, A., Kummasook, A., Pongpom, P., Kummasook, A., Pongpom, P., & Vanittanakom, N. (2007). Cloning, characterization and differential expression of an hsp70 gene from the pathogenic dimorphic fungus, *Penicillium marneffei*. *DNA Sequence, 18*, 385–394.

Lakshman, D. K., Roberts, D. P., Garrett, W. M., Natarajan, S. S., Darwish, O., Alkharouf, N., Pain, A., Khan, F., Jambhulkar, P. P., & Mitra, A. (2016). Proteomic investigation of *Rhizoctonia solani* AG 4 identifies secretome and mycelial proteins with roles in plant cell wall degradation and virulence. *Journal of Agricultural and Food Chemistry, 64*, 3101–3110.

Latouche, S., Totel, A., Lacube, P., Bolognini, J., Nevez, G., & Roux, A. (2001). Development of an RT-PCR on the heat shock protein 70 gene for viability detection of *Pneumocystis carinii* f. sp. hominis in patients with Pneumocystosis and in air sample. *Journal of Eukaryotic Microbiology, 48*, 176S–177S.

Lewis, J., Learmonth, R., & Watson, K. (1995). Induction of heat, freezing and salt tolerance by heat and salt shock in *Saccharomyces cerevisiae*. *Microbiology, 141*, 687–694.

Li, J., Qian, X., & Sha, B. (2003a). The crystal structure of the yeast Hsp40 Ydj1 complexed with its peptide substrate. *Structure, 11*, 1475–1483.

Li, X. S., Reddy, M. S., Baev, D., & Edgerton, M. (2003b). *Candida albicans* Ssa1/2p is the cell envelope binding protein for human salivary histatin 5. *The Journal of Biological Chemistry, 278*, 28553–28561.

Li, L., Li, Q., Liu, Z., & Wu, H. (2008). Immunological analysis and mass spectrometry identification of the major allergen from *Cladosporium cladosporioides*. *Journal of Hygiene Research, 37*, 50–52.

Liebman, S. W., & Chernoff, Y. O. (2012). Prions in yeast. *Genetics, 191*, 1041–1072.

Lindquist, S., & Craig, E. (1988). The heat-shock proteins. *Annual Review of Genetics, 22*, 631–677.

Lopez-Ribot, J. L., Alloush, H. M., Masten, B. J., & Chaffin, W. L. (1996). Evidence for presence in the cell wall of *Candida albicans* of a protein related to the hsp70 family. *Infection and Immunity, 64*, 3333–3340.

Lüders, J., Demand, J., & Höhfeld, J. (2000). The ubiquitin-related BAG-1 provides a link between the molecular chaperones Hsc70/Hsp70 and the proteasome. *The Journal of Biological Chemistry, 275*, 4613–4617.

Mayer, M., & Bukau, B. (2005). Hsp70 chaperones: Cellular functions and molecular mechanism. *Cellular and Molecular Life Sciences, 62*, 670.

McEwen, J., Taylor, J., Carter, D., Xu, J., Felipe, M., Vilgalys, R., Mitchell, T., Kasuga, T., White, T., & Bui, T. (2000). Molecular typing of pathogenic fungi. *Medical Mycology, 38*, 189–197.

Molinari, M., Galli, C., Piccaluga, V., Pieren, M., & Paganetti, P. (2002). Sequential assistance of molecular chaperones and transient formation of covalent complexes during protein degradation from the ER. *The Journal of Cell Biology, 158*, 247–257.

Monteiro, J. P., Clemons, K. V., Mirels, L. F., Coller, J. A., Jr., Wu, T. D., Shankar, J., Lopes, C. R., & Stevens, D. A. (2009). Genomic DNA microarray comparison of gene expression patterns in *Paracoccidioides brasiliensis* mycelia and yeasts in vitro. *Microbiology, 155*, 2795–2808.

Morano, K. A., Grant, C. M., & Moye-Rowley, W. S. (2012). The response to heat shock and oxidative stress in *Saccharomyces cerevisiae*. *Genetics, 190*, 1157–1195.

Newbury, J., & Peberdy, J. F. (1996). Characterization of the heat shock response in protoplasts of *Aspergillus nidulans*. *Mycological Research, 100*, 1325–1332.

Nikolaidis, N., & Nei, M. (2004). Concerted and nonconcerted evolution of the Hsp70 gene superfamily in two sibling species of nematodes. *Molecular Biology and Evolution, 21*, 498–505.

Odds, F. C., Brown, A. J., & Gow, N. A. (2003). Antifungal agents: Mechanisms of action. *Trends in Microbiology, 11*, 272–279.

Ouimet, P., & Kapoor, M. (1999). Nucleotide binding and hydrolysis properties of *Neurospora crassa* cytosolic molecular chaperones, Hsp70 and Hsp80, heat-inducible members of the eukaryotic stress-70 and stress-90 families. *Biochemistry and Cell Biology, 77*, 89–99.

Peñalva, M. A., & Arst, H. N. (2002). Regulation of gene expression by ambient pH in filamentous fungi and yeasts. *Microbiology and Molecular Biology Reviews, 66*, 426–446.

Perlman, J., & Feldman, J. F. (1982). Cycloheximide and heat shock induce new polypeptide synthesis in *Neurospora crassa*. *Molecular and Cellular Biology, 2*, 1167–1173.

Pfund, C., Huang, P., Lopez-Hoyo, N., & Craig, E. A. (2001). Divergent functional properties of the ribosome-associated molecular chaperone Ssb compared with other Hsp70s. *Molecular Biology of the Cell, 12*, 3773–3782.

Plesofsky-Vig, N., & Brambl, R. (1985a). The heat shock response of fungi. *Experimental Mycology, 9*, 7–94.

Plesofsky-Vig, N., & Brambl, R. (1985b). Heat shock response of *Neurospora crassa*: Protein synthesis and induced thermotolerance. *Journal of Bacteriology, 162*, 1083–1091.

Plesset, J., Palm, C., & McLaughlin, C. (1982). Induction of heat shock proteins and thermotolerance by ethanol in *Saccharomyces cerevisiae*. *Biochemical and Biophysical Research Communications, 108*, 1340–1345.

Qiu, X.-B., Shao, Y.-M., Miao, S., & Wang, L. (2006). The diversity of the DnaJ/Hsp40 family, the crucial partners for Hsp70 chaperones. *Cellular and Molecular Life Sciences, 63*, 2560–2570.

Raitt, D. C., Johnson, A. L., Erkine, A. M., Makino, K., Morgan, B., Gross, D. S., & Johnston, L. H. (2000). The Skn7 response regulator of *Saccharomyces cerevisiae* interacts with Hsf1 in vivo and is required for the induction of heat shock genes by oxidative stress. *Molecular Biology of the Cell, 11*, 2335–2347.

Röhl, A., Wengler, D., Madl, T., Lagleder, S., Tippel, F., Herrmann, M., Hendrix, J., Richter, K., Hack, G., & Schmid, A. B. (2015). Hsp90 regulates the dynamics of its cochaperone Sti1 and the transfer of Hsp70 between modules. *Nature Communications, 6*, 6655.

Roychowdhury, H., & Kapoor, M. (1988). Ethanol and carbon-source starvation enhance the accumulation of HSP80 in *Neurospora crassa*. *Canadian Journal of Microbiology, 34*, 162–168.

Salzer, H. J. F. (2008). Identification and characterization of heat shock protein 60 and 70 of common environmental fungi of medical interest. MedizinischeUniversität.

Sanchez, Y., Parsell, D., Taulien, J., Vogel, J., Craig, E., & Lindquist, S. (1993). Genetic evidence for a functional relationship between Hsp104 and Hsp70. *Journal of Bacteriology, 175*, 6484–6491.

Satyanarayana, C., Schröder-Köhne, S., Craig, E. A., Schu, P. V., & Horst, M. (2000). Cytosolic Hsp70s are involved in the transport of aminopeptidase 1 from the cytoplasm into the vacuole. *FEBS Letters, 470*, 232–238.

Schmidt, P., Walker, J., Selway, L., Stead, D., Yin, Z., Enjalbert, B., Weig, M., & Brown, A. J. (2008). Proteomic analysis of the pH response in the fungal pathogen *Candida glabrata*. *Proteomics, 8*, 534–544.

Setiadi, E. R., Doedt, T., Cottier, F., Noffz, C., & Ernst, J. F. (2006). Transcriptional response of *Candida albicans* to hypoxia: Linkage of oxygen sensing and Efg1p-regulatory networks. *Journal of Molecular Biology, 361*, 399–411.

Shaner, L., Sousa, R., & Morano, K. A. (2006). Characterization of Hsp70 binding and nucleotide exchange by the yeast Hsp110 chaperone Sse1. *Biochemistry, 45*, 15075–15084.

Shankar, J. (2013). An overview of toxins in Aspergillus associated with pathogenesis. *International Journal of Life Sciences Biotechnology and Pharma Research, 2*, 16–31.

Shankar, J., Nigam, S., Saxena, S., Madan, T., & Sarma, P. (2004). Identification and assignment of function to the genes of *Aspergillus fumigatus* expressed at 37 C. *The Journal of Eukaryotic Microbiology, 51*, 428–432.

Shankar, J., Restrepo, A., Clemons, K. V., & Stevens, D. A. (2011a). Hormones and the resistance of women to paracoccidioidomycosis. *Clinical Microbiology Reviews, 24*, 296–313.

Shankar, J., Wu, T. D., Clemons, K. V., Monteiro, J. P., Mirels, L. F., & Stevens, D. A. (2011b). Influence of 17β-estradiol on gene expression of Paracoccidioides during mycelia-to-yeast transition. *PLoS One, 6*, e28402.

Sharma, D., & Masison, D. C. (2009). Hsp70 structure, function, regulation and influence on yeast prions. *Protein and Peptide Letters, 16*, 571–581.

Sharma, D., Martineau, C. N., Le Dall, M.-T., Reidy, M., Masison, D. C., & Kabani, M. (2009a). Function of SSA subfamily of Hsp70 within and across species varies widely in complementing *Saccharomyces cerevisiae* cell growth and prion propagation. *PLoS One, 4*, e6644.
Sharma, S. K., Christen, P., & Goloubinoff, P. (2009b). Disaggregating chaperones: An unfolding story. *Current Protein and Peptide Science, 10*, 432–446.
Shearer, G., Jr., Birge, C. H., Yuckenberg, P. D., Kobayashi, G. S., & Medoff, G. (1987). Heat-shock proteins induced during the mycelial-to-yeast transitions of strains of *Histoplasma capsulatum*. *Microbiology, 133*, 3375–3382.
Shen, H. D., Au, L. C., Lin, W. L., Liaw, S. F., Tsai, J. J., & Han, S. H. (1997). Molecular cloning and expression of a *Penicillium citrinum* allergen with sequence homology and antigenic crossreactivity to a hsp 70 human heat shock protein. *Clinical and Experimental Allergy, 27*, 682–690.
Shen, Y., Meunier, L., & Hendershot, L. M. (2002). Identification and characterization of a novel endoplasmic reticulum (ER) DnaJ homologue, which stimulates ATPase activity of BiP in vitro and is induced by ER stress. *The Journal of Biological Chemistry, 277*, 15947–15956.
Shiber, A., & Ravid, T. (2014). Chaperoning proteins for destruction: Diverse roles of Hsp70 chaperones and their co-chaperones in targeting misfolded proteins to the proteasome. *Biomolecules, 4*, 704–724.
Shiber, A., Breuer, W., Brandeis, M., & Ravid, T. (2013). Ubiquitin conjugation triggers misfolded protein sequestration into quality control foci when Hsp70 chaperone levels are limiting. *Molecular Biology of the Cell, 24*, 2076–2087.
Silva, E. M., Freitas, J. S., Gras, D. E., Squina, F. M., Leal, J., Silveira, H. C., Martinez-Rossi, N. M., & Rossi, A. (2008). Identification of genes differentially expressed in a strain of the mold Aspergillus nidulans carrying a loss-of-function mutation in the palA gene. *Canadian Journal of Microbiology, 54*, 803–811.
Singh, K., Nizam, S., Sinha, M., & Verma, P. K. (2012). Comparative transcriptome analysis of the necrotrophic fungus *Ascochyta rabiei* during oxidative stress: Insight for fungal survival in the host plant. *PLoS One, 7*, e33128.
Sondermann, H., Ho, A. K., Listenberger, L. L., Siegers, K., Moarefi, I., Wente, S. R., Hartl, F.-U., & Young, J. C. (2002). Prediction of novel Bag-1 homologs based on structure/function analysis identifies Snl1p as an Hsp70 co-chaperone in *Saccharomyces cerevisiae*. *The Journal of Biological Chemistry, 277*, 33220–33227.
Song, Y., & Masison, D. C. (2005). Independent regulation of Hsp70 and Hsp90 chaperones by Hsp70/Hsp90-organizing protein Sti1 (Hop1). *The Journal of Biological Chemistry, 280*, 34178–34185.
Song, Y., Wu, Y.-x., Jung, G., Tutar, Y., Eisenberg, E., Greene, L. E., & Masison, D. C. (2005). Role for Hsp70 chaperone in *Saccharomyces cerevisiae* prion seed replication. *Eukaryotic Cell, 4*, 289–297.
Squina, F. M., Leal, J., Cipriano, V. T., Martinez-Rossi, N. M., & Rossi, A. (2010). Transcription of the *Neurospora crassa* 70-kDa class heat shock protein genes is modulated in response to extracellular pH changes. *Cell Stress & Chaperones, 15*, 225–231.
Stedman, T. T., Butler, D. R., & Buck, G. A. (1998). The HSP70 gene family in *Pneumocystis carinii*: Molecular and phylogenetic characterization of cytoplasmic members. *The Journal of Eukaryotic Microbiology, 45*, 589–599.
Steel, G. J., Fullerton, D. M., Tyson, J. R., & Stirling, C. J. (2004). Coordinated activation of Hsp70 chaperones. *Science, 303*, 98–101.
Stefani, R. M. P., & Gomes, S. L. (1995). A unique intron-containing hsp70 gene induced by heat shock and during sporulation in the aquatic fungus *Blastocladiella emersonii*. *Gene, 152*, 19–26.
Tereshina, V. (2005). Thermotolerance in fungi: The role of heat shock proteins and trehalose. *Microbiology, 74*, 247–257.

Thakur, R., Anand, R., Tiwari, S., Singh, A. P., Tiwary, B. N., & Shankar, J. (2015). Cytokines induce effector T-helper cells during invasive aspergillosis; what we have learned about T-helper cells? *Frontiers in Microbiology, 6*, 429.

Thakur, R., Tiwari, S., & Shankar, J. (2016). Differential expression pattern of heat shock protein genes in toxigenic and atoxigenic isolate of *Aspergillus flavus*. *British Microbiology Research Journal, 14*, 1–9.

Tibor Roberts, B., Moriyama, H., & Wickner, R. B. (2004). [URE3] prion propagation is abolished by a mutation of the primary cytosolic Hsp70 of budding yeast. *Yeast, 21*, 107–117.

Tilburn, J., Sarkar, S., Widdick, D., Espeso, E., Orejas, M., Mungroo, J., Penalva, M., & Arst, H., Jr. (1995). The Aspergillus PacC zinc finger transcription factor mediates regulation of both acid-and alkaline-expressed genes by ambient pH. *The EMBO Journal, 14*, 779.

Tiwari, S., Thakur, R., & Shankar, J. (2015). Role of heat-shock proteins in cellular function and in the biology of fungi. *Biotechnology Research International, 2015*, 1–11.

Tiwari, S., Thakur, R., Goel, G., & Shankar, J. (2016). Nano-LC-Q-TOF analysis of proteome revealed germination of *Aspergillus flavus* Conidia is accompanied by MAPK signalling and cell wall modulation. *Mycopathologia, 181*, 769–786.

Tiwari, S., & Shankar, J. (2018) Integrated proteome and HPLC analysis revealed quercetin-mediated inhibition of aflatoxin B1 biosynthesis in *Aspergillus flavus*. 3 Biotech 8 (1)

Tyson, J. R., & Stirling, C. J. (2000). LHS1 and SIL1 provide a lumenal function that is essential for protein translocation into the endoplasmic reticulum. *The EMBO Journal, 19*, 6440–6452.

Verghese, J., Abrams, J., Wang, Y., & Morano, K. A. (2012). Biology of the heat shock response and protein chaperones: Budding yeast (*Saccharomyces cerevisiae*) as a model system. *Microbiology and Molecular Biology Reviews, 76*, 115–158.

Wang, L., & Lin, X. (2012). Morphogenesis in fungal pathogenicity: Shape, size, and surface. *PLoS Pathogens, 8*, e1003027.

Wang, C., Duan, Z., & Leger, R. J. S. (2008). MOS1 osmosensor of *Metarhizium anisopliae* is required for adaptation to insect host hemolymph. *Eukaryotic Cell, 7*, 302–309.

Weitzel, G., Pilatus, U., & Rensing, L. (1985). Similar dose response of heat shock protein synthesis and intracellular pH change in yeast. *Experimental Cell Research, 159*, 252–256.

Werner-Washburne, M., & Craig, E. A. (1989). Expression of members of the Saccharomyces cerevisiae hsp70 multigene family. *Genome, 31*, 684–689.

Werner-Washburne, M., Stone, D. E., & Craig, E. A. (1987). Complex interactions among members of an essential subfamily of hsp70 genes in Saccharomyces cerevisiae. *Molecular and Cellular Biology, 7*, 2568–2577.

Wormley, F. L., Jr. (2011). Potential vaccine strategies for the management of cryptococcal disease. *Journal of Invasive Fungal Infection, 5*, 65.

Xavier, I. J. (1998). Environmental stress response of the Hyphomycetous Entomopathogenic fungi. http://hdl.handle.net/10388/etd-10212004-001033

Yi, M., Chi, M.-H., Khang, C. H., Park, S.-Y., Kang, S., Valent, B., & Lee, Y.-H. (2009). The ER chaperone LHS1 is involved in asexual development and rice infection by the blast fungus *Magnaporthe oryzae*. *Plant Cell, 21*, 681–695.

Zhu, C., Ai, L., Wang, L., Yin, P., Liu, C., Li, S., & Zeng, H. (2016). De novo transcriptome analysis of *Rhizoctonia solani* AG1 IA strain early invasion in Zoysia japonica root. *Frontiers in Microbiology, 7*, 708.

Index

A. A. A. Asea, P. Kaur (eds.), *HSP70 in Human Diseases and Disorders*, Heat Shock Proteins 14, https://doi.org/10.1007/978-3-319-89551-2

MIX
Papier aus verantwortungsvollen Quellen
Paper from responsible sources
FSC® C105338

If you have any concerns about our products,
you can contact us on
ProductSafety@springernature.com

In case Publisher is established outside the EU,
the EU authorized representative is:
Springer Nature Customer Service Center GmbH
Europaplatz 3, 69115 Heidelberg, Germany

Printed by Libri Plureos GmbH
in Hamburg, Germany